PATCH CLAMP TECHNIQUE

Edited by **Fatima Shad Kaneez**

Patch Clamp Technique
Edited by Fatima Shad Kaneez

Published by InTech
Janeza Trdine 9, 51000 Rijeka, Croatia

Publishing Process Manager Romana Vukelic

Technical Editor Teodora Smiljanic

Cover Designer InTech Design Team

Additional hard copies can be obtained from orders@intechopen.com

Patch Clamp Technique, Edited by Fatima Shad Kaneez
p. cm.
ISBN 978-953-51-0406-3

For EU product safety concerns:
IN TECH d.o.o., Prolaz Marije Krucifikse Kozulić 3, 51000 Rijeka, Croatia,
info@intechopen.com or visit our website at intechopen.com.

Meet the editor

Dr Fatima Shad Kaneez has been working with patch clamping cell membranes since 1989, and has been studying and recording the electrical currents passing through ionic channels in these membranes. She has used all four types of patch-clamp configurations while studying Ligand-gated ionic channels. She established and modified the patch clamp set up in Sydney and Newcastle in Australia, as well in Karachi, Pakistan at Dr Panjwani Centre for Molecular Medicine and Drug Research, University of Karachi, and she has also helped the UAE University in acquiring a similar set-up for their Central Laboratories Unit. In 1992, as a PhD student, she visited Professor Neher's laboratory in Gottingen, Germany and had the first-hand experience of seeing the development of Nobel Prize winning patch clamp technique in its early stages.

Contents

Preface

This book deals with the understanding of endogenous mechanisms of cells and their receptors as well as advantages of using this technique. It covers the basic principles and preparation types and also deals with the latest developments in the traditional patch clamp technique. Some chapters in this book take the technique to a next level of modulation and novel approach.

The patch-clamp technique was developed by Neher and Sakmann in 1976 as an effective refinement of the voltage-clamp-assay that demonstrated the function of single ion channels in fundamental cellular process. The patch-clamp technique has revolutionized the study of ion channels in a wide variety of cells.

This book has seventeen chapters and has been divided into two sections. First section deals with the exploration of underlying mechanisms of cellular activities using patch clamp technique. The second section deals with the advantages of this system.

The first section begins with a chapter on the black box properties of the system followed by how intracellular signalling pathways integrate the pore associated to P2X7 receptor with other large pores. Next chapter deals with combine usage of whole-cell patch and focal stimulation technique to study of single synapse of central nervous system. This is followed by the chapter on single-channel properties and pharmacological characteristics of KATP channels in primary afferent neurons. Chapters five and six explore how renal potassium channels are regulated by protein kinases and phosphatases and characteristic features of BK channels respectively. Then there are papers on "onion-peeling" technique, mitochondrial channels, membrane transport in red blood cells and electrical properties in Leishmania-macrophage membrane.

Seven chapters under section entitled "Advantages of using patch clamp technique" describe how this electrophysiological method is found to be useful in studying cardiac channelopathies, gating charge movements, off target effects of clinically used drugs, Non-Steroidal Anti-Inflammatory Drugs (NSAIDs) in Modulating Vascular Smooth Muscle Cells, Perforated Patch Clamp in non-neuronal cells, Drug screening and drug safety evaluation and antimicrobial peptides and viroporins.

This book will be of good value for students of physiology, neuroscience, cell biology and biophysics.

Dr. Fatima Shad Kaneez
Professor Human Physilogy
PAP RSB Institute of Health Sciences
Universiti Brunei Darussalam
Brunei Darussalam

Part 1

Exploring Cellular Mechanisms Using Patch Clamp Technique

1

Patch ClampTechnique for Looking at Serotonin Receptors in B103 Cell Lines: A Black Box Test

K. Fatima-Shad[1] and K. Bradley[2]
[1]*PAP RSB Institute of Health Sciences, Universiti Brunei Darussalam*
[2]*Faculty of Medicine and Health Sciences, University of Newcastle*
[1]*Brunei Darussalam*
[2]*Australia*

1. Introduction

In this chapter, we would like to describe black box testing phenomenon of patch clamp technique while looking at the serotonin receptors in B103 cell lines.

In a black box test, the tester only knows the inputs and what the expected outcomes should be and but not the mechanisms of those outputs. Patch clamp method is a great method for quantifying the research on Pico or femto scales, but most of the time even very controlled experiments will not give us the expected results. We will begin our chapter by introducing serotonin receptors and B103 cell lines.

In mammals, serotonin or 5-hydroxytryptamine (5-HT) behaves primarily as an inhibitory neurotransmitter of the central nervous system (CNS), decreasing neuronal activity and facilitating behavioural relaxation, while peripherally it has an excitatory role, promoting inflammatory responses, pain, and muscle spasm (Kirk et al 1997). Centrally this neurotransmitter is produced nearly exclusively by a group of neurons found in the rostro-ventral brainstem comprising the raphé nuclei from which project two major serotonergic pathways (Dahlstrom & Fuxe, 1960).

There are more than seventeen types of serotonin receptors and almost all are associated with G-proteins except 5-HT_3R, which is a member of the ligand-gated ion channel superfamily. The 5-HT_3R was initially identified as a monovalent cation channel by studies indicating that extra-cellularly recorded depolarising responses were diminished by removal of Na^+ from extracellular solution (Wallis & Woodward, 1975). The native 5-HT_3R is a cation-specific ion channel, but is otherwise relatively non-selective (demonstrating poor cation discrimination) allowing the passage of even large molecules, such as Ca^{2+} and Mg^{2+} (Maricq et al., 1991).

Serotonin type 3 receptors have been identified in the enteric nervous system (Branchek, et al, 1984), on sympathetic, parasympathetic, and sensory nerve fibres in the CNS (Kilpatrick et al, 1987), and on several mouse neuroblastoma cell lines, including the NCB-20 (Lambert et al., 1989, Maricq et al., 1991), N1E-115 (Lambert et al., 1989), and NG 108-15 (Freschi & Shain, 1982). All of these lines exhibits a rapid membrane depolarisation accompanied by increased membrane conductance in response to exogenously applied 5-HT (Peters & Lambert, 1989).

We are using B103 cell lines to study this fast acting receptor channel. The B103 rat neuroblastoma cell line was produced via transplacental exposure to nitroethylurea (Druckrey et al., 1967) and literature (Tyndale et al., 1994; Kasckow, et al., 1992) indicated the possibility that this line could be derived from cells of the raphé nuclei, and so might be representative of cells from the serotonergic pathway. The B103-line has been used as a model in a number of studies looking at GABA function, including GABA uptake (Schubert, 1975), and binding (Napias, et al., 1980). Studies looking at the functionality of $GABA_AR$s in a number of the lines initially generated (Schubert et al.,1974) via the patch-clamp technique indicated that while all lines were suitable for patch-clamp studies, none showed appreciable $GABA_A$-induced chloride conductance. Although the B103-line was not used in this study, it was reasonable to assume that it might exhibit similar characteristics and be suitable for electrophysiological studies (Hales & Tyndale, 1994). This was supported by the findings of (Kasckow et al., 1992) where patch clamping detected no functional $GABA_A$ chloride channels in the B103-line. Other studies involving the B103-line have centred around exploring the characteristics of Alzheimer's disease (specifically neuritic plaques) with particular focus paid to the β-amyloid peptide (Mook-Jung, 1997), and β/A4 protein precursor (Ninomiya et al, 1994).

Membrane excitability of the line was initially confirmed using anode-break stimulus, while ^{125}I-α-neurotoxin binding indicated the presence of AChRs. B103 cells were shown to contain the neurotransmitter GABA, and both choline acetyl transferase and glutamic acid decarboxylase activities - enzymes acting in ACh and glutamate anabolism (Schubert et al., 1974). This cell line has also been used for looking at the effects of extracellular Ca2+ influx on endothelin-1-induced mitogenesis, as B103 neuroblastoma cells predominantly express endothelin ETB receptors (Yoshifumi et al, 2001)

It has been shown previously that metastatic cells express high levels of voltage-gated Na+ channels (VGSCs) in prostate cancer (Laniado et al., 1997), breast cancer (Fraser et al., 2002; Roger, et al., 2003) and melanoma (Alien, et al, 1997).

Although, the cell line has previously proven suitable for patch clamp study, no work had yet been conducted about the presence of serotonin type 3 receptor channels and their relationship with the types of VGSCs for these cells.

The patch clamp technique has been applied to the B103 cell line in this experimental series in order to explore the native voltage-gated channels (VGCs) and serotonin sensitivity to type 3 receptors present in these cells. This project is aimed to explore whether these cells presented active/functional serotonin type 3 receptors (5-HT_3R) and voltage-gated sodium channels (VGSCs) and the link between each other.

2. Experimental procedures and methods

2.1 Cell culture

The B103 cells were donated by Dr Phil Rob (Cell Signalling Unit, Westmead). Stock aliquots were stored at -80°C and active stocks used for 20-25 passages before a new aliquot was revived - passage limitation decreased the incidence of cellular mutation (Figure 1).

Twice a week confluent active stocks were split and new flasks seeded in neuronal growth medium (NGM) (DMEM (TRACE), 10% foetal calf serum (FCS), 2% of 7.5% sodium bicarbonate, 200 mM L-glutamine, 2% 1 M HEPES). Five minute incubation in trypsin at 37°C, 5% CO_2, 90% humidity (Forma Scientific incubator) degraded the extra cellular matrix of the culture, releasing cells from flask adhesion (effective dislodging turned the trypsin cloudy).

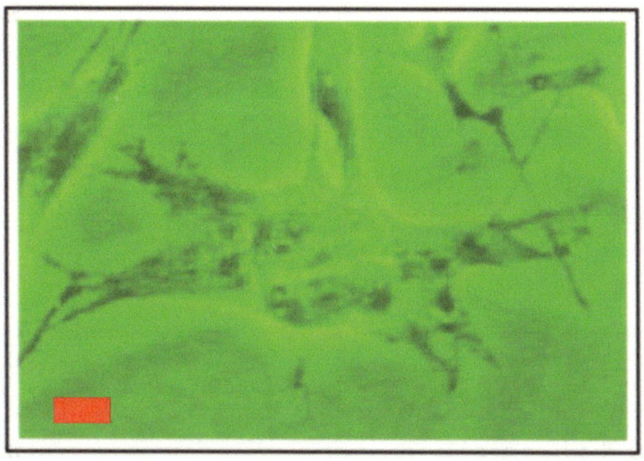

Fig. 1. A Sample Mutated Cell from the B103 Clonal-Line. Taken with an Olympus inverted microscope at 30× magnification showing dramatically altered morphology. These cells were typically seen to engulfing neighbouring cells.

Trypsin was inactivated by adding NGM, preventing continued digestion, which would have resulted in cell lysis. The suspension was spun at 400 rpm for 8-10 minutes in a megafuge (Heraeus Instruments). Supernatant was discarded and cell pellet gently resuspended in 10 ml NGM.

Later on cells were replated (Figure 2) and cover slips were prepared for patch clamp experiments.

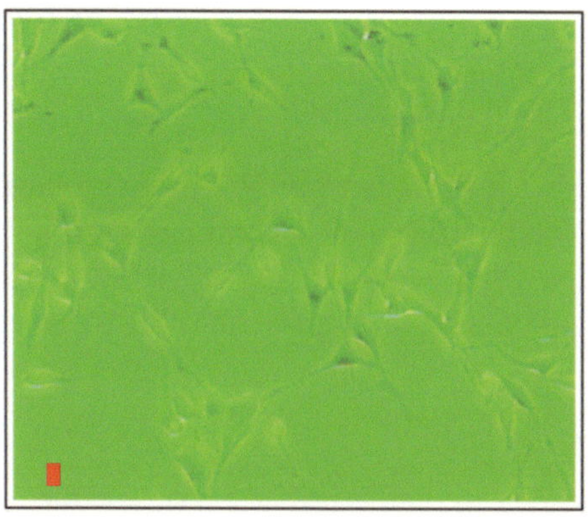

Fig. 2. A Typical B103 Cell Culture. Image at 10× magnification after 48 hours of incubation, showing a cellular concentration of 4.0×10^5. Note the extensive branching network generated.

2.1.1 Cell counting

Cells were counted from the outer four segments of a **hemocytometer** (Improve Neubave Weber) under **10×** magnification (using an Olympus CK2 microscope) and a total mean value was calculated. This value was used to determine the concentration of cells per millilitre in the diluted cell suspension by employing the formula:

mean cell count × 100 000 (gave a per ml value) = cells/ml

After cell concentration was calculated, the cell suspension was diluted to **1.0 × 10^5 cells/ml** and the cells plated at varying concentrations onto sterilised collagen-coated coverslips (see heading Collagen-Coating the Coverslips) in **35 × 10 mm tissue culture dishes** (Corning). The cellular concentration required for later work was **4.0 × 10^5** and because cells roughly **doubled** every **24 hours**, plates were seeded with four different cellular concentrations (Table#1).

Day of Use	Plating Cell Concentrations	Cell Culturing	
		FCS Media (ml)	Cell Suspension (ml)
Day 1	seeding performed	-	-
Day 2	2.0 × 10^5*	0	2
Day 3	1.0 × 10^5*	1	1
Day 4	5.0 × 10^4*	1.5	0.5
Day 5	2.5 × 10^4*	1.75	0.25

* Because of the doubling rate of neuronal cells, plates reached a concentration of 4.0 × 10^5 on their respective days of use.

Table 1. Cell Culturing Schedule

2.1.2 Collagen-coating the coverslips

Collagen provided a matrix for B103 cell adhesion when plated. Coverslips and culture dishes were coated with sterile **10 µg/ml** rat tail **collagen solution** (Roche) diluted in **phosphate buffered saline** (PBS), and incubated at **37°C** for **2 hours**. The collagen solution was removed and dishes washed with PBS to ensure complete removal of residual collagen.

2.2 Solutions

Cells were patched under two different sets of bath and pipette solutions. Initial results were obtained from physiologically normal solutions (normal pipette solution: 120 mM KCl, 3.7 mM NaCl, 1 mM $CaCl_2$, 2 mM $MgCl_2$, 20 mM TEACl, 10 mM HEPES, 11 mM EGTA (pH 7.4); normal bath solution: 137 mM NaCl, 5.4 mM KCl, 1.8 mM $CaCl_2$, 2 mM $MgCl_2$, 5 mM HEPES, 10 mM D-glucose (pH 7.4)) which were designed to mimic normal cellular conditions. Later recordings utilised solutions with symmetrical cation concentrations (normal pipette solution: 140 mM NaF, 1 mM $MgCl_2$, 10 mM HEPES, 10 mM EGTA (pH 7.4); experimental bath solution: 140 mM NaCl, 2 mM $CaCl_2$, 1 mM $MgCl_2$, 10 mM HEPES, 10mM

D-glucose (pH 7.4)). To promote long-term cell viability bath solution osmolarity was kept between 300-320 mOsm. A difference of 20 mOsm rendered cells non-viable for electrophysiological study (adversely affecting plasma membrane structure and function) either resulting in cell swelling (<300 mOsm) or shrinking (>320 mOsm) leading to premature cell death. The bath perfusion system was used to elute the cell cultures and was comprised of a solution reservoir connected to the bath via plastic tubing. A regulator was attached to the tubing allowing for control of solution flow - unrestricted flow was 0.38 ± 0.009 ml/sec.

2.2.1 Bath solution perfusion

Solution was removed from the bath and emptied into a waste reservoir via a system of tubing connected to a **miniport motor** (Neuberger). Between the waste reservoir and the motor was a second reservoir containing **silica gel crystals** which prevented moisture from reaching the motor.

The bath perfusion system was particularly prone to contamination, especially with bacteria which fed on the solution glucose. To prevent contamination the system was rinsed with distilled water after every use to remove any trace glucose. However when the inevitable contamination did occur **antibacterial solution** (Milton hospital-grade disinfectant) was used to flush the lines.

2.2.2 Technical difficulties

The technique employed for electrophysiological study of the B103 cell-line was not conducted under aseptic conditions therefore the cells were particularly prone to **bacterial infection**. Bacteria tended to attack the cellular **cytoplasm** forming small **vacuoles** (Figure#3) and rendering the cells unfit for study. Once an infection had been noted, in order to prevent further contamination (particularly of the surrounding equipment) the patch-clamp system had to be immediately decontaminated using **70% ethanol** and/or **antibacterial solution**. The coverslip had to be immediately discarded and the stage and bath had to be thoroughly disinfected to prevent contamination of subsequent coverslips.

2.3 Pharmacological agents

The following pharmacological agents were used: Serotonin, Ondansetron, Tetrodotoxin (TTX), Phenytoin, and d- Tubocurarine. All these were purchased from Sigma, except TTX (Alomone).

2.4 Patch clamp experiments

Cells were visualised with an Olympus IX70 inverted microscope and images recorded with a KOBI digital colour camera and the ASUS Live 3D Multimedia software. Electrophysiological manipulation and recordings were undertaken with a HEKA EPC9 amplifier and HEKA *Pulse* software package which supersedes older amplifier models by having a fully interactive, PC-compatible data retrieval and storage facility. The *PULSE* program allowed for automatic electronic noise adjustments such as fast and slow capacitative transients' nullifications.

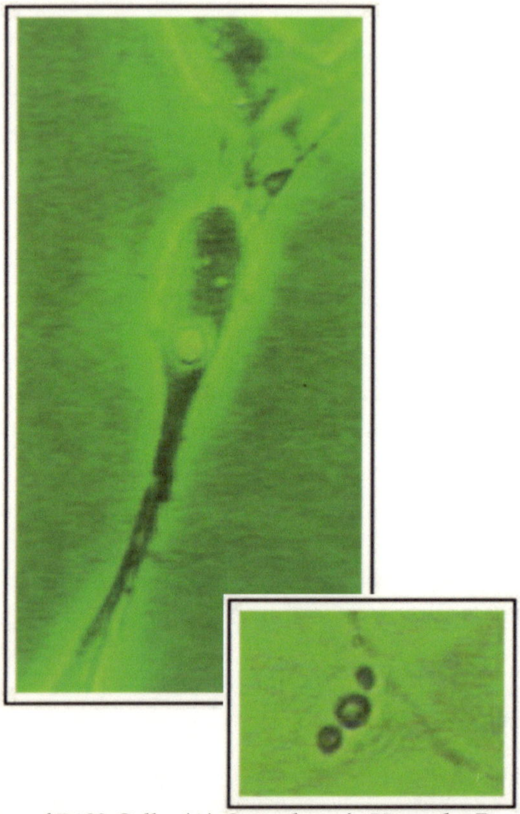

Fig. 3. Bacterial Infection of B103 Cells. **(A) Cytoplasmic Vacuole.** Bacteria entered the cells by generating holes in the cell membranes where they formed vacuoles in the cytoplasm. Image generated under phase-contrast filtering at 30× magnification. **(B) Bacterial Aggregate.** Image generated at magnification under bright-phase filtering at 60× magnification.

Thin walled borosilicate glass capillaries (1.5 mm O.D. × 1.17 mm I.D) were used to produce patch pipettes with a 3 MΩ resistance. Pipettes were half-filled using both the front- and back-filling techniques. Solution-filled glass pipettes were attached to an Ag/AgCl recording electrode and manipulated using a PCS-5000 series patch clamp micromanipulator (Burleigh Instruments). Cellular patching was performed according to the protocol outlined by (Hamil et al., 1981) Figure 4.

An appropriate B103 cell was chosen for patching on the basis of its general morphology: approximately 25 µm in diameter, well-defined clean cell membrane, and relatively isolated from contact with other cells. Morphological cellular standardisation was a critical component of the protocol. All cells were tested for their viability in the physiological saline before changing into symmetrical solutions (sodium on both side of the cell membrane) for measuring voltage activated sodium currents. 5-HT3 receptor channel currents were observed in B103 cells, when they were exposed to serotonin (endogenous currents of B103 cells were completely abolished by using TTX or Phenytoin).

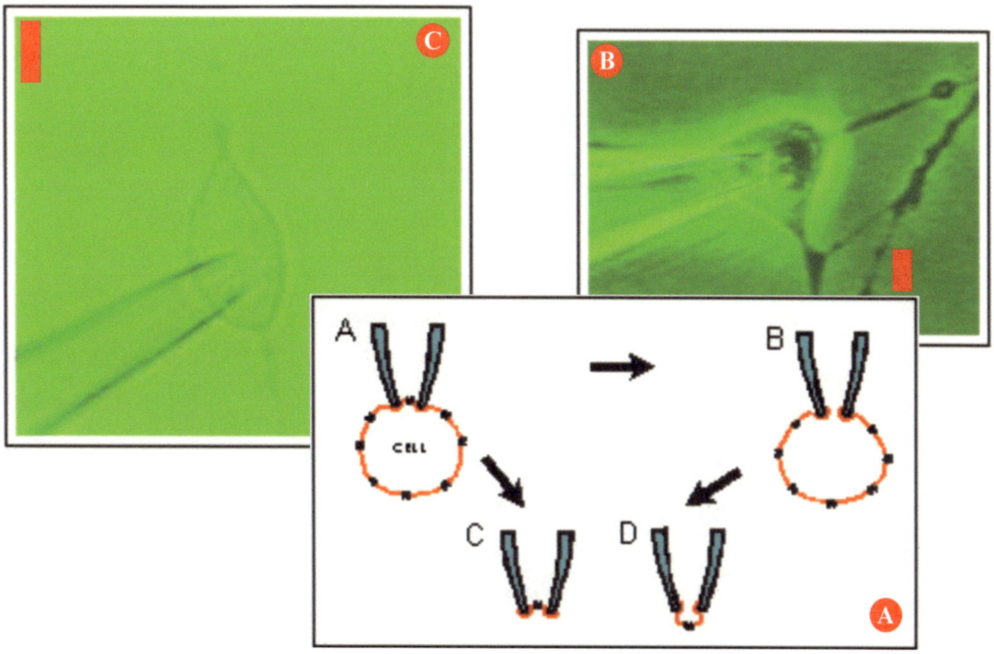

Fig. 4. (A) The Various Patch-Clamp Configurations. A indicates the cell-attached configuration where a pipette is attached to the outside of a cell with a GΩ resistance and effectively measures the conductance of a single channel. B shows the whole-cell patch-clamp configuration where the patch of membrane under the pipette tip has been ruptured allowing direct access to the cell interior so that pipette solution replaces the cytoplasmic contents of the cell. This configuration forms a continuous circuit with the electrode and the cell interior allowing for recordings of the conductance of channels from the entire membrane. Both of these configurations were used during this experimental series, while C (the inside-out) and D (outside-out) configurations were not used. **(B) A Cell-Attached Patched B103 Cell under Phase Contrast Filtering**. At 40× magnification **(C) A Whole-Cell Patched B103 Cell under Bright Phase Filtering**. At 30× magnification. Immediately after patch initiation cell will start to take on a slight spherical appearance.

A perfusion system was employed to introduce chemicals (both agonist and antagonist) onto a patched cell with application time being electronically controlled via solenoid valve. The agonist solutions used in this experimental series were a set of serotonin hydrochloride dilutions: 1 mM, 500 μM, and 10 μM. Patched cells were challenged with a 8000 ms exposure to agonist at 5 minute intervals - a transient method of agonist application avoided cellular desensitisation (Neijt et al., 1988), and results were recorded using the HEKA *PULSE* software. The solution used in our experiments to abolish serotonin activated current was Ondansetron a selective 5-HT3R antagonist. Cells were again challenged with 8000 ms exposure, both with and without agonist or antagonist solution.

Cells were stimulated using a Pulse Protocol facilitated via the HEKA *Pulse* software. Cellular stimulation ranged from -100 mV to +30 mV increasing in 10 mV steps with a resting period at 0 mV between each step (figure 5)

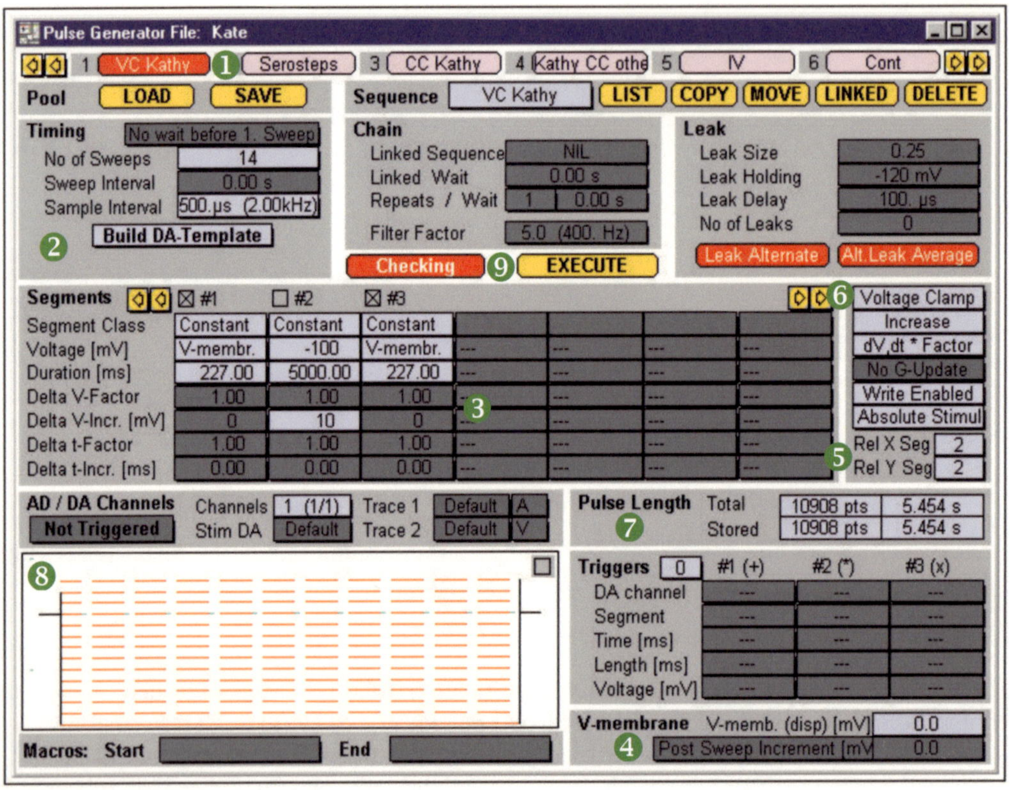

Fig. 5. The *Pulse Generator* Window showing the Pulse Protocol. This window was accessed by choosing Pulse Generator from the Pulse drop-down menu on *PULSE* main screen toolbar. In this window a Pulse Protocol is generated where the *PULSE* operator can predefine ❶ the desired cellular electrical stimulus so that it can later be used instantaneously during experimentation. The Timing section ❷ defined the number of stimulus Sweeps applied to the cell (14) and the frequency with which data is collected during each Sweep (once every 500 µs). Values in the Segments section ❸ defined the stimulus pattern internally for each Sweep, as well as the pattern between Sweeps. Here three Segments were defined, where Segments 1&3 were 227.0 ms Resting Phases with no electrical stimulation, while Segment 2 was the Stimulus Phase where for 5000 ms an electrical stimulus of -100 mV was initially applied to the cell. Subsequent Sweep Stimulation Phases increased by +10 mV so that the final Sweep stimulated at +30 mV. The holding membrane potential was defined as 0 mV ❹ because symmetrical Na^+ solutions were used during experimentation. The Relevant Segments ❺ for data retrieval were defined so that later data analysis used information collected from Segment 2 only, and the type of patch-clamping mode ❻ was selected here (i.e. either voltage-clamping or current-clamping). The total number of data points and the time for each Sweep was indicated in the Pulse Length Segment ❼ and the entire Protocol displayed diagrammatically ❽ for easy reference. Once the Protocol was defined was checked for errors by initiating the Checking sequence ❾ and the entire Protocol was complete and ready for use.

2.5 Data analysis

Each experiment in a given condition was carried out minimum of five times and the mean was determined as the representative result. Each condition was thus tested in at least 3 separate experiments. The average and the standard errors were calculated for the experimental values and analysed statistically by using Sigma Plot software (SDR Incorporation). Slopes of linear regressions were analysed by t-test.

3. Results

Electrophysiological heterogeneity of the B103 cell-line was observed where channel current responses divided the cells into three groups: with low, medium, and high conductance. There was no correlation between conductance and morphology because the cells used were morphologically identical as well as culture incubation time.

3.1 B103 currents in physiological solution

Cells were examined via the patch-clamp technique first in physiological solutions where K^+ was the primary cationic component of the pipette solution, imitating the internal and external conditions found *in vivo*. Throughout the course of the experimental series, all patch-clamp recordings were taken at a constant temperature of 22°C unless otherwise indicated. The average value of resting membrane potential for B103 cells in physiological saline was - 68 ± 3 mV close to potassium reversal potential expected for cells of neuronal origin.

Single-channel recordings in cell attached configurations in mammalian Ringer solution (Figure#6) gave a maximum conductance, of 0.44 nS, at 30 mV. The calculated ±30 mV slope conductance (the average conductance at +30 mV divided by the average conductance at -30 mV) was 1.02.

Subsequent Protocol applications showed a trend for decreasing current responses to the maximum applied potential from that initially recorded for each cell.

3.2 B103 Currents in symmetrical ionic concentration

The second set of solutions (with same sodium concentration on both sides) used during experiments gave a resting membrane potential of close to 0 mV. The presence of three subsets of conductances of B103 cells noted were based on their whole-cell current responses observed under symmetrical solutions.

3.2.1 The low conductance subset – Control in symmetrical solutions

These cells were categorised based on their current response to the maximum hyperpolarising step in the Protocol, that is at -100mV. Responses that were observed to be of 30 pA or less were categorised into this subset.

Whole-cell recordings were taken under symmetrical solutions (Figure#7) giving an average maximum conductance value of 0.28 nS at +30 mV. The calculated E_{rev} was -0.13 mV, while the calculated ±30 mV slope conductance was 1.08, indicating rather linear relationship between voltages and the current responses.

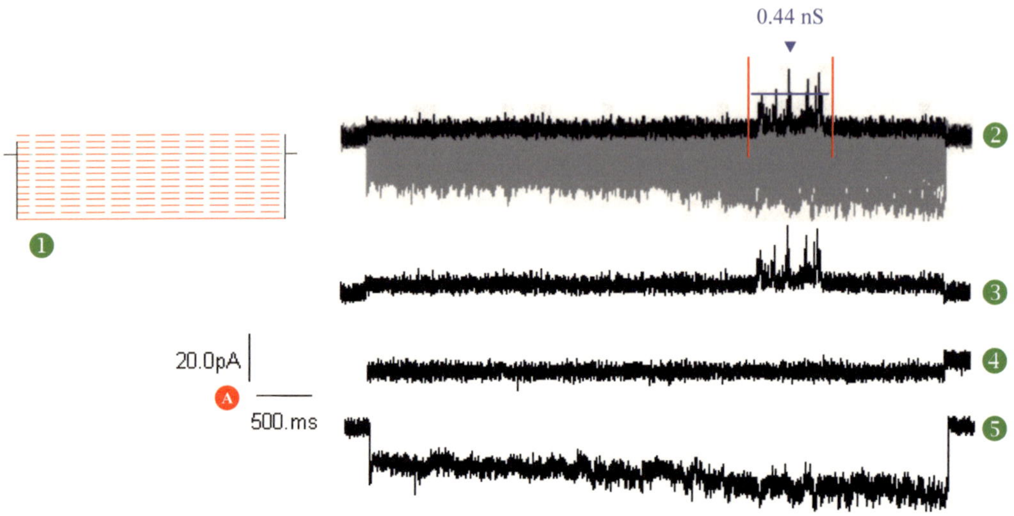

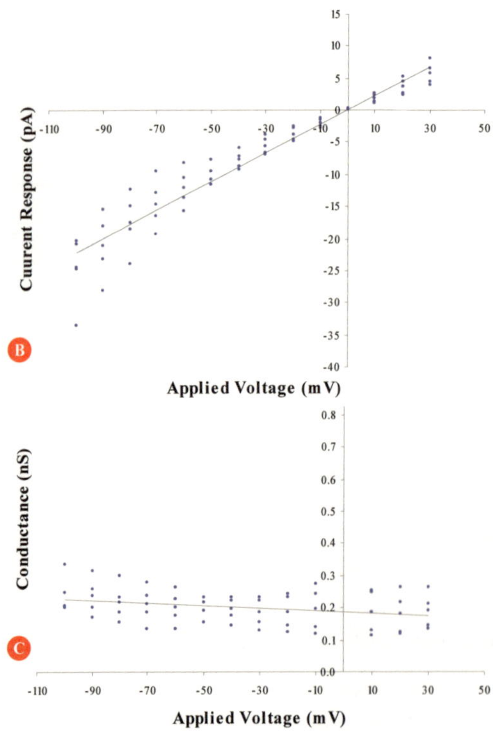

(A) Demonstrates a typical oscilloscope diagram of single-channel current response to the applied electrical stimulus, Pulse Protocol (shown in Figure#5), over time. ❶ demonstrates a schematic of the Pulse Protocol where vertically the steps are 10 mV each ranging from -100 to +30mV, and horizontally Segment2 shows a time of 5000 mS. ❷ indicates an entire Protocol response, ❸ the maximum (+30 mV) depolarising response, ❹ demonstrates the response to -30 mV stimulus, and ❺ the maximum (-100 mV) hyperpolarising response. The segment delineated by red lines indicates a single-channel time-delayed response to stimulation at +30mV, demonstrating multiple channel subconductance states, with a maximum conductance of 0.44 nS and an average of 0.25 nS.

(B) Shows the current response (pA) versus applied voltage (mV) plot for the single-channel recorded data in normal solution. Cells were in the cell-attached configuration and provided single-channel current recordings. The solid black line indicates the line of best fit for the averaged data points.

(C) Displays the response conductance (nS) versus applied voltage (mV) plot for the recorded single-channel data in normal solution conditions. The solid line is the line of best fit for the averaged conductances for each voltage step. The mean conductance was 201.43 pS, and the maximum conductance, recorded at -100 mV, was 248.25 pS. The calculated slope conductance at ±30 mV was 1.02.

Fig. 6. Single-Channel Control Results from B103 Cells Recorded in Normal Physiological Solutions: 137/3.7 $[Na^+]_o/[Na^+]_i$. All recordings were taken at a temperature of 20°C.

From this low conductance subset of B103 cells, two whole-cell current responses were observed: fast transient current (Figure#7*) and slow steady-state responses, where the amplitude and duration varied significantly. Fast transient currents were seen at the initiation of a voltage step and had a duration of 5-7 ms with a peak current, at maximum hyperpolarising potential, of -43.01 pA, while the slow steady-state current showed a greater duration of 4993-4995 ms. The average current recorded for the steady-state response was -20.64 pA. While subsequent current responses varied in amplitude, the durations were seen to remain constant unless otherwise indicated.

3.2.2 The medium conductance subset – Control in symmetrical solutions

The maximum average whole-cell conductance recorded from the B103 medium subset with experimental solutions (Figure#8) was 0.97 nS at + 30 mV. The calculated E_{rev} was -3.32 mV, while the calculated ±30 mV slope conductance was 1.4. Responses that were observed to be between 30-100 pA at -100 mV were categorised into the medium subset.

3.2.3 The high conductance subset – Control in symmetrical solutions

The average maximum control whole-cell conductance recorded for the high B103 subset with experimental solutions (Figure #9) was 1.39 nS at +30 mV. The calculated E_{rev} was 0.57 mV, while the calculated ±30 mV slope conductance was 1.09. Current response observed at -100 mV was greater than 100 pA in this high subset of B103 cells.

3.3 Serotonin receptor channel currents in B103 cell

Serotonin in different concentrations (10 μM , 500 μM & 1mM) was applied to low medium and high subsets of B103 cells. Serotonin gated currents were observed in B103 cells in the presence of 1 μM TTX.

3.3.1 Serotonin receptor channel currents in B103 cell (Low conductance subset)

The mean maximum whole-cell conductance recorded from low B103 cells in response to transient, externally applied serotonin (5-HT) in symmetrical sodium solutions (10 μM, Figure#10) was seen at +30 mV to be of 0.30 nS. The calculated E_{rev} was 0.34 mV, while the calculated ±30 mV slope conductance was 1.09. At maximum hyperpolarisation the fast transient peak was -86.30 pA and the steady-state response was -29.61 pA.

Where as in the presence of 500 μM (Figure#12) the maximum mean whole-cell conductance recorded from the low subset of B103 cells was 0.42 nS at +30 mV. The calculated Erev was 0.81 mV, while the calculated ±30 mV slope conductance was 1.24.

The maximal current value for the 440 ms fast transient was -40.28 pA and the average for then 4560 ms steady-state response was -25.0 pA.

3.3.2 Serotonin receptor channel currents in B103 cell (Medium conductance subset)

The maximal average whole-cell conductance recorded from the medium subset of B103 cells in response to external transiently applied 10 μM 5-HT with symmetrical solutions (Figure#11) was 3.09 nS at +30 mV. The calculated Erev was 13.91 mV, while the calculated ±30 mV slope conductance was 2.35.

Low B103 Subset Response to Transient Bath Application of 5-HT in Symmetrical Solutions

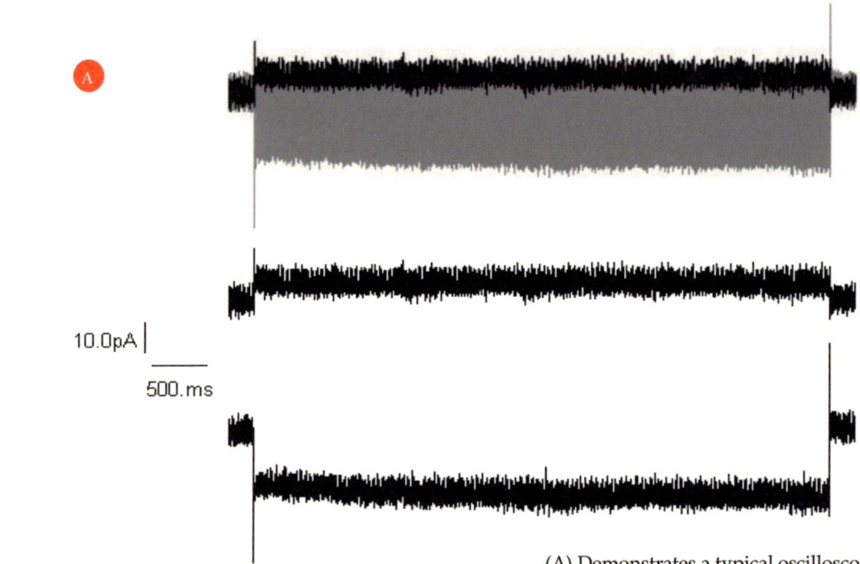

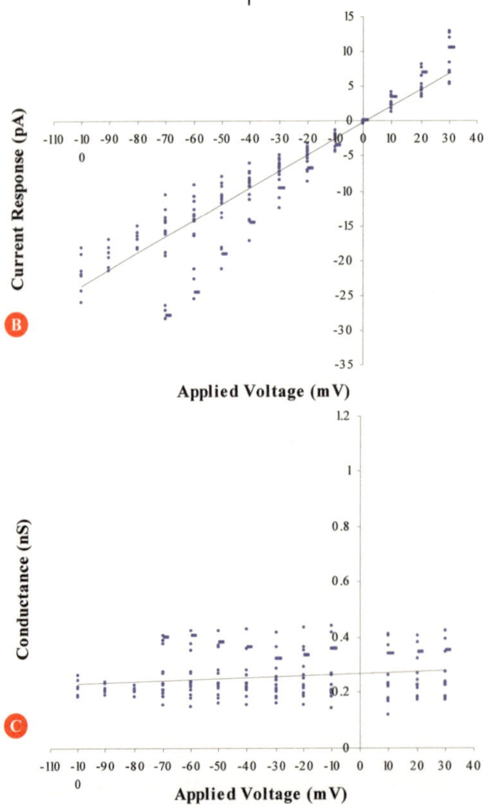

(A) Demonstrates a typical oscilloscope diagram of the whole-cell current response to the applied voltages. ❷ indicates an entire Protocol response, ❸ the maximum (+30 mV) depolarising response, and ❹ the maximum (-100 mV) hyperpolarising response. Importantly the spikes seen at the initiation and discontinuation of Segment2 are not capacitative transients as these values were rectified for during recording. From the recorded response data two types of B103 current response can be identified: fast transient current (*) and slow steady-state responses.
(B) Shows the current response (pA) versus applied voltage (mV) plot for the whole-cell recorded data in experimental solutions. The solid black line indicates the line of best fit for the averaged data points. The reversal potential calculated from the plotted data was -0.13 mV, with a mean current at +30 mV of -7.72 pA.
(C) Displays the response conductance (nS) versus applied voltage (mV) plot for the recorded whole-cell data in symmetrical solution conditions. The solid line is the line of best fit for the averaged conductances for each voltage step. The mean conductance at -30 mV was 0.26 nS, and the maximum conductance, recorded at +30 mV, was 0.28 nS. The calculated slope conductance at ±30 mV was 0.24.

Fig. 7. Whole-Cell Control Recordings from the Low Subset of B103 Cells in Symmetrical Solutions: 140/140 $[Na^+]_o/[Na^+]_i$.

Medium B103 Subset Response to Transient Bath Application of 5-HT in Symmetrical Solutions

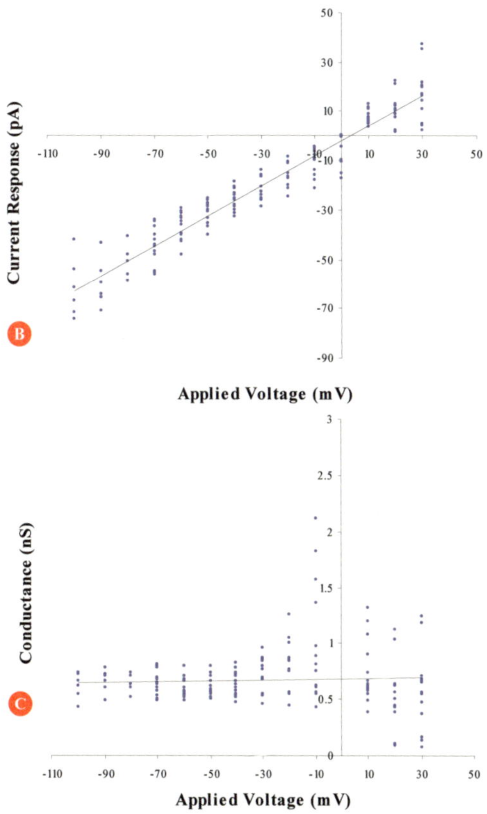

(A) Demonstrates a typical oscilloscope diagram of the whole-cell current response to the applied voltages, ❶ indicates an entire Protocol response, ❷ the maximum (+30 mV) depolarising response, and ❸ the maximum (-100 mV) hyperpolarising response. (B) Shows the current response (pA) versus applied voltage (mV) plot for the whole-cell recorded data in experimental solutions. The solid black line indicates the line of best fit for the averaged data points. The reversal potential calculated from the plotted data was -0.33 mV, with an average current at +30 mV of 16.36 pA. (C) Displays the response conductance (nS) versus applied voltage (mV) plot for the recorded whole-cell data in symmetrical solution conditions. The solid line is the line of best fit for the averaged conductances for each voltage step. The mean conductance at -30 mV was 0.69 pS, and the maximum conductance, recorded at -10 mV, was 0.97 nS. The calculated slope conductance was 0.79 for ±30 mV.

Fig. 8. Whole-Cell Control Recordings from the Medium Subset of B103 Cells in Symmetrical Solutions.

High B103 Subset Response to Transient Bath Application of 5-HT in Symmetrical Solutions

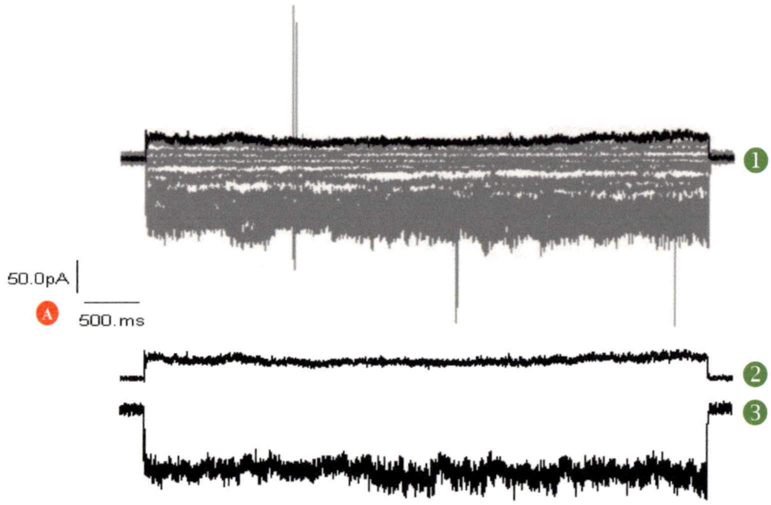

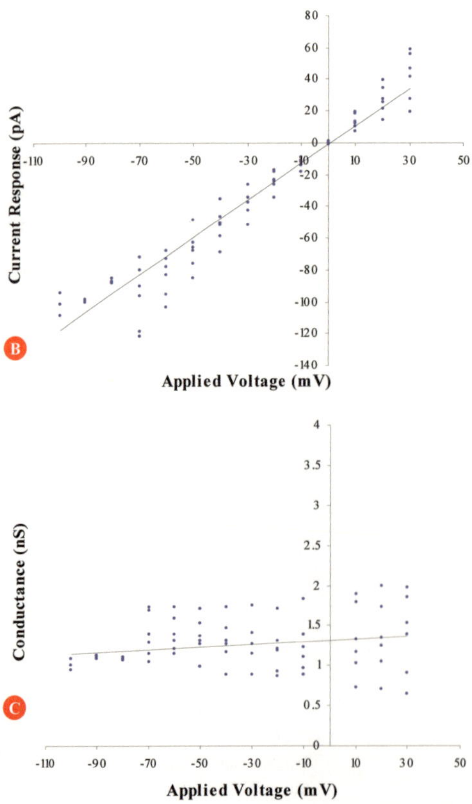

(A) Demonstrates a typical oscilloscope diagram of the whole-cell current response to the applied ❶ indicates an entire Protocol response, ❷ the maximum (+30 mV) depolarising response, and ❸ the maximum (-100 mV) hyperpolarising response.
(B) Shows the current response (pA) versus applied voltage (mV) plot for the whole-cell recorded data in symmetrical solutions. The solid black line indicates the line of best fit for the averaged data points. The reversal potential calculated from the plotted data was 0.57 mV, with an average current recording at +30 mV of 41.35 pA.
(C) Displays the response conductance (nS) versus applied voltage (mV) plot for the recorded whole-cell data in experimental solution conditions. The solid line is the line of best fit for the averaged conductances for each voltage step. The mean conductance at -30 mV was 1.27 nS, and the maximum conductance, recorded at -60 mV, was 1.39 nS. The calculated ±30 mV slope conductance was 1.09.

Fig. 9. Whole-Cell Control Recordings from the High Subset of B103 Cells in Symmetrical Solutions.

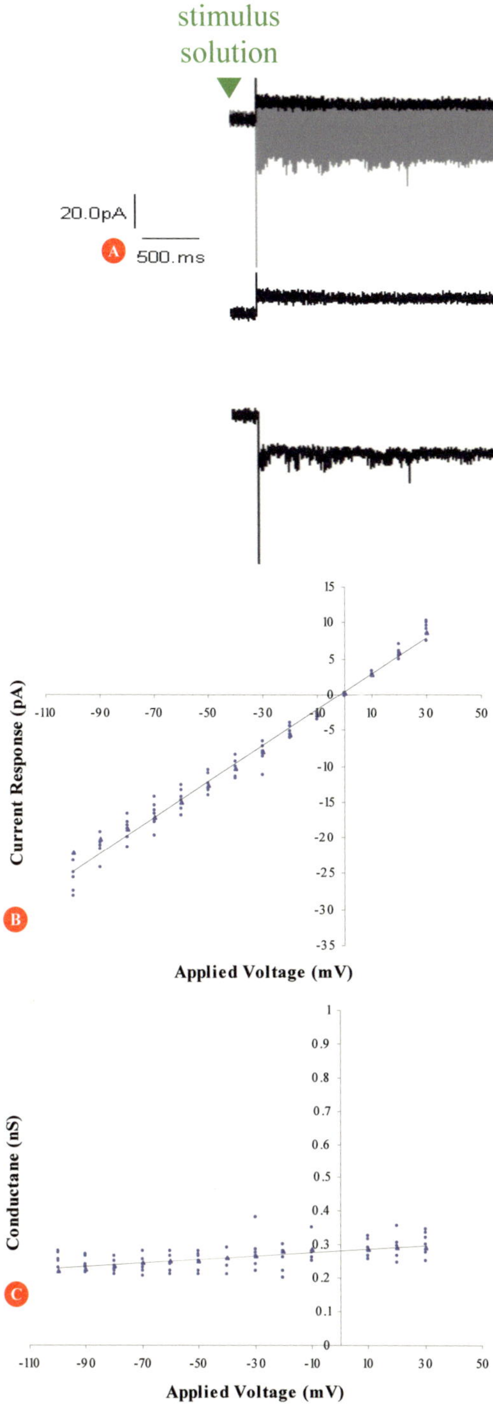

(A) Demonstrates a typical oscilloscope picture of the whole-cell current response to the applied pulse, ❶ indicates an entire Protocol response, ❷ the maximum (+30 mV) depolarising response, and ❸ the maximum (-100 mV) hyperpolarising response. Serotonin solution application was initiated at 0 ms and continued for the duration of the Protocol. This application pattern was continued throughout the experiments of agonist application.
(B) Shows the current response (pA) versus applied voltage (mV) plot for the whole-cell recorded data in symmetrical solutions. The solid black line indicates the line of best fit for the averaged data points. The reversal potential calculated from the plotted data was 0.34 mV, with an average current recording at +30 mV of 9.08 pA.
(C) Displays the response conductance (nS) versus applied voltage (mV) plot for the recorded whole-cell data in experimental solution conditions. The solid line is the line of best fit for the averaged conductances for each voltage step. The mean conductance at -30 mV was 0.26 nS, and the maximum conductance, recorded at +30 mV, was 0.30 nS. The calculated slope conductance at ±30 mV was 1.09.

Fig. 10. Whole-Cell Current Response of the Low B103 Subset to Transient Bath Application of 10 μM 5-HT in Symmetrical Solutions.

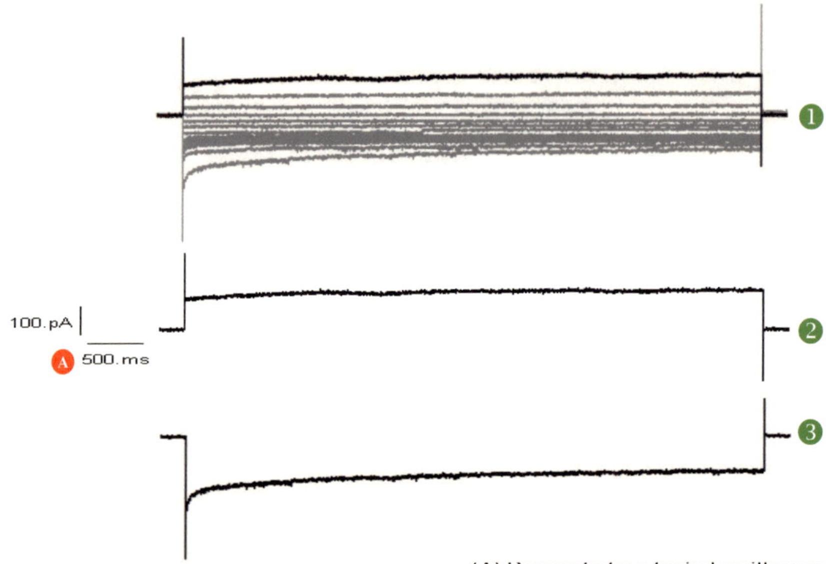

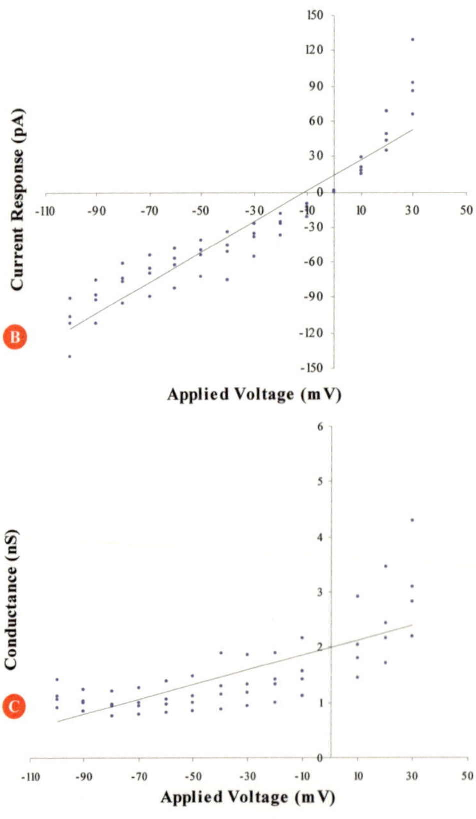

(A) Demonstrates a typical oscilloscope diagram of the whole-cell current response to the applied pulse protocol. ❶ indicates an entire Protocol response, ❷ the maximum (+30 mV) depolarising response, and ❸ the maximum (-100 mV) hyperpolarising response.
(B) Shows the current response (pA) versus applied voltage (mV) plot for the whole-cell recorded data in symmetrical solutions. The solid black line indicates the line of best fit for the averaged data points. The reversal potential calculated from the plotted data was 13.913 mV, with an average current recording of 92.78 pA at +30 mV.
(C) Displays the response conductance (nS) versus applied voltage (mV) plot for the recorded whole-cell data in experimental solution conditions. The solid line is the line of best fit for the averaged conductances for each voltage step. The mean conductance at -30 mV was 1.31 nS, and the maximum conductance, recorded at +30 mV, was 3.09 nS. The calculated slope conductance was 2.35.

Fig. 11. Whole-Cell Current Response of the Medium B103 Subset to Transient Bath Application of 10 μM 5-HT in Symmetrical Solutions.

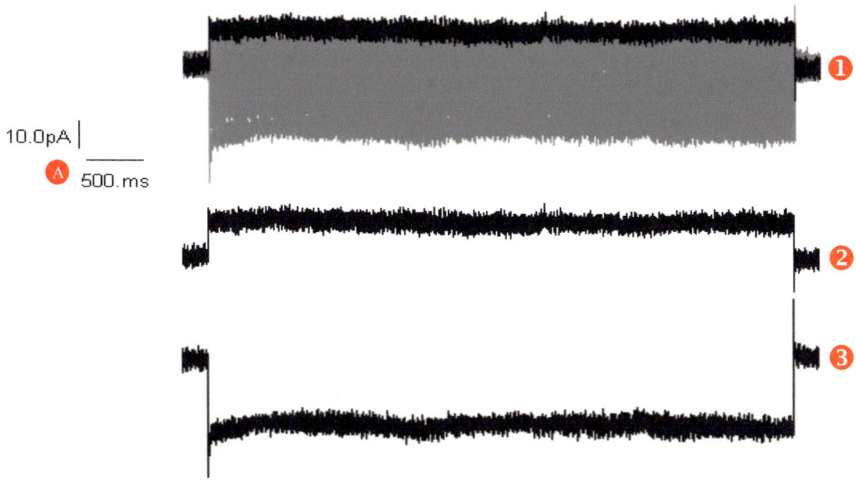

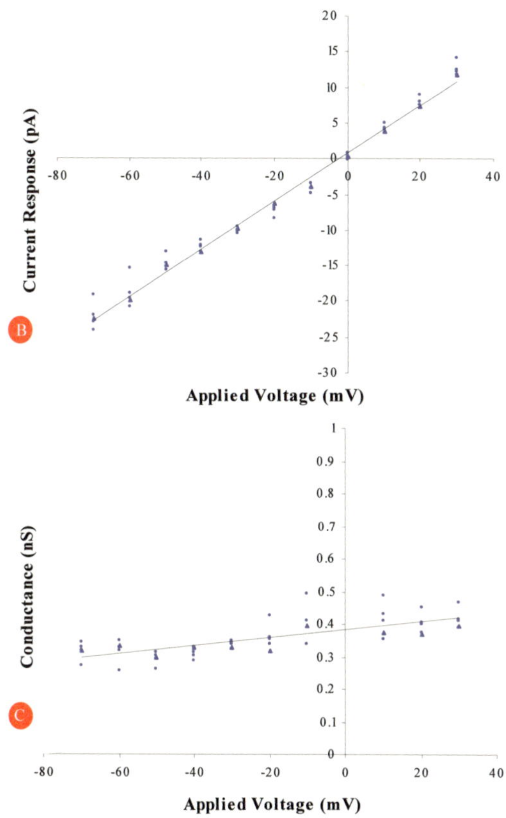

(A) Demonstrates a typical oscilloscope diagram of the whole-cell current response to the applied ❶ indicates an entire Protocol response, ❷ the maximum (+30 mV) depolarising response, and ❸ the maximum (-70 mV) hyperpolarising response.
(B) Shows the current response (pA) versus applied voltage (mV) plot for the whole-cell recorded data in experimental solutions. The solid black line indicates the line of best fit for the averaged data points. The reversal potential calculated from the plotted data was 0.81 mV, with an average current recording at +30 mV of 12.48 pA.
(C) Displays the response conductance (nS) versus applied voltage (mV) plot for the recorded whole-cell data in symmetrical solution conditions. The solid line is the line of best fit for the averaged conductances for each voltage step. The mean conductance for -30 mV was 0.34 nS, and the maximum conductance, recorded at +30 mV, was 0.42 nS. The calculated slope conductance at ±30 mV was 1.24.

Fig. 12. Whole-Cell Current Response of the Low B103 Subset to Transient Bath Application of 500 μM 5-HT in Symmetrical Solutions.

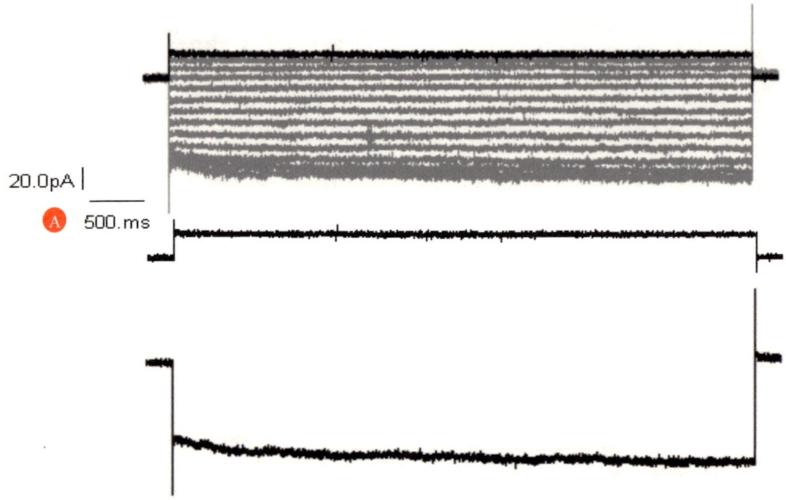

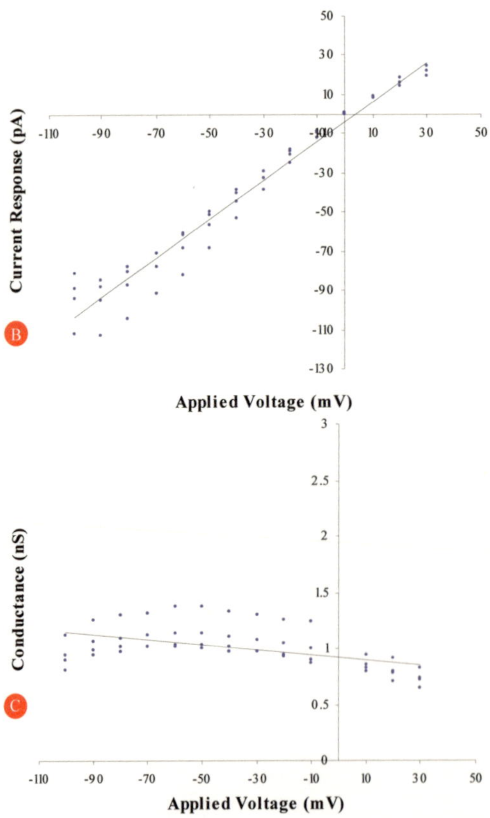

(A) Demonstrates a typical oscilloscope diagram of the whole-cell current response to the applied voltage protocol. ❶ indicates an entire Protocol response, ❷ the maximum (+30 mV) depolarising response, and ❸ the maximum (-100 mV) hyperpolarising response. (B) Shows the current response (pA) versus applied voltage (mV) plot for the whole-cell recorded data in symmetrical solutions. The solid black line indicates the line of best fit for the averaged data points. The reversal potential calculated from the plotted data was -3.64 mV, with an average current recording at +30 mV of 21.85 pA. (C) Displays the response conductance (nS) versus applied voltage (mV) plot for the recorded whole-cell data in experimental solution conditions. The solid line is the line of best fit for the averaged conductances for each voltage step. The mean conductance at -30 mV was 1.08 nS, and the maximum conductance, recorded at +30 mV, was 0.73 nS. The calculated ±30 mV slope conductance was 0.67.

Fig. 13. Whole-Cell Current Response of the Medium B103 Subset to Transient Bath Application of 500 μM 5-HT in Symmetrical Solutions.

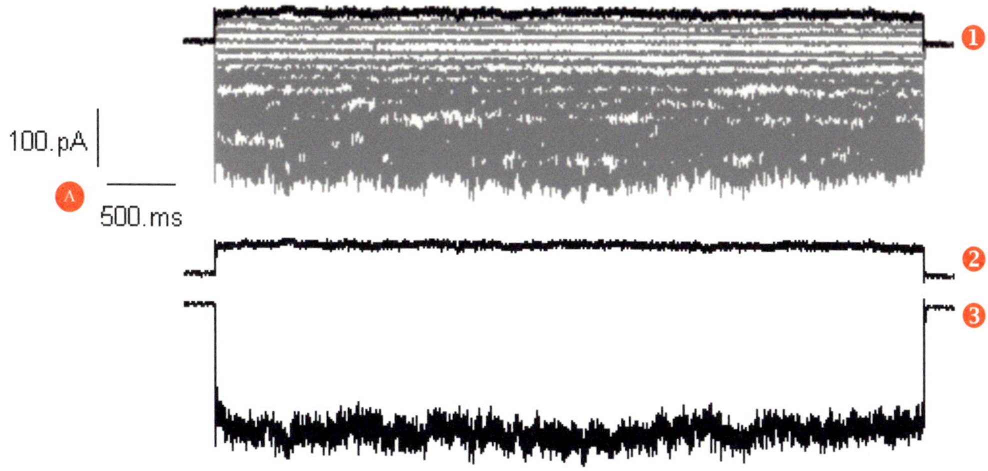

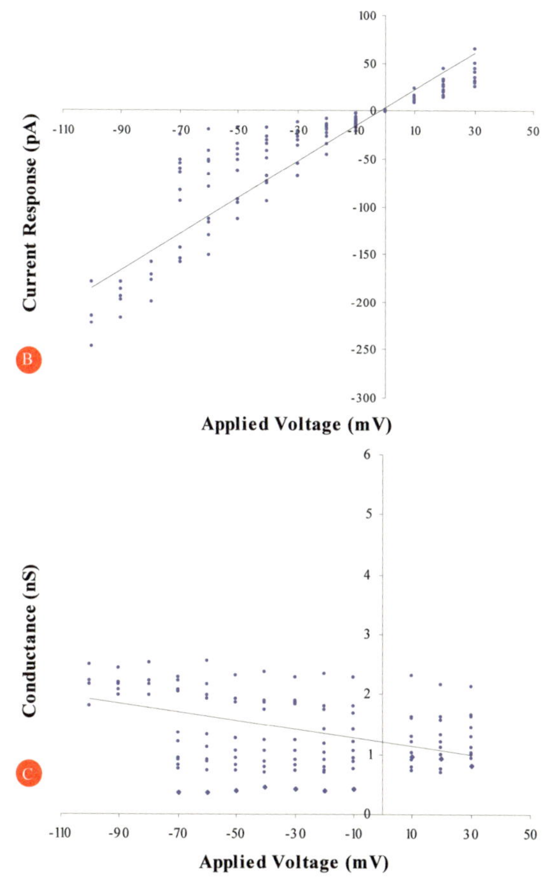

(A) Demonstrates a typical oscilloscope diagram of the whole-cell current response to the applied voltage protocol. ❶ indicates an entire Protocol response, ❷ the maximum (+30 mV) depolarising response, and ❸ the maximum (-100 mV) hyperpolarising response.

(B) Shows the current response (pA) versus applied voltage (mV) plot for the whole-cell recorded data in experimental solutions. The solid black line indicates the line of best fit for the averaged data points. The reversal potential calculated from the plotted data was 3.50 mV, with an average current recording at +30 mV of 39.09 pA.

(C) Displays the response conductance (nS) versus applied voltage (mV) plot for the recorded whole-cell data in symmetrical solution conditions. The solid line is the line of best fit for the averaged conductances for each voltage step. The mean conductance at -30 mV was 1.25 nS, and the mean conductance, recorded at +30 mV, was 1.30 nS. The calculated ±30 mV slope conductance was 1.05.

Fig. 14. Whole-Cell Current Response of the High B103 Subset to Transient Bath Application of 500 μM 5-HT in Symmetrical Solutions.

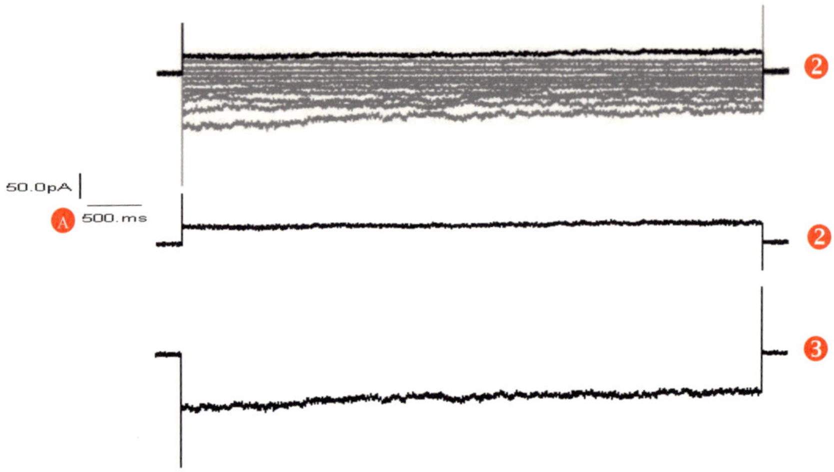

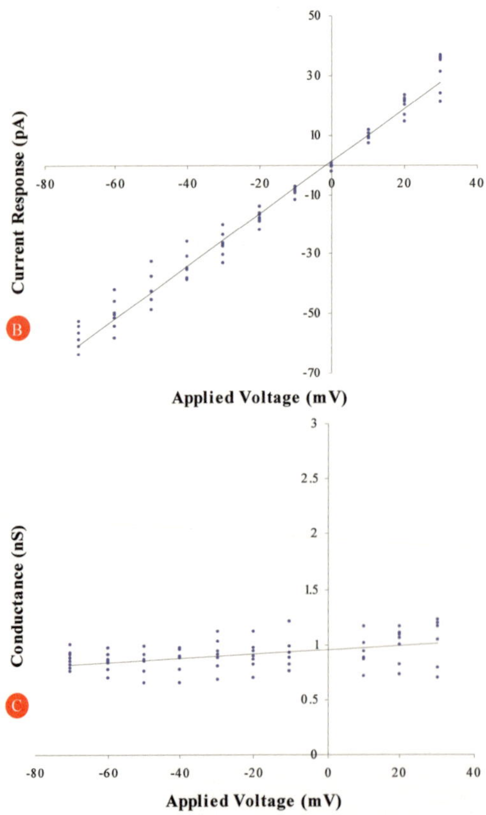

(A) Demonstrates a typical oscilloscope diagram of the whole-cell current response to the applied ❶ indicates an entire Protocol response, ❷ the maximum (+30 mV) depolarising response, and ❸ the maximum (-70 mV) hyperpolarising response.
(B) Shows the current response (pA) versus applied voltage (mV) plot for the whole-cell recorded data in experimental solutions. The solid black line indicates the line of best fit for the averaged data points. The reversal potential calculated from the plotted data was 1.26 mV, with an average current recording at +30 mV of 31.26 pA.
(C) Displays the response conductance (nS) versus applied voltage (mV) plot for the recorded whole-cell data in symmetrical solution conditions. The solid line is the line of best fit for the averaged conductances for each voltage step. The mean conductance at -30 mV was 0.90 nS, and the maximum conductance, recorded at +30 mV, was 1.04 nS. The calculated ±30 mV slope conductance was 1.16.

Fig. 15. Whole-Cell Current Response of the Low B103 Subset to Transient Bath Application of 1 mM 5-HT in Symmetrical Solutions.

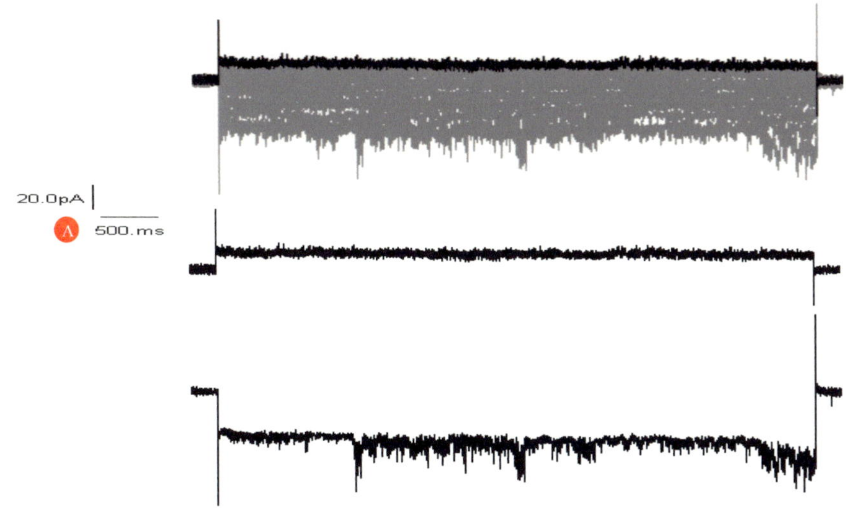

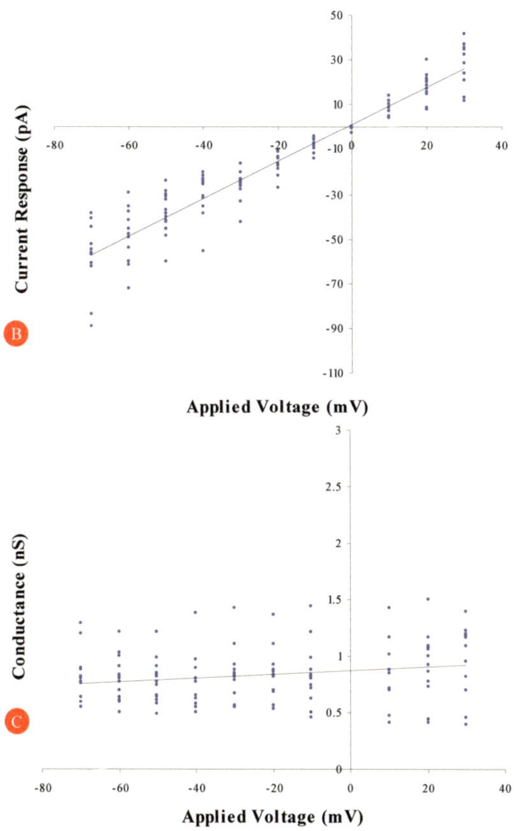

(A) Demonstrates a typical oscilloscope diagram of the whole-cell current response to the applied ❶ indicates an entire Protocol response, ❷ the maximum (+30 mV) depolarising response, and ❸ the maximum (-70 mV) hyperpolarising response.
(B) Shows the current response (pA) versus applied voltage (mV) plot for the whole-cell recorded data in symmetrical solutions. The solid black line indicates the line of best fit for the averaged data points. The reversal potential calculated from the plotted data was 1.33 mV, with an average +30 mV response current recording of 28.81 pA.
(C) Displays the response conductance (nS) versus applied voltage (mV) plot for the recorded whole-cell data in experimental solution conditions. The solid line is the line of best fit for the averaged conductances for each voltage step. The mean conductance at -30 mV was 0.82 nS, and the maximum conductance, recorded at +30 mV, was 0.96 nS. The calculated slope conductance was 1.17 for ±30 mV.

Fig. 16. Whole-Cell Current Response of the Medium B103 Subset to Transient Bath Application of 1 mM 5-HT in Symmetrical Solutions.

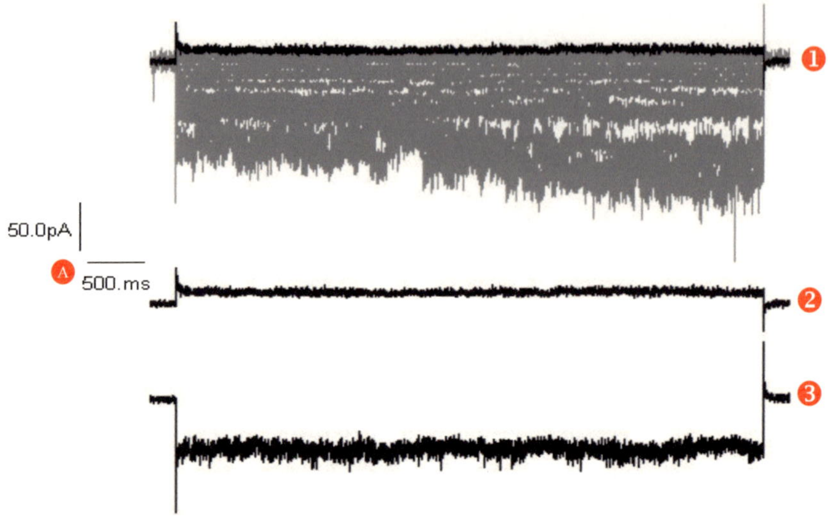

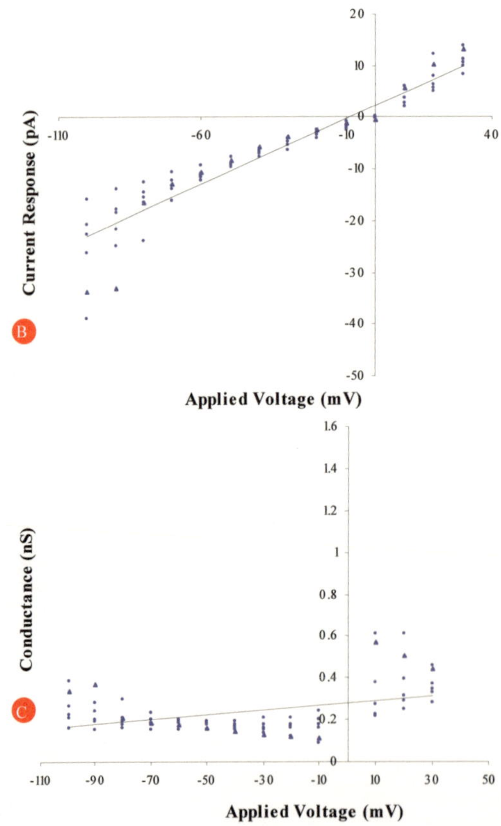

(A) Demonstrates a typical oscilloscope diagram of the whole-cell current response to the applied ❶ indicates an entire Protocol response, ❷ the maximum (+30 mV) depolarising response, and ❸ the maximum (-100 mV) hyperpolarising response.

(B) Shows the current response (pA) versus applied voltage (mV) plot for the whole-cell recorded data in experimental solutions. The solid black line indicates the line of best fit for the averaged data points. The reversal potential calculated from the plotted data was 2.46 mV, with an average current recording at +30 mV of 11.13 pA.

(C) Displays the response conductance (nS) versus applied voltage (mV) plot for the recorded whole-cell data in symmetrical solution conditions. The solid line is the line of best fit for the averaged conductances for each voltage step. The mean conductance at -30 mV was 0.16 nS, and the maximum conductance, recorded at +20 mV, was 0.39 nS. The calculated ±30 mV slope conductance was 2.3.

Fig. 17. Whole-Cell Current Response of the Low B103 Subset to Transient Bath Application of 5 μM D-Tubocurarine in Symmetrical Solutions.

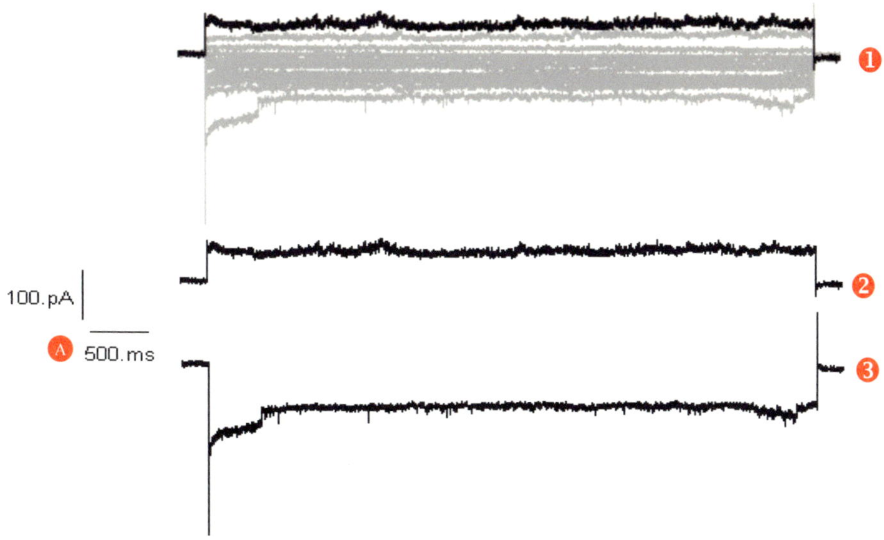

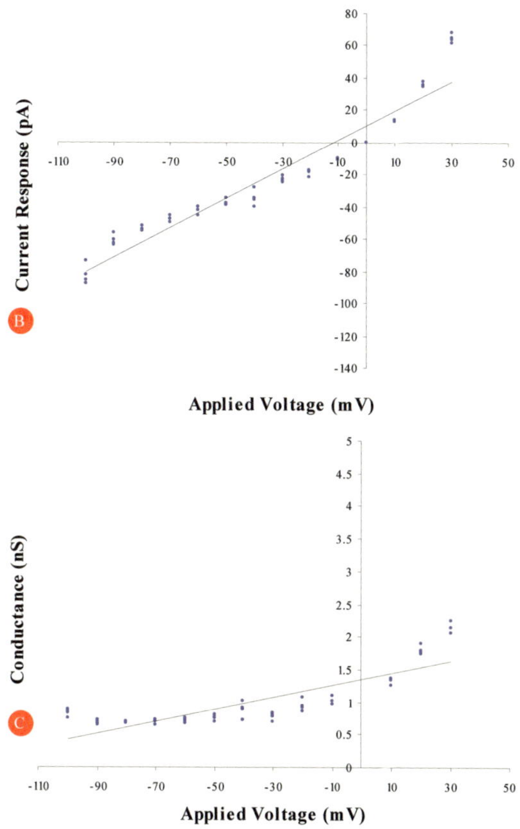

(A) Demonstrates a typical oscilloscope diagram of the whole-cell current response to the applied pulse. ❶ indicates an entire Protocol response, ❷ the maximum (+30 mV) depolarising response, and ❸ the maximum (-100 mV) hyperpolarising response.
(B) Shows the current response (pA) versus applied voltage (mV) plot for the whole-cell recorded data in symmetrical solutions. The solid black line indicates the line of best fit for the averaged data points. The reversal potential calculated from the plotted data was 10.50 mV, with an average current recording at +30 mV of 67.47 pA.
(C) Displays the response conductance (nS) versus applied voltage (mV) plot for the recorded whole-cell data in experimental solution conditions. The solid line is the line of best fit for the averaged conductances for each voltage step. The mean conductance at -30 mV was 0.78 nS, and the maximum conductance, recorded at +30 mV, was 2.25 pS. The calculated slope conductance at ±30 mV was 2.88.

Fig. 18. Whole-Cell Current Response of the Medium B103 Subset to Transient Bath Application of 5 μM D-Tubocurarine in Symmetrical Solutions.

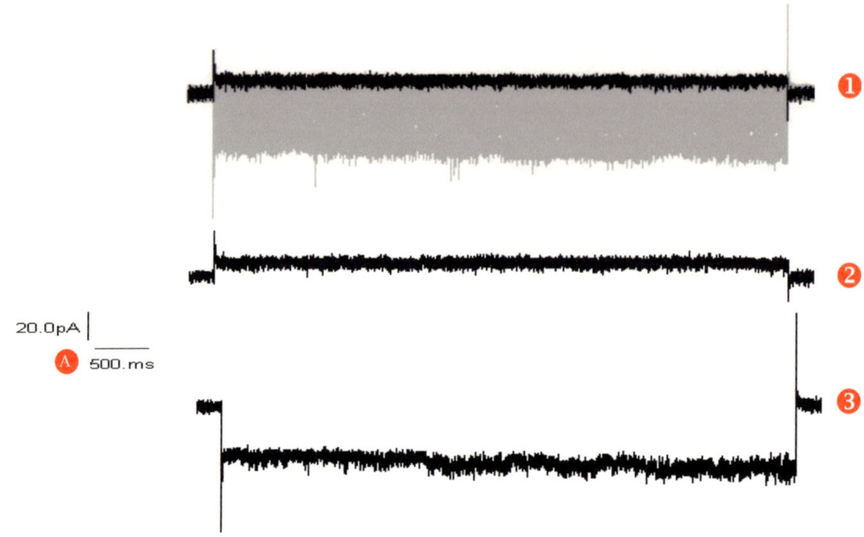

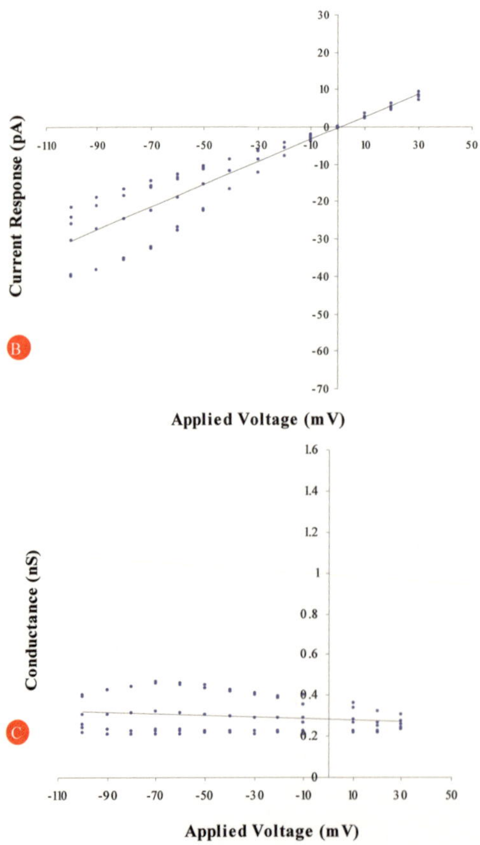

(A) Demonstrates a typical oscilloscope diagram of the whole-cell current response to the applied protocol. ❶ indicates an entire Protocol response, ❷ the maximum (+30 mV) depolarising response, and ❸ the maximum (-100 mV) hyperpolarising response.
(B) Shows the current response (pA) versus applied voltage (mV) plot for the whole-cell recorded data in symmetrical solutions. The solid black line indicates the line of best fit for the averaged data points. The reversal potential calculated from the plotted data was -0.36 mV, with an average +30 mV current response of 7.76 pA.
(C) Displays the response conductance (nS) versus applied voltage (mV) plot for the recorded whole-cell data in experimental solution conditions. The solid line is the line of best fit for the averaged conductances for each voltage step. The mean conductance at -30 mV was 0.29 nS, and the maximum conductance, recorded at -70 mV, was 0.32 nS. The calculated ±30 slope conductance was 0.89.

Fig. 19. Whole-Cell Current Response of the Low B103 Subset to Transient Bath Application of 5 μM D-Tubocurarine and 10 μM 5-HT in Symmetrical Solutions.

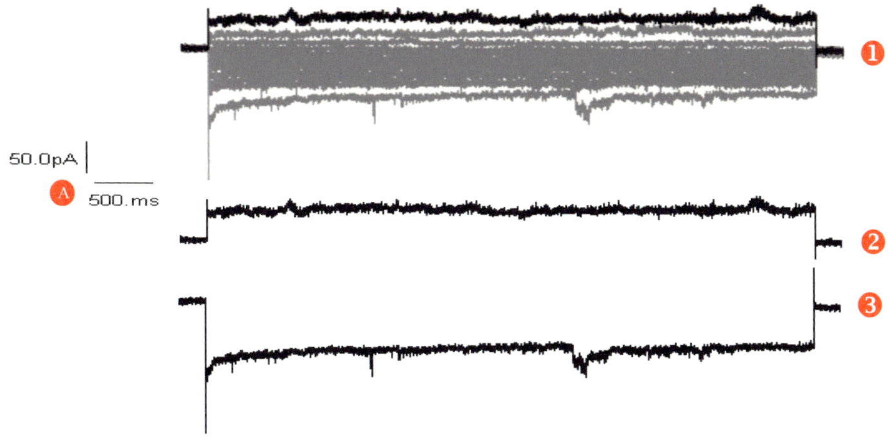

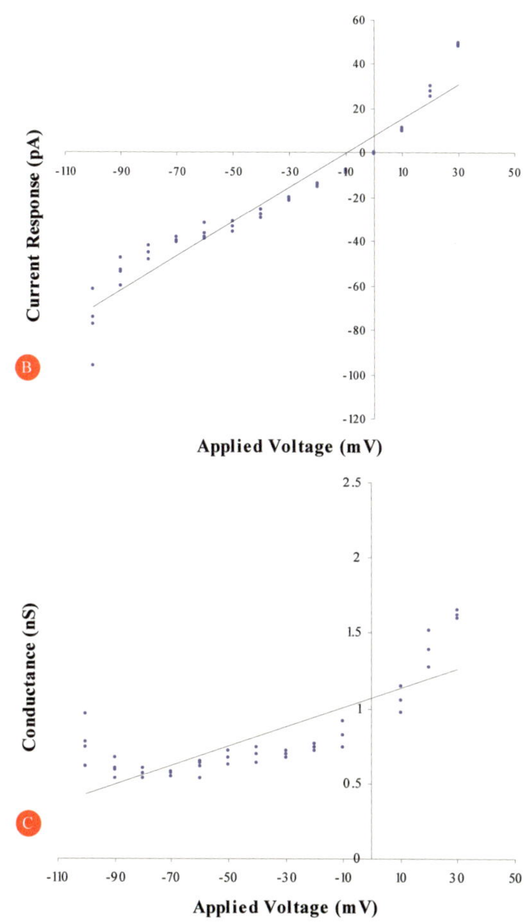

(A) Demonstrates a typical oscilloscope diagram of the whole-cell current response to the applied voltage. ❶ indicates an entire Protocol response, ❷ the maximum (+30 mV) depolarising response, and ❸ the maximum (-100 mV) hyperpolarising response.
(B) Shows the current response (pA) versus applied voltage (mV) plot for the whole-cell recorded data in experimental solutions. The solid black line indicates the line of best fit for the averaged data points. The reversal potential calculated from the plotted data was 7.76 mV, with an average +30 mV current recording of 48.70 pA.
(C) Displays the response conductance (nS) versus applied voltage (mV) plot for the recorded whole-cell data in symmetrical solution conditions. The solid line is the line of best fit for the averaged conductances for each voltage step. The mean -30 mV conductance was 0.70 nS, and the maximum conductance, recorded at +30 mV, was 1.62 nS. The calculated ±30 mV slope conductance was 2.33.

Fig. 20. Whole-Cell Current Response of the Medium B103 Subset to Transient Bath Application of 5 μM D-Tubocurarine and 10 μM 5-HT in Symmetrical Solutions.

		Mean Current (pA)		Mean g (nS)		Max g (nS)		Most	Slope g	E_{rev}	Transient Current (pA)		SVA
		@ -30 mV	@ +30 mV	@ -30 mV	@ +30 mV		@ mV	Frequent g	@ ±30mV		Fast	SS	
Normal Solutions		-5.57	5.67	0.19	0.19	0.25	-100	0.18	1.02	0.14	n/a	n/a	D
Symmetrical Solutions													
controls	Low	-7.72	8.31	0.26	0.28	0.28	+30	0.27	1.08	-0.13	-43.01	-20.64	B
	Medium	-20.61	16.36	0.69	0.55	0.97	-10	0.63	0.80	-3.32	-30.33	-24.55	I
	High	-37.98	41.35	1.27	1.38	1.39	-60	1.30	1.09	0.57	no peak	no peak	D
10 uM 5-HT	Low	-8.34	9.08	0.28	0.30	0.30	+30	0.24	1.09	0.34	-86.30	-29.61	D
	Medium	-39.41	92.78	1.31	3.09	3.09	+30	1.30	2.35	13.91	-244.40	-111.11	D
500 uM 5-HT	Low	-10.10	12.48	0.34	0.42	0.42	+30	0.32	1.24	0.81	-40.28	-25.0	B
	Medium	-32.43	21.85	1.08	0.73	1.14	-60	1.10	0.67	-3.64	-133.21	-86.67	I
	High	-37.35	39.09	1.25	1.30	2.22	-80	1.20	1.05	3.50	no peak	no peak	B
1 mM 5-HT	Low	-26.99	31.26	0.90	1.04	1.04	+30	0.94	1.16	1.260	-215.69	-110.07	I
	Medium	-24.52	28.81	0.82	0.96	0.96	+30	0.81	1.17	1.33	-91.53	-57.06	I
5 uM d-tubocurarine	Low	-4.87	11.13	0.16	0.37	0.39	+20	0.80	2.28	2.46	-115.18	-51.06	D
	Medium	-23.39	67.47	0.78	2.25	2.25	+30	0.70	2.88	10.50	-234.43	-127.00	B
5 uM d-tubocurarine + 10 uM 5-HT	Low	-8.75	7.76	0.29	0.26	0.32	-70	0.32	0.89	-0.36	-192.88	-52.88	D
	Medium	-20.86	48.70	0.70	1.62	1.62	+30	0.56	2.33	7.76	-202.54	-99.107	D

E_{rev} for solution components was derived using the Nernst Equation

Normal Solutions	E_K = -78.34 mV	Symmetrical Solutions	E_K = n/a
	E_{Na} = 91.24 mV		E_{Na} = 0.0 mV
	E_{Ca} = 0.007 mV		E_{Ca} = n/a
	E_{Cl} = -0.05 mV		E_{Cl} = -107.69 mV

g = conductance
SVA = successive voltage applications
where D = decrease in current response from initial, I = increase, & B = both increase and decrease noted.

CRT = Current Response Type
SS = steady-state

Note that data in this table is tabulated from averaged information and therefore some discrepencies might be noted when specific individual responses are viewed.

Table 2. B103 Electrophydiological Response Summary

Fast transients demonstrated a -423.20 pA response to -100 mV stimulation with a slower decay time than previously noted of 1000 ms. The steady state response then lasted for the remaining 4000 ms with an average current of -133.33 pA.

The maximal average whole-cell conductance recorded from the medium subset of B103 cells in response to transient externally applied 500 μM 5-HT with experimental solutions (Figure#13) was 1.39 nS at -60 mV. The calculated Erev was -3.64 mV and the calculated ±30 mV slope conductance was 0.67.

The 5-7 ms fast transient response to maximum hyperpolarisation was -133.21 pA with the steady-state component displaying an average -86.67 pA current response. However the steady-state transient displayed an initiation at approximately half-maximal then increased in response to reach the average current.

Consecutive Pulse Protocol applications showed a trend for maximum hyperpolarisation current response to decrease stepwise from that initially recorded for each cell.

3.3.3 Serotonin receptor channel currents in B103 cell (High conductance subset)

Exhibits Whole-Cell Current Response of the High B103 Subset to Transient Bath Application of 500 μM 5-HT (Figure#14) in Symmetrical Solutions, as 10 μM 5-HT was not able to produce any response in this sub set of B103 cells.

All serotonin concentrations except 500 μM and other drugs were applied to only low and medium subsets of B103 cells.

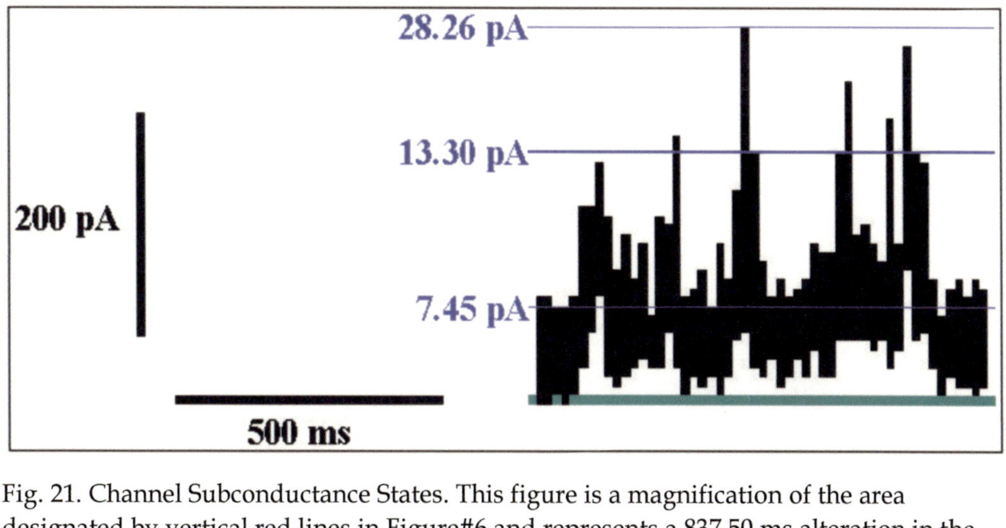

Fig. 21. Channel Subconductance States. This figure is a magnification of the area designated by vertical red lines in Figure#6 and represents a 837.50 ms alteration in the channel conducting state. As indicated in the figure by the horizontal blue lines, the max current recorded was 28.26 pA (0.88 nS), the average current was 7.45 (0.25nS), and the probable true maximum conductance state for the channel was when 13.30 pA of current was recorded (0.44 nS) - chosen on the basis of the number of peaks passing through the line.

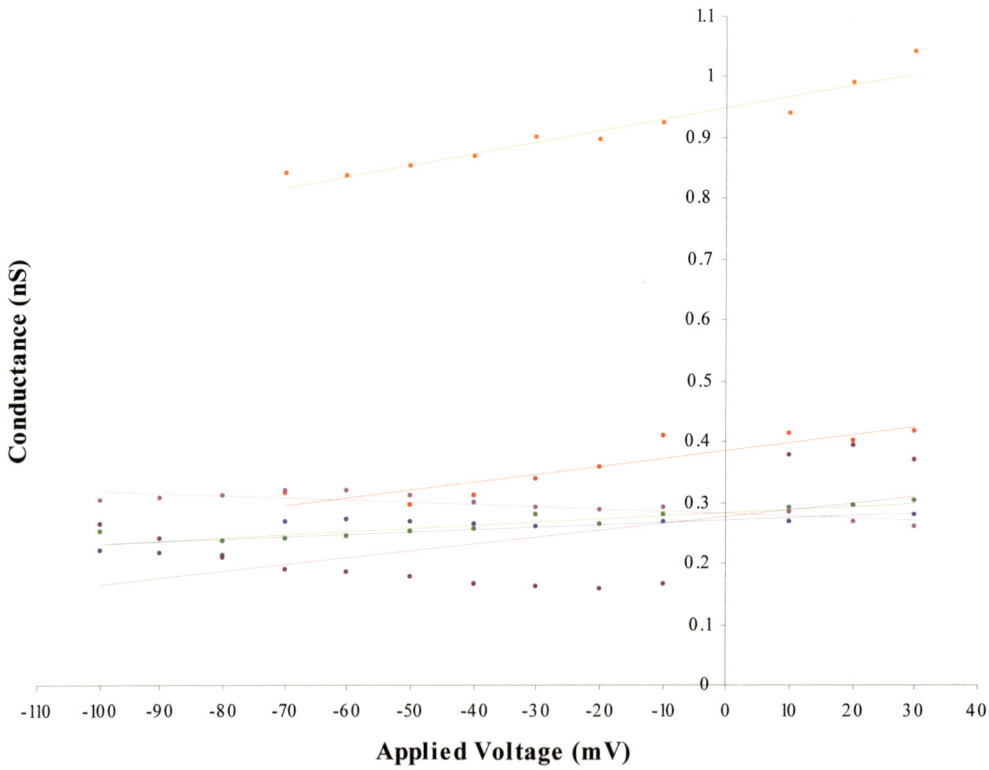

Fig. 22. Conductance Comparison for the Low B103 Subset. Mean values for control (blue), 10 μM 5-HT (green), 500 μM 5-HT (red), 1 mM 5-HT (orange), 5 μM d-tubocurarine (plum), and 5 μM d-tubocurarine plus 10 μM 5-HT (pink) are shown plotted against applied voltage. The solid lines represent the lines of best fit for each averaged data series. Clearly demonstrated is an increase in channel conductance associated with the addition of increasing concentrations of 5-HT so that 1 mM > 500 μM > 10 μM. Also shown is a decrease in conductance with the 5 μM d-tubocurarine at hyperpolarising potentials however stimulation with 5 μM d-tubocurarine and 10 μ M 5-HT at hyperpolarising potentials appears to increase the low subset conductance above that seen with 10 μM 5-HT alone. Further research will be required to isolate the cause of this increase.

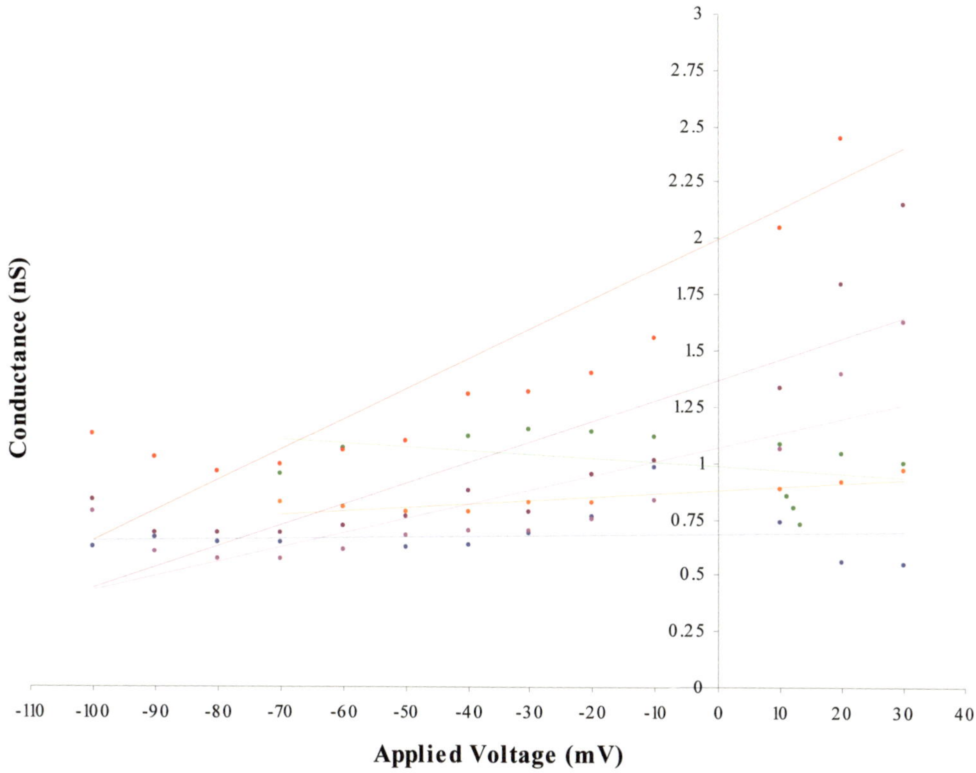

Fig. 23. Conductance Comparison for the Medium B103 Subset. Mean values for control (blue), 10 μM 5-HT (red), 500 μM 5-HT (green), 1 mM 5-HT (orange), 5 μM d-tubocurarine (plum), and 5 μM d-tubocurarine plus 10 μM 5-HT (pink) are shown plotted against applied voltage. The solid lines represent the lines of best fit for each averaged data series. Clearly demonstrated is a decrease in channel conductance associated with the addition of increasing concentrations of 5-HT so that 10 μM > 500 μM > 1 mM. Also shown is an apparently agonistic effect cause by the addition of 5 μM d-tubocurarine, interestingly the response to d-tubocurarine alone shows a greater conductance to that seen with 10 μM 5-HT as well. Further research will be required to isolate the cause of this phenomenon.

4. Discussion and conclusion

We describe the ionic movements in the physiological and symmetric solutions. The solutions used for physiological normal control results employed K^+ as the primary pipette solution cationic component, and Na^+ as the primary bath cation. This was because the normal conditions under which a cell exists demonstrates a higher internal K^+ than Na^+ concentration, thus recorded current results mimicked normal phenomenon. This meant that response currents were expected to be primarily carried mainly by K^+ efflux or Na^+ influx.

Ion movement is dependent on the electrochemical gradient produced only by its own subset of ions, thus it is independent of the concentration of other ions. This means that movement of Na^+ in solution is caused by the relative concentration of Na^+ alone and is not affected by the concentration of other ions in solution. In normal physiological conditions (*in vivo*) early transient currents that reverse their signs from inward current flow to outward current flow at values greater than around -60 mV (normal membrane resting potential) would be expected to be carried by Na^+ so that correspondingly our experimental results in normal physiological solutions ought to be reversing at around -91.2 mV (E_{Na+}). Alternatively late outward currents would be expected to be carried by K^+ with a E_{rev} more negative than -60 mV. A comparison of E_{rev} for Na^+ (91.2 mV), K^+ (-78.3 mV), Ca^{2+} (0.007), and Cl^- (-0.05) with the recorded E_{rev} (0.14 mV) clearly indicated that currents were passing through the point of origin demonstrating non-selective ion permeation.

The single-channel current responses noted for cells in physiologically normal solutions indicated the probability of multiple channel subconductance states (Figure#21). With a maximum conductance of 0.95 pS and an average of 0.25 nS. The probable true maximum conductance state for the channel was when 13.30 pA of current was recorded (0.44 nS). Subconductance states exist because while a channel might be open, and therefore conducting ions, it might not be fully activated or conducting at its full capacity.

While the normal physiological ionic concentration of a cell's environment is not symmetrical (i.e. the intracellular fluid and extracellular fluid do not have the same ionic composition) current recordings of a particular selective channel can be enhanced by using symmetrical solutions with a greater than normal concentration of the specific permeant ion. Na^+ was used in this experimental series in order to emphasis and characterise the kinetics of known channels.

The B103 cells were divided into three electrophysiological response groups based on the observed variation in cellular current response to -100 mV stimulus in symmetrical bathing and pipette solutions: low, medium, and high conductance B103 subsets.

The averaged maximum conductance control result (0.28 pS at +30 mV) indicates that this subset shows its highest voltage-determined conductance at **positive potentials** (>0 mV) thus displaying **outward current rectification** (positive ions move from the cellular cytoplasm into the surrounding solution).

A fast-activating increase in channel conductance in response to the addition of serotonin was observed, where an increase in 5-HT concentration resulted in a higher conductance level, so that conductance response for the low B103 subset was 1 mM > 500 μM > 10 μM (Figure#22).

E_{rev} for the low subset varied from control -0.13 mV to 2.46 mV for 5μM d-tubocurarine. While most of the values fell close enough to E_{Na+} (0.0 0± 0.5 mV) to indicate that Na^+ was the primary ionic contributor to current the values for 500 μM and 1 mM 5-HT and 5 μM d-tubocurarine were slightly higher suggesting other ions produced some component of these conductances. Further experiments (where ions are selectively removed from the bath and pipette) are required to identify the percentage of current response comprised by components other than Na^+. Only a fraction of the delayed steady-state current response could have been caused by Ca^{2+} permeation, however, as very little Ca^{2+} was presenting the symetrical solutions. This low concentration was deliberately produced as in the normal cellular resting state cytoplasmic free Ca^{2+} levels are held at extremely low concentrations lying in the range of 20-300 nM in living cells. This concentration is maintained by the combined action of the ATP-dependent pump and Na^+/Ca^{2+} exchanger systems on the surface of the membrane, as well as by ATP-dependent pumps present on intracellular organelles such as the endoplasmic reticulum.

In the low subset recordings (Figure#10) there was a high frequency response component to 10 μM 5-HT at -100 mV, just before the initiation of the steady-state current. This may indicate the 5-HT_3 receptor channel current component.

A heterogeneity of current responses was observed for 500 μM 5-HT applied to the medium subset demonstrating the presence of different receptor conductance states which keep on increasing even after 5000 ms. By 5000 ms the steady-state current amplitude get doubled as compared to the initial response. Receptor heterogeneity was again displayed in the presence of 1 mM 5-HT where some receptors were silent at hyperpolarising potentials while some were bursting. An increasing current after 5000 ms again indicated continued channel opening or increase in subconductance levels.

A comparison of the mean conductances recorded for the varying concentrations of 5-HT for medium B103 cells (Figure#23) shows that the cells demonstrated a decreasing current response to increasing 5-HT concentration where 10 μM > 500 μM > 1 mM. These results are comparable to the results previously obtained for N1E-115 cells, where maximal response was noted at 10 μM 5-HT (Neijt, Duits, & Vijverberg, 1988). Only 500 μM 5-HT stimulated the high subset of B103 cells, with the ±30 mV slope conductance showing that depolarising potentials demonstrate a higher conductance. Further investigation is warranted to clarify this decreased response.

D-tubocurarine was employed as a competitive antagonist to identify 5-HT_3Rs in B103 cell-lines. The recorded responses to d-tubocurarine indicated that rather than antagonising 5-HT_3R activity it was having a modulatory affect on the native B103 receptors for both low and medium subsets. The low cells had a more normal response with a decrease in conductance seen with d-tubocurarine at hyperpolarising potentials, however stimulation with both d-tubocurarine and 10 μM 5-HT at hyperpolarising potentials appeared to increase subset conductance to a level above that seen with 10 μM 5-HT alone.

Our results are indicative of either a change in the amino acid composition of the antagonist binding area of the 5-HT_3R (indicating different subunit composition of 5-HT_3R in these cell-lines as compared to native neuronal cells or isolated recombinant α and β subtypes), or that the same subunits are present with different amino acid compositions (splice variants). Also, at low concentrations some antagonist can act as positive modulators of receptors. Further research will be required to isolate the cause of this increase.

In summary, we describe the patch clamp experiments for B103 cells as **Black Box Test Known Inputs**

1. B103 cells were chosen for patching based on their general morphology: approximately 25 μm in diameter, well-defined clean cell membrane.
2. Only non-contaminated healthy B103 cells were used for patch clamp experiments.
3. Two sets of bath and pipette solutions were used through out the experiments. One which mimics the Extracellular and intracellular ionic composition and second with similar sodium concentration on both side of the cell membrane as to get close to zero reversal potential. The second set of solution was used to observe the serotonin gated currents.

4. Serotonin solutions of different known concentrations were used in the bath to see 5-HT3R currents. In these experiments TTX or phenytoin solution were used to abolish any endogenous currents
5. Pharmacological agents from the same companies were used through out the experiments.
6. Well-regulated bath perfusion system was used to challenge patched cells with serotonin hydrochloride solutions of 1 mM, 500 μM, and 10 μM concentrations.
7. Thin walled borosilicate glass capillaries (1.5 mm O.D. × 1.17 mm I.D) were used to produce patch pipettes with a 3 MΩ resistance. Pipettes were half-filled using both the front- and back-filling techniques
8. Same Patch clamp setup (HEKA EPC9 amplifier and HEKA *Pulse* software package) fully grounded without any noise was used through out the experiments, with daily calibration.
9. Constant Pulse Protocol facilitated via the HEKA *Pulse* software was used through out the experiments. Voltage procedure design for the voltage gated experiments ranges from -100 mV to +30 mV increasing in 10 mV steps with a resting period at 0 mV between each step.
10. B103 cells were categorized into three types based on their current response to the maximum hyperpolarizing step in the Protocol, which is at -100mV. Responses that were observed to be of 30 pA or less were categorized into low subset, between 30-100 pA were categorized into the medium subset and more than 100 pA in high subset.

Unknown outputs (some examples)

1. B103 cells with similar morphology and experimental conditions randomly generate three different sub sets of conductances.
2. An increasing steady state current even after 5000 ms in the medium subset
3. Only 500 μM 5-HT stimulated the high subset of B103 cells.
4. Action of d-tubocurarine as agonist to B103 currents of both low and medium subsets.
5. The low sub set cells had a more expected response with d-tubocurarine at hyperpolarising potentials.
6. D-tubocurarine in the presence of 10 μM 5-HT at hyperpolarising potentials increases 5-HT3 currents more than that seen with 10 μM 5-HT alone.

Looking for answers to the unknown outcomes and mechanisms of our experiments.

5. References

[1] Alien, D.H., Lepple-Wienhues, A., Cahalan, MD 1997. Ion channel phenotype of melanoma cell lines. J. Membr. Biol. 155:27–34

[2] Branchek T., Kates M., Gershon M.D. 1984. Enteric receptors for 5-hydroxytryptamine. *Brain Research.* 324(1):107-118.

[3] Dahlstrom A., Fuxe K. 1960. Evidence for the existence of monoamine-containing neurons in the central nervous system. I Demonstration of monoamines in the cell bodies of brain stem neurons. *Acta Physiologia Scandinava.* 62:1-55.

[4] Druckrey H., Preussmann R., Ivankovic S., Schmahl D. 1967. Organotropic carcinogenic effects of 65 various N-nitroso- compounds on BD rats. *Zeitschrift fur Krebsforschung.* 69(2):103-201

[5] Fraser, S.P., Diss, J.K.J., Mycielska, M.E., Coombes, R.C., Djamgoz, M.B.A 2002. Voltage-gated sodium channel expression in human breast cancer cells: Possible functional role in metastatis. Breast Cancer Research & Treatment 76 (Suppl 1):S142

[6] Freschi J.E., Shain W.G. 1982.Electrophysiological and pharmacological characteristics of the serotonin response on a vertebrate neuronal somatic cell hybrid. *The Journal of Neuroscience.* 2(1):106-112.

[7] Greenshaw A.J. 1993. Behavioural pharmacology of 5-HT3 receptor antagonists: a critical update on therapeutic potential. [Review]. *Trends in Pharmacological Sciences.* 14(7):265-270.

[8] Hales T.G., Tyndale R.F. 1994. Few cell lines with $GABA_A$ mRNAs have functional receptors. *Journal of Neuroscience.* 14(9):5429-5436.

[9] Hamil O. P., Marty A., Neher E.,Sakmann B., Sigworth F., 1981 Improved patch clamp techniques for high-resolution current recordings from cells and cell free membrane patches. Pfluger Archive. 391: 85-100

[10] Kasckow J.W., Tillakaratne N.J., Kim H., Strecker G.J., Tobin A.J., Olsen R.W. 1992. Expression of $GABA_A$ receptor polypeptides in clonal rat cell lines. *Brain Research.* 581(1):143-147.

[11] Kilpatrick G.J., Jones B.J., Tyers M.B. 1987. Identification and distribution of 5-HT3 receptors in rat brain using radioligand binding. *Nature.* 330 (6150):746-748.

[12] Kirk E.E., Giorano J., Anderson R.S. 1997. Serotonergic receptors as targets for pharmacotherapy. [Review]. *Journal of Neuroscience Nursing.* 29(3):191-197.

[13] Lambert J.J., Peters J.A., Hales T.G., Dempster J. 1989. The properties of 5-HT3 receptors in clonal cell lines studied by patch-clamp techniques. *British Journal of Pharmacology.* 97(1):27-40.

[14] Laniado, M.E., Lalani, E.N., Fraser, S.P., Grimes, J.A., Bhangal, G., Djamgoz, M.B.A., Abel, P.D 1997. Expression and functional analysis of voltage-activated Na+ channels in human prostate cancer cell lines and their contribution to invasion in vitro. Am. J. Pathol. 150:1213–1221

[15] Maricq A.V., Peterson A.S., Brake A.J., Myers R.M., Julius D. 1991. Primary structure and functional expression of the 5HT3 receptor, a serotonin-gated ion channel. *Science.* 254(5030):432-437.

[16] Mook-Jung I., Joo I., Sohn S., Kwon H.J., Huh K., Jung M.W. 1997. Estrogen blocks neurotoxic effects of beta-amyloid (1-42) and induces neurite extension on B103 cells. *Neuroscience Letters.* 235(3):101-104.

[17] Napias C., Olsen R.W., Schubert D. 1980. GABA and picrotoxinin receptors in clonal nerve cells. *Nature.* 283(5744):298-299.

[18] Neijt H.C., Te Duits I.J., Vijverberg H.P.M 1988. Pharmacological characterization of serotonin 5-HT3 receptor-mediated electrical response in cultured mouse neuroblastoma cells.. *Neuropharmacology.* 27(3):301-307.

[19] Ninomiya H., Roch J.M., Jin L.W., Saitoh T. 1994. Secreted form of amyloid beta/A4 protein precursor (APP) binds to two distinct APP binding sites on rat B103 neuron-like cells through two different domains, but only one site is involved in neuritotropic activity. *Journal of Neurochemistry.* 63 (2):495-500.

[20] Peters J.A., Lambert J.J. 1989. Electrophysiology of 5-HT3 receptors in neuronal cell lines. [Review]. *Trends in Pharmacological Sciences.* 10 (5):172-175.

[21] Roger, S., Besson, P., Le Guennec, J.Y 2003. Involvement of a novel fast inward sodium current in the invasion capacity of abreast cancer cell line. Biochim Biophys Acta. 1616:107–111

[22] Schubert D., Heinemann S., Carlisle W., Tarikas H., Kimes B., Patrick J., Steinbach J.H., Culp W., Brandt B.L. 1974. Clonal cell lines from the rat central nervous system. *Nature.* 249(454):224-227.

[23] Schubert D. 1975. The uptake of GABA by clonal nerve glia. *Brain Research.* 84(1):87-98.

[24] Segal M. M and Douglas, A. F. 1997. Late Sodium Channel Openings Underlying Epileptiform Activity Are Preferentially Diminished by the Anticonvulsant Phenytoin *J Neurophysiol* 77: 3021-3034

[25] Tyndale R.F., Hales T.G., Olsen R.W., Tobin A.J. 1994. Distinctive patterns of GABAA receptor subunit mRNAs in 13 cell lines. *Journal of Neuroscience.* 14(9):5417-5428.

[26] Wallis D.I., Woodward B. 1975. Membrane potential changes induced by 5-hydroxytryptamine in the rabbit superior cervical ganglion. *British Journal of Pharmacology.* 55(2):199-212.

[27] Yan D., Schulte M.K., Bloom K.E., White M.M. 1999. Structural features of the ligand-binding domain of the serotonin 5HT3 receptor. *Journal of Biological Chemistry.* 274(9):5537-5541.

[28] Yoshifumi Kawanabe , Nobuo Hashimoto, Tomoh Masaki 2001 B103 neuroblastoma cells predominantly express endothelin ET_B receptor; effects of extracellular Ca^{2+} influx on endothelin-1-induced mitogenesis Eur J of Pharmacol 425 (3), 173-179

2

Intracellular Signaling Pathways Integrating the Pore Associated with P2X7R Receptor with Other Large Pores

L.G.B. Ferreira[1], R.A.M. Reis[2], L.A. Alves[1] and R.X. Faria[1]
[1]*Laboratory of Cellular Communication, Oswaldo Cruz Institute, Oswaldo Cruz Foundation*
[2]*Laboratory of Neurochemistry, Biophysical Institute, University Federal of Rio de Janeiro*
Brazil

1. Introduction

The purinergic P2X$_7$ receptor (P2X7R) is a member of the family of ligand-gated ion channels composed of seven subtypes, P2X$_{1-7}$. These receptors possess three subunits assembled as homo- or heterotrimers to make functional receptors (Nicke et al., 1998, 2005), which has been confirmed by atomic force microscopy (Barrera et al, 2005) and by crystallography (Kawate et al, 2009). The P2X7R shares an overall membrane topology with the other members of this family of receptors; it contains two putative pore forming transmembrane segments, a large cysteine-rich ligand-binding extracellular domain, and intracellularly located N and C termini (Surprenant et al, 1996). This subtype is structurally distinguished from other members of P2XRs by its long intracellular C-terminal tail with multiple protein and lipid interaction motifs, besides a cysteine-rich 18 - amino acid segment. From a pharmacological point of view, the P2X7R requires at least a 100-fold higher ATP concentration for activation compared to other P2XRs (North, 2002). Extracellular divalent cation reduction increases agonist potency (Hibell et al, 2001; Michel et al, 1999). Moreover, extracellular cations and chloride have important effects on the channel gating (Gudipaty et al., 2001; Li et al., 2003, 2008, 2010; Riedel et al., 2007b; Virginio et al., 1997). In this context, P2X7R activation can sustainably induce a wide range of different intracellular signaling responses (Dubyak, 2007; North, 2002).

P2X7R, when activated by ATP or the potent agonist 3'-O-(4-benzoyl)benzoyl-ATP (BzATP), functions as a non-selective cation channel, permeant to small cations, such as Na^+, K^+, and Ca^{2+} and this activation mode is dependent on extracellular divalent cations (Ding & Sachs, 2000; Jiang, 2009; Ma et al, 2006; Sperlágh et al, 2006; Virginio et al, 1997). Upon repeated or prolonged application of agonist, the P2X7R becomes permeable to larger molecules like ethidium bromide, N-methyl-D-glucamine or neurotransmitters such as glutamate and ATP (Faria et al, 2005, 2010; Hamilton et al, 2008; Jiang et al, 2005; Marcoli et al, 2008), a process termed cell "permeabilization" or large conductance channel opening.

Up to now, several ideas have been proposed as a possible explanation to this pore opening (Alloisio et al, 2010; Coutinho-silva et al, 1997; Faria et al, 2005, 2009; Jiang et al, 2005; North,

2002; Pelegrin & Surprenant, 2006; Virginio et al, 1999; Yan et al, 2008). In a general manner, cell permeabilization can be observed in cell types transfected or natively expressing P2X7R, in contrast to Xenopus oocytes transfected cells (Petrou et al, 1997) or lymphocytes B cells of patients with Chronic lymphocytic leukaemia (Boldt et al, 2003; Gu et al, 2000).

Initially, some groups have proposed that the opening of the pore associated to P2X7R receptor occurs after the small channel (8 pS) allosterically changes and expands over time-dilatation (Chaumont & Khakh, 2008; Virginio et al, 1999; Yan et al, 2008). Others have reported that Pannexin-1 (panx-1), a hemichannel protein, is the large conductance channel that opens independently of the cationic P2X7R (Locovei et al, 2007; Pelegrin & Surprenant, 2006, 2007). On the other hand, some groups have demonstrated another putative explanation of the dye uptake. They have reported two possible charge selectively pathways that are activated by P2X7R: one for cationic and another for anionic dyes (Cankurtaran-Sayar et al, 2009; Schachter et al, 2008). Jiang and collaborators have proposed distinct pathways to permeate inorganic monovalent and divalent cations, organic cations ($NMDG^+$) and fluorescent dyes (Jiang et al, 2005). In addition, it has been suggested at least two different conductive pathways, one for Ca^{2+} and other for monovalent ions (Alloisio et al, 2010).

In this context, some groups have published that the P2X7R permeabilization depends on intracellular factors (Donnely-Roberts et al, 2004; Faria et al, 2005, 2010; Gu et al, 2009; Le Stunff & Raymond, 2007; Shemon et al, 2004; Zhao et al, 2007) to occur.

In this chapter, we come up with novel proposals of intracellular signaling regulation that would help us to understand about the several intriguing characteristics of the ATP-induced P2X7R.

2. Intracellular signalling associated with small conductance channel

In the P2X7R receptor gating mechanism (low conductance channel) the response to agonist challenge, allow rapid, non-selective passage of cations across the cell membrane. It is permeable to Na^+ and K^+ and presents high permeability to Ca^{2+} (North, 2002). The cellular response to P2X7R receptor activation after agonist exposition is generally very rapid and this reaction does not depend on the production and diffusion of second messengers within the cytosol (Burnstock, 2007; Ralevic & Burnstock, 1998). In contrast, an increase in the intracellular Ca^{2+} concentration and a consequent depolarization of the cell membrane are observed after the ionic channel opening, subsequently activating voltage-gated calcium channels. In addition, evidence indicates that the large Ca^{2+} ion concentration in the cytoplasm could activate intracellular kinases like protein kinase C (PKC), mitogen activated protein kinases (MAPKs), calcium-calmodulin-dependent protein kinase II (CaMKII) (Amstrup & Novak, 2003; Bradford & Soltoff, 2002; Heo & Han, 2006), caspases (Orinska et al, 2011), phosphoinositide 3-kinase (Pi3K) (Jaques-Silva et al, 2004) and phospholipases (Alzola et al, 1998; Perez-Andres et al, 2002; Pochet et al, 2007). Further, other functions may be mediated by P2X7R receptor activation such as IL-1β maturation (Ferrari et al, 1997), shedding of membrane proteins (Moon et al, 2006); Src (Denlinger et al, 2001), and glycogen synthase kinase 3 (Ortega et al, 2009); membrane blebbing (Morelli et al, 2003).

These different signaling pathways may contribute to a large complexity in the response of this receptor and it raises the question about how the correct coupling and fine tuning of the signaling in response to extracellular stimuli is achieved. To date, several pieces of evidence have been described about the small conductance P2X7R receptor and interactions with other proteins (Adinolfi et al, 2003; Antonio et al, 2011; Barbieri et al, 2008; Boumechache et al, 2009; Bradley et al, 2010; Denlinger et al, 2001; Guo et al, 2007; Lemaire et al, 2007; Liu et al, 2011; Wilson et al, 2002), mainly through its longer C-terminus region or lipids (Denlinger et al, 2001; Gonnord et al, 2009; Michel& Fonfria, 2007; Takenouchi et al, 2007; Zhao et al, 2007a, 2007b). However, there are only a few studies about the low conductance P2X7R receptor gating and the mechanism of transition to large conductance channel opening. This may be due to (i) a lack of selective agonists or antagonists only to small or large channel (North, 2002) or (ii) to P2X7R receptor polymorphisms or (iii) to artificial deletions in regions of this receptor resulting in distinct intracellular or extracellular regulation of this phenomenon. In this context, the concept that the C-terminal domain of P2X7R directly regulates a complex distinct from receptor-dependent pore activity was first introduced by El-Moatassim and Dubyak (El-Moatassim & Dubyak, 1992). They demonstrated that P2X7R receptor mediated phospholipase D (PLD) activity was dependent on GTP and independent of the large conductance channel opening. Another study found that human $P2X_7$ receptor currents were facilitated in response to repeated or prolonged agonist applications, via dynamic calmodulin binding (Roger et al, 2008). This Ca^{2+}-dependent component is related to the uptake of large compounds seen by the pore complex. These and other papers (Alloysio et al, 2010; Boldt et al, 2003; Le Stunff et al, 2004) suggest independent intracellular signaling pathways regulating the low and large conductance channel associated with P2X7R .

3. Intracellular signalling pathways associated with large conductance channel

The intracellular regulation of the P2X7R associated large conductance channel opening is still mostly unknown. The initial suggestions for dependency of cytoplasmic factors in this event were originated from electrophysiological data in outside or inside out configurations with no large conductance channel recordings (Coutinho-Silva & Persechini, 1997; Persechini et al, 1998; Petrou et al, 1997). Posteriorly, other groups have investigated the P2X7R pore formation induced by intracellular signaling and how mutated amino acids or truncated regions affect functional availability of the receptor. In this line, Smart used truncated and single-residue-mutated P2X7R receptors in HEK-293 cells and in Xenopus oocytes. Truncated P2X7R at residue 581 (of 595) were not able to dye uptake, but there was dye uptake similarly to the wild receptor in those cells expressing the truncated P2X7R at position 582. In contrast, the small channel function was only suppressed in the residues 380 (Smart et al, 2003). Two alternative splices variants were identified in the human P2X7R (one lacking the first transmembrane domain and the other the entire cytoplasmic tail, but they were compared to the full-length channel). The first variant exhibited a non-functional slow conductance channel, while the second did not affect the small ion channel activity, but affected the large conductance channel and caspase activation (Cheewatrakoolpong et al, 2005). In addition, threonine 283 (Thr283) has been described as a critical residue in the ectodomain for P2X7R receptor function and it has been suggested that the intracellular leucine residue (P451L) alters downstream signalling independently of ion channel activity (Young et al, 2006). Recently, Marques-da-Silva and collaborators (Marques-da-Silva et al, 2011) demonstrated that colchicine did not inhibit ATP-evoked currents in macrophages, but it decreased ATP-induced dye uptake. Large conductance channel opening on Xenopus oocytes and HEK293 cells expressing P2X7R were inhibited after colchicines treatment (Marques-da-Silva et al, 2011). Yan described that extracellular Ca^{2+} concentration is a physiological negative modulator of the P2X7R low conductance channel without affecting the large conductance channel opening (Yan et al, 2011).

There is some controversial data related to the biophysical, pharmacological or molecular tools that impair the actions of intracellular enzymes involved in the P2X7R pore formation. In one of the primary papers studying the intracellular signaling of the P2X7R large conductance channel, it was described that calmidazoliun, an inhibitor of the calmodulin protein, impaired the small channel activity, but had no effect in the large conductance channel opening (Virginio et al, 1997). This result was also confirmed by other groups (Donnelly-Roberts et al, 2004; Faria et al, 2010; Lundy et al, 2004). However, Roger and coworkers (Roger et al, 2008, 2010) reported that rat P2X7R induced large organic cation permeability ionic currents were dependent on critical residues of calmodulin binding domain when recorded in patch-clamp whole cell configuration. In this sense, the intracellular Ca^{2+} concentration dependence in the P2X7R pore formation is still unclear. We have shown in 2005 that intracellular Ca^{2+} acts as a second messenger in the large conductance channel opening in peritoneal macrophages and 2BH4 cells (Faria et al, 2005). Similar data were found by other groups (Cankurtaran-Sayar et al, 2009; Roger et al, 2008, 2010; Schachter et al, 2008), but others did not observe this effect (da Cruz et al, 2006; Iglesias et al, 2008; Schachter et al, 2008; Virginio et al, 1999). We continued to investigate the Ca^{2+} participation in more detail, and we described a major Ca^{2+} dependence in the P2X7R pore formation, but we also observed Ca^{2+} independent events in the same cell types (Faria et al, 2010). This variability in the responses may be due to preponderant expression of distinct P2X7R variants or activity of different large conductance channels, as was discussed above in the text. Phospholipase C (PLC) had no effect on P2X7R large conductance channel formation in THP-1 cells (Donnelly-Roberts et al, 2004), or in mouse 2BH4 cells or peritoneal macrophages cells (Faria et al, 2010). However, P2X7R pore formation was inhibited in mouse microglial cell line by PLC (Takenouchi et al, 2005). It was also shown that MAPK is associated with P2X7R activation (Donnelly-Roberts et al, 2004; Faria et al, 2005, 2010), but in other hands this was not confirmed (da Cruz et al, 2006; Michel et al, 2006). These discrepancies may be due to species variations, distinct intracellular machinery or differences in the protocol used to investigate a specific function in a same cell type (Faria et al, 2010). Other proteins have presented less divergent responses, such as: PKC (Donnelly-Roberts et al, 2004; Faria et al, 2010; Shemon et al, 2004), Ca2+-insensitive Phospholipase A PLA (Chaib et al, 2000), caspase-1 and-3 (Donnelly-Roberts et al, 2004; Faria et al, 2010), PLD (Stunff & Raymond, 2007), phosphatidylinositol 4,5-bisphosphate (PIP_2) (Zhao et al, 2007), cytoskeleton components (Marques-da-Silva et al, 2011), PI3K (Faria et al, 2010), src tyrosine phosphorylation (Iglesias et al, 2008), Peroxisome proliferator-activated receptor gamma (PPAR gamma) (Nagasawa et al, 2009a) antagonists and intracellular Ca^{2+} chelants (Faria et al, 2005, 2010).

As mentioned above, one possible drawback in relation to these results is due to the diversity of responses observed in $P2X_7$ species and cell types (Donnelly-Roberts et al, 2009; Michel et al, 2008). The variations may be proportional to natural P2X7R receptor polimorphisms and these may be, at least partially, functional promoting gain or loss of activity (Cheewatrakoolpong et al, 2005; Feng et al, 2006; Masin et al, 2011; Shemon et al, 2006). Another related factor to this matter is the native structural state of the P2X7R receptor. In this line, a monoclonal antibody (Ab) to P2X7R ectodomain was used to immunoprecipitate the receptor complex in central and peripheral immune cells (Kim et al, 2001b). Using western blotting, native P2X7R in peritoneal macrophage or bone marrow cells formed bound multimeric complex with numerous bands ranging in size from 25 up to 250 kDa, in contrast to P2X7R from brain glia and/or astrocytes that formed only monomeric subunits. This result suggests differential intracellular regulation of the P2X7R pore in distinct cell types (Kim et al, 2001b). Li and coworkers discovered in parotid acinar and duct cells a cell-specific assembly and gating of the P2X7R channels, in a way that upon exposure to ATP, P2X7Rs are assembled into functional channels with rapid gating. In contrast, P2X7Rs from duct cells are preassembled and continually subject to rapid gating by ATP (Li et al, 2003). Recently, other researches have found distinct pathways of dye uptake, mediated by P2X7R receptor activation after ATP treatment, possibly through different large conductance channels. Schachter and collaborators compared P2X7R-associated cation and anionic fluorescent dyes uptake of macrophages and HEK-293 cells transfected with P2X7R receptor (Schachter et al, 2008). Transfected cells did not take up anionic dyes and did not display single channel cell-attached recordings, in contrast to the native mice peritoneal macrophages. Anionic and cationic dye effluxes induced by ATP treatment were temperature independent and dependent, respectively (Schachter et al, 2008). In addition, another study examined the process of dye uptake by transfected or natively expressed P2X7R receptor leading to the pore formation. HEK-293 cells expressing rat P2X7R was permeable to cationic but not to anionic dyes in a way that intracellular Ca^{2+} concentration ($[Ca^{2+}]_i$) increase was not necessary to be activated (via 1). In the via 2, the pore was permeated only by lucifer yellow and it was completely dependent on $[Ca^{2+}]_i$ for activation. Also, RAW 264.7 cells presented both pathways similar to the transfected cells, but they did not require intracellular Ca^{2+} (Cankurtaran-Sayar et al, 2009).

Based on all these data from different groups suggesting that more than one pore might work simultaneously after ATP treatment, we describe the intracellular enzymes activated by the P2X7R associated large conductance channel opening compared to other large conductance channels.

4. Intracellular signaling cascades activated by these pores

Since the proteins responsible to the P2X7R large conductance channel opening are still largely unknown here we compare and discuss the intracellular signaling pathways and the possible candidates associated to the P2X7R pore formation. Among them are connexin hemichannels, pannexin-1, plasma membrane voltage dependent anion channel (pl-VDAC), maxi anion, transient receptor potential vaniloid-1 (TRPV1), transient receptor potential anquirin-1 (TRPA1), Maitotoxin-induced pore and Rising of intracellular Ca^{2+} concentration induced pore.

The connexin hemichannel, a hexameric protein composed of connexin subunits expressed in vertebrates, was the first large conductance channel studied with functional similarity to P2X7R receptor large conductance channels. An initial study used two types of J774 mouse macrophages, one sensitive and another ATP-insensitive. In the sensitive cells, connexin-43 (Cx43) gap junction mRNA and protein and P2X7R were expressed and the dye was taken up, but in the insensitive lineage there was not Cx43 expression and neither dye uptake. Therefore, they proposed that connexin 43 was the pore associated to P2X7R receptor (Beyer

& Steinberg, 1991). This concept was elegantly refuted by Alves and colleagues at least in peritoneal macrophages where they demonstrated that experimental conditions known to block hemichannels and Cx43 knockout mice maintained the P2X7R large conductance channel activity (Alves et al, 1996). Also, P2X7R and Cx43 are expressed in J774 macrophage lineage and are colocalized in the cell membrane (Fortes et al, 2004).

In relation to the intracellular signaling cascades, connexins can be modulated in the C-terminal domain by phosphorylation through PKC (Bao et al, 2007; Hawat & Baroudi, 2008), MAPK (Bao et al, 2007), S-nytrosylation with covalent biding of nitric oxide (NO) to cysteine (Cys) (De Vuyst et al, 2007; Retamal et al, 2009), protein kinase A (Liu et al, 2011), intracellular redox potential (Retamal et al, 2007) and intracellular Ca^{2+} concentration (De Vuyst et al, 2006; Schalper et al, 2008; Thimm et al, 2005). Compared to the P2X7R receptor pore, the connexin hemichannel may be dependent on intracellular Ca^{2+}, PKC and MAPK. Meanwhile, the unitary conductance value (20-250pS) for all known connexins are lower compared to the ones observed to P2X7R receptor (400pS). In addition, it has been shown that connexin hemichannel blockers have no effect on the P2X7R receptor pore formation (Faria et al, 2005) and this apparently ruled out the participation of connexins at least in cell types tested.

Maitotoxin (MTX), a marine toxin, described to increase calcium in GH4C1 rat pituitary cells (Young et al, 1995), increases intracellular Ca^{2+} concentration leading to the opening of a pore with biophysical properties similar to P2X7R large conductance channel (Schilling et al, 1999a, 1999b). The dye uptake observed after this pore opening may also be dependent on extracellular Ca^{2+} (Lundy et al, 2004; Wisnoskey et al, 2004), intracellular Ca^{2+} concentration (Wisnoskey et al, 2004), calmodulin (Donnelly-Roberts et al, 2004; Lundy et al, 2004) and PLC (Donnelly-Roberts et al, 2004). Although the maitotoxin pore may be functionally similar to the P2X7R pore, they might possess different intracellular pathways. A possible explanation to this fact may be that the same large conductance channel or different similar pores functioning in conjunction might be regulated by distinct signaling pathways

In 2009, our group described a large conductance channel stimulated by rising of intracellular Ca^{2+} concentration recorded in cell attached patches. This pore was blocked by calmodulin, Calcium-calmodulin kinase type II (CamKII), PLC, MAPK and caspase-1 and-3 antagonists and it was insensitive to PKC and P2X7R receptor antagonists (Faria et al, 2009). The intracellular signaling pathways modulating this pore and the one associated with P2X7R are distinct, but they possessed some common pathways. In addition, the pore induced by MTX presents large intracellular signalling similar to the pore described by us. Since the protein responsible for the opening of the MTX induced pore also is not indentified, the pore recorded in our conditions (Faria et al, 2009) may be the same as the MTX pore.

Another large conductance channel, that may be activated by rising of intracellular Ca^{2+} concentration (Locovei et al, 2006; Ma et al, 2006), is the pannexin hemichannel, which is a hexameric protein present in vertebrates and invertebrates. However, as reported by recent papers, the extracellular or intracellular Ca^{2+} did not interfere with the pannexin activity (Ma et al, 2009; Pelegrin & Surprenant, 2007). This large conductance channel might be activated by S-nitrosylation and Src kinase (Iglesias et al, 2009; Pelegrin & Surprenant, 2006; Suadicani et al, 2009). P2X7R receptor large conductance channel was inhibited by RNAi to pannexin-1, inhibitory peptide and pannexin antagonists (Pelegrin & Surprenant, 2006). In contrast, other groups did not observe inhibition of the P2X7R large conductance channel for pannexin-1 inhibitors or RNAi (Faria et al, 2005, 2010; Nagasawa et al, 2009b; Reyes et al, 2008; Schchater et al, 2008; Yan et al, 2008, 2011). Up to now, for some groups the pannexin-1 seems to be an important player in this phenomenon, but apparently this protein is working in conjunction with other protein(s). This information is based on the partial blockage of the P2X7R receptor induced dye uptake exhibited after pannexin-1 inhibition (Iglesias et al, 2009; Locovei et al, 2007; Pelegrin & Surprenant, 2007). Alternatively, other large conductance channel such as MTX-induced pore (Pelegrin & Surprenant, 2007), P2X2R large conductance channel (Marques-da-Silva et al, 2011) and the rising of intracellular Ca^{2+} induced pore (Faria et al, 2009) were not impaired by pannexin-1 inhibitors.

Maxi-anion channel possesses a wide nanoscopic pore suitable for nucleotide transport and an ATP-binding site in the middle of the pore lumen to facilitate the passage of the nucleotide (Sabirov & Okada, 2004). Physiologically, the same large conductance channel is operational in swelling-, ischemia-, and hypoxia-induced ATP release from neonatal rat cardiomyocytes (Dutta et al, 2004). In addition, raising the intracellular Ca^{2+} concentration (Bajnath et al, 1993; Groschner & Kukovetz, 1992; Hussy, 1992; Kawahara & Tawuka, 1991) as well as protein tyrosine dephosphorylation (Toychiev et al, 2009) can activate this pore. On the other hand, PKA (Okada et al, 1997), PKC (Kokubun et al, 1991; Saiguza & Kokubun, 1988; Vaca & Kunze, 1993), G proteins (Schwiebert et al, 1992; Sun et al, 1993) and Src kinase (Kajita et al, 1995) antagonists can inhibit it. This large conductance channel also has no protein constituents identified so far (Sabirov & Okada et al, 2009) and there are no studies comparing its effects with other pores, except to pl-VDAC (Sabirov et al, 2006).

Voltage-dependent anion channels (VDACs) were originally characterized as mitochondrial porins but other evidence began to accumulate that VDACs could also be expressed in the plasma membrane (pl-VDAC). VDAC may be activated changing the applied voltage in the presence of NADH (Zizi et al, 1994) and under apoptotic conditions (Elinder et al, 2005). In relation to the intracellular signaling pathways, there are few data up to now, but some groups have shown the involvement of this pore with lipid rafts (Ferrer, 2009; Herrera et al, 2011). Moreover, the large conductance channel and pl-VDAC may be activated by excised patch, indicating an independence of the intracellular signals to open the large conductance channel (Guibert et al, 1998; Sun et al,1993). Relevant information is about the nucleotide-binding sites in the C-terminus of the mitochondrial VDAC, which presents the same C-terminal sequence of the pl-VDAC (Yehezkel et al, 2006). The main discrepancy in relation to both pores compared to the P2X7R pore is due to the lack of single channel recordings of the P2X7R pore in excised patches (Riedel et al, 2007). This fact may indicate that maxi anion and pl-VDAC pores are different compared to the P2X7R receptor pore, since they did not depend on cytoplasmic factors to open. But, this does not rule out the participation of this pore in this phenomenon.

The capsaicin induced receptor, transient receptor potential vanilloid 1 (TRPV1), is activated not only by capsaicin but also by heat (>43°C), acid and various lipids (Moran et al, 2011). Since capsaicin and its analogues, such as resiniferatoxin (RTX), are lipophilic, it is quite possible that they pass through the cell membrane and act on the binding sites present in the intracellular surface of TRPV1. It is a Ca^{2+} permeable non-specific cation channel. It was demonstrated that activation of native or recombinant rat TRPV1 leads to time- and agonist concentration-dependent increase in the relative permeability of large cations and changes in Ca^{2+} permeability (Chung et al, 2008). TRPV1 induced small channel can be modulated by calmodulin, PKC, PKA, intracellular Ca^{2+}, PLC, G protein and PIP_2/ Src (Bhave et al, 2002;

Chuang et al, 2001; Dai et al, 2004; Moriyama et al, 2003, 2005; Sugiura et al, 2002; Tominaga et al, 2001). TRPV1 induced a large conductance channel similar to the P2X7R since it depends on the C-terminus and it is modulated by PKC phosphorylation (Chung et al, 2008). Although biophysically this pore is similar to the P2X7R associated pore, the intracellular signaling is poorly understood up to now.

TRPA1 is a nonselective cation channel that belongs to the superfamily of mammalian TRP ion channels and is unique since it possesses a large number of ankyrin repeats in its N-terminal domain (Montell, 2005). TRPA1, when activated, are permeable to small cations such as Ca^{2+}, K^{+}, Na^{+}; simultaneously it depolarizes the plasma membrane and raises intracellular Ca^{2+}, which subsequently triggers a variety of physiological responses. Recently, it was described that TRPA1 activation induces dye uptake, which is blocked by selective TRPA1 antagonists. In addition, outside-out patch recordings using N-methyl-D-glucamine ($NMDG^{+}$) as the sole external cation and Na^{+} as the internal cation, TRPA1

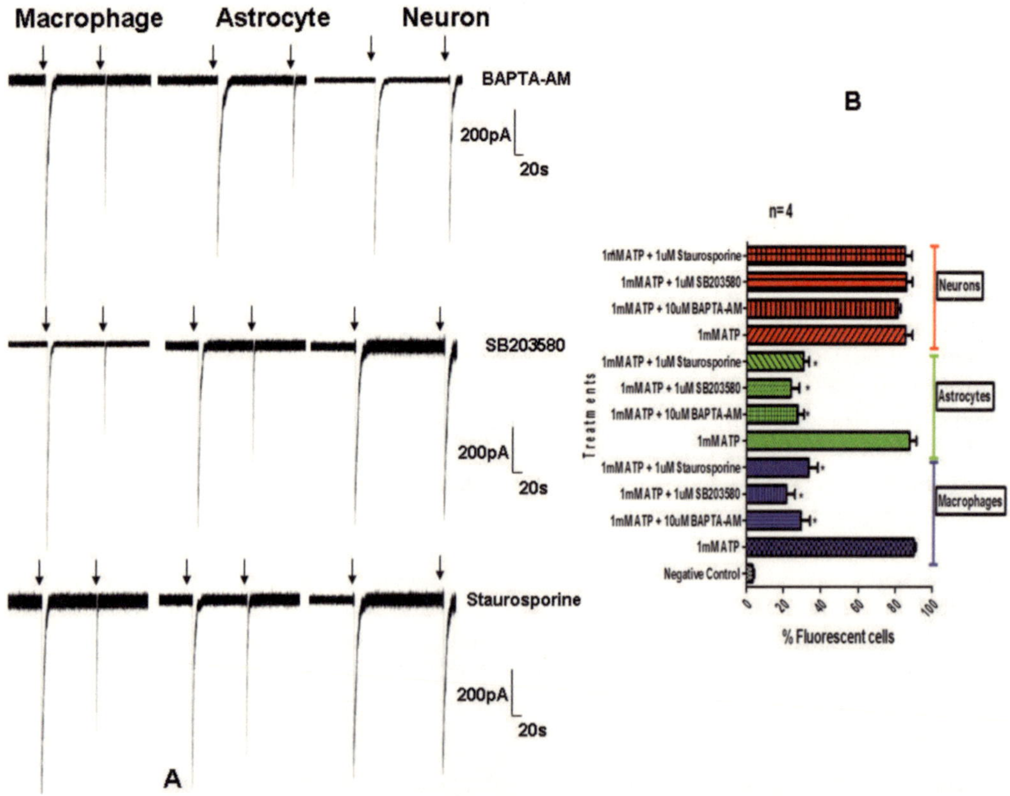

Fig. 1. Pharmacological comparison of the intracellular pathway of the pore associated with $P2X_7$ receptor in different cell types. A- Whole cell experiments in mice peritoneal macrophages, mice cortical astrocytes or mice mesencephalic. We applied 1mM ATP, after the incubation of the cells with 10µM BAPTA-AM, 1µM Sb203580 or 1µM Staurosporine for 5 minutes at 37°C. B- The graphic represents the quantification of dye uptake experiments in the cell types cited above. The values represent the mean ± SD of three to four experiments performed on different days. *$p<0.05$ compared with the ATP treatment of each group.

activation results in dynamic changes in permeability to $NMDG^{+}$ (Chen et al, 2008). Other groups have reproduced this data (Banke et al, 2010, 2011), but in every cell studied the intracellular signaling was not investigated. Moreover, the fact that TRPA1 associated large conductance channel permeates large cations in outside out configuration suggests a possible independence of intracellular factors.

As we can observe above, there are diverse common intracellular signaling proteins that may be used in the activation of these large conductance pores. Moreover, these pores might be biophysically and functionally similar. Thus, when we performed assays which preincubated cells with intracellular signaling pathway blockers were stimulated with 1mM ATP, there was an activation of the large conductance channel associated to P2X7R (Figure 1), as previously shown in our published papers (Faria et al, 2005, 2010). Using whole cell configuration, we used (i) neurons to evaluate the P2X7R pore formation in cells expressing TRPV1 and TRPA1; (ii) astrocytes to study the expression of Maxi anion, pl-VDAC, Connexin 43 and Pannexin-1; and (iii) macrophages to study Maitotoxin and intracellular Ca^{2+} increase induced pores.

5. Conclusions

Finally, based on data discussed here, several issues might explain why a common gate mechanism for the P2X7R pore is not yet understood: (1) a large conductance channel is activated by different signaling pathways; (2) these signaling cascades might be related to the activation of distinct pores; (3) both [(1) and (2)] mechanisms might act together in certain cells; (4) it might exist a gate modulator that is cell-type specific, (5) $P2X_7$ might be part of a macromolecular protein complex or a protein-lipid complex. In summary, more studies are necessary in order to comprehend the functional mechanism of the $P2X_7$ receptor.

6. References

Adinolfi E, Kim M, Young MT, Di Virgilio F, Surprenant A (2003) Tyrosine phosphorylation of HSP90 within the P2X7 receptor complex negatively regulates P2X7 receptors. J Biol Chem 278:37344–37351.

Alloisio S, Di Garbo A, Barbieri R, Bozzo L, Ferroni S, Nobile M (2010) Evidence for two conductive pathways in P2X receptor: differences in modulation and selectivity. J Neurochem. 113, 796-806.

Alves LA, Coutinho-Silva R, Persechini PM, Spray DC, Savino W, Campos de Carvalho AC (1996) Are there functional gap junctions or junctional hemichannels in macrophages? Blood. 88, 328-334.

Alzola E, Pérez-Etxebarria A, Kabré E, Fogarty DJ, Métioui M, Chaïb N, Macarulla JM, Matute C, Dehaye JP, Marino A (1998) Activation by P2X7 agonists of two phospholipases A2 (PLA2) in ductal cells of rat submandibular gland. Coupling of the calcium-independent PLA2 with kallikrein secretion. J Biol Chem. 273, 30208-30217.

Amstrup J, Novak I (2003) P2X7 receptor activates extracellular signal-regulated kinases ERK1 and ERK2 independently of Ca2+ influx. Biochem J. 374, 51-61.

Antonio LS, Stewart AP, Xu XJ, Varanda WA, Murrell-Lagnado RD, Edwardson JM (2011) P2X4 receptors interact with both P2X2 and P2X7 receptors in the form of homotrimers. Br J Pharmacol. 163, 1069-1077.

Bajnath RB, Groot JA, de Jonge HR, Kansen M, Bijman J (1993) Calcium ionophore plus excision induce a large conductance chloride channel in membrane patches of human colon carcinoma cells HT-29cl.19A. Experientia 1993 49, 313-316.

Banke TG, Chaplan SR, Wickenden AD (2010) Dynamic changes in the TRPA1 selectivity filter lead to progressive but reversible pore dilation. Am J Physiol Cell Physiol. 298, C1457-68.

Banke TG (2011) The dilated TRPA1 channel pore state is blocked by amiloride and analogues. Brain Res 1381, 21-30.

Bao X, Lee SC, Reuss L, Altenberg GA (2007) Change in permeant size selectivity by phosphorylation of connexin 43 gap-junctional hemichannels by PKC. Proc Natl Acad Sci U S A. 104, 4919-4924.

Barbieri R, Alloisio S, Ferroni S, Nobile M (2008) Differential crosstalk between P2X7 and arachidonic acid in activation of mitogen-activated protein kinases. Neurochem Int. 53, 255-262.

Barrera NP, Ormond SJ, Henderson RM, Murrell-Lagnado RD, Edwardson JM. (2005) Atomic force microscopy imaging demonstrates that P2X2 receptors are trimers but that P2X6 receptor subunits do not oligomerize. J Biol Chem. 280, 10759-65.

Beyer EC, Steinberg TH (1991) Evidence that the gap junction protein connexin-43 is the ATP-induced pore of mouse macrophages. J Biol Chem. 266, 7971-7974.

Bhave G, Zhu W, Wang H, Brasier DJ, Oxford GS, Gereau RW 4th (2002) cAMP-dependent protein kinase regulates desensitization of the capsaicin receptor (VR1) by direct phosphorylation. Neuron. 2002 35, 721-731.

Boldt W, Klapperstück M, Büttner C, Sadtler S, Schmalzing G, Markwardt F (2003) Glu496Ala polymorphism of human P2X7 receptor does not affect its electrophysiological phenotype. Am J Physiol Cell Physiol. 284, C749-56.

Boumechache M, Masin M, Edwardson JM, Górecki DC, Murrell-Lagnado R (2009) Analysis of assembly and trafficking of native P2X4 and P2X7 receptor complexes in rodent immune cells. J Biol Chem 284, 13446-13454.

Bradford MD, Soltoff SP (2002) P2X7 receptors activate protein kinase D and p42/p44 mitogen-activated protein kinase (MAPK) downstream of protein kinase C. Biochem J 366, 745-755.

Bradley HJ, Liu X, Collins V, Owide J, Goli GR, Smith M, Surprenant A, White SJ, Jiang LH (2010) Identification of an intracellular microdomain of the P2X7 receptor that is crucial in basolateral membrane targeting in epithelial cells. FEBS Lett. 584, 4740-4744.

Burnstock G (2007) Purine and pyrimidine receptors. Cell Mol Life Sci. 64, 1471-1483.

Cankurtaran-Sayar S, Sayar K, Ugur M (2009) P2X7 receptor activates multiple selective dye-permeation pathways in RAW 264.7 and human embryonic kidney 293 cells. Mol Pharmacol. 76, 1323-1332.

Chaib N, Kabré E, Alzola E, Pochet S, Dehaye JP (2000) Bromoenol lactone enhances the permeabilization of rat submandibular acinar cells by P2X7 agonists. Br J Pharmacol. 129, 703-708.

Chaumont S, Khakh BS (2008) Patch-clamp coordinated spectroscopy shows P2X2 receptor permeability dynamics require cytosolic domain rearrangements but not Panx-1 channels. Proc Natl Acad Sci U S A. 105, 12063-12068.

Cheewatrakoolpong B, Gilchrest H, Anthes JC, Greenfeder S (2005) Identification and characterization of splice variants of the human P2X7 ATP channel. Biochem Biophys Res Commun. 332, 17-27.

Chen J, Zhang XF, Kort ME, Huth JR, Sun C, Miesbauer LJ, Cassar SC, Neelands T, Scott VE, Moreland RB, Reilly RM, Hajduk PJ, Kym PR, Hutchins CW, Faltynek CR (2008) Molecular determinants of species-specific activation or blockade of TRPA1 channels. J Neurosci. 28, 5063-5071.

Chuang HH, Prescott ED, Kong H, Shields S, Jordt SE, Basbaum AI, Chao MV, Julius D (2001) Bradykinin and nerve growth factor release the capsaicin receptor from PtdIns(4,5)P2-mediated inhibition. Nature 411, 957-962.

Chung MK, Güler AD, Caterina MJ (2008) TRPV1 shows dynamic ionic selectivity during agonist stimulation. Nat Neurosci. 11, 555-564.

Coddou C, Yan C, Obsil T, Huidobro-Toro JP, Stojilkovic SS (2011) Activation and Regulation of Purinergic P2X Receptor Channels. Pharmacol Rev print July 7, 2011, doi: 10.1124/pr.110.003129

Coutinho-Silva R, Persechini PM (1997) P2Z purinoceptor-associated pores induced by extracellular ATP in macrophages and J774 cells. Am J Physiol. 273, C1793-800.

da Cruz CM, Ventura AL, Schachter J, Costa-Junior HM, da Silva Souza HA, Gomes FR, Coutinho-Silva R, Ojcius DM, Persechini PM (2006) Activation of ERK1/2 by extracellular nucleotides in macrophages is mediated by multiple P2 receptors independently of P2X(7)-associated pore or channel formation. Br J Pharmacol 147, 324–334.

Dai Y, Moriyama T, Higashi T, Togashi K, Kobayashi K, Yamanaka H, Tominaga M, Noguchi K (2004) Proteinase-activated receptor 2-mediated potentiation of transient receptor potential vanilloid subfamily 1 activity reveals a mechanism for proteinase-induced inflammatory pain. J Neurosci. 24, 4293-4299.

De Vuyst E, Decrock E, Cabooter L, Dubyak GR, Naus CC, Evans WH, Leybaert L (2006) Intracellular calcium changes trigger connexin 32 hemichannel opening. EMBO J. 25, 34-44.

De Vuyst E, Decrock E, De Bock M, Yamasaki H, Naus CC, Evans WH, Leybaert L (2007) Connexin hemichannels and gap junction channels are differentially influenced by lipopolysaccharide and basic fibroblast growth factor. Mol Biol Cell. 18, 34-46.

Denlinger LC, Fisette PL, Sommer JA, Watters JJ, Prabhu U, Dubyak GR, Proctor RA, Bertics PJ.

Cutting edge: the nucleotide receptor P2X7 contains multiple protein- and lipid-interaction motifs including a potential binding site for bacterial lipopolysaccharide (2001) J Immunol. 167, 1871-1876.

Denlinger LC, Sommer JA, Parker K, Gudipaty L, Fisette PL, Watters JW, Proctor RA, Dubyak GR, Bertics PJ (2003) Mutation of a dibasic amino acid motif within the C terminus of the P2X7 nucleotide receptor results in trafficking defects and impaired function. J Immunol 171, 1304–1311

Ding S, Sachs F (2000) Inactivation of P2X2 purinoceptors by divalent cations. J Physiol 522, 199–214.

Donnelly-Roberts DL, Namovic MT, Faltynek CR, Jarvis MF (2004) Mitogen-activated protein kinase and caspase signaling pathways are required for P2X7 receptor (P2X7R)-induced pore formation in human THP-1 cells. J Pharmacol Exp Ther 308, 1053–1061.

Donnelly-Roberts DL, Namovic MT, Han P, Jarvis MF (2009) Mammalian P2X7 receptor pharmacology: comparison of recombinant mouse, rat and human P2X7 receptors. Br J Pharmacol. 157, 1203-1214.

Dubyak GR (2007). Go it alone no more--P2X7 joins the society of heteromeric ATP-gated receptor channels. Mol Pharmacol. 72, 1402-1405.

Dutta AK, Sabirov RZ, Uramoto H, Okada Y (2004) Role of ATP-conductive anion channel in ATP release from neonatal rat cardiomyocytes in ischaemic or hypoxic conditions. J Physiol 559, 799-812.

El-Moatassim C, Dubyak GR (1992) A novel pathway for the activation of phospholipase D by P2z purinergic receptors in BAC1.2F5 macrophages. J Biol Chem 267, 23664–23673.

Elinder F, Akanda N, Tofighi R, Shimizu S, Tsujimoto Y, Orrenius S, Ceccatelli S (2005) Opening of plasma membrane voltage-dependent anion channels (VDAC) precedes caspase activation in neuronal apoptosis induced by toxic stimuli. Cell Death Differ 12, 1134-1140.

Faria RX, Defarias FP, Alves LA (2005) Are second messengers crucial for opening the pore associated with P2X7 receptor? Am J Physiol Cell Physiol. 288, C260-71.

Faria RX, Reis RA, Casabulho CM, Alberto AV, de Farias FP, Henriques-Pons A, Alves LA (2009) Pharmacological properties of a pore induced by raising intracellular Ca2+. Am J Physiol Cell Physiol. 297, C28-42.

Faria RX, Cascabulho CM, Reis RA, Alves LA (2010) Large-conductance channel formation mediated by P2X7 receptor activation is regulated through distinct intracellular signaling pathways in peritoneal macrophages and 2BH4 cells. Naunyn Schmiedebergs Arch Pharmacol. 382, 73-87.

Feng YH, Li X, Wang L, Zhou L, Gorodeski GI (2006) A truncated P2X7 receptor variant (P2X7-j) endogenously expressed in cervical cancer cells antagonizes the full-length P2X7 receptor through hetero-oligomerization. J Biol Chem. 281, 17228-17237.

Ferrari D, Chiozzi P, Falzoni S, Dal Susino M, Melchiorri L, Baricordi OR, Di Virgilio F (1997) Extracellular ATP triggers IL-1 beta release by activating the purinergic P2Z receptor of human macrophages. J Immunol 159, 1451–1458.

Ferrer I. (2009) Altered mitochondria, energy metabolism, voltagedependent anion channel, and lipid rafts converge to exhaust neurons in Alzheimer's disease. J. Bioenerg. Biomembr. 41, 425–431.

Fortes FS, Pecora IL, Persechini PM, Hurtado S, Costa V, Coutinho-Silva R, Braga MB, Silva-Filho FC, Bisaggio RC, De Farias FP, Scemes E, De Carvalho AC, Goldenberg RC (2004) Modulation of intercellular communication in macrophages: possible interactions between GAP junctions and P2 receptors. J Cell Sci. 117, 4717-4726.

Gonnord P, Delarasse C, Auger R, Benihoud K, Prigent M, Cuif MH, Lamaze C, Kanellopoulos JM (2009) Palmitoylation of the P2X7 receptor, an ATP-gated channel, controls its expression and association with lipid rafts. FASEB J. 23, 795-805.

Groschner K, Kukovetz WR (1992) Voltage-sensitive chloride channels of large conductance in the membrane of pig aortic endothelial cells. Pflugers Arch. 421, 209-217. Erratum in: Pflugers Arch 1992 Sep;421, 613.

Gu B, Bendall LJ, Wiley JS (1998) Adenosine triphosphate-induced shedding of CD23 and L-selectin (CD62L) from lymphocytes is mediated by the same receptor but different metalloproteases. Blood 92, 946–951.

Gudipaty L, Humphreys BD, Buell G, Dubyak GR (2001) Regulation of P2X(7) nucleotide receptor function in human monocytes by extracellular ions and receptor density. Am J Physiol Cell Physiol 280, C943–C953.

Guibert B, Dermietzel R, Siemen D (1998) Large conductance channel in plasma membranes of astrocytic cells is functionally related to mitochondrial VDAC-channels. Int J Biochem Cell Biol. 30, 379-391.

Guo C, Masin M, Qureshi OS, Murrell-Lagnado RD (2007) Evidence for functional P2X4/P2X7 heteromeric receptors. Mol Pharmacol. 72, 1447-1456.

Hamilton N, Vayro S, Kirchhoff F, Verkhratsky A, Robbins J, Gorecki DC, Butt AM (2008) Mechanisms of ATP- and glutamate-mediated calcium signaling in white matter astrocytes. Glia 56, 734-749.

Hawat G, Baroudi G (2008) Differential modulation of unapposed connexin 43 hemichannel electrical conductance by protein kinase C isoforms. Pflugers Arch. 456, 519-527.

Heo JS, Han HJ (2006) ATP stimulates mouse embryonic stem cell proliferation via protein kinase C, phosphatidylinositol 3-kinase/Akt, and mitogen-activated protein kinase signaling pathways.Stem Cells 24, 2637-2648.

Herrera JL, Diaz M, Hernández-Fernaud JR, Salido E, Alonso R, Fernández C, Morales A, Marin R (2011) Voltage-dependent anion channel as a resident protein of lipid rafts: post-transductional regulation by estrogens and involvement in neuronal preservation against Alzheimer's disease. J Neurochem. 116, 820-827. doi: 10.1111/j.1471-4159.2010.06987.x.

Hibell AD, Thompson KM, Xing M, Humphrey PP, Michel AD (2001) Complexities of measuring antagonist potency at P2X(7) receptor orthologs. J Pharmacol Exp Ther. 296, 947-57.

Hussy N (1992) Calcium-activated chloride channels in cultured embryonic Xenopus spinal neurons. J Neurophysiol. 68, 2042-2050.

Iglesias R, Locovei S, Roque A, Alberto AP, Dahl G, Spray DC, Scemes E (2008) P2X7 receptor-Pannexin1 complex: pharmacology and signaling. Am J Physiol Cell Physiol. 295, C752-60.

Iglesias R, Dahl G, Qiu F, Spray DC, Scemes E (2009) Pannexin 1: the molecular substrate of astrocyte "hemichannels". J Neurosci. 29, 7092-7097.

Kawahara K, Takuwa N (1991) Bombesin activates large-conductance chloride channels in Swiss 3T3 fibroblasts. Biochem Biophys Res Commun. 177, 292-298.

Kawate T, Michel JC, Birdsong WT, Gouaux E (2009) Crystal structure of the ATP-gated P2X(4) ion channel in the closed state. Nature. 460, 592-598.

Kim M, Jiang LH, Wilson HL, North RA, Surprenant A (2001a) Proteomic and functional evidence for a P2X7 receptor signaling complex. EMBO J 20:6347–6358.

Kim M, Spelta V, Sim J, North RA and Surprenant A (2001b) Differential assembly of rat purinergic P2X7 receptor in immune cells of the brain and periphery. J Biol Chem, 276, 23262–23267.

Kokubun S, Saigusa A, Tamura T (1991) Blockade of Cl channels by organic and inorganic blockers in vascular smooth muscle cells. Pflugers Arch. 418, 204-213.

Locovei S, Scemes E, Qiu F, Spray DC, Dahl G (2007) Pannexin1 is part of the pore forming unit of the P2X(7) receptor death complex. FEBS Lett 581, 483-488.

Lundy PM, Nelson P, Mi L, Frew R, Minaker S, Vair C, Sawyer TW (2004) Pharmacological differentiation of the P2X7 receptor and the maitotoxin-activated cationic channel. Eur J Pharmacol 487, 17-28.

Jacques-Silva MC, Rodnight R, Lenz G, Liao Z, Kong Q, Tran M, Kang Y, Gonzalez FA, Weisman GA, Neary JT (2004) P2X7 receptors stimulate AKT phosphorylation in astrocytes. Br J Pharmacol 141, 1106-1117.

Jiang LH (2009) Inhibition of P2X(7) receptors by divalent cations: old action and new insight. Eur Biophys J. 38, 339-346.

Jiang LH, Rassendren F, Mackenzie A, Zhang YH, Surprenant A, North RA (2005) N-methyl-D-glucamine and propidium dyes utilize different permeation pathways at rat P2X(7) receptors. Am J Physiol Cell Physiol. 289, C1295-302.

Le Stunff H, Auger R, Kanellopoulos J, Raymond MN (2004) The Pro-451 to Leu polymorphism within the C-terminal tail of P2X7 receptor impairs cell death but not phospholipase D activation in murine thymocytes. J Biol Chem. 279, 16918-16926.

Le Stunff H, Raymond MN (2007) P2X7 receptor-mediated phosphatidic acid production delays ATP-induced pore opening and cytolysis of RAW 264.7 macrophages. Cell Signal 19, 1909–1918.

Lemaire I, Falzoni S, Leduc N, Zhang B, Pellegatti P, Adinolfi E, Chiozzi P, Di Virgilio F (2006) Involvement of the purinergic P2X7 receptor in the formation of multinucleated giant cells. J Immunol. 177, 7257-7265.

Li Q, Luo X, Zeng W, Muallem S (2003) Cell-specific behavior of P2X7 receptors in mouse parotid acinar and duct cells. J Biol Chem. 278, 47554-47561.

Li M, Chang TH, Silberberg SD, Swartz KJ (2008) Gating the pore of P2X receptor channels. Nat Neurosci 11, 883–887.

Li M, Kawate T, Silberberg SD, Swartz KJ (2010) Pore-opening mechanism in trimeric P2X receptor channels. Nat Commun 1, 44.

Li Q, Luo X, Muallem S (2005) Regulation of the P2X7 receptor permeability to large molecules by extracellular Cl- and Na+. J Biol Chem 280, 26922–26927.

Liu Y, Xiao Y, Li Z (2011) P2X7 receptor positively regulates MyD88-dependent NF-κB activation. Cytokine 55, 229-236.

Liu J, Ek Vitorin JF, Weintraub ST, Gu S, Shi Q, Burt JM, Jiang JX (2011) Phosphorylation of connexin 50 by protein kinase A enhances gap junction and hemichannel function. J Biol Chem. 286, 16914-16928.

Locovei S, Wang J, Dahl G (2006) Activation of pannexin 1 channels by ATP through P2Y receptors and by cytoplasmic Ca2+. FEBS Lett 580, 239–244.

Ma W, Korngreen A, Weil S, Cohen EB, Priel A, Kuzin L, Silberberg SD (2006) Pore properties and pharmacological features of the P2X receptor channel in airway ciliated cells. J Physiol 571, 503–517.

Ma W, Hui H, Pelegrin P, Surprenant A (2009) Pharmacological characterization of pannexin-1 currents expressed in mammalian cells. J Pharmacol Exp Ther. 328, 409-418.

Marcoli M, Cervetto C, Paluzzi P, Guarnieri S, Alloisio S, Thellung S, Nobile M, Maura G (2008) P2X7 pre-synaptic receptors in adult rat cerebrocortical nerve terminals: a role in ATP-induced glutamate release. J Neurochem. 105, 2330-2342.

Marques-da-Silva C, Chaves MM, Castro NG, Coutinho-Silva R, Guimaraes MZ (2011) Colchicine inhibits cationic dye uptake induced by ATP in P2X2 and P2X7 receptor-expressing cells: implications for its therapeutic action Br J Pharmacol. 163, 912-926

Masin M, Young C, Lim K, Barnes SJ, Xu XJ, Marschall V, Brutkowski W, Mooney ER, Gorecki DC, Murrell-Lagnado R (2011) Expression, assembly and function of novel C-terminal truncated variants of the mouse P2X7 receptor: Re-evaluation of P2X7 knockouts. Br J Pharmacol. Aug 12. doi: 10.1111/j.1476-5381.2011.01624.x.

Michel AD, Chessell IP, Humphrey PP (1999) Ionic effects on human recombinant P2X7 receptor function. Naunyn Schmiedebergs Arch Pharmacol. 359, 102-109.

Michel AD, Thompson KM, Simon J, Boyfield I, Fonfria E, Humphrey PP (2006) Species and response dependent differences in the effects of MAPK inhibitors on P2X(7) receptor function. Br J Pharmacol 149, 948-957

Michel AD, Fonfria E (2007) Agonist potency at P2X7 receptors is modulated by structurally diverse lipids. Br J Pharmacol. 152, 523-537.

Michel AD, Clay WC, Ng SW, Roman S, Thompson K, Condreay JP, Hall M, Holbrook J, Livermore D, Senger S (2008) Identification of regions of the P2X(7) receptor that contribute to human and rat species differences in antagonist effects. Br J Pharmacol. 155, 738-751.

Montell C (2005) The TRP superfamily of cation channels. Sci STKE 2005(272):re3.

Moon H, Na HY, Chong KH, Kim TJ (2006) P2X7 receptor-dependent ATP-induced shedding of CD27 in mouse lymphocytes. Immunol Lett. 102, 98-105.

Moran MM, McAlexander MA, Bíró T, Szallasi A (2011) Transient receptor potential channels as therapeutic targets. Nat Rev Drug Discov. 2011 Aug 1;10(8):601-20. doi: 10.1038/nrd3456.

Morelli A, Chiozzi P, Chiesa A, Ferrari D, Sanz JM, Falzoni S, Pinton P, Rizzuto R, Olson MF, Di Virgilio F (2003) Extracellular ATP causes ROCK I-dependent bleb formation in P2X7-transfected HEK293 cells. Mol Biol Cell 14, 2655-2664.

Moriyama T, Iida T, Kobayashi K, Higashi T, Fukuoka T, Tsumura H, Leon C, Suzuki N, Inoue K, Gachet C, Noguchi K, Tominaga M (2003) Possible involvement of P2Y2 metabotropic receptors in ATP-induced transient receptor potential vanilloid receptor 1-mediated thermal hypersensitivity. J Neurosci. 23, 6058-6062.

Moriyama T, Higashi T, Togashi K, Iida T, Segi E, Sugimoto Y, Tominaga T, Narumiya S, Tominaga M (2005) Sensitization of TRPV1 by EP1 and IP reveals peripheral nociceptive mechanism of prostaglandins. Mol Pain 17, 1:3.

Nagasawa K, Miyaki J, Kido Y, Higashi Y, Nishida K, Fujimoto S (2009a) Possible involvement of PPAR gamma in the regulation of basal channel opening of P2X7 receptor in cultured mouse astrocytes. Life Sci 84, 825-831.

Nagasawa K, Escartin C, Swanson RA (2009b) Astrocyte cultures exhibit P2X7 receptor channel opening in the absence of exogenous ligands. Glia. 57, 622-633.

Nicke A, Bäumert HG, Rettinger J, Eichele A, Lambrecht G, Mutschler E, Schmalzing G (1998) P2X1 and P2X3 receptors form stable trimers: a novel structural motif of ligand-gated ion channels. EMBO J 17, 3016–3028

Nicke A, Kerschensteiner D, Soto F (2005) Biochemical and functional evidence for heteromeric assembly of P2X1 and P2X4 subunits. J Neurochem 92, 925–933.

Nicke A (2008) Homotrimeric complexes are the dominant assembly state of native P2X7 subunits. Biochem Biophys Res Commun. 377, 803-808.

North RA (2002) Molecular physiology of P2X receptors. Physiol Rev 82, 1013–1067.

Orinska Z, Hein M, Petersen F, Bulfone-Paus S, Thon L, Adam D (2011) Retraction: Extracellular ATP induces cytokine expression and apoptosis through P2X7 receptor in murine mast cells. J Immunol. 186, 2683.

Ortega F, Pérez-Sen R, Delicado EG, Miras-Portugal MT (2009) P2X7 nucleotide receptor is coupled to GSK-3 inhibition and neuroprotection in cerebellar granule neurons. Neurotox Res. 15(3):193-204.

Pelegrin P, Surprenant A (2006) Pannexin-1 mediates large pore formation and interleukin-1beta release by the ATP-gated P2X7 receptor. EMBO J 25, 5071–5082.

Pelegrin P, Surprenant A (2007) Pannexin-1 couples to maitotoxin- and nigericin-induced interleukin-1beta release through a dye uptake-independent pathway. J Biol Chem. 282, 2386-2394.

Pérez-Andrés E, Fernández-Rodriguez M, González M, Zubiaga A, Vallejo A, García I, Matute C, Pochet S, Dehaye JP, Trueba M, Marino A, Gómez-Muñoz A (2002) Activation of phospholipase D-2 by P2X(7) agonists in rat submandibular gland acini. J Lipid Res. 43, 1244-1255.

Persechini PM, Bisaggio RC, Alves-Neto JL, Coutinho-Silva R (1998) Extracellular ATP in the lymphohematopoietic system: P2Z purinoceptors off membrane permeabilization. Braz J Med Biol Res. 31, 25-34.

Petrou S, Ugur M, Drummond RM, Singer JJ, Walsh JV Jr (1997) P2X7 purinoceptor expression in Xenopus oocytes is not sufficient to produce a pore-forming P2Z-like phenotype. FEBS Lett. 411, 339-345.

Pochet S, Garcia-Marcos M, Seil M, Otto A, Marino A, Dehaye JP (2007) Contribution of two ionotropic purinergic receptors to ATP responses in submandibular gland ductal cells. Cell Signal. 19, 2155-2164.

Ralevic V, Burnstock G (1998) Receptors for purines and pyrimidines. Pharmacol Rev. 50, 413-92

Retamal MA, Yin S, Altenberg GA, Reuss L (2009) Modulation of Cx46 hemichannels by nitric oxide. Am J Physiol Cell Physiol. 296, C1356-63.

Reyes JP, Pérez-Cornejo P, Hernández-Carballo CY, Srivastava A, Romanenko VG, Gonzalez-Begne M, Melvin JE, Arreola J (2008) Na+ modulates anion permeation and block of P2X7 receptors from mouse parotid glands. J Membr Biol. 223, 73-85.

Riedel T, Schmalzing G, Markwardt F (2007) Influence of extracellular monovalent cations on pore and gating properties of P2X7 receptor-operated single-channel currents. Biophys J. 93, 846-58.

Roger S, Pelegrin P, Surprenant A (2008) Facilitation of P2X7 receptor currents and membrane blebbing via constitutive and dynamic calmodulin binding. J Neurosci 28, 6393–6401.

Roger S, Gillet L, Baroja-Mazo A, Surprenant A, Pelegrin P (2010) C-terminal calmodulin-binding motif differentially controls human and rat P2X7 receptor current facilitation. J Biol Chem. 285, 17514-17524.

Saigusa A, Kokubun S (1988) Protein kinase C may regulate resting anion conductance in vascular smooth muscle cells. Biochem Biophys Res Commun. 155, 882-889.

Sabirov RZ, Okada Y (2004) Wide nanoscopic pore of maxi-anion channel suits its function as an ATP-conductive pathway. Biophys J. 87, 1672-1685.

Sabirov RZ, Sheiko T, Liu H, Deng D, Okada Y, Craigen WJ (2006) Genetic demonstration that the plasma membrane maxianion channel and voltage-dependent anion channels are unrelated proteins. J Biol Chem281, 1897-1904.

Schachter J, Motta AP, de Souza Zamorano A, da Silva-Souza HA, Guimarães MZ, Persechini PM (2008) ATP-induced P2X7-associated uptake of large molecules involves distinct mechanisms for cations and anions in macrophages. J Cell Sci. 121, 3261-3270.

Schalper KA, Palacios-Prado N, Retamal MA, Shoji KF, Martínez AD, Sáez JC (2008) Connexin hemichannel composition determines the FGF-1-induced membrane permeability and free [Ca2+]i responses. Mol Biol Cell. 19, 3501-3513.

Schilling WP, Wasylyna T, Dubyak GR, Humphreys BD, Sinkins WG (1999a) Maitotoxin and P2Z/P2X(7) purinergic receptor stimulation activate a common cytolytic pore. Am J Physiol. 277, C766-76.

Schilling WP, Sinkins WG, Estacion M (1999b) Maitotoxin activates a nonselective cation channel and a P2Z/P2X(7)-like cytolytic pore in human skin fibroblasts. Am J Physiol. 277, C755-65.

Schwiebert EM, Kizer N, Gruenert DC, Stanton BA (1992) GTP-binding proteins inhibit cAMP activation of chloride channels in cystic fibrosis airway epithelial cells. Proc Natl Acad Sci U S A. 89, 10623-10627.

Shemon AN, Sluyter R, Conigrave AD, Wiley JS (2004) Chelerythrine and other benzophenanthridine alkaloids block the human P2X7 receptor. Br J Pharmacol 142, 1015–1019.

Shemon AN, Sluyter R, Fernando SL, Clarke AL, Dao-Ung LP, Skarratt KK, Saunders BM, Tan KS, Gu BJ, Fuller SJ, Britton WJ, Petrou S, Wiley JS (2006) A Thr357 to Ser polymorphism in homozygous and compound heterozygous subjects causes absent or reduced P2X7 function and impairs ATP-induced mycobacterial killing by macrophages. J Biol Chem. 281, 2079-2086.

Smart ML, Gu B, Panchal RG, Wiley J, Cromer B, Williams DA, Petrou S (2003) P2X7 receptor cell surface expression and cytolytic pore formation are regulated by a distal C-terminal region. J Biol Chem 278, 8853–8860.

Sperlágh B, Vizi ES, Wirkner K, Illes P (2006) P2X7 receptors in the nervous system. Prog Neurobiol. 78, 327-346.

Suadicani SO, Iglesias R, Spray DC, Scemes E (2009) Point mutation in the mouse P2X7 receptor affects intercellular calcium waves in astrocytes. ASN Neuro. 2009 Apr 14;1(1). pii: e00005. doi: 10.1042/AN20090001.

Sugiura T, Tominaga M, Katsuya H, Mizumura K (2002) Bradykinin lowers the threshold temperature for heat activation of vanilloid receptor 1. J Neurophysiol. 88, 544-548.

Sun XP, Supplisson S, Mayer E (1993) Chloride channels in myocytes from rabbit colon are regulated by a pertussis toxin-sensitive G protein. Am J Physiol264, G774-85.

Surprenant A (1996) Functional properties of native and cloned P2X receptors. Ciba Found Symp 198:208–219; discussion 219–222.

Takenouchi T, Ogihara K, Sato M, Kitani H (2005) Inhibitory effects of U73122 and U73343 on Ca2+ influx and pore formation induced by the activation of P2X7 nucleotide receptors in mouse microglial cell line. Biochim Biophys Acta. 1726, 177-186.

Takenouchi T, Sato M, Kitani H (2007) Lysophosphatidylcholine potentiates Ca2+ influx, pore formation and p44/42 MAP kinase phosphorylation mediated by P2X7 receptor activation in mouse microglial cells. J Neurochem. 102, 1518-1532.

Thimm J, Mechler A, Lin H, Rhee S, Lal R (2005) Calcium-dependent open/closed conformations and interfacial energy maps of reconstituted hemichannels. J Biol Chem. 280, 10646-10654.

Tominaga M, Wada M, Masu M (2001) Potentiation of capsaicin receptor activity by metabotropic ATP receptors as a possible mechanism for ATP-evoked pain and hyperalgesia. Proc Natl Acad Sci U S A. 98, 6951-6956.

Toychiev AH, Sabirov RZ, Takahashi N, Ando-Akatsuka Y, Liu H, Shintani T, Noda M, Okada Y (2009) Activation of maxi-anion channel by protein tyrosine dephosphorylation. Am J Physiol Cell Physiol. 297, C990-1000.

Vaca L, Kunze DL (1993) Depletion and refilling of intracellular Ca2+ stores induce oscillations of Ca2+ current. Am J Physiol. 264, H1319-22.

Virginio C, Church D, North RA, Surprenant A (1997) Effects of divalent cations, protons and calmidazolium at the rat P2X7 receptor. Neuropharmacology 36, 1285–1294.

Virginio C, MacKenzie A, Rassendren FA, North RA, Surprenant A (1999) Pore dilation of neuronal P2X receptor channels. Nat Neurosci 2, 315–321.

Wilson HL, Wilson SA, Surprenant A, North RA (2002) Epithelial membrane proteins induce membrane blebbing and interact with the P2X7 receptor C terminus. J Biol Chem 277:34017–34023.

Wisnoskey BJ, Estacion M, Schilling WP (2004) Maitotoxin-induced cell death cascade in bovine aortic endothelial cells: divalent cation specificity and selectivity. Am J Physiol Cell Physiol. 287, C345-56.

Yan Z, Li S, Liang Z, Tomić M, Stojilkovic SS (2008) The P2X7 receptor channel pore dilates under physiological ion conditions. J Gen Physiol. 132, 563-573.

Yan Z, Khadra A, Sherman A, Stojilkovic SS (2011) Calcium-dependent block of P2X7 receptor channel function is allosteric. J Gen Physiol. 138, 437-452.

Yehezkel G, Hadad N, Zaid H, Sivan S, Shoshan-Barmatz V (2006) Nucleotide-binding sites in the voltage-dependent anion channel: characterization and localization. J Biol Chem. 281, 5938-5946.

Young RC, McLaren M, Ramsdell JS (1995) Maitotoxin increases voltage independent chloride and sodium currents in GH4C1 rat pituitary cells. Nat Toxins. 3, 419-427.

Young MT, Pelegrin P, Surprenant A (2006) Identification of Thr283 as a key determinant of P2X7 receptor function. Br J Pharmacol 149, 261–268.

Zhao Q, Yang M, Ting AT, Logothetis DE (2007) PIP(2) regulates the ionic current of P2X receptors and P2X(7) receptor-mediated cell death. Channels (Austin) 1:46–55.

Zizi M, Forte M, Blachly-Dyson E, Colombini M (1994) NADH regulates the gating of VDAC, the mitochondrial outer membrane channel. J Biol Chem 269, 1614-1626.

3

Patch Clamp Study of Neurotransmission at Single Mammalian CNS Synapses

Norio Akaike
Kumamoto Health Science University
Japan

1. Introduction

Single mammalian CNS neurons can be acutely isolated with adherent and functional excitatory and inhibitory synaptic nerve terminals (boutons) using a mechanical dissociation procedure without any enzyme treatment (Vorobjev, 1991; Haage & Johansson, 1998; Rhee et al., 1999). This 'synaptic bouton' preparation is particularly suitable for physiological and pharmacological investigations of mammalian CNS synaptic transduction mechanisms, and the properties of both the receptors, transporters, and 2nd messengers present in the presynaptic terminals (boutons), and the synaptic and extrasynaptic receptors on the postsynaptic membrane, can be studied.

The truncated dissociated potsynaptic neurons are well space-clamped allowing accurate measurements of synaptic currents, and the isolated neurons are devoid of complications arising from surrounding cells such as other neurons or astrocytes (glial cells). The acute, mechanical dissociation avoids possible changes in protein distribution and/or function as result of either enzyme treatment or *in vitro* culture. Yet, the extremely small size of typical mammalian presynaptic terminals (< 1μm) have presented a challenge for functional studies on neurotransmitter release. In the synaptic bouton preparation, neurotransmitter is released from these adherent terminals giving rise to spontaneous synaptic potentials. Furthermore, a single presynaptic nerve terminal (bouton) can be focally stimulated with electrical pulses (Akaike et al., 2002; Murakami et al., 2002; Akaike & Moorhouse, 2003) to result in evoked synaptic potentials. Therefore, this 'synapse bouton' preparation has helped to unravel the mechanisms and modulation of synaptic transmission in the mammalian CNS. In this article, the general properties of this preparation are described, along with some typical examples of its applications to the study of synaptic transmission.

2. Mechanical dissociation of mammalian CNS neurons

The synaptic bouton has to date been prepared from mature (1~3 months old) and immature (12~18 days old) rats, immature and juvenile mice, and guinea-pigs (10 days~1 month old) (e.g. Akaike et al, 2002) using the following approach. Animals were decapitated under pentobarbital anesthesia (50 mg kg^{-1} *i.p.*), and the brain quickly removed and immersed in an ice-cold physiological incubation solution, saturated with 95 % O_2 and 5 % CO_2. Brain or spinal cord slices containing the region of interest were cut using a vibrating microtome (VT 1200S; Leica, Nussloch, Germany) at a thickness of ~400 μm. The brain slices were then incubated in a medium oxygenated with 95 % O_2 and 5 % CO_2 at room temperature (21-24 °C), for at least 1h before mechanical dissociation. For mechanical dissociation, the brain slice was transferred into a 35 mm culture dish (Primaria 3801, Becton Dickinson, Rutherford, NJ, USA) containing a HEPES-buffered standard external solution, and fixed by an anchor made from a platinum frame and nylon thread (Fig. 1A). The region of interest was identified under a binocular microscope (SMZ645, Nikon Tokyo) and the tip of a fire-polished glass pipette was lightly placed on the slice surface above the target neurons and vibrated horizontally (0.2-2 mm displacement) at 50-60 Hz using the manufactured device (S1-10 cell isolator, K.T. Labs, Tokyo) (Fig. 1B). The dish is typically also moved horizontally along the target region by hand to enable isolation of neurons from more than a single spot. Thereafter, the slices were removed from the dish, and the mechanically dissociated neurons left to settle and adhere to the bottom of the dish for at least 15 min before electrophysiological measurements.

An electron microscopy study was done to confirm that boutons are attached to dissociated CNS neurons. Fifteen boutons were observed on a single dissociated rat hippocampal neuron with an averaged diameter of 0.6±0.04 μm and were clearly separated from the closest neighboring bouton, as shown in Figure 1C (Akaike et al., 2002). These adherent synaptic boutons contain functional voltage-dependent ionic channels, various chemo-receptors, transporters, and 2nd messenger pathways, as also found in the postsynaptic neuron. To date, this 'synaptic bouton' preparation has been obtained from various brain regions in our laboratory, including: the Meynert's nuclei (Rhee et al., 1999 ; Arima et al., 2001), the basolateral amygdala (Koyama et al., 1999), the hippocampal CA1 (Matsumoto et al., 2002) and CA3 (Yamamoto et al., 2011) regions, the periaqueductal gray (Kishimoto et al., 2001), the ventromedial hypothalamus (Jang et al, 2001), and the spinal sacral dorsal commissural nucleus (Akaike et al., 2010) and many other regions.

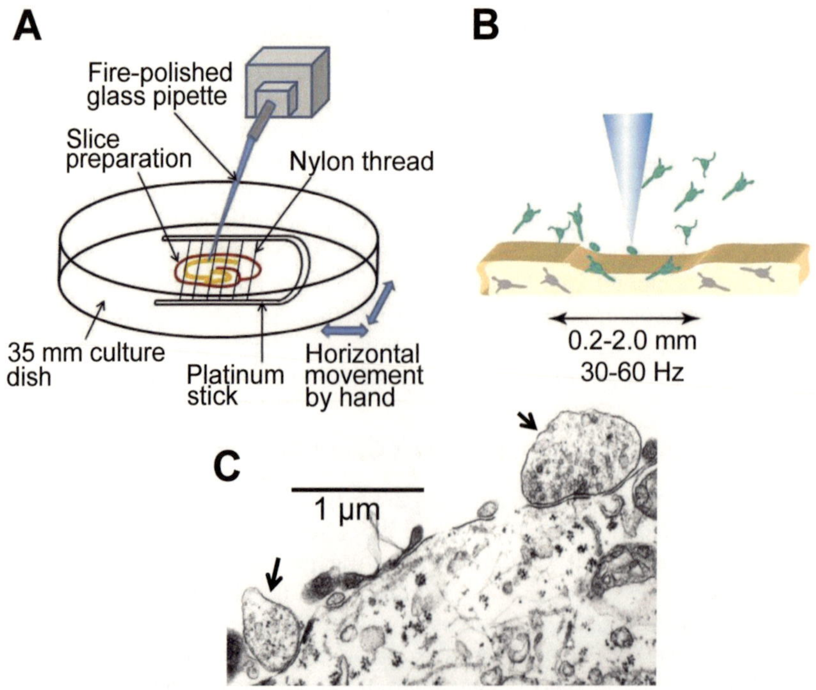

Fig. 1. Mechanical isolation of single CNS neurons

A, B: Schematic illustration of the mechanical dissociation of rat hippocampal CA1 pyramidal neurons. A fire-polished glass pipette is vibrated horizontally across the surface of the CA1 area at 50~60 Hz. Successful liberation of viable neurons results in a fine mist originating from the dissociation site. The treated slice is then removed and the liberated neurons are left to adhere to the base of the culture dish. C: Electron microscopy image showing two presynaptic boutons (arrowheads) adherent to a dissociated hippocampal CA1 neuron. Part C was used, with permission, from (Akaike et al.2002).

3. 'Synaptic bouton' preparations preserve functional presynaptic terminals (boutons)

Neurons mechanically isolated as described above show spontaneous synaptic potentials, as shown and described below. Fluorescence was also used to visualize functional presynaptic boutons. FM1-43 (Molecular probes, OR, USA) was applied to dissociated rat sacral dorsal commissural nucleus (SDCN) neurons, at a concentration of 10µM and in a depolarizing external solution containing high (45mM) K^+ for 30 sec, before the neurons were well washed with a standard external solution. FM1-43 flourescent spots, representing putative presynaptic boutons, were seen under the inverted microscope attached to an SDCN neuron at the soma and proximal dendrites (Fig. 2Aa). These fluorescent spots quickly distained when a second high $[K^+]_o$ (15~20mM) external solution was applied for 30 sec (Fig. 2Ab), indicating these are functional boutons undergoing endocytosis and exocytosis of the

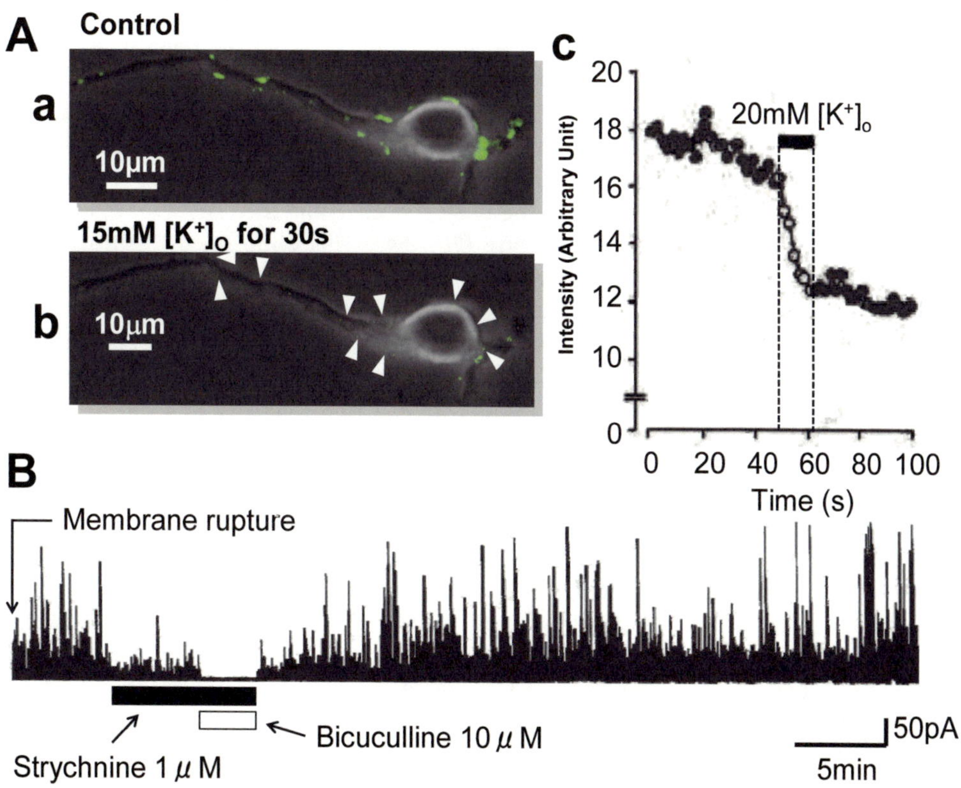

Fig. 2. The 'synaptic bouton' preparation.

fluorescent dye. Figure 2Ac shows the time course of one of the fluorescent spots before, during, and after application of a 20mM K^+ external solution.

Spinal SDCN neurons receive the projections of two kinds of inhibitory nerves, glycinergic and GABAergic ones, and hence the sIPSC is completely ceased by cumulative application of strychinine (a selective glycine receptor antagonist) and bicuculline (a selective $GABA_A$ receptor antagonist) (Fig. 2B). These representative results suggest that mechanically dissociated spinal neurons maintain functional presynaptic nerve terminals, which are sensitive to $[K^+]_o$ dipolarization, and useful for pharmacological studies of release and/or synaptic receptors.

Aa~c: A dissociated hippocampal CA1 neuron with presynaptic boutons stained by FM1-43 fluorescence (arrowheads). The intensity of the fluorescent signal in the boutons disappeared after adding of 15 mM K^+ to external solution (a, b). (c) A typical time course of the fluorescence intensity of a presynaptic bouton during the addition of 20mM K^+. B: Effects of strychnine and bicuculline on spontaneous inhibitory (glycinergic and GABAergic) currents recorded from a single rat spinal interneuron from the sacral dorsal commissural nucleus (SDCN) region. The result indicates that these neurons receive both glycinergic and GABAergic projections. Parts A and B were adapted with permission, from (Akaike et al., 2002 ; Jang et al., 2002)

4. Focal electrical stimuli of a single bouton using *a "θ"* glass pipette

The stimulating pipette for focal electrical stimulation of a single bouton adherent to mechanically dissociated CNS neurons was made fromθglass tube (φ= OD 2mm, ID 1.4mm, WPI) filled with normal external test solution. A θ glass pipette is separated down the centre by a wall to give rise to two adjacent and separated compartments. Both compartments are filled with external solution and electrical wires, thereby acting like a bipolar electrode. The pipette was placed closer as possible to the postsynaptic soma membrane of a single CNS neuron during a whole-cell patch recording (Fig. 3A). The stimulating pipette was then carefully moved along the surface membrane of the soma or dendrites while applying stimulation pulses and monitoring for responses. Paired pulses are typically used if investigating presynaptic mechanisms, and stimuli applied typically once every 5-10 sec and applied via a stimulus isolator (SS-202 J, Nihon Koden, Tokyo). In individual neurons, the stimulus paradigms used are 100 μs duration, 0.1–0.3 mA intensity and 30–60 ms inter-stimulus intervals for evoked IPSCs (eIPSCs), and 100 μs duration, 0.05–0.08 mA intensity and 20–30 ms inter-stimulus intervals for evoked EPSCs (eEPSCs).

To determine whether GABAergic IPSCs were really evoked from a single bouton or, alternatively, from multiple separate boutons, the stimulus-amplitude and stimulus-distance relationships were examined. When a GABAergic eIPSC was identified, it appeared in an all-or-none fashion as stimulus strength increased or decreased (Fig.3B), indicating that the stimulating pipette was positioned just above a single GABAergic bouton. Furthermore, when the stimulus pipette was moved horizontally along the surface of a dissociated neuron, the eIPSC again appeared or disappeared in an all-or-none fashion. With shifts in distance of less than 0.4μm, the eIPSCs were maintained in the majority of boutons tested. The shift in the electrode did not affect the mean amplitude of eIPSCs but increased the failure rate of eIPSCs (Fig. 3Bb). In the case of #4 in Figure 3Bb, however, the eIPSCs were still elicited even when the stimulus electrode was shifted ±0.4μm (totally about 0.8μm), suggesting that the eIPSCs were elicited from several boutons. Hence, studies on 'single boutons' seem to require stimuli locations that fail to elicit the eIPSC response if shifted by more than 0.4μm.

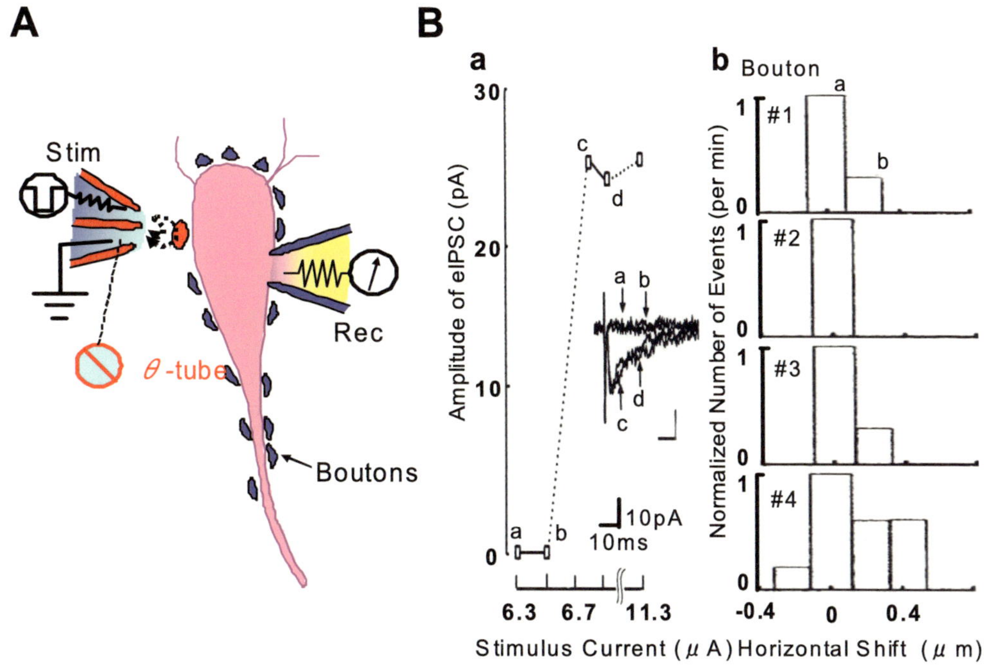

Fig. 3. Focal electrical stimulation of a single bouton

A: Schematic representation of a 'synaptic bouton' preparation and focal electrical stimulation of a single nerve ending (bouton). Ba: The relationship between stimulus strength and the amplitude of GABAergic eIPSCs, which appeared in an 'all or none' nature as stimulus strength varied. Inset shows two failure responses and two successfully evoked responses. Bb: The relationship between the extent of lateral displacement of the stimulating pipette (from a starting position, 0μm, that induced a minimum failure rate) and the proportion of successfully evoked eIPSCs. Shifts greater than 0.4~0.6μm abolished the response. Each of the four panels (#1~4) comes from four different boutons. The Y-axis shows the number of failures relative to the maximum incidence of successfully evoked responses, obtained at position 0μm. Part B was adapted with permission, from (Akaike et al., 2002).

5. General properties of spontaneous and evoked transmitter release in the 'synaptic bouton' preparation

The frequency of spontaneous IPSCs (sIPSC) and spontaneous EPSCs (sEPSCs) recorded from different CNS regions is between 1 and 10Hz, and the variability presumably reflects the differences in the number and excitability and adherent boutons. The spontaneous synaptic currents are both action potential-dependent and -independent (TTX; tetrodotoxin resistant). The addition of TTX, a selective Na channel blocker, decreases the frequency of

GABAergic or glycinergic sIPSC (Rhee et al., 1999) and glutamatergic sEPSC (Jang et al., 2001) by about 50 % at least. In the presence of Ca channel blockers (Rhee et al., 1999; Koyama et al., 1999; Shoudai et al., 2007) or in nominal Ca-free solution (Maeda et al., 2009), glycinergic sIPSC frequency decreases 30 ~40% of control. A large proportion of these TTX-resistant miniature IPSCs (mIPSCs) are independent from Ca^{2+} influx (Emptage et al., 2001; Miller et al., 1998; Scholtz et al., 1992), as reported by others. Both Ca^{2+} release from internal Ca stores and store - depleted Ca^{2+} influx contribute to these mIPSCs (Emptage et al., 2001). Consequently, many 'minis' remain in the absence of external Ca^{2+} influx. The ability to dissect transmitter release into such spontaneous and miniature postsynaptic currents, Ca^{2+} influx resistant or sensitive, is useful for examining the locus of action of presynaptic neuromodulators.

As indicated above, the selective activation of a single excitatory glutamatergic (Yamamoto et al., 2011; Akaike et al., 2010) or inhibitory GABAergic (Akaike et al., 2002; Murakami et al., 2002; Akaike et al., 2010; Ogawa et al., 2011) and glycinergic (Akaike et al., 2010; Nonaka et al., 2010) nerve terminals (boutons) are possible in the 'synaptic bouton' preparation by using focal electrical stimulation technique. The current amplitudes of glutamatergic eEPSCs, glycinergic eIPSCs, and GABAergic eIPSCs recorded from SDCN neurons are on average about 140, 250, and 80 pA, respectively. On the other hand, the average failure rates of glutamatergic, glycinergic and GABAergic single bouton synaptic responses are 18, 37, and 32%, respectively.

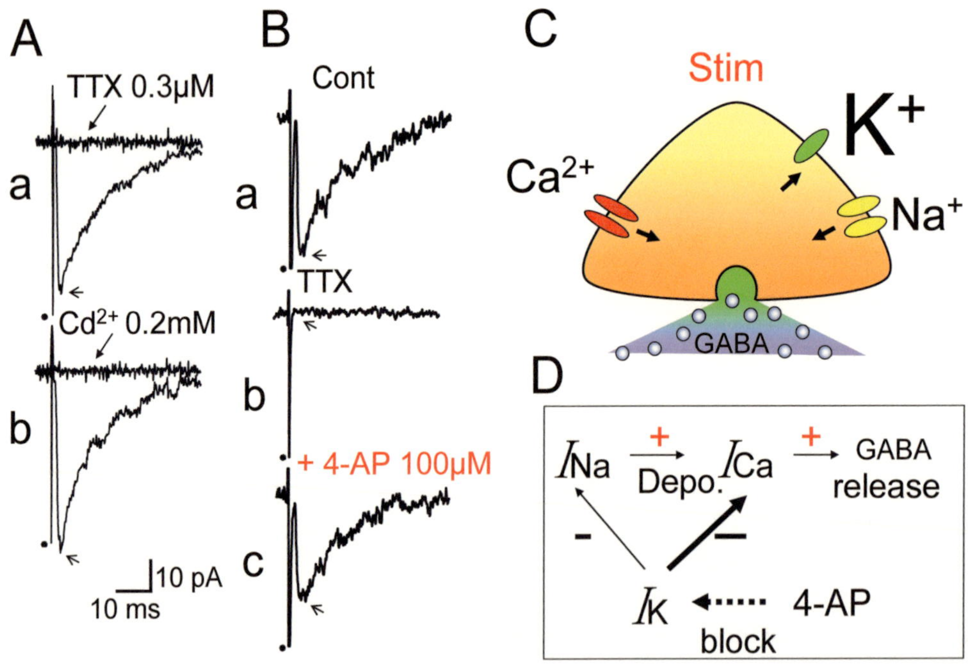

Fig. 4. Focal stimulation of a single GABAergic bouton

A,B: GABAergic eIPSCs in the presence and absence of 0.3μM tetrodotoxin (TTX), 0.2mM Cd^{2+}, and 0.3μM TTX + 100μM 4-AP as indicated. All traces were recorded from the same hippocampal CA1 pyramidal neuron. C: A schematic illustration of voltage-dependent Na^+, K^+ and Ca^{2+} channels in single GABAergic boutons. D: Diagram of the functional interaction among Na^+, K^+ and Ca^{2+} currents in single boutons. Parts A and B were adapted with permission, from (Akaike et al., 2002).

The depolarization of the nerve terminals triggered by Na channel activation results in subsequent Ca^{2+} influx passing through voltage-dependent Ca channels (Nonaka et al., 2010; Jackson & Zhang, 1995). The application of 0.3μM TTX reversibly abolished GABAergic eIPSCs. In the presence of TTX, eIPSCs could not be evoked even after increases in the stimulus strength. However, in the presence of TTX, eIPSC reappears when 100μM 4-AP (a nonselective K channel blocker) is applied (Fig.4A,B). The application of the voltage-dependent Ca^{2+} channel blocker, Cd^{2+}, reversibly blocks eIPSCs *in either in control conditions or in the presence of TTX and 4-AP*. This suggests that eIPSCs are evoked by Ca^{2+} influx through voltage-dependent Ca^{2+} channels that typically activate due to the opening of voltage-dependent Na^+ channels. In the presence of TTX, electrical stimulation are unable to directly depolarize the terminals sufficiently to activate Ca^{2+} influx and release, seemingly due to a high density and/or lower activation threshold of presynaptic K channels. Once these K^+ channels are blocked (by adding 4-AP) the nerve terminals can be directly activated by focal stimulation (Akaike et al., 2002) (Fig. 4C,D).

6. Voltage-dependent Ca channel subtypes on single GABAergic and glycinergic nerve terminals

Varying the external Ca^{2+} concentration ($[Ca^{2+}]_o$) changes both the current amplitude and the failure rate (Rf) of eIPSCs elicited by focal stimulation of a single glycinergic bouton (Fig.5). Raising $[Ca^{2+}]_o$ from 2 to 5mM slightly increases the eIPSC current amplitude while decreasing Rf. Decreasing $[Ca^{2+}]_o$ from 2 mM to 1mM substantially decreases the current amplitude and increases Rf. The eIPSCs are abolished at 0.3mM $[Ca^{2+}]_o$ (Fig.5). These results indicate glycine release is highly dependent on Ca^{2+} influx into single glycinergic nerve terminals, via both changes in release probability (changing Rf), and changes in the amount of glycine released (changing eIPSC amplitude).

Voltage-dependent Ca^{2+} channels are distributed throughout the CNS and play a key role in many neuronal functions including synaptic transmission. Ca^{2+} entering into the presynaptic terminal through different Ca^{2+} channel subtypes result in local intra-terminal "hot-spots" in which Ca^{2+} binds to various Ca^{2+} -binding proteins located at the release sites to trigger exocytosis of neurotransmitter vesicles (Borst & Sakmann, 1996; Seager et al., 1999). In these brain slice and cultured neuronal preparations, different Ca^{2+} channel subtypes coexist and co-regulate transmitter release.

The precise functional arrangement of Ca channel subtypes on small CNS nerve terminals is technically challenging, and focal electrical stimulation of single GABAergic and glycinergic nerve terminals in nerve-bouton preparations of hippocampal CA1 and spinal SDCN neurons, respectively, has addressed this question. The L-, N- and P/Q subtypes of Ca^{2+} channels were identified on the CA1 GABAergic nerve terminals (Murakami et al., 2002) and P- and R Ca^{2+} channel subtypes on the SDCN glycinergic terminals (Nonaka et al., 2010). There is some Ca^{2+} channel cooperativity in the individual terminals, and the different subtypes present all contribute to determining the total Ca^{2+} influx associated with synaptic vesicle release.

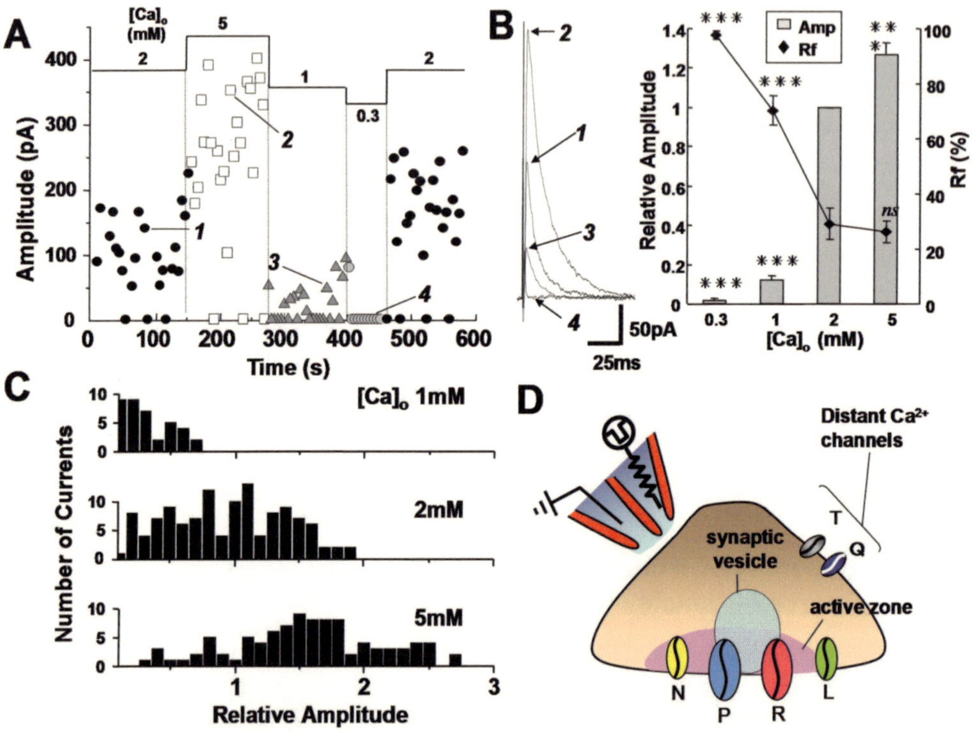

Fig. 5. Effects of changes in the external Ca^{2+} concentration ($[Ca^{2+}]_o$) on glycinergic eIPSCs from the spinal SDCN neuron

A: A plot of eIPSC amplitude (evoked at a frequency of 0.2 Hz) in different $[Ca^{2+}]_o$ concentrations as indicated (0.3~5mM). B: Left panel, Representative current traces (1~4) appeared at times and in A. Right panel, Relationship between the averaged current amplitude and failure rate (Rf), and $[Ca^{2+}]_o$. ** $P<0.01$, *** $P<0.001$, ns, no significant. Each data point represents the mean ± SEM of 6 neurons. C: Histogram of the relative current amplitudes of eIPSCs at 1, 2 and 5mM $[Ca^{2+}]_o$. All current amplitudes in different $[Ca^{2+}]_o$ were normalized to the average current amplitude obtained from each bouton in control external solution with 2mM Ca^{2+}. D: Schematic diagram of the proposed distribution of high- and low-voltage threshold Ca^{2+} channels on a single glycinergic bouton projecting to rat spinal SDCN neurons. Dominant control of glycine release depends on P- and R-type Ca^{2+} channels. Parts A~C were obtained, with permission, from (Nonaka et al., 2010).

7. GABA$_A$ receptor-mediated 'autoinhibition' and 'presynaptic inhibition' in single CNS nerve terminals

GABA is accepted as a major inhibitory neurotransmitter, and can act at GABA$_A$ receptors to cause both postsynaptic and/or presynaptic inhibition. Presynaptic GABA$_A$ receptors can inhibit GABA release from GABAergic nerve terminals, as an example of classical auto-inhibition, or can presynaptic inhibition in glutamatergic nerve terminals including the classically studied primary-afferent depolarization in the spinal cord. To investigate how presynaptic $GABA_A$ receptors modulate spontaneous and action potential mediated GABA release, 'synaptic bouton' preparations isolated from hipppocampal CA3 region were used. Muscimol, a selective $GABA_A$ receptor agonist, increased spontaneous GABAergic IPSCs (sIPSCs) in a concentration-dependent manner, without affecting the current amplitude, indicating that muscimol acts on $GABA_A$ receptors in the presynaptic GABAergic nerve terminals. The increase in sIPSC frequency is reversibly prevented by the addition of Cd^{2+}, or in Ca^{2+}-free external solution, suggesting that muscimol depolarizes nerve terminals to induce Ca^{2+} influx through voltage-dependent Ca channels (Jang et al., 2002; Jang et al., 2006; Yamamoto et al., 2011). The depolarization indicates a GABA-induced Cl^- efflux and hence a higher Cl^- concentration in presynaptic nerve terminals than prediction by a passive distribution. In neuronal soma and in nerve terminals, this results from the activity of the Na^+, K^+, $2Cl^-$ cotransporter type1 (NKCC-1) (Jang et al., 2001; Kakazu et al., 1999) and in fact blocking this transporter with bumetanide prevents the GABA-induced increase in sIPSC frequency (Jang et al., 2001). The functional role of presynaptic $GABA_A$ receptors on eIPCS at GABAergic hippocampal CA1 synapses was also studied. Muscimol (3μM) decreased the eIPSC amplitude, and increased the Rf (Fig.6), with this inhibitory effect being completely abolished by bicuculline, confirming the role of $GABA_A$ receptors.

Similar inhibitory effects were also seen for the presynaptic inhibition of excitatory responses at Glutamatergic hippocampal CA3 synapses by muscimol. At a concentration range of between about 0.3~10 μM, muscimol decreased eEPSC amplitude, and increased the Rf (Fig 6A), and this effect was also sensitve to bicuculline (Fig. 6C). At a lower concentration (0.03μM), muscimol had an excitatory effects, increase of eEPSC amplitude and decrease of the Rf (Fig. 6Ab). This presynaptic action also depends on NKCC-1 mediated Cl- uptake (Kakazu et al., 1999; Payne et al., 2003) into the glutamatergic nerve terminals, as bumetanide (10μM), a blocker of NKCC-1, completely blocks the muscimol effects on eEPSCs (Figure 6D). Consequently, activation of presynaptic $GABA_A$ receptors induces a small or large depolarization of the terminals, which induces a sustained increase in eEPSCs or a decrease in eEPSC, respectively. The decrease in evoked glutamate release may result from either or both of the blockade of action potentials as a consequence of inactivation of voltage-dependent Na channels (Sasaki et al., 2008) or a depolarization-induced shunt of the membrane conductance (Yamamoto et al., 2011; Jang et al., 2002; Cattaert et al., 1994; Graham et al., 1994).

We have also examined the effects of muscimol on short-term synaptic plasticity, including paired-pulse facilitation responses (PPF). The response to the second stimulus in a paired stimulus paradigm depends on residual intracellular Ca^{2+} and remaining vesicles available for release after the initial stimulus evoked response. Hence paired pulse responses are a measure of PPF is considered to be a presynaptic phenomenon which is regulated by presynaptic vesicle and intracellular Ca^{2+} homeostasis (Zucker, 2002). Muscimol caused a significant enhancement of the paired-pluse ratio (PPR = response P_2/ response P_1) suggesting either an alteration of presynaptic Ca^{2+} homeostasis or an increase in the number of vesicles available for release induced by the cation (Fig. 6B). The result supports the presynaptic locus of effect, with the probable likely mechanism being reason is that muscimol of high concentration causes some inactivation of voltage-dependent Na^+ channels, hence a less effective action potential, a reduced activation could inhibit Ca^{2+} influx into presynaptic terminals through voltage-dependent Ca channels (VDCC) to a reduced Ca^{2+} influx into presynaptic terminals by causing presynaptic inhibition resulting from the inactivation of voltage-dependent Na channels, then under an intraterminal lower Ca^{2+} condition and a reduced presynaptic release probability (P_r). Therefore, relatively more vesicles are available for the second response, which would potentiate PPR (Fig. 6B). This is consistent with previous studies showing an inverse relationship between P_r and PPR (Asztely et al., 1996; Murthy et al., 1997).

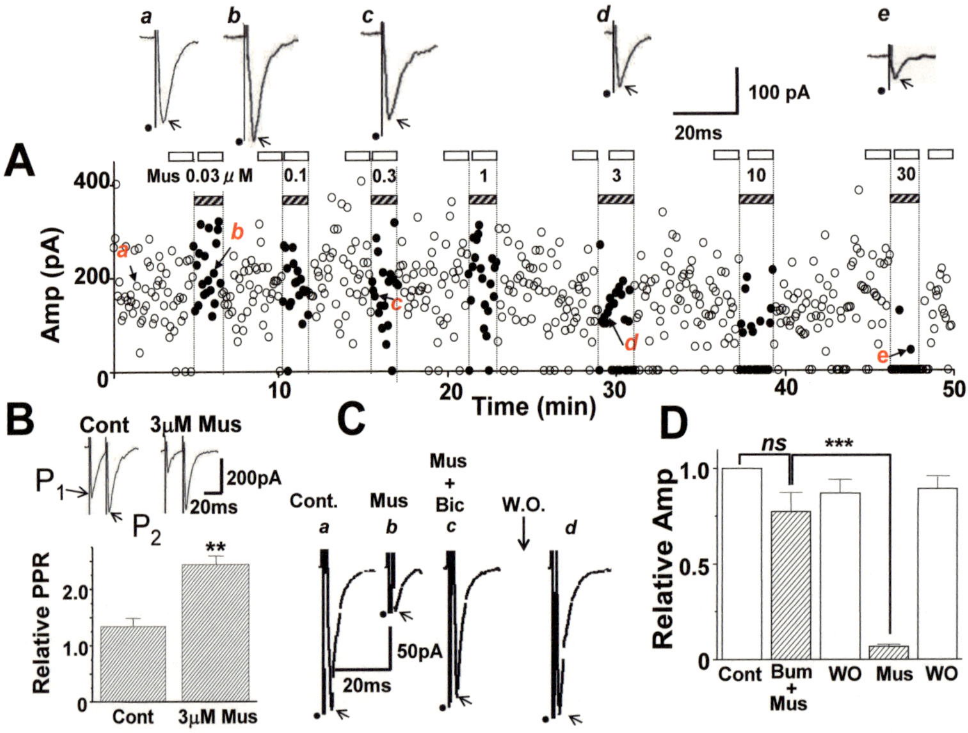

Fig. 6. Glutamatergic evoked EPSCs

Effects of muscimol on glutamatergic eEPSCs recorded from CA3 pyramidal neurons. A: All eEPSCs were evoked from a single glutamatergic bouton stimulated every 5s while the postsynaptic hippocampal CA3 neuron was voltage-clamped at a holding potential (V_H) of −65mV. The amplitude of the eEPSCs gradually decreased with increasing muscimol concentration. The insets a~e show sample single eEPSCs at the time points a~e indicated in the graph.

B: The effects of muscimol on the paired pulse responses. The paired pulse ratio, PPR (P2/P1), increases in the presence of 3μM muscimol. ** P<0.01. C: Muscimol (10μM) -induced inhibition of eEPSC is reversibly abolished by bicuculline (10μM). D: Bumetanide (10μM) prevents the muscimol (3μM) -induced inhibition of eEPSC. *** P<0.001, ns, no significant. Parts A~D were quoted with permission, from (Yamamoto et al., 2011)

8. Investigating the effects of neuromodulators on presynaptic nerve terminals using the 'synaptic bouton' preparation

8.1 5-HT action on GABAergic transmission

Serotonin (5-HT) is an important neurotransmitter in CNS and can modulate neuronal activities via 5-HT receptors (Bruns et al., 2000; Levkovitz & Segal, 1997), which consist of seven families of membrane proteins comprising a total of fourteen subtypes (Barns & Sharp et al., 1999). In the rat hippocampal CA1 region, both the pyramidal neurons and the GABAergic interneurons are innervated by serotoninergic neurons originating from the midbrain raphe nuclei (Azmitia & Segal, 1978). Activation of 5-HT_{1A} and 5-HT_3 receptors in the hippocampal slice *in vitro* reduces and enhances GABAergic transmission, respectively (McMahon & Kauer , 1997; Schmitz et al., 1995). Figure 7Ab shows such serotoninergic modulation of GABAergic eIPSCs recorded from CA1 pyramidal neurons in a rat hippocampal slice preparation. Interestingly, the eIPSC amplitude is initially reduced, then followed by a gradual increase with increasing 5-HT concentration. This biphasic action is consistent with 5-HT activating at least two different receptor subtypes. In fact, a selective 5-HT_{1A} receptor agonist, 8-OH DPAT, increases eIPSC amplitude while a selective 5-HT_3 receptor agonist, mCPBG, inhibits eIPSC amplitude, both acting in a concentration-dependent fashion (Fig.7Ac). The results indicate two subtypes at least present at GABAergic terminals, but it is difficult in the slice preparation, with numerous synaptic connections on to a single neuron, to determine if both subtypes exist on the same terminal, and if they interact.

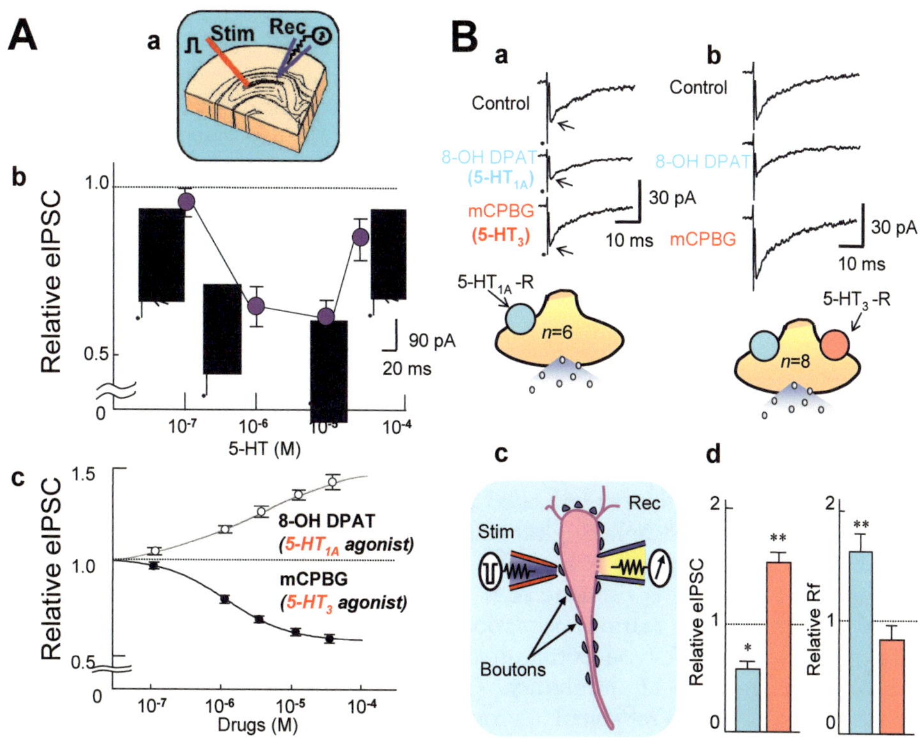

Fig. 7. Modulation of GABAergic evoked IPSCs by 5-HT

A: 5-HTergic modulation of GABAergic eIPSCs recorded in rat hippocampal CA1 pyramidal neurons in the brain slice preparation, illustrated schematically in Aa. (Ab): Concentration-response curves and sample GABAergic eIPSCs, all recorded in the presence of glutamatergic antagonists, 10μM CNQX and 10μM AP-V. 5-HT initially decreases the responses, and then increases the response as the 5-HT concentration is increased. (Ac): The effects of selective agonist of 5-HT_{1A} receptor, 8-OH DPAT, and a selective agonist of 5-HT_3 receptor, mCPBG on the mean amplitude of GABAergic eIPSCs. Each point shows the mean

±SEM of data from eight neurons. B: 5-HTergic modulation of eIPSCs elicited focally in single GABAergic boutons. Single boutons could be classified as either Type 1 boutons (Ba) that have only $5HT_{1A}$ receptors and respond only to 8-OH DPAT (1μM), or type 2 boutons (Bb) have both $5\text{-}HT_{1A}$ and $5\text{-}HT_3$ receptors and also respond to mCPBG (1μM). (Bc): Schematic illustration of 'synaptic bouton' preparation and focal electrical stimulation. (Bd): Effects of $5\text{-}HT_{1A}$ (blue) and $5\text{-}HT_3$ agonists (red) on the mean amplitude and Rf of GABAergic eIPSCs. The data in each column are expressed relative to the initial control value, and represent the mean ±SEM of 8~14 neurons. * P<005、 ** P<0.01 Parts A, Ba, b, d, were adapted, with permission, from (Katsurabayashi et al.,2003).

Using the 'synaptic bouton' preparations of single hipocampal CA1 neurons, a single GABAergic nerve terminal (bouton) was activated by focal electrical stimuli (Fig. 7Bc). At six boutons tested, 8-OH DPAT (1μM) decreased the GABAergic eIPSC amplitude and increased the Rf 1.8 fold, while mCPBG (1μM) to these cells had no effect, indicating that these six boutons have no $5\text{-}HT_3$ receptors but only functional $5\text{-}HT_{1A}$ receptors (Fig.7 Ba,d,). However, at other eight boutons, both 8-OH DPAT and mCPBG had effects, decreasing and increasing eIPSC amplitude, respectively. Hence, the results indicate that these boutons had both $5\text{-}HT_{1A}$ and $5\text{-}HT_3$ receptors (Fig. 7Bb,d). Interestingly, mCPBG only decreased Rf modestly at these boutons. Furthermore, there were no boutons which had only $5\text{-}HT_3$ receptors, or which had neither $5\text{-}HT_{1A}$ nor $5\text{-}HT_3$ receptors. The physiological consequences of this co-localisation of $5\text{-}HT_{1A}$ and $5\text{-}HT_3$ receptors on single boutons is that 5-HT may cause an initial transient enhancement of GABA release, progressing into a reduction of GABA release as the $5\text{-}HT_3$ receptors become desensitized and the slower actions of metabotropic $5\text{-}HT_{1A}$ receptors takes over. Such combination of transient excitatory receptors and persistent inhibitory receptors may have some benefits for more rapid and sustained signaling, respectively (Koyama et al., 1999; Koyama et al., 2000, Koyama et al., 2002; Katsurabayashi et al, 2003).

8.2 Modulation of excitatory and inhibitory presynaptic terminals by A type botulinum toxin

Botulinum toxins (BoNTs) are currently widely used to study the molecular events that are involved in exocytosis (Schiavo et al., 2000; Sudhof, 2004). Studies on brain slices, cultured neurons and synaptosomes have indicated that BoNTs can impede the release of various transmitters such as acetylcholine, glutamate, glycine, noradrenalin, dopamine and ATP (Ashton & Dolly, 1988; Capogna et al., 1997), in addition to the well documented actions on neuro-muscular transmission (Schiavo et al., 2000). Therefore, it is of interest to study the effects of BoNTs on fast neurotransmission at mammalian CNS terminals. Below, we describe the effects of A2 type botulinum toxin (A2NTX) on spontaneous and evoked neurotransmitter release at inhibitory (glycinergic or GABAergic) and excitatory (glutamatergic) synapses in rat spinal neurons using 'synaptic bouton' preparations (Akaike et al., 2010; Sakaguchi et al., 1981).

The rank order of the sensitivity of these different synapses to the inhibitory effects of A2NTX (0.1~10pM) on spontaneous transmitter release was glycinergic > GABAergic≫glutamatergic synapses (Fig. 8A). Using focal electrical stimulation to evoke eIPSCs or eEPSCs of large amplitude and with low Rf, we showed that A2NTX (0.01~1pM) completely abolishes the eIPSC and eEPSC in a time-dependent fashion and with partial reversibility. The rank order of this inhibitory effect was glycinergic eIPSC≧GABAergic eIPSC ≧glutamatergic eEPSC (Fig. 8 Ba,b). The neurotoxin sensitivity for the evoked transmitter release of three transmitters was greater than the spontaneous one. The other striking feature of this study was that the spontaneous or evoked release of the inhibitory transmitters was 10-100 times more sensitive to A2NTX, as compared to those of the excitatory glutamate release. This observation suggests that the precise molecular events underlying excitatory and inhibitory, and spontaneous and evoked neurotransmitter release, may be different. We have also seen differences between spontaneous and evoked release in their sensitivity of divalent cations, and had previously suggested that spontaneous and evoked glycine release in SDCN neurons involved Ca^{2+} binding to different synaptotagmins (Maeda et al., 2009). Recent studies have now in fact indicated that > 95% spontaneous release is induced by Ca^{2+}-binding to synaptotagmin 1 in murine cortical neurons (Xu et al., 2009), while synaptotagmins 1, 2, and/or 9 are involved in evoked neurotransmitter release (Sollner, 2003; Rizo & Sudhof, 1998). This involvement of different synaptotagmins in spontaneous and evoked neurotransmitter release could also explain the different sensitivities of spontaneous and evoked release of glycine, GABA and glutamate to A2NTX. In addition, transmitter vesicles at CNS terminals are divided into two general pools, a ready-to-release pool, and a reserve pool (Sudhof, 2004; Schikorski & Stevens, 2001). The different vesicle pools may contribute differently to spontaneous and evoked release, and A2NTX may also acts differentially on these processes.

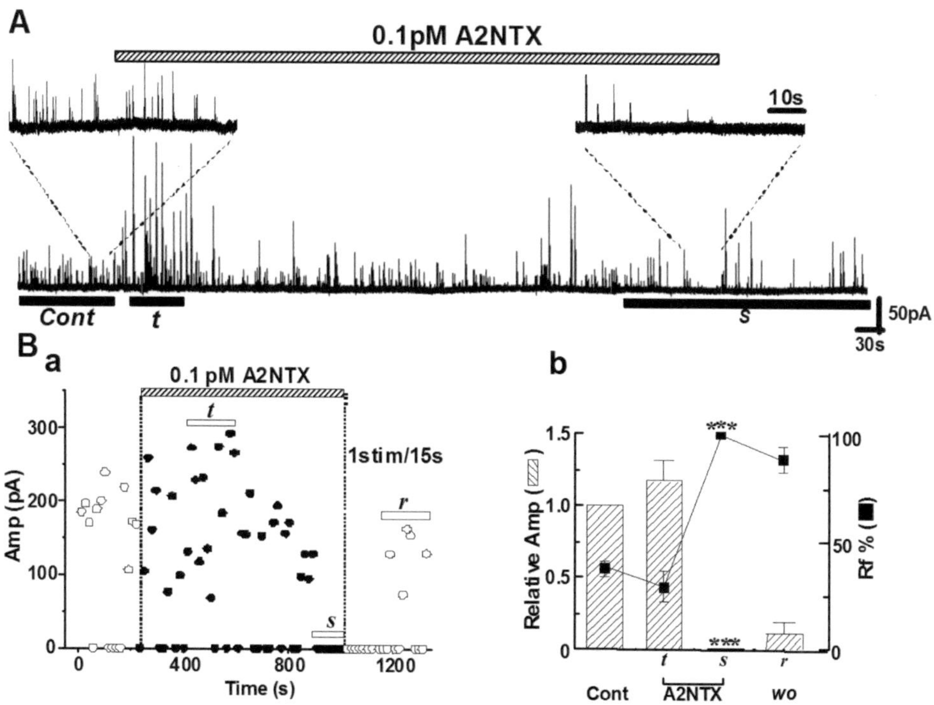

Fig. 8. Modulation of spontaneous and evoked glycinergic IPSCs by A2 type botulinum toxin

Effects of A2NTX on glycinergic spontaneous inhibitory postsynaptic currents (sIPSCs). A: Typical sIPSCs recorded from a mechanically isolated spinal SDCN neuron. The glycinergic sIPSCs are isolated by allowing the GABAergic sIPSCs to run down because the internal patch pipette solution is without ATP. Application of 0.1pM A2NTX transiently enhanced both the frequency and amplitude of glycinergic sIPSCs, before gradually decreasing them. The periods in the current trace indicated by "cont", "t", and "s" represent the control currents, transient facilitation and steady-state inhibition of currents in the presence of A2NTX. Expanded current recordings are shown in the upper panel. B: Effects of 0.1pM A2NTX on evoked glycinergic eIPSCs. (a): A single glycinergic bouton was activated by focal electrical stimuli every 15s. Stimuli that failed to evoke a response (i.e. current amplitude=0) were referred to as failure and used to calculate the failure rate (Rf). (b): Data were analyzed during the control, transient (t), steady-state (s), and recovery (r) periods, as indicated in panel A. Each column and filled square shows the mean value of 7 experiments. *** P< 0.001 Parts A and B were quoted with permission from (Akaike et al., 2010).

8.3 Action mechanisms of volatile anesthetics

Volatile anesthetics inhibit neuronal activity throughout the CNS, causing complex behavioral effects including sedation, analgesia, hypnosis, unconsciousness, and immobility. Many previous studies using brain slice preparations and primary cultured neurons indicated that the volatile anesthetics enhance GABAergic inhibitory transmission at both synaptic and extrasynaptic sites (Jones et al., 1992; Zimmerman et al., 1994; Nishikawa &

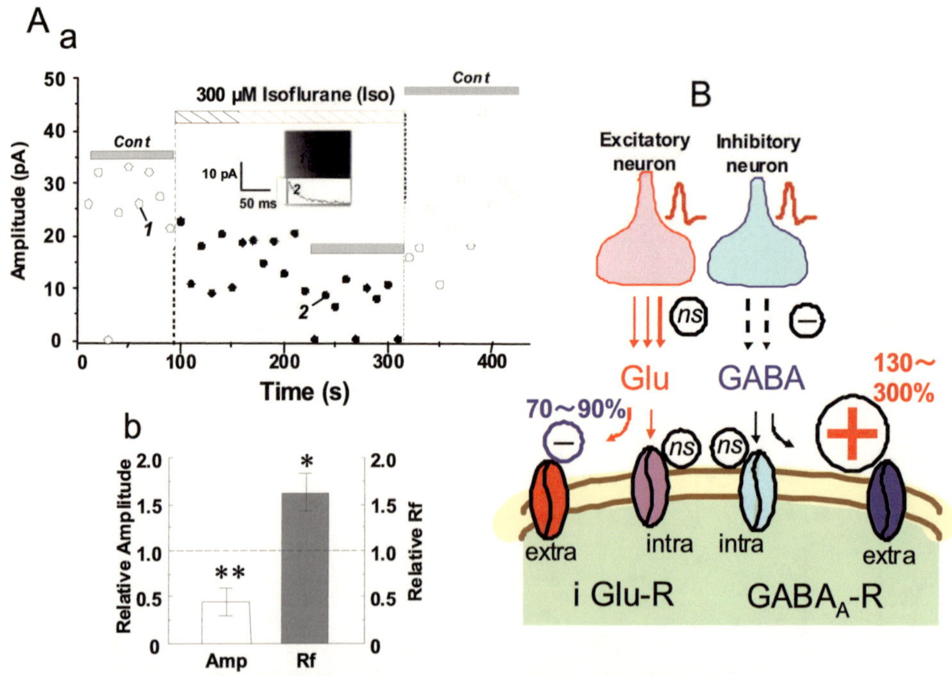

Fig. 9. Modulation of GABAergic evoked IPSCs by volatile anesthetics

A: Effects of isoflurane on GABAergic eIPSCs. (a): Plots of the eIPSC amplitude against time in the absence and presence of 300μM isoflurane. Focal electrical stimulation was applied to a single GABAergic bouton projecting to rat hippocampal CA1 neuron every 10s. (b): Relative current amplitude and Rf of eIPSCs in the presence of isoflurane. Each column shows the mean value of 4~8 neurons. Error bar represents ±SEM. * P < 0.05, ** P < 0.01. B: Schematic illustration of how volatile anesthetics modulate excitatory and inhibitory presynaptic terminals, synaptic receptors, and extrasynaptic receptors. Parts A was obtained with permission, from (Ogawa et al., 2011).

Maclver, 2005; Nishikawa et al., 2005; Bai et al., 2001; Bonin & Orser, 2008; Bieda et al., 2009). In an attempt to more clearly delineate the sites of actions of volatile anaestheics at different pre- and postsynaptic levels at single synapses, devoid of complications from surrounding cells, we used the 'synaptic bouton' preparation of rat hippocampal CA1 pyramidal neurons. As shown in Figure 9Aa, b, isoflurane (300μM) inhibited the amplitude of eIPSCs induced by focal stimuli of a single GABAergic bouton, and increased the Rf. This result, obtained at the single GABAergic synapse level, clearly indicates that volatile anesthetics such as enflurane, isoflurane and sovoflurane also act presynaptically to inhibit GABA release, and this dominates any potentiation of the postsynaptic $GABA_A$ receptors, which is often been thought to be the main site of action of these drugs. As shown in Figure9B, the volatile anaesthetics also had no effect at glutamatergic synapses. As has been reported frequently by others, the extrasynaptic $GABA_A$ receptor-mediated response (by exogenous GABA application) was greatly enhanced by the volatile anaesthetics, while the extrasynaptic glutamatergic receptor-mediated response was significantly inhibited. Thus, the behavioral effects of volatile anesthetics may result from both the enhancement of extrasynaptic $GABA_A$ responses and the suppression of extrasynaptic glutamate responses (Ogawa et al., 2011), although the synaptic responses are quite differently affected.

9. Concluding remarks

The 'synaptic bouton' preparation is a simple and convenient methodology to investigate the pharmacology, physiology and transduction mechanisms of neurotransmission at mammalian nerve terminals (boutons), the vast majority of which are small (diameter less than a few μm) and difficult to access for functional studies by other means. Spontaneous IPSCs and EPSCs mediated by the classical fast neurotransmitters can be recorded in acute preparations from many brain regions, with accurate space-clamp of the postsynaptic membrane voltage and with good control of both the cytoplasmic constitutions and the test solutions bathing single neurons. The preparation is devoid of complications arising from surrounding other neurons and glia cells, and from possible changes in protein distribution and function resulting from enzyme treatment and *in vitro* culture. A single bouton in this preparation can also be selectively activated by focal electrical stimulation and visualized by fluorescent signals. The 'synaptic bouton' preparation could be helpful to reveal further the repertoire of receptors, ion channels, transporters, and second messengers that mediate and regulate synaptic transmission in mammalian presynaptic terminals.

10. Acknowledgements

The author wish to thank Dr. A. Moorhouse for his helpful discussion and T. Yamaga for his assistance with the figures.

11. References

Akaike, N., Murakami, N., Katsurabayashi, S., Jin, Y. H. & Imazawa, T. (2002). Focal stimulation of single GABAergic presynaptic boutons on the rat hippocampal neuron., Neurosci. Res. 42, pp. 187-195

Akaike, N. & Moorhouse, AJ. (2003) ,Techniques; applications of the nerve-bouton preparation in neuropharmaclogy, TREND in Pharmacological Science 24.1, pp. 44-47

Akaike, N., Ito, Y., Shin, M.C., Nonaka, K., Torii, Y., Harakawa, T., Ginnaga, A., Kozaki, S. & Kaji, R. (2010). Effects of A2type botulinum toxin on spotaneous miniature and evoked transmitter release from the rat spinal excitatory and inhibitory synapses, Toxicon 56, pp. 1315-1326

Arima, J., Matsumoto, N., Kishimoto, K. & Akaike, N. (2001). Spontaneous miniature outward currents in mechanically dissociated rat Meynert neurons. J. Physiol. (Lond.) 1, pp. 99-107

Ashton, A.C., & Dolly, J.O. (1988). Characterization of the inhibitory action of botulinum neurotoxin type A on the release of several transmitters from rat cerebrocortical synaptosomes. J. Neurochem. 50, pp.1808-1816

Asztely, F., Xiao, M.Y. and Gustafsson, B. (1996). Long-term potentiation and paired-pulse facilitation in the hippocampal CA1 region. Neuroreport 7, pp. 1609-1612

Azmitia, E. & Segal, M. (1978). An autoradiographic analysis of the differential ascending projections of the dorsal and median raphenuclei in the rat. J. Comp. Neurology 179, pp. 641-668

Bai, D., Zhu, G., Pennefather, P., Jackson, M.F., MacDonald, J.F. & Orser, B.A. (2001). Distinct functional and pharmacological properties of tonic and quantal inhibitory postsynaptic currents mediated by gamma-aminobutyric $acid_A$ receptors in hippocampal neurons. Mol. Pharmacol. 59, pp. 814-824

Barns, N. & Sharp, T. (1999). A review of central 5-HT receptors and their function. Neuropharmacol. 38, pp. 1083-1152

Bieda, M.C., Su, H., & Maciver, M.B. (2009). Anesthetics discriminate between tonic and phasic gamma-aminobutyric acid receptors on hippocampal CA1 neurons. Anesth. Analg.108, pp. 484-490

Bonin, R.P. & Orser, B.A. (2008). $GABA_A$ receptor subtypes underlying general anesthesia. Pharmacol. Biochem. Behav. 90, pp. 105-112

Borst,J. & Sakmann,B. (1996). Calcium influx and transmitter release in a fast CNS synapse. Nature 383, pp. 431-434

Bruns, D., Riedel, D., Klingauf, J. & Jahn, R. (2000). Quantal release of serotonin. Neuron 28, pp. 205-220

Capogna, M., mckinney, R.A., O'Connor, V., Gahwiler, B.H. & Thompson, S.M. (1997). Ca^{2+} or Sr^{2+} partially rescues synaptic transmission in hippocampal cultures treated with botulinum toxin A and C, but not tetanus toxin. J. Neurosci. 17, pp. 7190-7202

Cattaert, D., El, M.A. & Clarac F. (1994). Chloride conductance produces both presynaptic inhibition and antidromic spikes in primary afferents, Brain Res. 666, pp. 109-112

Emptage, N.J., Reid, and C.A. and Fine, A. (2001). Calcium stores in hippocampal synaptic boutons mediate short-term plasticity, store-operated Ca^{2+} entry, and spontaneous transmitter release. Neuron 29, pp. 197-208

Graham, B. & Redman, S. (1994). A simulation of action potentials in synaptic boutons during presynaptic inhibition. J. Neurophysiol 71, pp. 538-549

Hage, D, Karlsson, U. & Johansson, S. (1998). Heterogenous presynaptic Ca^{2+} channel types triggering GABA release onto medial preoptic neurons from rat. J.Physiol.(Lond.)507.1, pp. 77-91

Jackson, M.B. & Zhang, S.J. (1995). Action potential propagation and propagation block by GABA in rat posterior pituitary nerve terminals. J.Physiol. (Lond) 483, pp. 597-611

Jang IS., Jeong, HJ. & Akaike, N. (2001). Contribution of the Na-K-Cl cotransporter on GABAA receptor-mediated presynaptic depolarization in excitatory nerve terminals., J. Neurosci. 21, pp. 5962-5972

Jang, I.S., Jeoung, H.J., Katsurabayashi, S. & Akaike, N. (2002). Functional roles of presynaptic GABA(A) receptors on glycinergic nerve terminals in rat spinal cord. J. Physiol (Lond) 541, pp. 423-434

Jang, I.S., Nakamura, M., Ito, Y. & Akaike, N. (2006). Presynaptic $GABA_A$ receptors facilitate spontaneous glutamate release from presynaptic terminals on mechanically dissociated rat CA3 pyramidal neurons. Neurosci. 138, pp. 25-35

Jones, M.V., Brooks, P.A., & Harrison, N.L. (1992). Enhancement of gamma-aminobutyric acid-activated Cl^- currents in cultured rat hippocampal neurons by three volatile anaesthetics. J.Physiol.(Lond.)449, pp. 279-293

Kakazu, N., Akaike, N., Komiyama, S. & Nabekura, J. (1999). Regulation of intracellular chloride by cotransporters in developing lateral superior olive neurons. J. Neurosci. 19, pp. 2843-2851

Katsurabayashi, S., Kubota, H., Tokutomi, N. & Akaike, N. (2003). A distinct distribution of functional presynaptic 5-HT receptor subtypes on GABAergic nerve terminals projecting to single hippocampal CA1 pyramidal neurons. Neuropharmacol. 44, pp. 1022-1030

Kishimoto, K., Matsuo, S., Kanemoto, Y., Ishibashi, H., Oyama, Y. & Akaike, N. (2001). Nanomolar concentraions of tri-n-butyltin facilitate γ-aminobutyric acidergic synaptic trasmission in rat hypothalamic neurons., J. Pharmacol. Exp. Ther. 299, pp. 171-177

Koyama, S., Kubo, C., Rhee, J-S. & Akaike, N. (1999) Presynaptic serotonergic inhibition of GABAergic synaptic transmission in mechanically dissociated rat basolateral amygdala neurones., J. Physiol. (Lond.) 518, pp. 525-538

Koyama, S., Matsumoto, N., Kubo, C. & Akaike, N. (2000). Presynaptic $5\text{-}HT_3$ receptor-mediated modulation of synaptic GABA release in the mechanically dissociated rat amygdale neurons. J. Physiol (Lond) 529, pp.373-383

Koyama, S., Matsumoto,N., Murakami, N., Kubo, C., Nabekura, J. & Akaike, N. (2002). Role of presynaptic $5\text{-}HT_{1A}$ and $5\text{-}HT_3$ receptors in modulation of synaptic GABA transmission in dissociated rat basolateral amygdale neurons. Life Sci. 72, pp. 375-387

Levkovitz, Y. and Segal, M. (1997). Serotonin $5\text{-}HT_{1A}$ receptors modulate hippocampal reactivity to afferent stimulation. J. Neurosci. 17, pp. 5591-5598

Maeda, M., Tanaka, E., Shoudai, K., Nonaka, K., Murayama N., Ito Y. &Akaike, N. (2009). Differential effects of divalent cations on spontaneous and evoked glycine release from spinal interneurons, J. Neurophysiol. 101, pp. 1103-1113

Matsumoto, N., Komiyama, S. & Akaike, N. (2002). Pre-and postsynaptic ATP-sensitive potassium channels during metabolic inhibition of rat hippocampal CA1 neurons, J. Physiol. (Lond.) 541.1, pp. 511-520

McMahon, L. & Kauer, J. (1997). Hippocampal interneurons are excited via serotonin-gated ion channels. J. Neurophysiol. 78(5), pp.2493-2502

Miller, R.J. (1998). Presynaptic receptors. Annu. Rev. Pharmacol. Toxicol. 38, pp. 201-207

Murakami, N., Ishibashi, H., Katsurabayashi, S. & Akaike, N. (2002). Calcium channel subtypes on single GABAergic presynaptic terminal projecting to rat hippocampal neurons., Brain Res. 951, pp. 121-129

Murthy V.N., Sejnowski, T.J. & Stevens, C.F. (1997). Heterogeneous release properties of visualized individual hippocampal synapses. Neuron 18, pp. 599-612

Nishikawa, K. & Maclver, M.B. (2001). Agent-selective effects of volatile anesthetics on $GABA_A$ receptor-mediated synaptic inhibition in hippocampal interneurons. Anesthesiology 94, pp. 340-347

Nishikawa, K., Kubo, K., Ishizeki, J., Takazawa, T., Saito, S. & Goto, F. (2005). The interaction of noradrenaline with sevoflurane on $GABA_A$ receptor-mediated inhibitory postsynaptic currents in the rat hippocampus. Brain. Res. 1039, pp. 153-161

Nonaka, K., Murayama, N., Maeda, M., Shoudai, K., Shin, M.C. & Akaike N. (2010). P- and R- type Ca^{2+} channels regulating spinal glycinergic nerve terminals. Toxicon 55(7), pp. 1283-1290

Ogawa S., Tanaka, E., Shin, M.C., Kotani, N. & Akaike, N. (2011). Volatile anesthetic effects on isolated GABA synapses and extrasynaptic receptors., Neuropharmacol. 60, pp. 701-710

Payne, J.A., Rivera, C., Voipio, J. & Kaila, K. (2003). Cation-chloride cotransporter in neuronal communication, development and trauma. Trends Neurosci. 26, pp. 199-206

Rhee, J.-S., Ishibashi, H. & Akaike, N. (1999). Calcium channels in the GABAergic presynaptic nerve terminals projecting to Meynert neurons of the rat. J. Neurochem. 72, pp. 800-807

Rizo, J. & Sudhof, T.C. (1998). Progress in membrane fusion: from structure to function. Nat. Struc. Biol. Chem. 275, pp. 6328-6336

Sakaguchi, G., Oishi, I. & Kozaki, S. (1981). Purification and oral toxicities of clostridium botulinum progenitor toxins. In: Lewis, G.E. (Ed.), Biomedical Aspects of Botulism. Academic Press, New York, pp. 21-34

Sasaki, K., Takayama, Y., Tahara, T., Anraku, K., Ito, Y. & Akaike, N. (2008). Quantitative analysis of toxin extracts from various tissues of wild and cultured puffer fish by an electrophy siological method. Toxicon 51, pp. 606-614

Schiavo, G., Matteoli, M. & Montecucco, C. (2000). Neurotoxins affecting neuroexocytosis. Ohysiol. Rev. 80; 717-766

Schikorski, T. & Stevens, C.F. (2001). Morphological correlates of functionally defined synaptic vesicle populations. Nat. Neurosci. 4, pp.391-395

Scholtz, K.P. & Miller, R.J. (1992). Inhibition of quantal transmitter release in the absence of calcium influx by a G-protein linked adenosine receptor at hippocampal synapse. Neuron 8, pp. 1139-1150

Schmitz, D., Empson, R. & Heinemann, U. (1995). Serotonin reduces inhibition via 5-HT_{1A} receptors in area CA1 of rat hippocampal slices in vitro. J. Neurosci. 15, pp. 7217-7225

Seagar, M., Leveque, C., Charvin, N., Marqueze, B., Martin-Moutot, N., boudier, J.A., Boudier, J.L., Shoji-Kasai, Y., Sato, K. & Takahashi, M. (1999). Interactions between proteins implicated in exocytosis and voltage-gated calcium channel. Philos. Trans. R. Soc. Lond. B Biol. Sci. 354, pp. 289-297

Shoudai,K., Nonaka, K., Maeda, M., Wang, Z.M., Jeong, H.J., Higashi, H., Murayama, N. & Akaike, N. (2007). Effects of various K^+ channel blockers on spontaneous glycine release at rat spinal neurons. Brain Res. 1157, pp. 11-22

Sollner, T.H. (2003). Regulated excocytosis and SNARE fusion. Mol. Membr. Bio. 20, pp. 209- 220

Sudhof, T.C. (2004). The synaptic vesicle cycle. Annu. Rev. Neurosci. 27; 509-547

Vorobjev, V.S. (1991). Vibrodissociation of sliced mammalian nervous tissue. J. Neurosci. Meth. 38, pp. 141-150

Xu, J., Pang, Z.P., Shin, O.H. & Sudhof, T.C. (2009). Synaptotagmin-1 functions as the Ca^{2+} -sensor for spontaneous release. Nat. Neurosci. 12, pp. 756-766

Yamamoto,S., Yoshimura, M., Shin, M.C., Wakita, M., Nonaka, K. & Akaike, N. (2011). $GABA_A$ receptor-mediated presynaptic inhibition on glutamatergic transmission, Brain Res. Bull. 84, pp. 22-30

Zimmerman, S.A., Jones, M.V. & Harrison, N.L. (1994). Potentiation, of gamma-amino butyric acid A receptor Cl^- current correlates with in vivo anesthetic potency. J. Pharmacol. Exp. Ther. 270, pp.987-991

Zucker, R.S. & Regehr, W.G. (2002). Short-term synaptic plasticity. Annu. Rev. Physiol. 64, pp. 355-405

4

Single-Channel Properties and Pharmacological Characteristics of K_{ATP} Channels in Primary Afferent Neurons

Takashi Kawano
Department of Anesthesiology and Critical Care Medicine, Kochi Medical School
Japan

1. Introduction

ATP- sensitive potassium (K_{ATP}) channels first discovered in cardiac myocytes in 1983 (Noma, 1983), and then found in many other metabolically active tissues, including central nervous systems (Babenko et al., 1998). K_{ATP} channels are inhibited by physiological concentration of intracellular ATP, and are activated when the intracellular ADP/ATP ratio increases secondary to hypoxia, ischemia, or metabolic stress (Babenko et al., 1998; Miki & Seino, 2005; Nichols, 2006). The activation of K_{ATP} channels results in an enhanced outward repolarizing flow of K^+ and cell membrane hyperpolarization, and thus they regulate cell excitability and mediate cellular responses that determine cell survival during metabolic stress (Miki & Seino, 2005).

Recent electrophysiological and molecular genetic studies of K_{ATP} channels have provided insights into their physiological and pathophysiological roles, such as insulin secretion (Seghers et al., 2000; Miki et al., 1999), cardioprotection (Saito et al., 2005; Suzuki et al., 2001), vasodilatation (Chutkow et al., 2002; Miki et al., 2002), and neuroprotection (Sun et al., 2006; Yamada et al., 2001). Indeed, K_{ATP} channels are the target for a number of pharmacological agents, including inhibitors such as the sulfonylureas and a structurally unrelated group of K^+ channel openers (Babenko et al., 2000; Nichols, 2006).

In central nervous systems, K_{ATP} channels are predominantly expressed in basal ganglia, thalamus, hippocampus, and cerebral cortex (Mourre et al., 1989). Although detailed functional roles of K_{ATP} cahnnel in central nervous systems still remain to be clarified, their activation results in K^+ efflux, leading to membrane hyperpolarization, decreased excitability, attenuation of transmitter release, and protection from cell death (Yamada et al., 2001).

In addition to central nervous systems, we have recently reported that functional K_{ATP} channels are expressed in rat sensory neurons (Kawano et al., 2009a, 2009b, 2009c). The primary sensory neuron is known to be an important site of pathogenesis for neuropathic pain, which is a common clinical condition that is difficult to treat by current methods (Amir et al., 2005; Gold, 2000; Zimmermann, 2001). Since our observations also reveal that currents through K_{ATP} channels are significantly decreased by painful nerve injury (Sarantopoulos et al., 2003; Kawano et al., 2009a, 2009b, 2009c, Zoga et al., 2010), identification of their roles in sensory neurons, particularly regarding excitability, may reveal a contribution to the genesis of neuropathic pain. These researches may lead to development of novel therapies for neuropathic pain. Thus, the overall objective of this chapter is to discuss the characteristics and role of K_{ATP} channels in rat sensory neurons.

2. Tissue-specific moleculer structure of K_{ATP} channel

K_{ATP} channels are a widely distributed family of potassium-selective ion channels, whose structure consists of an inwardly rectifying, pore-forming, K^+ channel (Kir6.x) subunit, each coupled to a regulatory sulfonylurea receptor (SUR) subunit (Babenko et al., 1998; Miki et al., 1999; Yokoshiki et al., 1998). The Kir 6.x subunits belong to the Kir family, and determine the inward rectification, ATP-sensitivity, and unitary single-channel conductance of K_{ATP} channel (Babenko et al., 1998). On the other hand, SUR subunits belong to the ATP-binding cassette superfamily, and confer responsiveness to K_{ATP} channel openers and sulfonylureas (Miki et al., 1999). The functional K_{ATP} channel is assembled as an octamer with a 4:4 stoichiometry of Kir6.x and SUR subunit (Fig. 1 and 2).

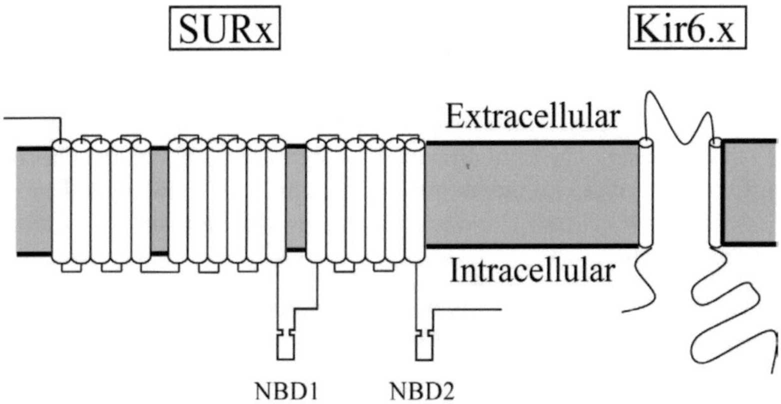

Fig. 1. Molecular structure of the K_{ATP} channel. Schematic representation of the transmembrane topology of a single sulfonylurea receptor (SUR) or inwardly rectifying K^+ channel (Kir6.x) subunit. SUR is an ABC protein bearing transmembrane domains and two nucleotide-binding domains (NBD1 and NBD2). The Kir6.x subunits presumably form the pore of the channel and determine the sensitivity of the channel to inhibition by ATP (Nichols, 2006).

Native K_{ATP} channels in different tissues show distinct single channel properties, modulating potency of nucleotides, and varying sensitivity to drugs that act as channel openers (such as diazoxide, pinacidil, nicorandil, etc) or blockers (sulphonylureas, like glibenclamide, tolbutamide, etc). These tissue-specific biophysical and pharmacological properties of K_{ATP} channels (summarized in Table 1) are thought to be endowed by their different molecular composition of Kir6.0 and SUR subunits (Yokoshiki et al., 1998). For instance, affinity for sulfonylureas is high for SUR1 but low for SUR2A and SUR2B subunits. Similarly, there are differences in response to openers, with the SUR1 and SUR2B channels responding more potently to diazoxide in contrast to the response of SUR2A channels.

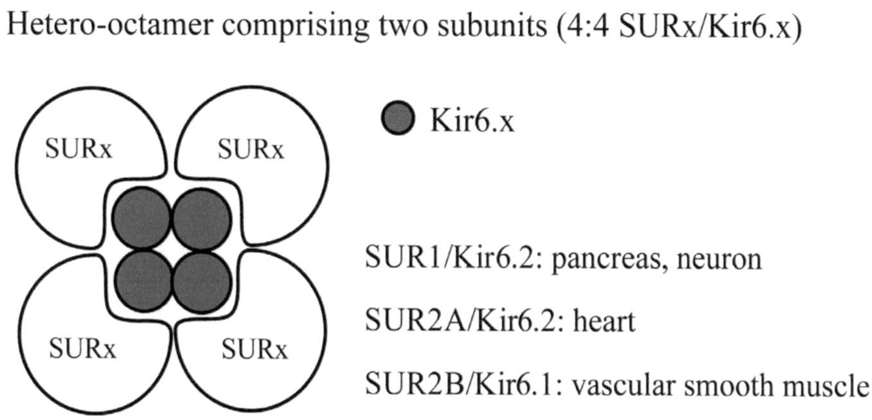

Fig. 2. Schematic representation of the octameric K_{ATP} channel complex viewed in cross section. Four Kir6.x subunits come together to form the K^+ channel pore, and each is associated with a regulatory SURx subunit. Several subtypes have been identified based on subunit combinations; co-expressing SUR1 and Kir6.2 forms the pancreatic β-cell and neuronal type K_{ATP} channel, SUR2A and Kir6.2 form the cardiac type K_{ATP} channel, and SUR2B and Kir6.1 form the vascular smooth muscle type K_{ATP} channel (Babenko et al., 1998; Miki et al., 1999; Yokoshiki et al., 1998).

	Pancreas (SUR1/Kir6.2)	Heart (SUR2A/Kir6.2)	VSM (SUR2B/Kir6.2)	Neuron (SUR1-2/Kir6.2)
Metabolic activity				
$[ATP]_i$	(- - -)	(- - -)	(- ~ 0)	(- - -)
$[ADP]_i$	(+)	(+)	(+)	(+)
$[Acidosis]_i$	(+ +)	(+ +)	(+ + +)	(+ +)
Channel openers				
Diazoxide	(+ + +)	(0)	(+ + +)	(+ + +)
Pinacidil	(+)	(+ + +)	(+ + +)	(+)
Nicorandil	(0)	(+)	(+ + +)	(+)
Sulfonylurea				
Glibenclamide	(- - -)	(- -)	(- -)	(- - -)
Tolbutamide	(- - -)	(- ~ 0)	(- ~ 0)	(- - -)

VSM: vascular smooth muscle.
$[ATP]_i$: Intracellular ATP, $[ADP]_i$: Intracellular ADP, $[Acidosis]_i$: Intracellular acidosis
(+): activation, (0): no effect, (-): inhibition.

Table 1. Tissue-specific properties of K_{ATP} channels.

3. Biophysical properties of K_{ATP} channel in primary afferent neurons

3.1 Distribution of the K_{ATP} channel subunits in primary afferent neurons

Western blot analysis and immunostaining show that peripheral sensory neurons express SUR1, SUR2 and Kir6.2 protein, but not Kir6.1 protein (Kawano et al., 2009b, 2009c). Co-localization of Kir6.2 with SUR1 subunits is also demonstrated by staining with antibody against Kir6.2 and BODIPY-Glibenclamide, which specifically binds to SUR1 with high affinity (Zoga et al., 2010). These findings indicate that K_{ATP} channels in peripheral sensory neuron are composed of SUR1/Kir6.2 (pancreas/neuronal type) or SUR2/Kir6.2 (cardiac type) subunits.

In addition to neuronal somata, image analysis results of immunostaining show that K_{ATP} channels are present in glial satellite and Schwann cells (Zoga et al., 2010), which are known to express K^+ currents (Chiu et al., 1984). K_{ATP} channels in these sites are thought to convey glial cell-mediated clearance of extracellular K^+, often termed "K^+ spatial buffering" (Kofuji et al., 2002).

3.2 Single-channel characteristics of K_{ATP} channel in primary afferent neurons

3.2.1 Cell isolation and plating

The L5 and L4 dorsal root ganglia (DRG) was harvested after normal adult Sprague-Dawley rats were decapitated under isoflurane anesthesia. DRG neurons were enzymatically dissociated in a solution containing 0.25 ml Liberalize Blendzyme 2 (0.05%) and 0.25 ml Dulbecco's modified Eagle's medium (DMEM) for 30 min at 37 °C.

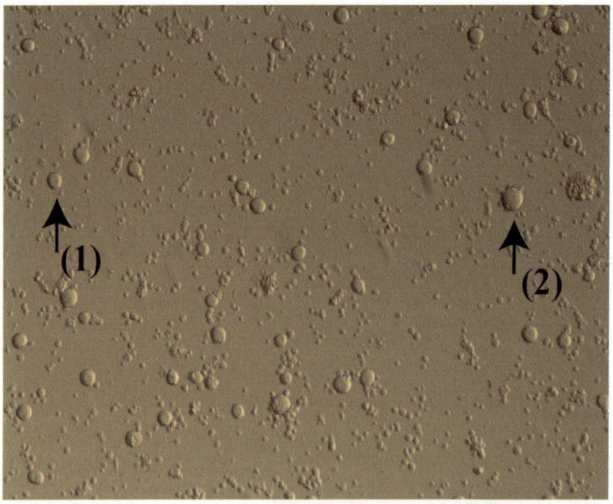

Fig. 3. Isolated DRG neurons. Neurons were viewed using Hoffman modulation optic systems under an inverted microscope (Nikon Diaphot 300). DRG neuronal somata observed after isolation varied in size about 15-50 μm in diameter, and were stratified by diameter into either large (≥ 40 μm) or small (<30 μm) neurons. These sizes correlates roughly with electrophysiological characteristics corresponding to either Aβ or C fibers, respectively (Lawson, 2002; Harper & Lawson, 1985). Arrowhead (1) and (2) point to small (24 μm in diameter) and large (43 μm in diameter) neuron, respectively.

After centrifugation and removal of the supernatant, a second incubation at 37 °C followed for another 30 min in 0.2 ml trypsin (0.0625%) and deoxyribonuclease 1 (0.0125%) in 0.25 ml DMEM. Cells were then isolated by centrifugation (600 rpm for 5 min) after adding 0.25 ml trypsin inhibitor (0.1%), and re-suspended in a medium consisting of 0.5 mM glutamine, 0.02 mg/ml gentamicin, 100 ng/ml nerve growth factor 7 S , 2% B27 supplement, and 98% neural basal medium A. Neurons were plated onto poly-L-lysine-coated 12-mm glass coverslips, and kept in a humidified incubator at 37 °C with 95% air and 5% CO_2. Patch-clamp experiments were performed within 3–8 h after the cell dissociation (Fig. 3).

3.2.2 Single DRG K_{ATP} channel currents from cell-attached patches

Biophysical and pharmacological characteristics of DRG K_{ATP} channels were first studied in cell-attached recordings at a sampling frequency of 5 kHz with 1 kHz low-pass filter. Cell-attached patch configuration is a non-invasive approach which is used to describe the endogenous properties of ion channels without disturbance of the intracellular milieu (Fig 4).

Both bath and pipette (extracellular) solutions were composed of the following (in mM): 140 KCl, 10 HEPES, 10 D-glucose, and 0.5 EGTA. The pH of all solutions was adjusted to 7.4 with KOH. Patch micropipettes were made from borosilicate glass capillaries using a Flaming/Brown micropipette puller, model P-97 (Sutter, San Rafael, CA) and flame polished with a microforge polisher (Narishige, Tokyo, Japan) prior to use. Their resistance ranged between 3 and 6 MΩ when filled with the internal solution, and placed into the recording solutions.

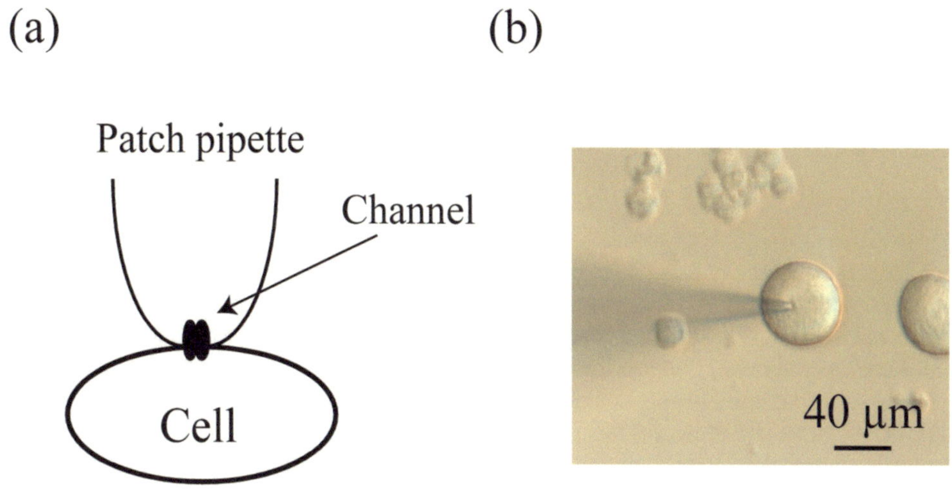

Fig. 4. Cell-attached patch clamp configuration. (a) Diagram illustrating the methods of making cell-attached patches. (b) Cell-attached pipette on a large DRG somata.

In cell-attached patches (Fig 5), infrequent but significant spontaneous channel activity was recorded. These basal channel activities were observed in cell-size independent manner. However, bath application of the uncoupler of mitochondrial ATP synthesis, 2,4-Dinitrophenol (DNP, 100 µM), gradually activated these baseline currents.

Subsequent bath application of glibenclamide 1 µM, a specific K_{ATP} channel inhibitor, completely blocked DNP-induced currents in both groups, indicating that these currents are conveyed via K_{ATP} channels.

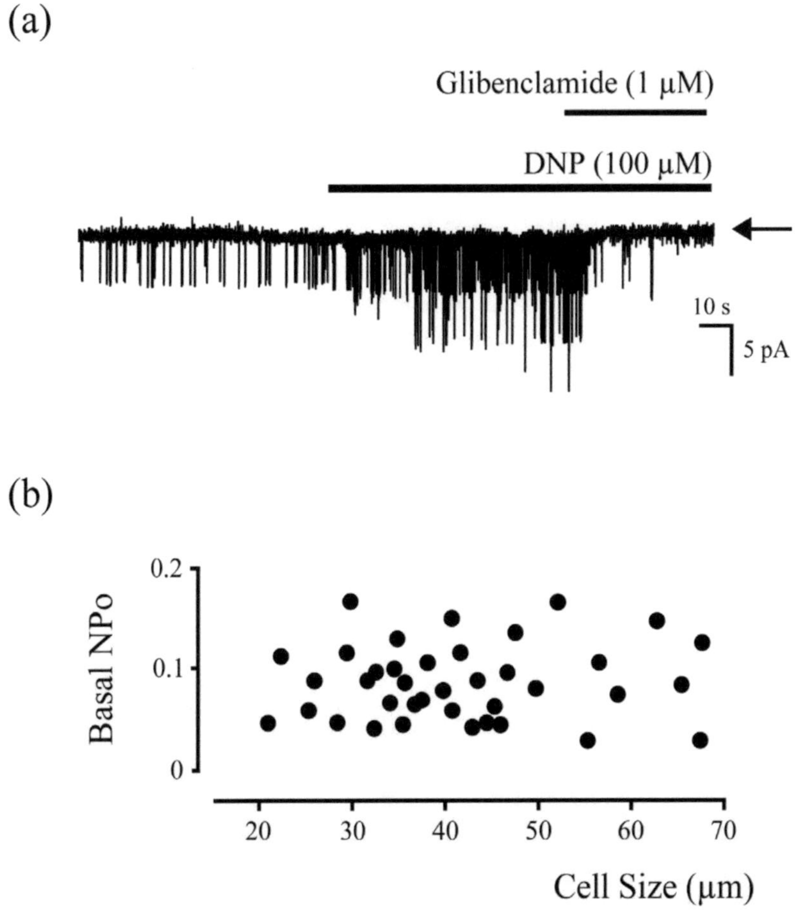

Fig. 5. Single-channel characteristics of K_{ATP} channel in DRG neurons from cell-attached patch clamp configuration. (a) Representative current trace of K_{ATP} channels in isolated DRG neurons recorded in cell-attached configuration at a holding potential of -60mV. These patches typically showed one predominant channel-type at a holding potential of -60 mV, whereas a second channel-type was observed infrequently in 5-10% of all patches. Arrows indicate closed channel state. (b) Basal channel open probability (Po) in individual DRG neurons. Po was determined from the ratios of the area under the peaks in the amplitude histograms fitted by a Gaussian distribution. Channel activity was calculated as NPo, where N is the number of observed channels in the patch.

3.2.3 Single DRG K_{ATP} channel currents from excised inside-out patches

In order to investigate the relative contribution of the intracellular milieu regulation of the intrinsic channel properties in DRG neurons, the K_{ATP} channel behavior was next examined in excised inside-out membrane patches. Inside-out patches were made by pulling the membrane patch off the cell into the bath solution.

The bath (intracellular) solution contained 140 mM KCl, 1.2 mM $MgCl_2$, 10 mM HEPES, 1.5 mM EGTA and 5.5 mM dextrose. The pipette (extracellular) solution was composed of the following: 140 mM KCl, 10 mM HEPES, 5.5 mM dextrose, and 1 mM EGTA. The pH of all solutions was adjusted to 7.4 with KOH. Osmolality was adjusted approximately to 300 mOsm/l by adding sucrose if necessary.

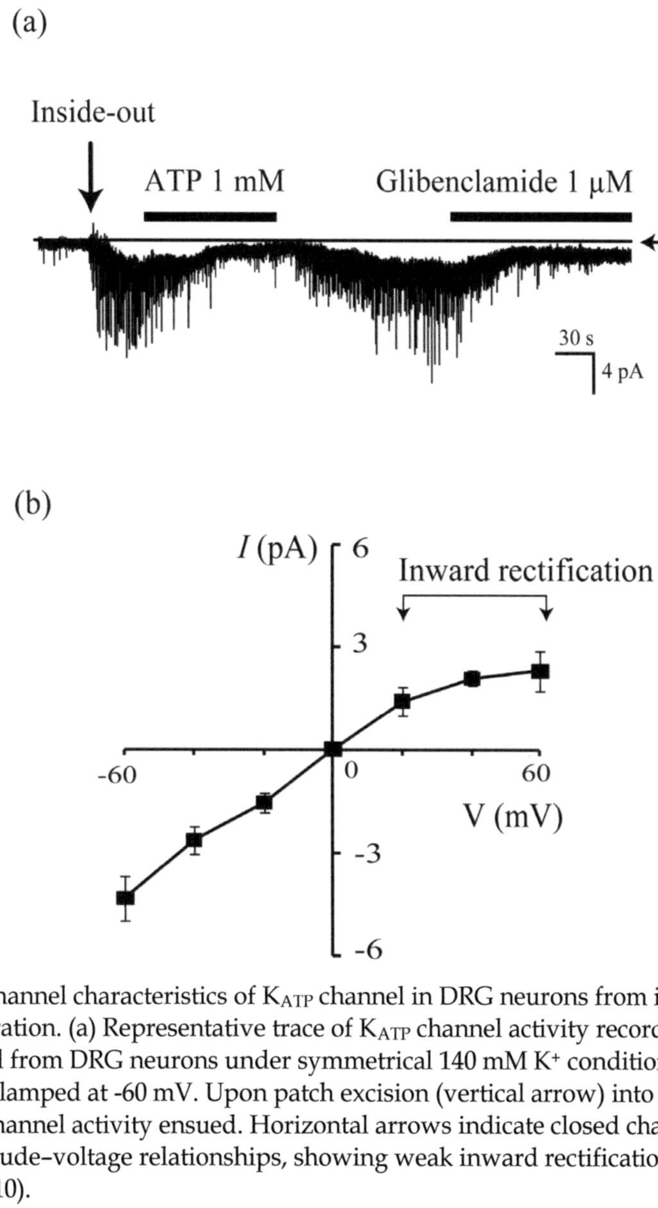

Fig. 6. Single-channel characteristics of K_{ATP} channel in DRG neurons from inside-out patch clamp configuration. (a) Representative trace of K_{ATP} channel activity recorded in cell-free patches excised from DRG neurons under symmetrical 140 mM K^+ conditions. Membrane potential was clamped at -60 mV. Upon patch excision (vertical arrow) into an ATP-free bath marked channel activity ensued. Horizontal arrows indicate closed channel state. (b) Current amplitude-voltage relationships, showing weak inward rectification. Means ± SD are shown (n=10).

When inside-out patch recordings at a holding potential of -60 mV were obtained in ATP-free solution, intense channel activity was observed in all patches without any differences between groups of neurons classified by size. The current-voltage relationship showed weak inward rectification with single channel conductance of 70-80 pS without any differences between neuron sizes. This channel activity was reversibly blocked by 1 mM of ATP. In the same recordings, subsequent superfusion with glibenclamide (1 μM) eliminated channel activity in ATP-free solution. Channel activation by ATP-free solution, as well as the inhibition by ATP and glibenclamide, rapidly occurred within a few seconds (Fig. 6).

Sensitivity to ATP and glibenclamide distinguishes these currents as conveyed through K_{ATP} channels (Babenko et al., 1998; Edwards & Weston, 1993).

4. Functional roles of K_{ATP} channel in primary afferent neurons

4.1 Basal K_{ATP} channel activity contributes to the resting membrane potential

To test the functional importance of K_{ATP} channel currents, their effect on resting membrane potential (RMP) in DRG neurons was examined using current-clamp recordings from β-escin perforated whole-cell patches.

The extracellular Tyrode's solution consisted of the following (in mM): 140 NaCl, 4 KCl, 2 $CaCl_2$, 2 $MgCl_2$, 10 D-glucose and 10 HEPES, at pH 7.4 and osmolality 300 mOsm/L. Internal pipette solution contained (in mM) 120 KCl, 20 HEPES, 5 EGTA, 5 $MgCl_2$, 1 Na-ATP, and $CaCl_2$ 2.25. pH was adjusted to 7.4 and osmolality to 300 mOsm/L. In this experiment, 50 μM β-escin was added into the pipette solution as a perforating agent (Sarantopoulos, et al., 2004) and recording was carried out when the access resistance was less than 15 MΩ.

RMP was recorded at baseline for at least 1 min. After stability was confirmed, glibenclamide (1 μM) was superfused in the bath by a gravity dependent flow system, which depolarized the RMP in DRG neurons. These results imply that basal K_{ATP} channel opening physiologically regulates the RMP in DRG neurons (Fig 7).

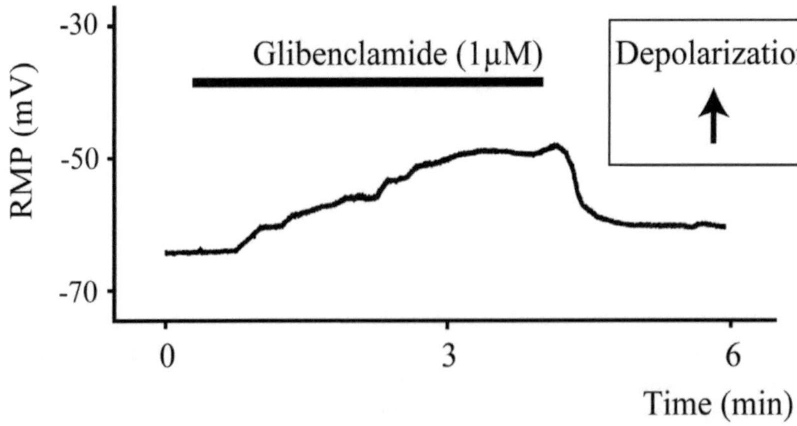

Fig. 7. Changes in resting membrane potential (RMP) induced by glibenclamide in DRG neurons. Representative RMP traces recorded in the current-clamp whole-cell patch configuration during the bath application of glibenclamide (1 μM).

4.2 K_{ATP} channel activity regulates neurotransmitter release

K_{ATP} channel is known to modulate neurotransmission (Stefani & Gold, 2001; Steinkamp et al., 2007). To examine whether K_{ATP} channel in DRG neuron also modulate neurotransmitter release, carbon-fiber amperometry was used to detect exocytosis from single DRG neuron in real time (Kawagoe, et al., 1993).

Amperometry provides high resolution to detect molecules released from single secretory vesicles. Amperometric measurements are normally limited to cells that package and secrete an endogenous oxidizable molecule such as catecholamines or serotonin; however, in some cases oxidizable molecules can be introduced artificially (Smith et al., 1995; Zhou & Misler, 1996). So, amperometric analysis from DRG neurons was conducted by measured release of the pseudo-transmitter dopamine that had been loaded in DRG neurons (Fig. 8).

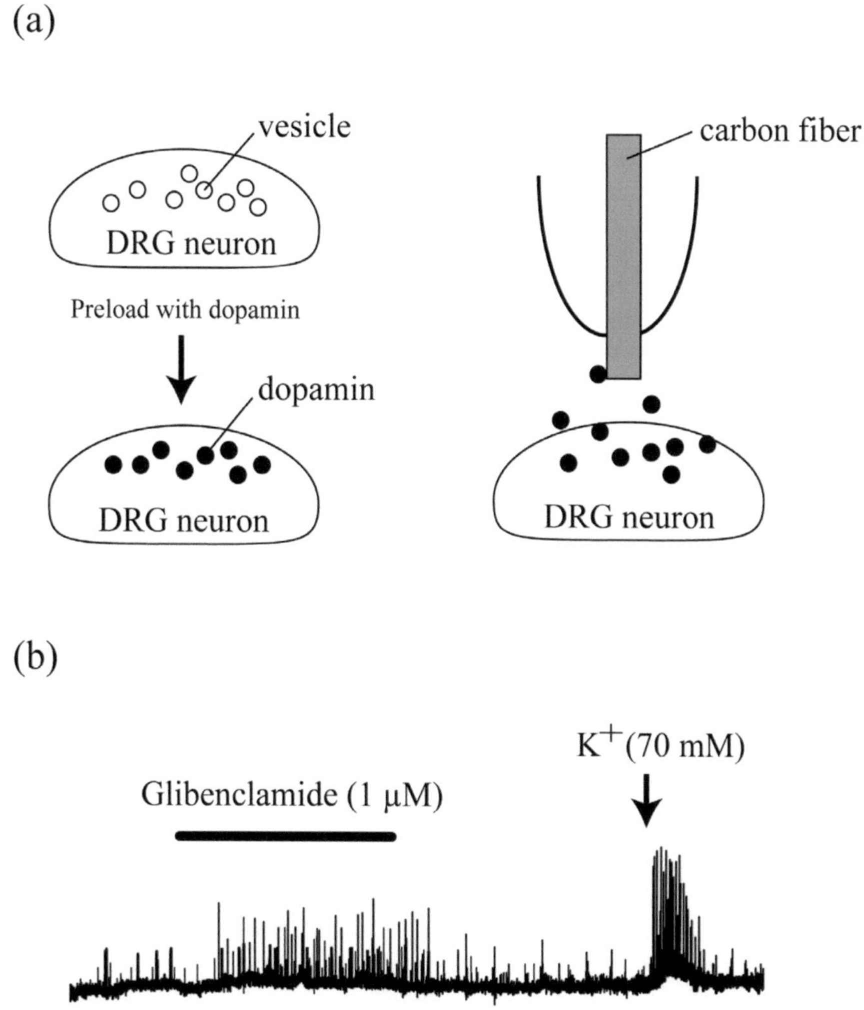

Fig. 8. Amperometric recordings from dopamine-loaded DRG neurons indicating the false transmitter release. Superfusion with glibenclamide (1 μM) enhanced basal transmitter release. Arrow indicates onset of massive transmitter release after depolarization with KCl (70 mM).

In this experiment, DRG neurons were incubated for 40 min in solution containing (in mM): 70 dopamine, 68 NaCl, 2.5 KCl, 2 $CaCl_2$, 1 $MgCl_2$, 10 D-glucose, and 10 HEPES at pH 7.4

with NaOH at 37°C. Recordings were performed at room temperature in amine-free external Tyrode's solution. To record amperometric events, 5 μm carbon fiber electrodes were backfilled with 3 M KCl. A carbon fiber electrode connected to a patch clamp amplifier was attached to the plasma membrane of the cell held at +800 mV for all experiments. The tip of the carbon fiber was manipulated onto the cell surface without disturbing cellular morphology. If infrequent response to high K^+ (70 mM KCl) was observed, these data results were discarded. Currents were low-pass filtered at 5 kHz and sampled at 1 kHz by a Digidata 1440 Series interface (Axon Instruments).

As shown in Fig. 8b, carbon-fiber amperometry showed spikes (< 5 events/min) that were amplified by depolarization with external 70 mM K^+ (200-300 events/min). Similarly, glibenclamide (1 μM) reversibly increased spike rate (20-50 events/min) These results indicate that K_{ATP} channels regulate vesicle release from DRG neurons.

4.3 Neuroprtection

Increased neuronal survival through activation of K_{ATP} channels has been demonstrated in association with membrane hyperpolarization, and reduction of excitability in response to hypoxia, ischemia or metabolic stress (Amoroso et al., 1990; Ballanyi, 2004; Sun et al., 2007; Yamada & Inagaki, 2005). K_{ATP} channels also act as transducers and effectors of neuronal preconditioning (Blondeau et al., 2000; Heurteaux et al., 1995). Pharmacological induced preconditioning with the K_{ATP} channel opener diazoxide may offer effective neuroprotection during hypothermic circulatory arrest (Shake et al., 2001). In addition, opening of K_{ATP} channels in the hippocampus or neocortex stabilizes the resting potential against anoxic stress, and protects neurons (Sun et al., 2006, 2007). In whole animals, the absence of Kir6.2 was associated with dramatically increased damage following ischemia induced by middle cerebral artery occlusion (Sun et al., 2006).

5. K_{ATP} channel in neuropathic pain

5.1 Pathophysiology of neuropathic pain

The specific cellular and molecular mechanisms underlying neuropathic pain remains largely unknown, but membrane hyperexcitability in those neurons that have lost their normal synaptic, physiological or electrical patterns is a common feature of most conditions leading to neuropathic pain (Gold, 2000; Woolf, 2010; Zimmermann, 2001).

Following nerve injury, damaged peripheral nerves become more excitable, with regards to their capacity to generate action potentials, leading to spontaneous, ongoing, ectopic electrical activity (Chung et al., 2002; Raouf et al., 2010; Woolf, 2010). In addition to the injury site, neuronal somata in the dorsal root ganglia are recognized as an important focus of ectopic electrical activity (Devor, 2009; Sapunar et al., 2005). This increase in primary afferent traffic to the dorsal horn is thought to induce raw pain signal (Devor, 2009) as well as central sensitization (Woolf, 2010). Substantial evidence indicates that altered expressions of ion channels on peripheral afferent neuronal somata contribute to abnormal sensory function following nerve injury (Raouf et al., 2010). Specifically, important changes have been noted in sodium (Amir et al., 2006), calcium (Gemes et al., 2011; McCallum et al., 2006) and potassium channels (Abdulla & Smith, 2001).

5.2 Animal models of neuropathic pain

Examination of the pathogenesis of neuropathic pain has been aided by the development of increasingly sophisticated rodent models of nerve injury that produce behavior indicative of on-going and evoked pain (Hogan, 2002).

A complete section of a nerve produces spontaneous pain, whereas they also lead to an anesthetic limb. On the other hand, partial injury retains a subset of afferent fibers and results in altered sensory function. Therefore, the latter is currently widely used for the study of neuropathic pain. These models involve: chronic constriction injury by loose ligation of the sciatic nerve (CCI) model (Bennett & Xie, 1988), tight ligation of the partial sciatic nerve (PSL) model (Shir & Seltzer, 1990), and tight ligation of spinal nerves (SNL) model (Kim & Chung, 1992).

5.2.1 CCI surgery

The right common sciatic nerve is exposed at the level of the middle of the thigh. Four loose ligatures of 4-0 chromic gut are placed around the sciatic nerve , and are loosely tied such that the diameter of the nerve was barely constricted.

5.2.2 PSL surgery

The right sciatic nerve was exposed near the trochanter. An 8-0 silk suture was inserted in the middle of the nerve, trapping in a tight ligation.

5.2.3 SNL surgery

The right paravertebral region was exposed via a lumbar incision, and the L6 transverse process was removed. The L5 and L6 spinal nerves were tightly ligated with a 6–0 silk suture and transected distal to the ligature.

5.3 Sensory testing

The purpose of behavioral sensory testing is to identify rats in which nerve injury has successfully produced behavior consistent with neuropathic pain. Operated rats are being tested for mechanical hyperalgesia, which provides a more consistent feature of the SNL model. Testing includes preoperative familiarization and acclimatization to testing environment, and subsequent repeated testing sessions.

Recently, Hogan et al. demonstrated the novel sensory testing that identifies hyperalgesia after SNL with high specificity (Hogan et al., 2004). Briefly, the plantar surface of each hind paw was touched with the tip of a Quincke 22–gauge spinal needle, which was applied with pressure adequate to indent but not penetrate the skin. Five needle applications were delivered in random order to each paw and repeated 5 min later for a total of 10 applications per session. These mechanical stimuli produced either a normal brief reflexive withdrawal or a hyperalgesia-type response that included sustained (> 1 s) paw lifting, shaking, and grooming (Fig. 9). The latter response occurs only after true SNL, and thus this may be accepted as an indication of a neuropathic pain. The intensity of hyperalgesia was assessed by the probability (%) of hyperalgesia-type responses out of ten trials of needle stimulation.

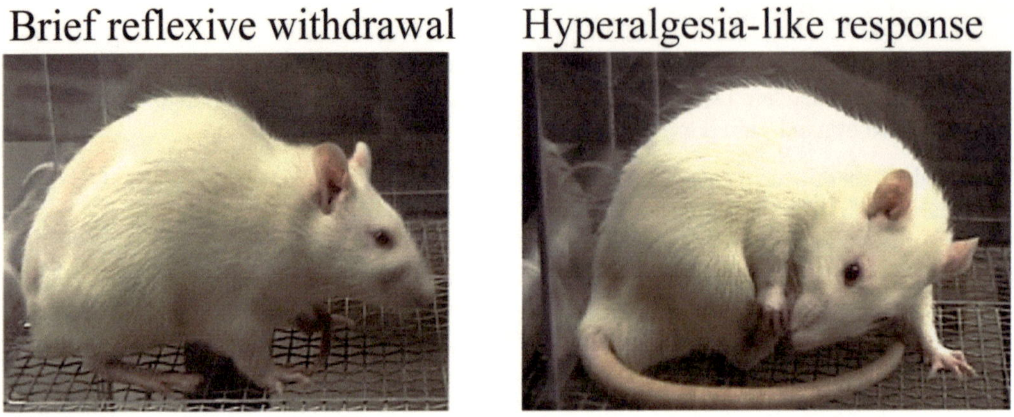

Fig. 9. Behavioral testing using a Quincke 22–gauge spinal needle. When a pin is applied to rat planter, the response is either a brief reflex withdrawal or a hyperalgesic reaction characterized by sustained lifting, shaking, and licking of the paw.

5.4 K_{ATP} channel activity in cell-attached patches depends on nerve injury status

To tested whether K_{ATP} channels in primary afferent neuron contribute to the pathogenesis of neuropathic pain, basal K_{ATP} channel openings in either control (non-surgery) or SNL neurons were measured at -60 mV membrane holding potential using cell-attached patch clamp configurations.

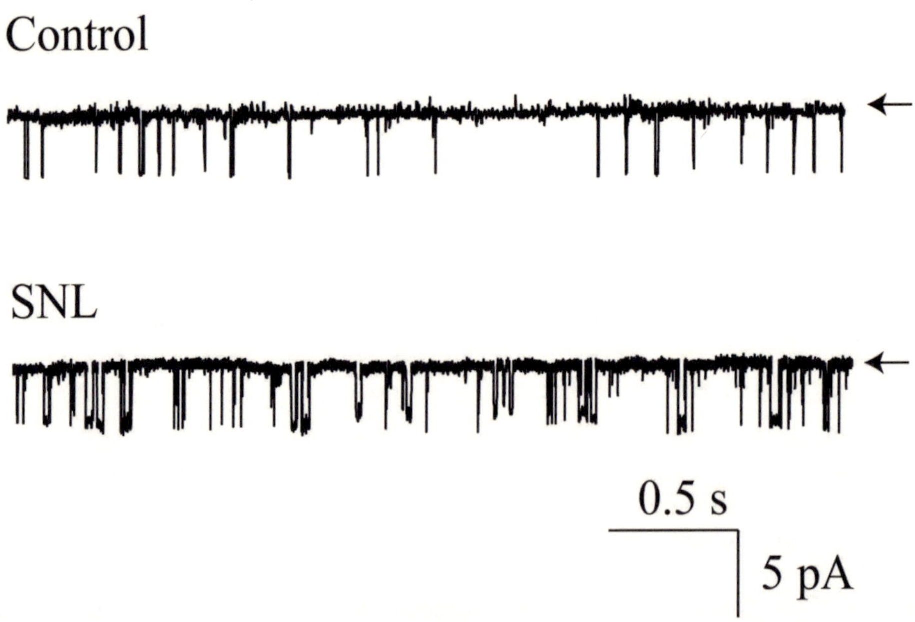

Fig. 10. Basal K_{ATP} channel activity from cell-attached patches in control and SNL neurons. Representative cell-attached recording traces in control (non surgery) and SNL neurons. Basal single channel currents were recorded at -60 mV. Horizontal arrows indicate closed channel state.

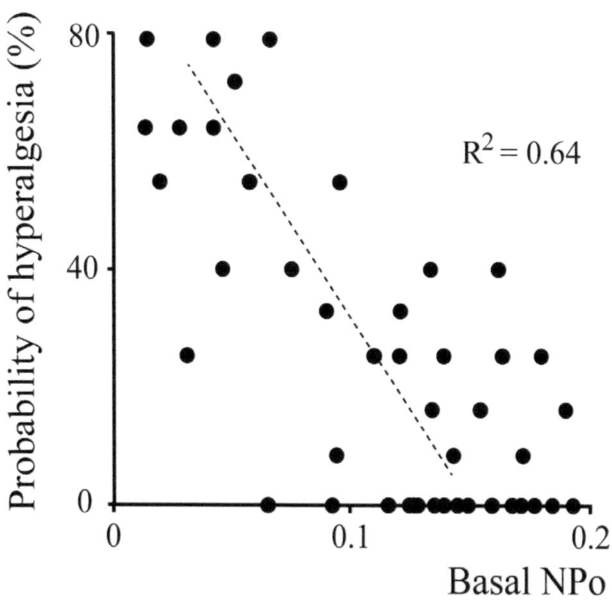

Fig. 11. Correlation of basal K_{ATP} channel NPo with each donor rat's probability of hyperalgesia, showing an inverse relationship. Simple linear regression curve and confidence intervals are shown.

In these recordings, basal channel opening was observed in both control and SNL neurons. In SNL neurons, however, basal K_{ATP} channel activity was diminished compared to controls (Fig. 10). NPo values in SNL neuron were also significantly reduced compared to control neurons. Analysis of single channel kinetics indicated that mean open time was shorter in SNL group compared to control group. Furthermore, basal K_{ATP} channel NPo correlated inversely with the probability of the donor animal responding to punctuate mechanical stimulus with a sustained, complex hyperalgesia-type behavior (Fig. 11.)

These results suggest that loss of current through these channels contributes to the pathogenesis of neuropathic pain. This hypothesis is further supported by a previous report that non-specific K^+ channel blockade evokes spontaneous firing in large Aβ fibers after SNL (Liu et al., 2001). In addition, other study reported that hyperexcitability following peripheral nerve injury is mediated by loss of various K^+ currents (Chung & Chung, 2002).

5.5 K_{ATP} channel activity in inside-out patches are not altered by nerve injury

To examine whether unitary K_{ATP} channel currents is altered by axotomy, single channel properties in cell-free patches in either control (non-surgery) or SNL neurons were measured at -60 mV membrane holding potential.

In these recordings, marked current activity was observed in inside-out patches excised from either control or SNL neurons into ATP-free solution with symmetrical 140 mM K^+ condition (Fig. 12). In both groups, inside-out patches showed only one type of K^+ channel current, especially at negative potentials. In addition, single-channel conductance was the same in control and SNL neurons (70-80 pS). Furthermore, sensitivity to ATP and diazoxide, a selective SUR-1 containing K_{ATP} channel opener, also did not differ between groups (Fig. 12).

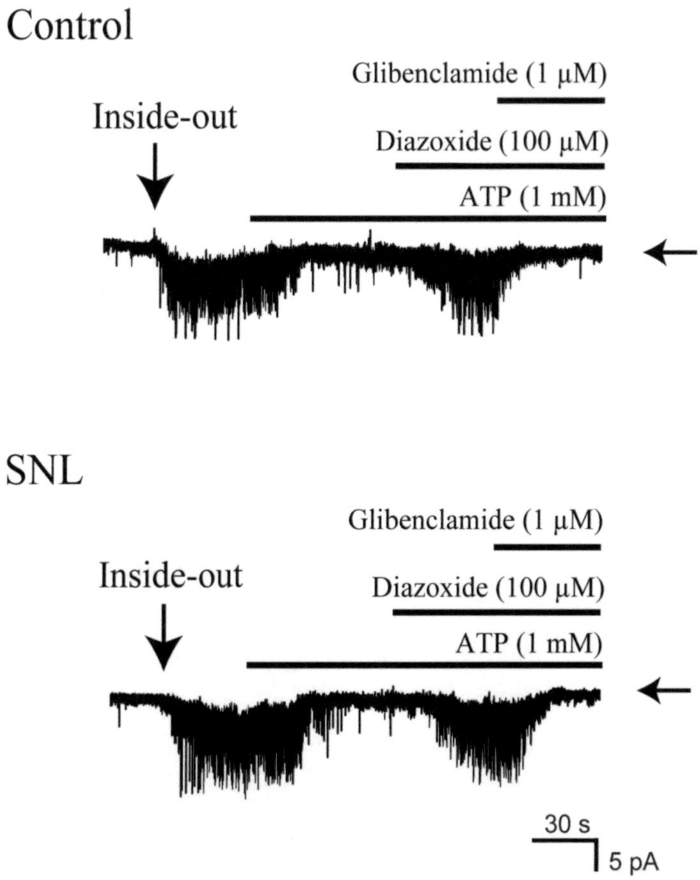

Fig. 12. Single-channel characteristics of K_{ATP} channels from inside-out recordings from control or SNL neurons. Membrane potential was clamped at -60 mV. Upon patch excision (vertical arrow) into an ATP-free bath marked channel activity ensued. ATP (1 mM), diazoxide (100 μM), and glibenclamide (1 μM) was added to the intracellular (bath) solution as indicated by the horizontal solid bar. Horizontal arrows indicate closed channel state.

These results suggest that SNL does not affect the K_{ATP} channel per se or any associated membrane-resident regulatory proteins. Our findings further imply that the molecular composition of the K_{ATP} channels is not affected by axotomy.

Therefore, the suppressed K_{ATP} channel activity observed from cell attached recordings may be attributed to alterations in the cytosolic signaling following painful nerve injury.

6. Conclusion

K_{ATP} channels couple cellular electrical activity to cytosolic metabolic status in various excitable tissues. These channels are widely expressed in central neurons, wherein they regulate membrane excitability and neurotransmitter release, and they provide neuroprotection. In addition to the functional K_{ATP} channels in the central nervous system, we have identified these channels in rat primary afferent neurons, dissociated from the rat DRG.

Altered sensory function contributes to the pathogenesis of neuropathic pain via hyperexcitability in injured axons and the corresponding somata in the DRG, increased synaptic transmission at the dorsal horns, and loss of DRG neurons. We have identified loss of K_{ATP} currents in DRG somata from rats that demonstrated sustained hyperalgesia-type response to nociceptive stimulation after axotomy. Thus, reduced K_{ATP} currents may be a factor in generating neuropathic pain through increased excitability, amplified excitatory neurotransmission, and enhanced susceptibility to neuronal cell death. In addition, intrinsic single-K_{ATP} channel characteristics are preserved even after painful nerve injury. Therefore, intact biophysical and pharmacological properties provide opportunities for therapeutic targeting with K_{ATP} channel openers against neuropathic pain.

7. References

Abdulla, FA. & Smith, PA. (2001) Axotomy- and autotomy-induced changes in Ca^{2+} and K^+ channel currents of rat dorsal root ganglion neurons. *J Neurophysiol.* Vol.85, No.2, (February 2001), pp. 644-58, ISSN 0022-3077

Amir, R., Argoff, CE., Bennett, GJ., Cummins, TR., Durieux, ME., Gerner, P., Gold, MS., Porreca, F. & Strichartz, GR. (2006) The role of sodium channels in chronic inflammatory and neuropathic pain. *J Pain.* Vol.7, No.5, (May 2006), pp. S1-29, ISSN 1526-5900

Amir, R., Kocsis, JD. & Devor, M. (2005) Multiple interacting sites of ectopic spike electrogenesis in primary sensory neurons. *J Neurosci.* Vol.25, No.10, (March 2005), pp. 2576-85, ISSN 0270-6474

Amoroso, S., Schmid-Antomarchi, H., Fosset, M. & Lazdunski, M. (1990) Glucose, sulfonylureas, and neurotransmitter release: role of ATP-sensitive K^+ channels. *Science.* Vol.247, No.4944, (February 1990), pp. 852-4, ISSN 0036-8075

Babenko, AP., Aguilar-Bryan, L. & Bryan, J. (1998) A view of sur/KIR6.X, K_{ATP} channels. *Annu Rev Physiol.* Vol.60, pp. 667-87, ISSN 0066-4278

Babenko, AP., Gonzalez, G. & Bryan, J. (2000) Pharmaco-topology of sulfonylurea receptors. Separate domains of the regulatory subunits of K_{ATP} channel isoforms are required for selective interaction with K^+ channel openers. *J Biol Chem.* Vol.275, No.2, (Jannuary 2000), pp. 717-20, ISSN 0021-9258

Ballanyi, K. (2004) Protective role of neuronal K_{ATP} channels in brain hypoxia. *J Exp Biol.* Vol.207, No.18, (August 2004), pp. 3201-12, ISSN 1477-9145

Bennett, GJ. & Xie, YK. (1988) A peripheral mononeuropathy in rat that produces disorders of pain sensation like those seen in man. *Pain.* Vol.33, No.1, (April 1988), pp. 87-107, ISSN 0304-3959

Blondeau, N., Plamondon, H., Richelme, C., Heurteaux, C. & Lazdunski, M. (2000) K_{ATP} channel openers, adenosine agonists and epileptic preconditioning are stress signals inducing hippocampal neuroprotection. *Neuroscience.* Vol.100, No.3, (Setember 2000), pp. 465-74, ISSN 0306-4522

Chiu, SY., Schrager, P. & Ritchie, JM. (1984) Neuronal-type Na^+ and K^+ channels in rabbit cultured Schwann cells. *Nature.* Vol.311, No.5982, (Sepember 1984), pp. 156-7, ISSN 0028-0836

Chung, JM. & Chung, K. (2002) Importance of hyperexcitability of DRG neurons in neuropathic pain. *Pain Pract.* Vol.2, No.2, (June 2002), pp. 87-97, ISSN 1530-7085

Chutkow, WA., Pu, J., Wheeler, MT., Wada, T., Makielski, JC., Burant, CF. & McNally, EM. (2002) Episodic coronary artery vasospasm and hypertension develop in the absence of Sur2 K_{ATP} channels. *J Clin Invest.* Vol.110, No.2, (July 2002), pp. 203-8, ISSN 0021-9738

Devor, M. (2009) Ectopic discharge in Aβ afferents as a source of neuropathic pain. *Exp Brain Res.* Vol.196, No.1, (June 2009), pp. 115-28, ISSN 0014-4819

Edwards, G. & Weston, AH. (1993) The pharmacology of ATP-sensitive potassium channels. *Annu Rev Pharmacol Toxicol.* Vol.33, (April 1993), pp. 597-637, ISSN 0362-1642

Gemes, G., Bangaru, ML., Wu, HE., Tang, Q., Weihrauch, D., Koopmeiners, AS., Cruikshank, JM., Kwok, WM. & Hogan, QH. (2011) Store-operated Ca^{2+} entry in sensory neurons: functional role and the effect of painful nerve injury. *J Neurosci.* Vol.31, No.10, (March 2011), pp. 3536-49, ISSN 0270-6474

Gold, MS. (2000) Spinal nerve ligation: what to blame for the pain and why. *Pain.* Vol.84, No.2-3, (February 2000), pp. 117-20, ISSN 0304-3959

Harper, AA. & Lawson, SN. (1985) Electrical properties of rat dorsal root ganglion neurones with different peripheral nerve conduction velocities. *J Physiol.* Vol.359, (February 1985), pp. 47-63, ISSN 0022-375

Heurteaux, C., Lauritzen, I., Widmann, C. & Lazdunski, M. (1995) Essential role of adenosine, adenosine A1 receptors, and ATP-sensitive K^+ channels in cerebral ischemic preconditioning. *Proc Natl Acad Sci U S A.* Vol.92, No.10, (May 1995), pp. 4666-70, ISSN 0027-8424

Hogan, Q. (2002) Animal pain models. *Reg Anesth Pain Med.* Vol.27, No.4, (July-August 2002), pp. 385-401, ISSN 1098-7339

Hogan, Q., Sapunar, D., Modric-Jednacak, K. & McCallum, JB. (2004) Detection of neuropathic pain in a rat model of peripheral nerve injury. *Anesthesiology.* Vol.101, No.2, (August 2004), pp. 476-87, ISSN 0003-3022

Kawagoe, KT., Zimmerman, JB. & Wightman, RM. (1993) Principles of voltammetry and microelectrode surface states. *J Neurosci Methods.* Vol.48, No.3, (July 1993), pp. 225-40, ISSN 0165-0270

Kawano, T., Zoga, V., Kimura, M., Liang, MY., Wu, HE., Gemes, G., McCallum, JB., Kwok, WM., Hogan, QH. & Sarantopoulos, CD. (2009a) Nitric oxide activates ATP-sensitive potassium channels in mammalian sensory neurons: action by direct S-nitrosylation. *Mol Pain.* Vol.5, No.12, (March 2009b), ISSN 1744-8069

Kawano, T., Zoga, V., Gemes, G., McCallum, JB., Wu, HE., Pravdic, D., Liang, MY., Kwok, WM., Hogan, Q. & Sarantopoulos, C. (2009b) Suppressed Ca^{2+}/CaM/CaMKII-dependent K_{ATP} channel activity in primary afferent neurons mediates hyperalgesia after axotomy. *Proc Natl Acad Sci U S A.* Vol.106, No.21, (May 2009), pp. 8725-30, ISSN 0027-8424

Kawano,T., Zoga, V., McCallum, JB., Wu, HE., Gemes, G., Liang, MY., Abram, S., Kwok, WM., Hogan, QH. & Sarantopoulos, CD. (2009c) ATP-sensitive potassium currents in rat primary afferent neurons: biophysical, pharmacological properties, and alterations by painful nerve injury. *Neuroscience.* Vol.162, No.2, (August 2009), pp. 431-43, ISSN 0306-4522

Kim, SH. & Chung, JM. (1992) An experimental model for peripheral neuropathy producedby segmental spinal nerve ligation in the rat. *Pain.* Vol.50, No.3, (Sepember 1992), pp. 355-63, ISSN 0304-3959

Kofuji, P., Biedermann, B., Siddharthan, V., Raap, M., Iandiev, I., Milenkovic, I., Thomzig, A., Veh, RW., Bringmann, A. & Reichenbach, A. (2002) Kir potassium channel subunit expression in retinal glial cells: implications for spatial potassium buffering. *Glia.* Vol.39, No.3, (Sepember 2002), pp. 292-303, ISSN 0894-1491

Lawson, SN. (2002) Phenotype and function of somatic primary afferent nociceptive neurones with C-, Aδ- or Aα/β-fibres. *Exp Physiol.* Vol.87, No.2, (March 2002), pp. 239-44, ISSN 0958-0670

McCallum, JB., Kwok, WM., Sapunar, D., Fuchs, A. & Hogan, QH. (2006) Painful peripheral nerve injury decreases calcium current in axotomized sensory neurons. *Anesthesiology.* Vol.105, No.1, (July 2006), pp. 160-8, ISSN 0003-3022

Miki, T., Nagashima, K. & Seino, S. (1999) The structure and function of the ATP-sensitive K^+ channel in insulin-secreting pancreatic beta-cells. *J Mol Endocrinol.* Vol.22, No.2, (April 1999), pp. 113-23, ISSN 0952-5041

Miki, T. & Seino, S. (2005) Roles of K_{ATP} channels as metabolic sensors in acute metabolic changes. *J Mol Cell Cardiol.* Vol.38, No.6, (June 2005), pp. 917-25, ISSN 0022-2828

Miki, T., Suzuki, M., Shibasaki, T., Uemura, H., Sato, T., Yamaguchi, K., Koseki, H., Iwanaga, T.,Nakaya, H. & Seino, S. (2002) Mouse model of Prinzmetal angina by disruption of the inward rectifier Kir6.1. *Nat Med.* Vol.8, No.5, (May 2002), pp. 466-72, ISSN 1078-8956

Mourre, C., Ben Ari, Y., Bernardi, H., Fosset, M. & Lazdunski, M. (1989) Antidiabetic sulfonylureas: localization of binding sites in the brain and effects on the hyperpolarization induced by anoxia in hippocampal slices. *Brain Res.* Vol.486, No.1, (May 1989), pp. 159-64, ISSN 0006-8993

Nichols, CG. (2006) K_{ATP} channels as molecular sensors of cellular metabolism. *Nature.* Vol.440, No.7083, (March 2006), pp. 470-6, ISSN 0028-0836

Noma, A. (1983) ATP-regulated K^+ channels in cardiac muscle. *Nature.* Vol. 305, No.5930, (Sepember 1983), pp. 147-8, ISSN 0028-0836

Raouf, R., Quick, K. & Wood, JN. (2010) Pain as a channelopathy. *J Clin Invest.* Vol.120, No.11, (November 2010), pp. 3745-52, ISSN 0021-9738

Saito, T., Sato, T., Miki, T., Seino, S. & Nakaya, H. (2005) Role of ATP-sensitive K^+ channels in electrophysiological alterations during myocardial ischemia: a study using Kir6.2-null mice. *Am J Physiol Heart Circ Physiol.* Vol.288, No.1, (Jannuary 2005), pp. H352-7, ISSN 0363-6135

Sapunar, D., Ljubkovic, M., Lirk, P., McCallum, JB. & Hogan, QH. (2005) Distinct membrane effects of spinal nerve ligation on injured and adjacent dorsal root ganglion neurons in rats. *Anesthesiology.* Vol.103, No.2, (August 2005), pp. 360-76, ISSN 0003-3022

Sarantopoulos, C., McCallum, JB., Kwok, WM. & Hogan, Q. (2004) β-escin diminishes voltage-gated calcium current rundown in perforated patch-clamp recordings from rat primary afferent neurons. *J Neurosci Methods.* Vol.139, No.1, (October 2004), pp. 61-8, ISSN 0165-0270

Sarantopoulos, C., McCallum, B., Sapunar, D., Kwok, WM. & Hogan, Q. (2003) ATP-sensitive potassium channels in rat primary afferent neurons: the effect of neuropathic injury and gabapentin. *Neurosci Lett.* Vol.343, No.3, (June 2003), pp. 185-9, ISSN 0304-3940

Seghers, V., Nakazaki, M., DeMayo, F., Aguilar-Bryan, L. & Bryan, J. (2000) Sur1 knockout mice . A model for K_{ATP} channel-independent regulation of insulin secretion. *J Biol Chem.* Vol.275, No.13, (March 2000), pp. 9270-7, ISSN 0021-9258

Shake, JG., Peck, EA., Marban, E., Gott, VL., Johnston, MV., Troncoso, JC., Redmond, JM. & Baumgartner, WA. (2001) Pharmacologically induced preconditioning with diazoxide: a novel approach to brain protection. *Ann Thorac Surg.* Vol.72, No.6, (December 2001), pp. 1849-54, ISSN 1552-6259

Shir, Y. & Seltzer, Z. (1990) A-fibers mediate mechanical hyperesthesia and allodynia and C-fibers mediate thermal hyperalgesia in a new model of causalgiform pain disorders in rats. *Neurosci Lett.* Vol.115, No.1, (July 1990), pp. 62-7, ISSN 0304-3940

Smith, PA., Duchen, MR. & Ashcroft, FM. (1995) A fluorimetric and amperometric study of calcium and secretion in isolated mouse pancreatic β-cells. *Pflugers Arch.* Vol.430, No.5, (Sepember 1995), pp. 808-18, ISSN 1432-2013

Stefani, MR. & Gold, PE. (2001) Intrahippocampal infusions of K-ATP channel modulators influence spontaneous alternation performance: relationships to acetylcholine release in the hippocampus. *J Neurosci.* Vol.21, NO.2, (Jannuary 2001), pp. 609-14, ISSN 0270-6474

Steinkamp, M., Li, T., Fuellgraf, H. & Moser, A. (2007) K_{ATP}-dependent neurotransmitter release in the neuronal network of the rat caudate nucleus. *Neurochem Int.* Vol.50, No.1, (Jannuary 2007), pp. 159-63, ISSN 0197-0186

Sun, HS., Feng, ZP., Barber, PA., Buchan, AM. & French, RJ. (2007) Kir6.2-containing ATP-sensitive potassium channels protect cortical neurons from ischemic/anoxic injury in vitro and in vivo. *Neuroscience.* Vol.144, No.4, (February 2007), pp. 1509-15, ISSN 0306-4522

Sun, HS., Feng, ZP., Miki, T., Seino, S. & French, RJ. (2006) Enhanced neuronal damage after ischemic insults in mice lacking Kir6.2-containing ATP-sensitive K^+ channels. *J Neurophysiol.* Vol.95, No.4, (April 2006), pp. 2590-601, ISSN 0022-3077

Suzuki, M., Li, RA., Miki, T., Uemura, H., Sakamoto, N., Ohmoto-Sekine, Y., Tamagawa, M., Ogura, T., Seino, S., Marbán, E. & Nakaya, H. (2001) Functional roles of cardiac and vascular ATP-sensitive potassium channels clarified by Kir6.2-knockout mice. *Circ Res.* Vol.88, No.6, (March 2001), pp.570-7, ISSN 0009-7330

Woolf, CJ. (2010) What is this thing called pain? *J Clin Invest.* Vol.120, No.11, (November 2010), pp. 3742-4, ISSN 0021-9738

Yamada, K. & Inagaki, N. (2005) Neuroprotection by K_{ATP} channels. *J Mol Cell Cardiol.* Vol.38, No.6, (June 2005), pp. 945-9, ISSN 0022-2828

Yamada, K., Ji, JJ., Yuan, H., Miki, T., Sato, S., Horimoto, N., Shimizu, T., Seino, S. & Inagaki, N. (2001) Protective role of ATP-sensitive potassium channels in hypoxia-induced generalized seizure. *Science.* Vol.292, No.5521, (May 2001), pp. 1543-6, ISSN 0036-8075

Yokoshiki, H., Sunagawa, M., Seki., T. & Sperelakis, N. (1998) ATP-sensitive K^+ channels in pancreatic, cardiac, and vascular smooth muscle cells. *Am J Physiol.* Vol.274, (Jannuary 1998), pp. C25-37, ISSN 0363-6135

Zhou, Z. & Misler, S. (1996) Amperometric detection of quantal secretion from patch-clamped rat pancreatic β-cells. *J Biol Chem.* Vol.271, No.1, (Jannuary 1996), pp. 270-7, ISSN 0021-9258

Zimmermann, M. (2001) Pathobiology of neuropathic pain. *Eur J Pharmacol.* Vol.429, No.1-3,(October 2001), pp. 23-37, ISSN 0014-2999

Zoga, V., Kawano, T., Liang, MY., Bienengraeber, M., Weihrauch, D., McCallum, B., Gemes, G., Hogan, Q. & Sarantopoulos, C. (2010) K_{ATP} channel subunits in rat dorsal root ganglia: alterations by painful axotomy. *Mol Pain.* Vol.6, No.6, (Jannuary 2010), ISSN 1744-8069

5

Regulation of Renal Potassium Channels by Protein Kinases and Phosphatases

Manabu Kubokawa
Department of Physiology, School of Medicine,
Iwate Medical University, Yahaba, Iwate
Japan

1. Introduction

Ion channels are well-known functional proteins, and their activity can be directly estimated by monitoring ion currents through channel proteins using the patch-clamp technique (Hamill et al., 1981), which is a fine tool for the investigation of not only ion channels, but also the membrane current and membrane potential of the whole cell. Many investigators have applied the patch-clamp technique to various cells, and found that many kinds of ion channel are present in the cell membrane.

Classical electrophysiological studies have demonstrated that the membrane potential of epithelial cells is indispensable to drive electrolytes and solute transport across cell membranes (Bello-Reuss & Weber, 1986; Beck et al., 1991; Fujimoto et al., 1991). It has also been demonstrated that the major part of the membrane potential consists of ion conductance (Kubota et al., 1983; Kubokawa et al., 1990), and that several factors regulate the membrane potential by changing the ion conductance of the cell membrane (Kubota et al., 1983; Hagiwara et al., 1990). The development of the patch-clamp technique has revealed that ion conductance of the cell membrane is mainly facilitated by current passing through the ion channels present in the membrane (Edelman et al., 1986; Fujimoto et al., 1991; Kubokawa et al., 1998). Irrespective of the recent advances in molecular technology, investigation of the functional significance of channel proteins is largely dependent on experiments using the patch-clamp technique.

Among the many kinds of ion channel, potassium selective channels (K^+ channels) are the most abundant channels in both excitable and non-excitable cells, and they play important roles in the formation of the membrane potential. It is generally accepted that the membrane potential of epithelial cells is indispensable as the driving force for electrogenic transport, such as Na^+-coupled solute co-transport. Since the driving force is largely dependent on the membrane potential, electrogenic transport is mediated, at least in part, by its potential. Namely, it is conceivable that membrane hyperpolarization stimulates and depolarization suppresses such transport. Moreover, since the membrane potential of the epithelial cells is mainly formed by the activity of K^+ channels and Na^+-K^+-ATPase, investigation of the mechanisms leading to changes in K^+ channel activity using the patch-clamp technique would provide important knowledge for the regulation of epithelial transport systems.

2. Distribution of renal K^+ channels along the nephron

In the kidney tubules, epithelial transport plays important roles in body fluid homeostasis, such as the electrolyte and acid-base balance (Giebisch, 1998; Gennari & Maddox, 2005). It has been demonstrated that several types of K^+ channel are present in the apical and basolateral membranes of tubular epithelia, and the functional importance of these K^+ channels in epithelial transport has also been reported in individual nephron segments (Guggino et al., 1987; Gögelein, 1990; Palmer, 1992; Giebisch, 1995; Quast, 1996). The most frequently observed types of K^+ channel with the patch-clamp technique are Ca^{2+}-activated channels and inwardly rectifying channels regulated by cytosolic ATP (Ohno-Shosaku, et al., 1989; Giebisch, 1995; Quast, 1996; W. Wang et al., 1997). The former are mainly present in the apical membranes (Merot et al., 1989; Hirano et al., 2001) and the latter are in the basolateral membranes (Ohno-Shosaku et al., 1990; Robson & Hunter, 1997; Kubokawa et al., 1998; Nakamura et al., 2001) of the proximal tubule cells, the principal cells of the cortical collecting duct (CCD) (Hirsch et al., 1993), the apical membranes of the thick ascending limb of Henle (TAL) (W. Wang, 1994), and the principal cells of CCD (Giebisch, 1995; W. Wang et al., 1997; Kawahara & Anzai, 1997). Studies using molecular techniques have also demonstrated the structures (Ho et al., 1993) and tissue distribution of cloned K^+ channels in several nephron segments of the kidney (Boim et al., 1995; McNicholas et al., 1996; Derst et al., 2001). It has also been reported that K^+ channels requiring ATP to maintain their activity possess protein kinase-mediated phosphorylation sites in cloned K^+ channels (Ho et al., 1993; Hebert et al., 2005). Although ATP-sensitive K^+ (K_{ATP}) channels in the kidney are usually inhibited by an increase in cytosolic ATP (W. Wang & Giebisch, 1991a; Tsuchiya et al., 1992; Welling, 1995; Kawahara & Anzai, 1997; Mauere et al., 1998a), several lines of evidence strongly suggest that ATP-regulated channels with ATP-dependent activation are mediated by protein kinases (W. Wang & Giebisch, 1991b; McNicholas et al., 1994; Levitan, 1994; Kubokawa et al., 1997; Nakamura et al., 2001) and phosphatases (Kubokawa et al., 1995a; 2000). Namely, the ATP-regulated K^+ channels are mainly mediated by protein kinases and phosphatases. To date, ATP-regulated K^+ channels have been found in the proximal tubule cells, TAL, and principal cells of CCD. Thus, we have focused on ATP-regulated K^+ channels in these tubule cells, and reviewed the involvement of protein kinases and phosphatases in the regulation of these channels and their importance in kidney functions.

2.1 Roles of K^+ channels in the proximal tubule cells

A model of ion transport in the proximal tubule is shown in Fig. 1A. An explanation of the individual transporters, pump, and ion channels in the apical and basolateral cell membranes is presented in the lower part of the figure. The filtered Na^+ and many solutes such as glucose and amino acids are reabsorbed into the proximal tubule cells by co-transporters according to the electrochemical gradient for Na^+ across the apical membrane (Weinstein, 2000). The basolateral Na^+-K^+-ATPase (Na^+-K^+ pump) plays a crucial role in the formation of the Na^+ and K^+ gradient across the apical and basolateral membranes. The basolateral ATP-regulated K^+ channels are indispensable for the negative cell potential, which is largely dependent on the K^+ gradient across the cell membrane and K^+ channel activity, and serves as the driving force for the apical Na^+-coupled transporters. The Na^+ that enters the cell is excluded to the interstitial space by basolateral Na^+-K^+ pump (Na^+-K^+

pump). Thus, the major part of the ATP produced in the cell is used for transport by Na^+-K^+ pump and ATP-regulated K^+ channels in renal tubule cells (Beck et al., 1991). Several protein kinases are involved in the regulation of the basolateral ATP-regulated K^+ channels, as described below (Kubokawa et al., 1997; Mauerer et al., 1998b; Mori., et al, 2001, Nakamura et al., 2002). The transport of other ions, such as H^+ or HCO_3^-, is mediated mainly by Na^+-coupled processes. It has also been reported that some K^+ channels are present in the apical membrane of proximal tubule cells (Merot et al., 1989). Although the precise role of the apical K^+ channel is still unknown, the K^+ channel requires a high intracellular Ca^{2+} concentration $[Ca^{2+}]_i$ to open (Hirano et al., 2001). Thus, the open probability (P_o) of the apical K^+ channel would usually be low under the normal conditions.

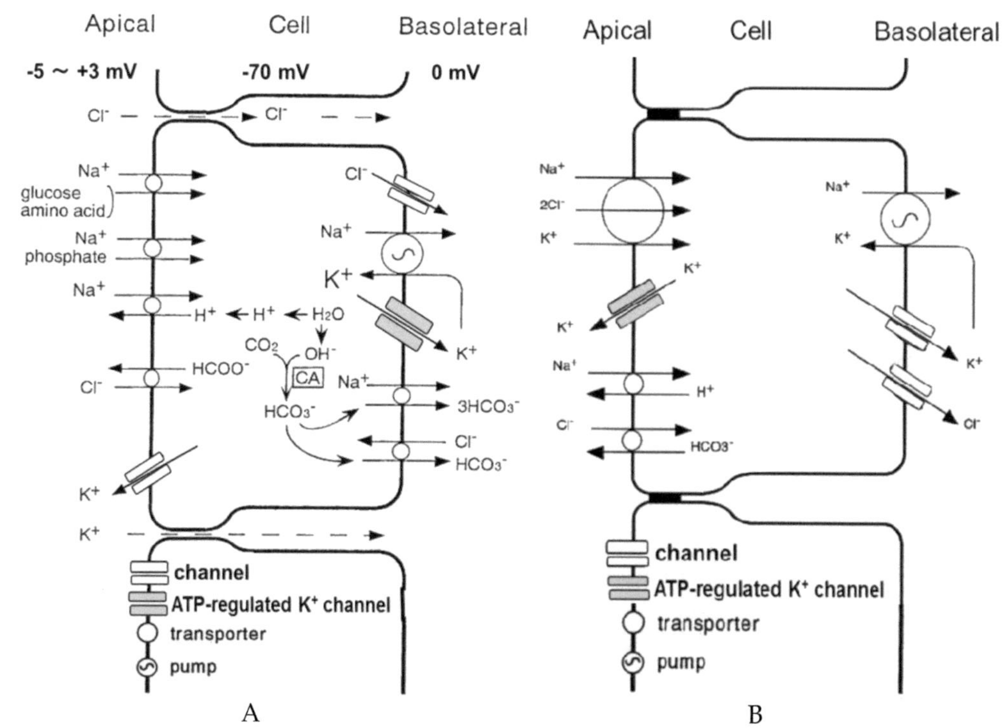

Fig. 1. Simplified models of the ion transport processes in the proximal tubule (A) and TAL (B). The double rectangles indicate ion channels (the gray double rectangles are ATP-regulated K^+ channels). The circles are transporters and the pump, as depicted in the lower part of the figures.

2.2 Roles of K^+ channels in TAL

The functional significance of TAL is to dilute the luminal fluid by the re-absorption of Na^+ and Cl^- without H_2O, which results in elevation of the interstitial osmolarity. As shown in Fig. 1B, the $Na^+/K^+/2Cl^-$ co-transporter plays a major role in the entry of these ions into cells. For the sufficient entry of these ions via the $Na^+/K^+/Cl^-$ co-transporter, a K^+ supply to the lumen is required. Thus, the role of the K^+ channel in the apical membrane of TAL is to move K^+ into the lumen from the cytosol. Two types of apical K^+ channel were reported in

TAL (W. Wang, 1994). Both of these channels were inhibited by a high concentration of ATP, but one is an ATP-regulated channel with a high P_o, which is activated by protein kinase A (W.Wang, 1994). These results suggest that the apical ATP-regulated K^+ channel with a high open probability is a major candidate for the K^+ channel which supplies the lumen with K^+.

In the basolateral membrane of TAL, an Na^+-K^+ pump and K^+ conductance were observed, as has been shown in normal transporting epithelia. However, little information is available regarding the basolateral K^+ channel in TAL, although the properties of basolateral Cl^- channels have been reported (Winters et al., 1999).

2.3 Roles of K^+ channels in the principal cells of CCD

The renal collecting duct including CCD, and inner and outer medullary collecting duct (IMCD and OMCD, respectively) are the most important nephron segments for determining the final urine conditions to maintain body fluid homeostasis. It has been demonstrated that the Na^+ and K^+ concentrations of blood are largely dependent on Na^+ re-absorption and K^+ secretion in CCD (Malnic et al., 2000). Na^+ is reabsorbed from the lumen to cell mainly through the apical epithelial Na^+ channel (ENaC) (Palmer & Carty, 2000). It is also generally accepted that K^+ is secreted from the cell to lumen mainly through the apical ATP-regulated K^+ channel (Giebisch, 1995) (Fig.2), although a few reports have demonstrated that the Ca^{2+}-activated maxi-K^+ channel is a major candidate for K^+ secretion in CCD (Taniguchi & Imai, 1998; Woda et al., 2001). The apical K^+ channels are very important to maintain the normal K^+ level of blood (Giebisch, 1998). If the apical K^+ channel is suppressed, the blood K^+ level is elevated, which often suppresses the cardiac function. In contrast, massive K^+ secretion through K^+ channels in CCD results in hypokalemia.

A K^+ channel similar to the apical ATP-regulated K^+ channel in CCD was cloned from the outer medulla of the rat kidney, which is named ROMK (Ho et al., 1993). Analyses of the cloned K^+ channel yielded several important findings regarding the apical K^+ channel. One of the most significant findings was that this ATP-regulated channel possessed separate PKA- and PKC-mediated phosphorylation sites. As reported previously, the apical K^+ channel in CCD was stimulated by PKA and suppressed by PKC (W. Wang & Giebisch, 1991b), suggesting that the roles of PKA and PKC in the modulation of the channel were different. Molecular analyses of the cloned K^+ channel supported the view obtained from observing the native K^+ channel that the PKA-mediated site is clearly distinct from PKC-mediated sites (Ho et al., 1993). Moreover, several subtypes of the ATP-regulated K^+ channel were cloned from the distal nephron segment including TAL (Hebert et al., 2005). These channels possess similar properties, such as conductance of about 30 pS, being ATP-regulated, and PKA-activated channels (Hebert et al., 2005). ATP-regulated K^+ channels in mammalian proximal tubule cells are also PKA-activated channels, although the conductance of these channels was greater (40 -90 pS) than that of the cloned channels (Kubokawa et al., 1997; Nakamura et al., 2001).

In the basolateral membrane of the principal cells, two types of K^+ channel have been reported. One is the Ca^{2+}-activated channel (Hirsch et al., 1993) and the other is the PKG-mediated channel (Hirsch & Schlatter, 1995). The former requires a high $[Ca^{2+}]_i$ to open, but the latter is able to open under normal conditions. Furthermore, the latter channel is considered to be an ATP-regulated channel, since it was modulated by a protein kinase,

PKG (Hirsch & Schlatter, 1995). Thus, it is suggested that the ATP-regulated (PKG-mediated) K⁺ channel is one of the major candidates for the formation of the membrane potential of the principal cells.

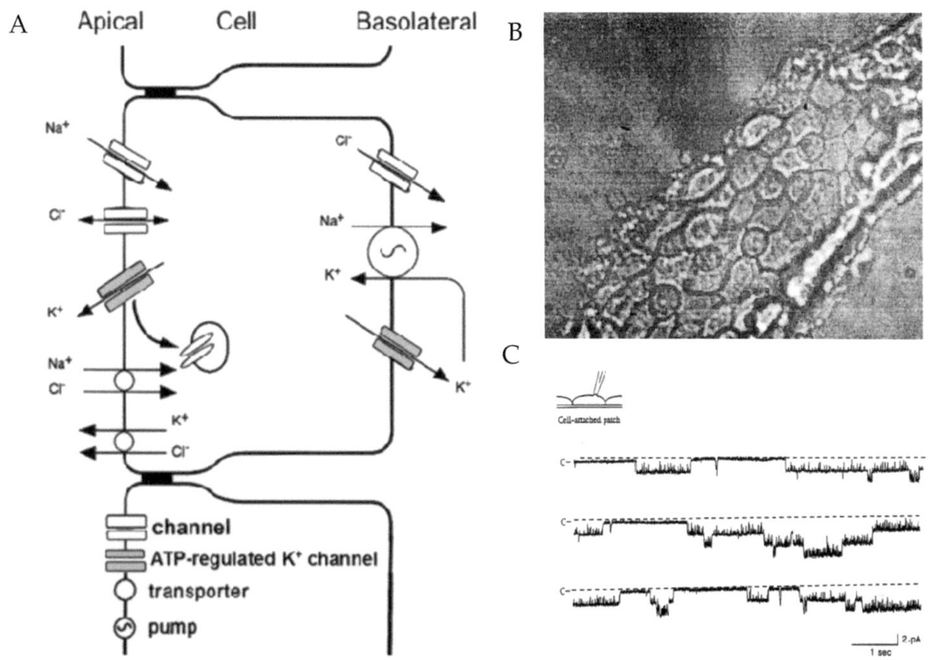

Fig. 2. A model of ion transport in the principal cell of CCD (A), the apical surface of CCD (B), and channel currents obtained from the apical membrane of the principal cell. ATP-regulated K⁺ channels are present in both apical and basolateral membranes (A). A microscopic view of the apical surface of CCD revealed that two types of cell, the principal cell (large cells) and intercalated cell (small, round cells), are present (B). The represented current trace was the apical ATP-regulated K⁺ channel observed in a cell-attached patch (C).

3. Mechanisms of activation and suppression of renal ATP-regulated K⁺ channels by protein kinases and phosphatases

3.1 ATP-dependency of the renal ATP-regulated K⁺ channels

In general, ATP-regulated K⁺ channels in renal tubule cells were previously believed to be ATP-sensitive (ATP-inhibitable) channels (W. Wang & Giebisch, 1991a; Tsuchiya et al., 1992, Welling, 1995; Robson & Hunter, 1997; Mauerer et al., 1998a), since a high concentration of internal ATP inhibited the activity of these channels. However, most of these channels require relatively low concentrations of ATP to maintain their activity (Kubokawa et al., 1995a; Mauerer et al., 1998b; Nakamura et al., 2001). The ATP-regulated K⁺ channels cloned from the renal outer medulla also require ATP to maintain their activity (Ho et al., 1993). Although high concentrations of ATP may suppress the activity of the ATP-regulated channels including a cloned renal K⁺ channel (McNicholas et al., 1996), a relatively low concentration of ATP is indispensable for maintenance of the channel activity. Thus, the renal K⁺ channels regulated by ATP are usually ATP-dependent channels.

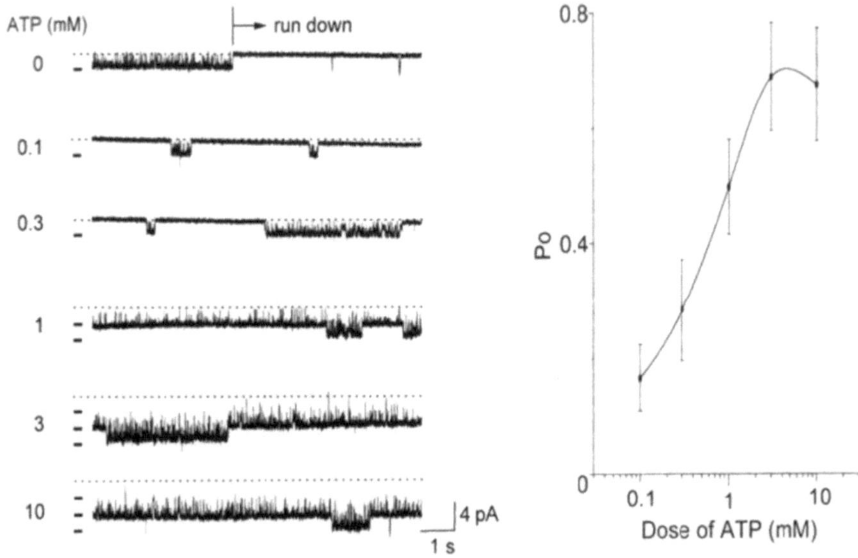

A B

Fig. 3. ATP-dependency of the ATP-regulated K+ channel in cultured human proximal tubule cells. ATP is required to maintain channel activity, and an increase in ATP on the cytosolic surface of the patch membrane enhanced channel activity (A). The dose-response curve indicates that the maximal P_o was observed around 3 mM ATP (B). [Reprinted from Nakamura, K., Hirano, J., & Kubokawa, M. (2001). An ATP-regulated and pH-sensitive inwardly rectifying K+ channel in cultured human proximal tubule cells. *Jpn. J. Physiol.*, Vol. 51, No. 4, pp. 523-530, ISSN: 0021-521X with permission, Copyright 2001]

In the inside-out mode of the patch-clamp technique, ATP-dependent channel activity was suppressed by protein kinase inhibitors (Kubokawa et al., 1997; Nakamura et al., 2001). Since protein kinase induces protein phosphorylation, inhibitors of protein kinase suppress the phosphorylation processes. Moreover, it is well-known that the cell membranes bind to many protein kinases. Thus, ATP is required to induce the protein kinase-mediated phosphorylation of the channel or associated protein in inside-out patches. Cytosolic ATP would act as a donor of phosphate for protein phosphorylation. After the removal of ATP, channel activity is rapidly reduced as shown in the top trace of Fig. 3A, which is usually called "run-down". Such channel run-down would result from phosphatase-mediated protein dephosphorylation, since the run-down observed in the inside-out patch can be blocked by phosphatase inhibitors (Kubokawa et al., 1995a). Moreover, closed channels after run-down can be re-activated by the addition of ATP. Thus, it is suggested that not only protein kinases but also phosphatases are bound to the inside of the cell membrane (Kubokawa et al., 1995a, 2000).

3.2 Protein kinase-mediated activation and phosphatases-mediated suppression of the renal ATP-regulated K^+ channels

Among several kinds of protein kinase, cAMP-dependent PKA is known to stimulate the ATP-regulated K+ channels in proximal tubule cells (Kubokawa et al., 1997; Nakamura et al.,

2001), TAL (W. Wang et al., 1994), and principal cells of CCD (W. Wang & Giebisch, 1991a; Kubokawa et al., 1995a), including the cloned renal K^+ channel (McNicholas et al, 1994). Thus, PKA-mediated channel activation seems to be a common characteristic of the renal ATP-regulated K^+ channels.

As shown in Fig. 4A, the application of membrane-permeant 8Br-cAMP to the bath elevated the K^+ channel activity of the proximal tubule cells in a cell-attached patch (Kubokawa et al., 1997; Nakamura et al., 2001). This result suggested that an increase in cytosolic cAMP stimulated PKA, and then PKA-mediated phosphorylation resulted in channel activation. Indeed, as shown in Fig. 4B, the direct application of PKA to the cytosolic surface of the patch-membrane in an inside-out patch elevated the activity of the ATP-regulated K^+ channel in the proximal tubule cells (Nakamura et al., 2001). As described above, PKA-

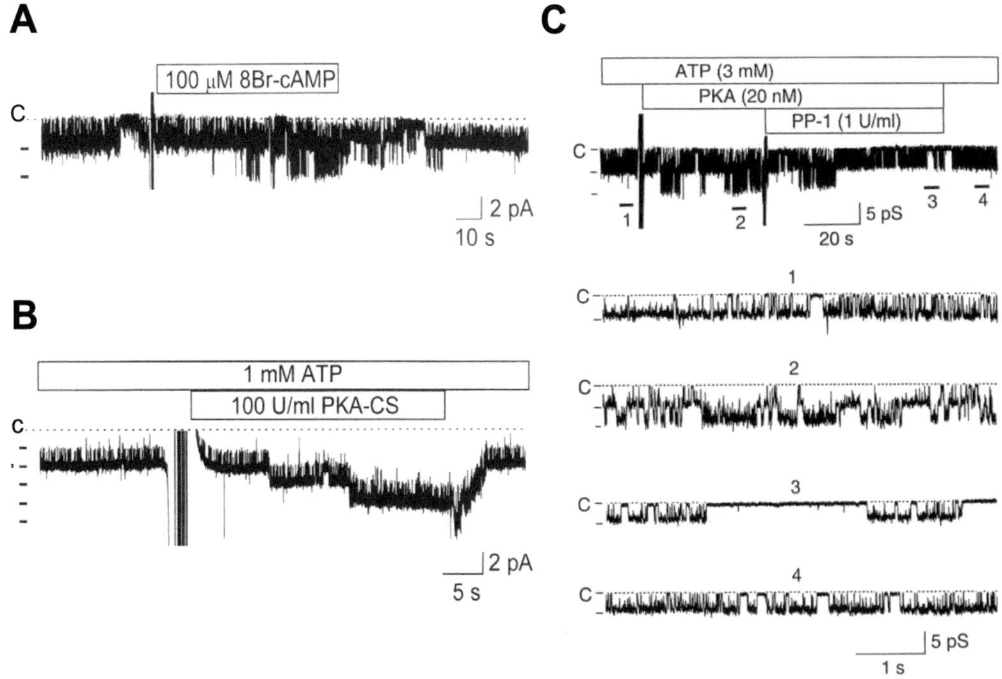

Fig. 4. Examples of the effects of cAMP-dependent PKA and protein phosphatase type 1 (PP-1) on the activity of the renal ATP-regulated K^+ channels in mammalian proximal tubule cells. The addition of membrane-permeant 8Br-cAMP to the bath enhanced channel activity in the cell-attached patch (A), and a catalytic subunit of PKA (PKA-CS) in the presence of ATP stimulated activity in the inside-out patch (B). In contrast, enhanced channel activity induced by PKA was inhibited by the addition of PP-1 in an inside-out patch (C). [A and B are reprinted from Nakamura, K., Hirano, J. & Kubokawa, M. (2001). An ATP-regulated and pH-sensitive inwardly rectifying K^+ channel in cultured human proximal tubule cells. *Jpn. J. Physiol.*, Vol. 51, No. 4, pp. 523-530, ISSN: 0021-521X with permission, Copyright 2001, and C is reprinted from Kubokawa, M., Nakamura, K., Hirano, J., Yoshioka, Y., Nakaya, S., Mori, Y., & Kubota, T. (2000). Regulation of inwardly rectifying K^+ channel in opossum proximal tubule cells by protein phosphatases 1 and 2A. *Jpn. J. Physiol.*, Vol. 50, No. 2, pp. 249-256, ISSN: 0021-521X, Copyright 2000]

induced channel activation is a common property of the renal ATP-regulated K^+ channel in both proximal and distal nephron segments. Moreover, a similar property was observed not only in human (Nakamura et al., 2001) but also opossum kidney proximal tubule cells (Kubokawa et al., 1997). These results suggest that the regulatory mechanisms of renal ATP-regulated K^+ channels along the nephron are similar.

In addition to PKA, cGMP-dependent protein kinase (PKG) was reported to enhance the activity of the ATP-regulated K^+ channel in the basolateral membrane of the proximal tubule (Kubokawa et al., 1998; Nakamura et al., 2002) and principal cells of CCD (Hirsch & Schlatter, 1995). Although the effect of cGMP or PKG on the activity of other renal K^+ channels has not been examined, PKG would also be an important protein kinase in the regulation of channel activity.

As mentioned above, phosphorylated protein would be dephosphorylated by some phosphatases. The importance of some protein phosphatases in regulating channel activity has been demonstrated in neuronal NMDA receptor channels (L. Wang et al., 1994; Lieberman & Mody, 1994). Also, in renal tubules, the effects of a protein phosphatase on the activity of ATP-regulated K^+ channels have been demonstrated, as shown in Fig. 4C. In this experiment, PP-1 inhibited the enhanced channel activity brought about by PKA in the proximal tubule cells of the opossum kidney (OK) in an inside-out patch (Kubokawa et al., 2000). Moreover, PP-2A also has an inhibitory effect on channel activity in OK proximal tubule cells (Kubokawa et al., 2000). However, the ATP-regulated K^+ channel in the apical membrane of the principal cells of CCD was not inhibited by PP-1, but inhibited by PP-2A (Kubokawa et al., 1995a). Thus, the types of protein kinase and phosphatase affecting the activity of renal ATP-regulated K^+ channel are not always the same.

Roles of protein kinases and phosphatases in modulating the renal ATP-regulated K^+ channel are schematically shown in Fig. 5. Namely, as mentioned above, the ATP-regulated K^+ channels in the proximal tubule cells are activated by PKA- and PKG-mediated phosphorylation of the channel or associated protein. The apical ATP-regulated K^+ channel in TAL and the principal cell of CCD is activated by PKA, and the basolateral channel in the

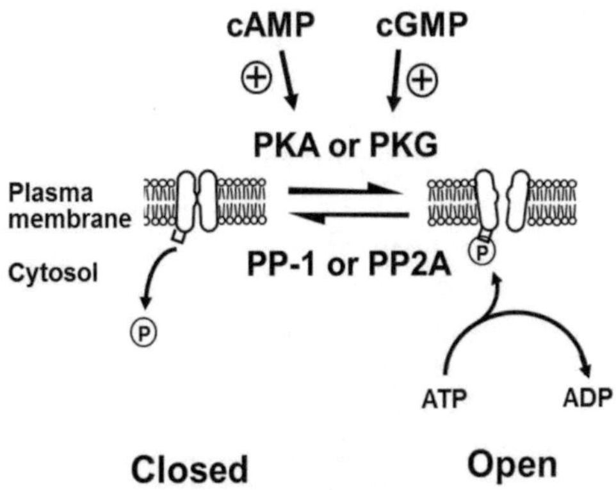

Fig. 5. A schematic representation of the roles of protein kinase-mediated phosphorylation and phosphatase-mediated dephosphorylation in modulation of the renal ATP-regulated K^+ channel. PKA- and/or PKG-mediated phosphorylation lead the channel to open, whereas PP-1 and/or PP-2A-mediated dephosphorylation promote the closed state of the channel.

principal cell is activated by PKG. Taken together, cAMP-dependent PKA or cGMP-dependent PKG induced phosphorylation which elevates the channel activity, and PP-1 or PP-2A induced dephosphorylation which lowers the channel activity. Thus, the phosphorylation and dephosphorylation induce the reversible alteration of channel activity.

3.3 Effects of Ca^{2+}-dependent protein kinase and phosphatase on the activity of renal ATP-regulated K^+ channels

It has been demonstrated that an increase in $[Ca^{2+}]_i$ suppresses the renal ATP-regulated K^+ channels, which is mainly induced by the activation of Ca^{2+}-dependent protein kinase C (PKC) in the proximal tubule cells (Mauerer et al., 1998; Mori et al., 2001) and the apical membrane of principal cells of CCD (W. Wang & Giebisch, 1991). Moreover, it was reported that Ca^{2+}/calmodulin-dependent protein kinase II (CaMKII) also suppressed the activity of the ATP-regulated K^+ channels in the basolateral membrane of the proximal tubule cells (Kubokawa et al., 2009) and the apical membrane of the principal cells in CCD (Kubokawa et al., 1995b). Indeed, the Ca^{2+}-induced K^+ channel inhibition was not blocked by PKC inhibitor alone, but was almost completely blocked by KN-62 (Tokumitsu et al., 1990), a CaMKII inhibitor, in addition to PKC inhibitor (Kubokawa et al., 1995b, 2009). Thus, it is possible that both PKC and CaMKII are involved in K^+ channel regulation.

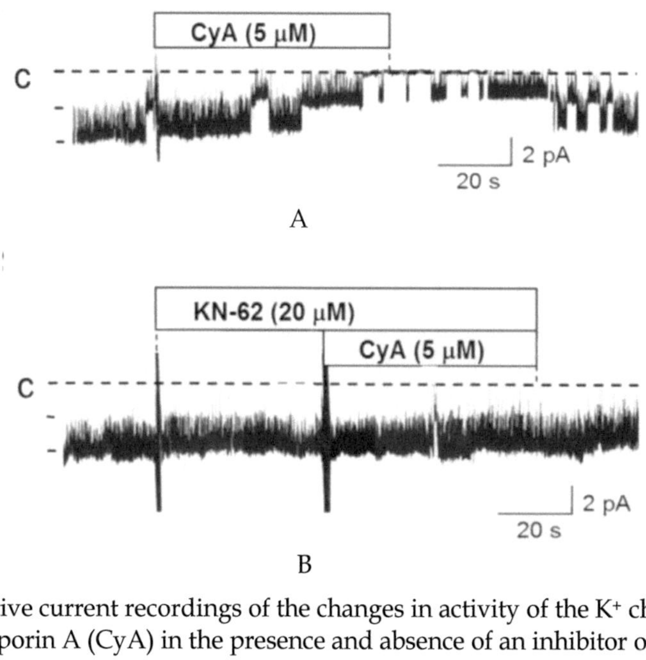

Fig. 6. Representative current recordings of the changes in activity of the K^+ channel in response to cyclosporin A (CyA) in the presence and absence of an inhibitor of CaMKII, KN-64, in cell-attached patches. Channel activity was suppressed by CyA alone (A), but the suppression was blocked in the presence of KN62 (B). [Modified from Kubokawa, M., Kojo. T., Komagiri Y. & Nakamura, K. (2009). Role of calcineurin-mediated dephosphorylation in modulation of an inwardly rectifying K^+ channel in human proximal tubule cells. *J. Membr. Biol.*, Vol. 231, No 2-3, pp. 79-92, ISSN: 0022-2631, with kind permission from Springer Science, Copyright 2009]

It has also been demonstrated that cyclosporin A (CyA), an inhibitor of Ca^{2+}/calmodulin-dependent phosphatase, calcineurin (CaN) (Hemenway & Heitman, 1999), inhibited the activity of the K^+ channel in both proximal tubule (Ye et al., 2006, Kubokawa et al., 2009) and principal cells of CCD (Ling & Eaton, 1993). Cyclosporin A is a a well-known immunosuppressive agent (Morris, 1981; White & Calne, 1982; Cohen, 1984), and affects various functions of the kidney other than K^+ channels (Epting et al., 2006; Tumlin, 1993; Grinyó & Cruzado, 2004; J. Wang et al., 2009; Damiano et al., 2010). Since the inhibition of CaN may induce a phosphorylation-dominant state, a Ca^{2+}/calmodulin-dependent kinase, such as CaMKII may be stimulated in this condition. This was supported by experiments using cultured human renal proximal tubule cells. That is, CyA-induced channel suppression was blocked by a CaMKII inhibitor, KN-62, as shown in Fig. 6 (Kubokawa et al., 2009). Moreover, Western blot analysis also revealed that CyA increased phospho-CaMKII, an active form of CaMKII (Gerges et al., 2005), as shown in Fig. 7 (Kubokawa et al., 2009). Taken together, these results strongly suggest that CaMKII-mediated phosphorylation suppresses channel activity, and CaN-mediated dephosphorylation reversibly elevates activity of the renal ATP-regulated K^+ channel. It was also demonstrated that CyA increased $[Ca^{2+}]_i$ in various cells, including renal tubule cells (Gordjani et al., 2000; Frapier et al., 2001; Bultynck et al., 2003; Kubokawa et al., 2009). Thus, the CyA-induced increase in $[Ca^{2+}]_i$ may enhance the activity of CaMKII.

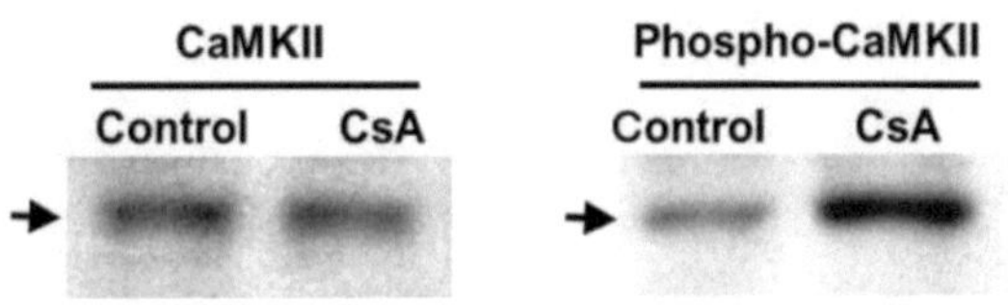

Fig. 7. Effects of CyA on CaMKII and phospho-CaMKII (Thr 286) in human kidney tubule cells. Western blot showing the effect of CyA (5 μM) on CaMKII detected with CaMKII antibody (A). CyA induced no appreciable change in CaMKII protein. Western blot showing the effect of CyA (5 μM) on phospho-CaMKII detected with a specific phospho-CaMKII antibody (Thr 286) (B). CyA clearly increased 50-kDa phospho-CaMKII (Thr 286). [Reprinted from Kubokawa, M., Kojo, T., Komagiri, Y. & Nakamura, K. (2009). Role of calcineurin-mediated dephosphorylation in modulation of an inwardly rectifying K^+ channel in human proximal tubule cells. *J. Membr. Biol.*, Vol. 231, pp. 79-92, ISSN: 0022-2631, with kind permission from Springer Science, Copyright 2009]

3.4 Possible relationship between CaMKII and CaN in modulation of the renal ATP-regulated K^+ channels

Protein kinase-mediated phosphorylation and phosphatase-mediated dephosphorylation are coupled mediators for the regulation of functional protein. It has been reported that CaN is an important phosphatase in regulating several cell functions (Rusnak & Merts, 2000). The above data suggest that CaMKII and CaN are coupled mediators in modulating the ATP-regulated K^+ channels. However, it has been reported that the target phosphorylation site which was dephosphorylated by CaN is mainly mediated by PKA (Santana et al., 2002). This report demonstrated that the functional coupling of CaN and PKA modulated Ca^{2+} release in ventricular myocytes (Santana et al., 2002). It has also been demonstrated that Na^+-K^+

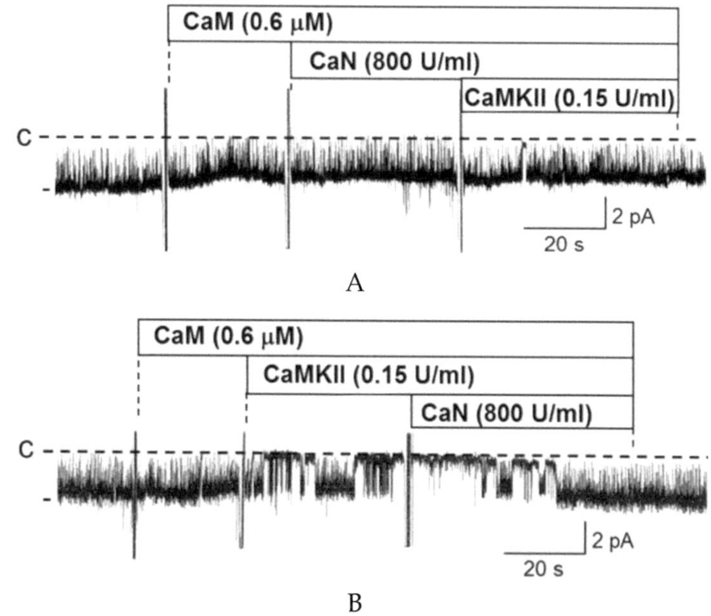

Fig. 8. Effects of CaMKII and CaN on activity of the ATP-regulated K^+ channel in the human proximal tubule cells in inside-out patches. [Reprinted from Kubokawa, M., Kojo, T., Komagiri, Y. & Nakamura, K. (2009). Role of calcineurin-mediated dephosphorylation in modulation of an inwardly rectifying K^+ channel in human proximal tubule cells. J. Membr. Biol., Vol. 231, pp. 79-92, ISSN: 0022-2631, with kind permission from Springer Science, Copyright 2009]

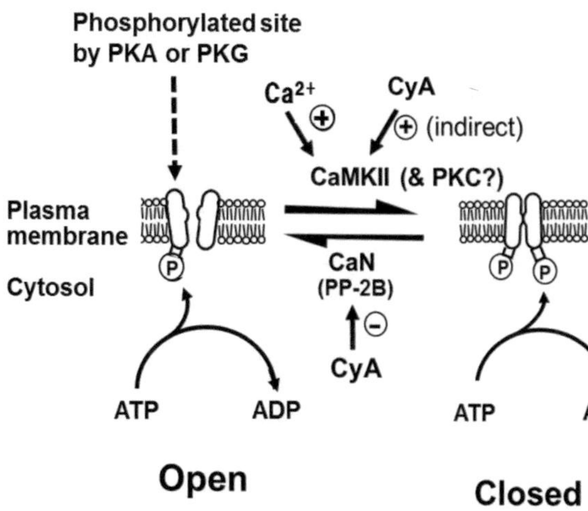

Fig. 9. A schematic representation of Ca^{2+} (/calmodulin)-dependent phosphorylation and dephosphorylation in modulation of the renal ATP-regulated K^+ channel. CaMKII- (PKC-) mediated phosphorylation leads the channel close, whereas CaN-mediated dephosphorylation causes the channel to open.

pump at the basolateral membrane of kidney tubular epithelia was inhibited by CyA (Tumlin & Sands, 1993) and stimulated by PKA (Carranza et al., 1998). On the other hand, a cardiac Na^+/Ca^{2+} exchanger was reported to be regulated by CaN and PKC (Shigekawa et al., 2007). A cellular process which depended on mitogen-activated protein kinase was reported to be negatively regulated by CaN (Tian & Karin, 1991). Thus, the protein kinases opposing CaN-mediated processes would not act in unity. Only a few reports have suggested that CaMKII-mediated processes are abolished by CaN (Wu et al., 2002; Gerges et al., 2005). Despite the variety of coupled mediators for CaMKII or CaN, experiments showing the effects of CaMKII on channel activity were abolished in the presence of CaN and that CaMKII-induced channel inhibition was restored by CaN in inside-out patches (Fig. 8) indicate the coupling of CaMKII and CaN in modulating the activity of the ATP-regulated K^+ channel. A model of the modulation of the ATP-regulated K^+ channel is schematically represented in Fig.9.

4. Conclusion

Among the many kinds of protein kinase, PKA- and/or PKG-mediated phosphorylation induces opening (an active state) of the renal ATP-regulated K^+ channel in proximal tubule cells (Kubokawa et al., 1997; Mauerer et al., 1998; Kubokawa et al., 1998), TAL (W. Wang,

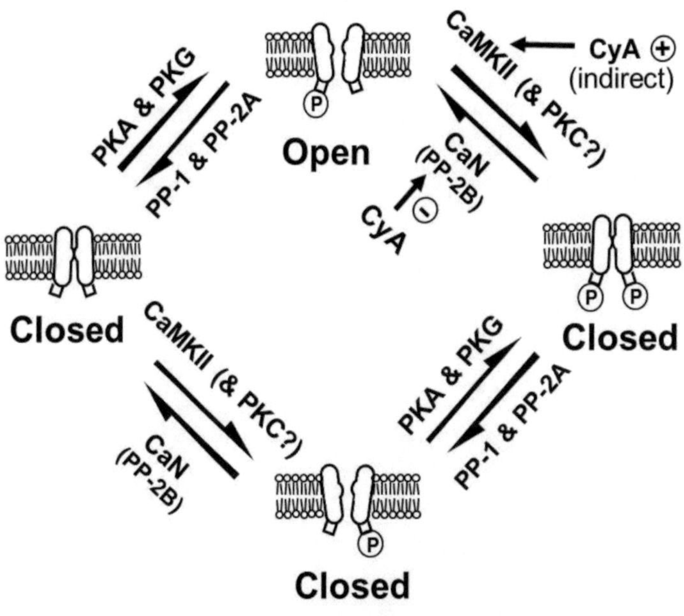

Fig. 10. A schematic representation of the 4-state phosphorylation model of the K^+ channel observed in the proximal tubule cells. CaN is often called protein phosphatase-2B (PP-2B). The circled "P" indicates phosphate. The circled "+" and " − " indicate stimulation and inhibition, respectively. [Reprinted from Kubokawa, M., Nakamura, K. & Komagiri, Y. (2011) Functional relationships between Ca^{2+}/calmodulin-dependent kinase and phosphatase in the regulation of K^+ channel activity and intracellular Ca^{2+} in kidney tubule cells. J. Iwate med. Assoc, Vol. 63, No4, pp. 209-218, with kind permission from Iwate Medical Association, Copyright 2011]

1994), and CCD principal cells (W. Wang & Giebisch, 1991; Kubokawa et al., 1995). In contrast, the open channels are closed by PP-1 and/or PP-2A (Kubokawa et al., 1995; Kubokawa et al., 2000), suggesting that PKA- or PKG-mediated phosphorylation was dephosphorylated by PP-1 or PP-2A. In addition, it is strongly suggested that CaMKII or PKC phosphorylates the other site, resulting in a closed state, while CaN-mediated dephosphorylation results in channel opening (Kubokawa et al., 2009). CyA directly inhibits CaN, and indirectly increases the active type of CaMKII, phospho-CaMKII, through the inhibition of CaN-mediated dephosphorylation processes (Kubokawa et al., 2009). A certain CaM-kinase kinase may be involved in the phosphorylation of CaMKII. Taken together, at least 2 phosphorylation sites are present at the cytosolic site of the channel or its associated protein, and individual phosphorylation sites are independently phosphorylated or dephosphorylated. Thus, the 4 different states of phosphorylation and dephosphorylation conditions would occur in the channel or the associated proteins in the regulatory processes of channel activity (Kubokawa et al., 2011). A putative 4-state model for the mechanism of K^+ channel regulation by phosphorylation and dephosphorylation is shown in Fig. 10. Although the channel state at the CaMKII-mediated phosphorylation alone is still unclear, these states would determine the channel activity at least in phosphorylation and dephosphorylation levels.

5. Acknowledgement

The author wishes to thank Drs. K. Nakamura and Y. Komagiri for their assistance in the preparation of the manuscript. This work was supported in part by a Grant-in-Aid for Scientific Research from the Japan Society for the Promotion of Science (to M. K., 23590264).

6. References

Beck, J. S.; Breton. S.; Mairbaurl, H.; Laprade, R. & Giebisch, G. (1991). Relationship between sodium transport and intracellular ATP in isolated perfused rabbit proximal convoluted tubule. *Am. J. Physiol.*, Vol. 261, No. 4, pp. F634-F639, ISSN: 0002-9513.

Bello-Reuss, E. & Weber, M. R. (1986). Electrophysiological studies on primary cultures of proximal tubule cells. *Am J Physiol.*, Vol. 251, No. 3, pp. F490-F498, ISSN: 0002-9513.

Boim, M. A.; Ho, K.; Shuck, M. E.; Bienkwski, M. J.; Block, J. H.; Slightom, J. L.; Yang, Y.; Brenner, B. M. & Hebert, S. C. (1995). ROMK inwardly rectifying ATP-sensitive K^+ channel. II. Cloning and distribution of alternative forms. *Am. J. Physiol.*, Vol. 268, No. 6, pp. F1132-F1140, ISSN: 0002-9513.

Bultynck, G.; Vermassen. E.; Szlufcik, K.; De Smet, P.; Fissore, R. A.; Callewaert, G.; Missiaen, L.; De Smedt, H. & Parys, J. B. (2003) Calcineurin and intracellular Ca^{2+}-release channels: regulation or association. *Biochem. Biophys. Res. Commun.*, Vol. 311, No. 4, pp. 1181-1193, ISSN: 0006-291X.

Carranza, M. L.; Rousselot, M.; Chibalin, A. V.; Bertorello, A. M.; Favre, H. & Feraille, E. (1998). Protein kinase A induces recruitment of active Na^+, K^+-ATPase units to the plasma membrane of rat proximal convoluted tubule cells. *J. Physiol. (Lond.).*, Vol. 511, No. 1, pp. 235-243, ISSN: 0022-3751.

Cohen, D. J.; Loertscher, R.; Rubin, M. F.; Tilney, N. L.; Carpenter, C. B. & Strom, T. B. (1984). Cyclosporine: a new immunosuppressive agent for organ transplantation. *Ann. Intern. Med.*, Vol. 101, No. 5, pp. 667-682, ISSN: 0003-4819

Damiano, S.; Scanni, R.; Ciarcia, R.; Florio, S. & Capasso, G. (2010). Regulation of sodium transporters in the kidney during cyclosporine treatment. *J. Nephrol.*, Vol. 23, No. 16, pp. S191-S198, ISSN: 1121-8428.

Derst, C.; Hirsch, J. R.; Preisig-Muller, R.; Wischmeyer, E.; Karschin, A.; Doring, F.; Thomzig, A.; Veh, R.W.; Schlatter, E.; Kummer, W. & Daut, J. (2001). Cellular localization of the potassium channel Kir7.1 in guinea pig and human kidney. *Kidney Int.*, Vol. 59, No. 6, pp. 2197-205, ISSN: 0085-2538.

Donella-Deana, A.; Krinks, M. H.; Ruzzene, M.; Klee, C. & Pinna, L. A. (1994). Dephosphorylation of phosphopeptides by calcineurin (protein phosphatase 2B). *Eur. J. Biochem.*, Vol. 219, No. 1-2, pp. 109-17, ISSN 0014-2956.

Edelman, A.; Garabedian, M. & Anagnostopoulos, T. (1986). Mechanisms of $1,25(OH)_2D_3$-induced rapid changes of membrane potential in proximal tubule: role of Ca^{2+}-dependent K^+ channels. *J. Membr. Biol.*, Vol. 90, No. 2, pp. :137-143, ISSN: 0022-2631.

Epting, T.; Hartmann, K.; Sandqvist, A.; Nitschke, R. & Gordjani, N. (2006). Cyclosporin A stimulates apical Na^+/H^+ exchange in LLC-PK1/PKE20 proximal tubular cells. *Pediatr. Nephrol.*, Vol. 21, No. 7, pp. 939-946, ISSN: 0931-041X..

Frapier, J. M.; Choby, C.; Mangoni, M.; Nargeot, J.; Albat, B. & Richard, S.(2001). Cyclosporin A increases basal intracellular calcium and calcium responses to endothelin and vasopressin in human coronary myocytes. *FEBS Lett.*, Vol. 493, No. 1, pp. :57-62, ISSN: 0014-5793.

Fujimoto, M.; Kubota, T.; Hagiwara, N.; Ohno-Shosaku, T.; Kubokawa, M. & Kotera, K. (1991). Ion-selective microelectrodes to study proton and bicarbonate transport in the renal epithelium. *Kidney Int.*, Suppl., Vol. 33, pp. S23-528, ISSN: 0098-6577.

Gennari, F. J. & Maddox, D. A. (2000). Renal regulation of acid-base homeostasis: integrated response. In: *The kidney*, Seldin D. W., & Giebisch. G (Eds:), pp. 2015-2053, Lippincott Williams & Wilkins, ISBN: 0-397-58784-8, Philadelphia.

Gerges, N.Z.; Alzoubi, K. H. & Alkadhi, K. A. (2005). Role of phosphorylated CaMKII and calcineurin in the differential effect of hypothyroidism on LTP of CA1 and dentate gyrus. *Hippocampus*, Vol. 15, No. 4, pp. 480-490, ISSN: 1050-9631.

Giebisch, G. (1995). Renal potassium channels: an overview. *Kidney Int.* Vol. 48, No. 4, pp. 1004-1009, ISSN: 0085-2538.

Giebisch, G. (1998). Renal potassium transport: mechanisms and regulation. *Am. J. Physiol.*, Vol. 274, No. 5, pp. F817-F833, ISSN: 0002-9513.

Gordjani, N.; Epting, T.; Fischer-Riepe, P.; Greger, R. F.; Brandis, M.; Leipziger, J. & Nitschke R. (2000). Cyclosporin-A-induced effects on the free Ca^{2+} concentration in LLC-PK1-cells and their mechanisms. *Pflügers Arch.*, Vol. 439, No. 5, pp. 627-33, ISSN: 0031-6768.

Grinyó, J. M. & Cruzado, J. M. (2004). Cyclosporine nephrotoxicity. *Transplant Proc.*, Vol. 36, No. 2 (Suppl.), pp. 240S-242S, ISSN: 0041-1345.

Guggino, S. E.; Guggino, W. B.; Green, N. & Sacktor, B. (1987). Ca^{2+}-activated K^+ channels in cultured medullary thick ascending limb cells. *Am. J. Physiol.*, Vol. 252, No. 2, pp. C121-C127, ISSN: 0002-9513.

Gögelein, H. (1990). Ion channels in mammalian proximal renal tubules. *Renal Physiol. Biochem.*, Vol. 13, No. 1-2, pp. 8-25, ISSN: 1011-6524.

Hagiwara, N.; Kubota, T.; Kubokawa, M. & Fujimoto, M. (1990). Effects of dopamine on the transport of Na, H, and Ca in the bullfrog proximal tubule. *Jpn. J. Physiol.*, Vol. 40, No. 3, pp. 351-368, ISSN: 0021-521X.

Hamill, O. P.; Marty, A.; Neher, E.; Sakmann, B. & Sigworth, F. J. (1981). Improved patch-clamp techniques for high-resolution current recording from cells and cell-free membrane patches. *Pflugers Arch.*, Vol. 391, No. 2, pp. 85-100, ISSN: 0031-6768.

Hebert, S. C.; Desir, G.; Giebisch, G. & Wang, W. (2005). Molecular diversity and regulation of renal potassium channels. *Physiol. Rev.*, Vol. 85, No. 1, pp. 319-71, ISSN: 0031-9333.

Hemenway, C. S. & Heitman, J. (1999). Calcineurin. Structure, function, and inhibition. *Cell Biochem. Biophys.*, Vol. 30, No. 1, pp. 115-151, ISSN: 1085-9195.

Hirano, J.; Nakamura, K. & Kubokawa, M. (2001). Properties of a Ca^{2+}-activated large conductance K^+ channel with ATP sensitivity in human renal proximal tubule cells. *Jpn. J. Physiol.*, Vol. 51, No. 4, pp. 481-489, ISSN: 0021-521X.

Hirsch, J.; Leipziger, J.; Fröbe, U. & Schlatter, E. (1993). Regulation and possible physiological role of the Ca^{2+}-dependent K^+ channel of cortical collecting ducts of the rat. *Pflügers Arch.*, Vol. 422, No. 5, pp. 492-498, ISSN: 0031-6768.

Hirsch, J. & Schlatter, E. (1995). K^+ channels in the basolateral membrane of rat cortical collecting duct. *Kidney Int.*, Vol. 48, No. 4, pp. 1036-1046, ISSN: 0085-2538.

Hirsch, J. R.; Weber, G.; Kleta, I. & Schlatter, E. (1999). A novel cGMP-regulated K^+ channel in immortalized human kidney epithelial cells (IHKE-1). *J. Physiol. (Lond.).*, Vol. 519, No. 3, pp. 645-655, ISSN: 0022-3751.

Ho, K.; Nichols, C. G.; Lederer, W. J.; Lytton, J.; Vassilev, P. M;. Kanazirska, M. V. & Hebert, S. C. (1993). Cloning and expression of an inwardly rectifying ATP-regulated potassium channel. *Nature* Vol. 362, No. 6415, Mar. pp. 31-38, ISSN: 0028-0836.

Kawahara, K. & Anzai, N. (1997). Potassium transport and potassium channels in the kidney tubules. *Jpn. J. Physiol.*, Vol. 47, No. 1, pp. 1-10, ISSN: 0021-521X.

Kubokawa, M.; Kubota, T. and Fujimoto, M: (1990). Potassium permeability of luminal and peritubular membranes in the proximal tubule of bullfrog kidneys. *Jpn J Physiol.* Vol. 40, No.5 :613-634, ISSN 0021-521X.

Kubokawa, M.; McNicholas, C. M.; Higgins, M. A.; Wang, W. & Giebisch, G. (1995a). Regulation of ATP-sensitive K^+ channel by membrane-bound protein phosphatases in rat principal tubule cell. *Am. J. Physiol.*, Vol. 269, No. 3, pp. F355-F362, ISSN: 0002-9513.

Kubokawa, M.; Wang, W.; McNicholas, C. M. & Giebisch, G. (1995b). Role of Ca^{2+}/CaMK II in Ca^{2+}-induced K^+ channel inhibition in rat CCD principal cell. *Am. J. Physiol.*, Vol. 268, No. 2, pp. F211-229, ISSN: 0002-9513.

Kubokawa, M.; Mori.; Y. & Kubota, T. (1997). Modulation of inwardly rectifying ATP-regulated K^+ channel by phosphorylation process in opossum kidney cells. *Jpn. J. Physiol.*, Vol. 47, No. 1, pp. 111-119, ISSN: 0021-521X.

Kubokawa, M.; Mori, Y.; Fujimoto, K. & Kubota, T. (1998). Basolateral pH-sensitive K^+ channels mediate membrane potential of proximal tubule cells in bullfrog kidney. *Jpn. J. Physiol.*, Vol. 48, No. 1, pp. :1-8, ISSN : 0021-521X.

Kubokawa, M.; Nakaya, S.; Yoshioka, Y.; Nakamura, K.; Sato, F.; Mori, Y. & Kubota, T. (1998). Activation of inwardly rectifying K^+ channel in OK proximal tubule cell involves cGMP-dependent phosphorylation process. *Jpn. J. Physiol.*, Vol. 48, No. 6, pp. 467-76, ISSN: 0021-521X.

Kubokawa, M.; Nakamura, K.; Hirano, J.; Yoshioka, Y.; Nakaya, S.; Mori, Y. & Kubota, T. (2000). Regulation of inwardly rectifying K^+ channel in opossum proximal tubule cells by protein phosphatases 1 and 2A. *Jpn. J. Physiol.*, Vol. 50, No. 2, Apr. pp. 249-56, ISSN: 0021-521X.

Kubokawa, M.; Kojo. T.; Komagiri Y. & Nakamura, K. (2009). Role of calcineurin-mediated dephosphorylation in modulation of an inwardly rectifying K^+ channel in human proximal tubule cells. *J. Membr. Biol.*, Vol. 231, No 2-3, pp. 79-92, ISSN: 0022-2631.

Kubokawa, M.; Nakamura, K. & Komagiri, Y. (2011). Interaction between calcineurin and Ca^{2+}/calmodulin kinase-II in modulating cellular functions. *Enzyme Res.*, DOI:10.4061/2011/587359.

Kubota, T.; Biagi, B. A. & Giebisch, G. (1983). Effects of acid-base disturbances on basolateral membrane potential and intracellular potassium activity in the proximal tubule of Necturus. *J. Membr. Biol.*, Vol. 73, No. 1, pp. :61-68, ISSN: 0022-2631.

Levitan, I. B. (1994). Modulation of ion channels by protein phosphorylation and dephosphorylation. *Annu. Rev. Physiol.*, Vol. 56, pp.193-212, ISSN: 0066-4278.

Lieberman, D. N. & Mody, I. (1994). Regulation of NMDA channel function by endogenous Ca^{2+}-dependent phosphatase, *Nature*, Vol. 369, No. 6477, pp. 235-239, ISSN: 0028-0836.

Ling, B. N. & Eaton, D. C. (1993). Cyclosporin A inhibits apical secretory K^+ channels in rabbit cortical collecting tubule principal cells. *Kidney Int.*, Vol. 44, No. 5, pp. 974-984, ISSN: 0085-2538.

Malnic, G.; Muto, S. & Giebisch, G. (2000). Regulation of potassium excretion. In: *The kidney*, Seldin D.W., & Giebisch, G. (Eds:), pp. 1575-1613, Lippincott Williams & Wilkins, ISBN: 0-397-58784-8, Philadelphia.

Matifat F. Hague F. Brule G. & Collin T. (2001). Regulation of InsP3-mediated Ca^{2+} release by CaMKII in Xenopus oocytes. *Pflügers Arch.*, Vol. 441, No. 6, Mar. pp. 796-801, ISSN: 0031-676.

Mauerer, U. R.; Boulpaep, E. L. & Segel, A. S. (1998a). Properties of an inward rectifying ATP-sensitive K^+ channel in the basolateral membrane of renal proximal tubule. *J. Gen. Physiol.*, Vol. 111, No. 1, pp. 139-160, ISSN: 0022-1295.

Mauerer, U. R.; Boulpaep, E. L. & Segel, A. S. (1998b). Regulation of an inward rectifying ATP-sensitive K^+ channel in the basolateral membrane of renal proximal tubule. *J. Gen. Physiol.*, Vol. 111, No. 1, pp. 161-180, ISSN: 0022-1295

McNicholas, C M.; Wang, W.; Ho, K.; Hebert, S C. & Giebisch, G. (1994) Regulation of ROMK1 K^+ channel activity involves phosphorylation processes. *Proc. Natl. Acad. Sci. U S A.* Vol. 91, No. 17, pp. 8077-8081, ISSN 0027-8424.

McNicholas, C. M.; Yang, Y.; Giebisch, G. & Hebert, S. C. (1996). Molecular site for nucleotide binding on an ATP-sensitive renal K^+ channel (ROMK2). *Am. J. Physiol. Renal Physiol.*, Vol. 271, No. 2, pp. F275-F285, ISSN: 0002-9513.

Merot, J.; Bidet, M.; Lemaout, S.; Tauc, M. & Poujeol, P. (1989). Two types of K^+ channels in the apical membrane of rabbit proximal tubule in primary culture. *Biochim. Biophys. Acta.*, Vol. 978, No. 1, pp. :134-144, ISSN: 0006-3002.

Mori, Y.; Kawasaki, A.; Takamaki, A.; Kitano, I.; Yoshida, R.; Kubokawa, M. & Kubota, T. (2001). Ca^{2+}-dependent inhibition of inwardly rectifying K^+ channel in opossum kidney cells *Jpn. J. Physiol.*, Vol. 51, No. 3, pp. 371-380, ISSN: 0021-521X.

Morris, P. J. (1981). Cyclosporin A. *Transplantation.* Vol. 32, No. 5, Nov. pp. 49-54, ISSN: 0041-1337.

Naesens, M.; Kuypers, D. R. & Sarwal, M. (2009). Calcineurin inhibitor nephrotoxicity. *Clin. J. Am. Soc. Nephrol.*, Vol. 4, No. 2, Feb. pp. 481-508, ISSN: 1555-905X.

Nakamura, K.; Hirano, J. & Kubokawa M. (2001). An ATP-regulated and pH-sensitive inwardly rectifying K^+ channel in cultured human proximal tubule cells. *Jpn. J. Physiol.*, Vol. 51, No. 4, pp. 523-30, ISSN: 0021-521X.

Nakamura, K.; Hirano, J.; Itazawa, S. & Kubokawa, M. (2002). Protein kinase G activates inwardly rectifying K^+ channel in cultured human proximal tubule cells. *Am. J. Physiol. Renal Physiol.*, Vol. 283, No. 4, pp. F784-F791, ISSN: 1931-857X.

Ohno-Shosaku, T.; Kubota, T.; Yamaguchi, J.; Fukase, M.; Fujita, T. & Fujimoto, M. (1989). Reciprocal effects of Ca^{2+} and Mg-ATP on the 'run-down' of the K^+ channels in opossum kidney cells. *Pflugers Arch.* Vol. 413, No. 5, pp: 562-564, ISSN: 0031-6768.

Ohno-Shosaku, T.; Kubota, T.; Yamaguchi, J. & Fujimoto, M. (1990). Regulation of inwardly rectifying K^+ channels by intracellular pH in opossum kidney cells. *Pflügers Arch.*, Vol. 416, No. 1-2, pp. 138-143, ISSN: 0031-6768.

Palmer, L.G. (1992). Renal potassium channels. In: *Handbook of Physiology: Renal Physiology*, Windhager E. E. (Ed:), pp. 715-738, Oxford University Press, ISBN: 0-19-506006-7, New York.

Palmer, L.G. & Carty, H. (2000). Epihtelial Na^+ channels. In: *The kidney, Physiology and pathophysiology*, Seldin D.W., & Giebisch, G. (Eds:), pp. 1575-1613, Lippincott Williams & Wilkins, ISBN: 0-397-58784-8, Philadelphia.

Quast, U. (1996). ATP-sensitive K^+ channels in the kidney. *Naunyn Schmiedebergs Arch Pharmacol.*, Vol. 354, No. 3, pp. 213-25, ISSN: 0028-1298.

Robson, L. & Hunter, M. (1997). Two K(+)-selective conductances in single proximal tubule cells isolated from frog kidney are regulated by ATP. *J. Physiol. (Lond.).*, Vol. 500, No. 3, pp. 605-616, ISSN: 0022-3751.

Rusnak, F. & Merts, F. (2000). Calcineurin: Form and function. *Physiol. Rev.*, Vol. 80, No. 4, Oct. pp. 1483-521, ISSN 0031-9333.

Santana, L F.; Chase, E.G.; Votaw. V.S.; Nelson, M. T. & Greven, R. (2002). Functional coupling of calcineurin and protein kinase A in mouse ventricular myocytes. *J. Physiol. (Lond.).*, Vol. 544, No. 1, pp. 57-69, ISSN: 0022-3751.

Shigekawa, M.; Katanosaka, Y. and Wakabayashi, S. (2007). Regulation of the cardiac Na^+/Ca^{2+} exchanger by calcineurin and protein kinase C. *Ann N Y Acad Sci.*, Vol. 1099, pp. 53-63, ISSN: 0077-8923.

Stoner, L.C. and Morley, G. E. (1995). Effect of basolateral or apical hyposmolarity on apical maxi K channels of everted rat collecting tubule. *Am J Physiol.*, Vol. 268, No. 4, pp. F569-F680, ISSN: 0002-9513.

Taniguchi, J. and Imai, M. (1998). Flow-dependent activation of maxi K^+ channels in apical membrane of rabbit connecting tubule. *J. Membr. Biol.*, Vol. 164, No. 1, pp. 35-45, ISSN: 0022-2631.

Tian, J. & Karin, M . (1991). Stimulation of Elk1 transcriptional activity by mitogen-activated protein kinases is negatively regulated by protein phosphatase 2B (calcineurin). *J. Biol. Chem.*, Vol. 274, No. 21, pp. 15173-15180, ISSN: 0021-9258.

Tokumitsu, H.; Chijiwa, T.; Hagiwara, M.; Mizutani, A.; Terasawa, M. & Hidaka, H. (1990). KN-62, 1-[N,O-bis(5-isoquinolinesulfonyl)-N-methyl-L-tyrosyl]-4-phenylpiperazine, a specific inhibitor of Ca^{2+}/calmodulin-dependent protein kinase II. *J. Biol. Chem.*, Vol. 265, No. 8, pp. 4315-4320, ISSN: 0021-9258.

Tsuchiya, K.; Wang, W.; Giebisch, G. & Welling, P. A. (1992). ATP is a coupling modulator of parallel Na, K-ATPase-K-channel activity in the renal proximal tubule. *Proc. Natl. Acad. Sci. USA*, Vol. 89, No. 14, pp. 6418-6422, ISSN: 0027-8424.

Tumlin, J. A. (1997). Expression and function of calcineurin in the mammalian nephron: Physiological roles, receptor signaling, and ion transport. *Am. J. Kidney. Dis.*, Vol. 30, No. 6, pp. 884-895, ISSN: 0272-6386.

Tumlin, J. A. & Sands, J. M. (1993). Nephron segment-specific inhibition of Na^+/K^+-ATPase activity by cyclosporin A. *Kidney Int.*, Vol. 43, No. 1, pp. 246-251, ISSN: 0085-2538.

Wang. J.; Zhang. Z. R.; Chou. C. F.; Liang, Y. Y.; Gu. Y. & Ma, H. P. (2009) . Cyclosporine stimulates the renal sodium channel by elevating cholesterol. *Am. J. Physiol. Renal Physiol.*, Vol. 296, No. 2, pp. F284-F290, ISSN: 0363-6127

Wang L. Y.; Orser, B. A.; Brautigan, D. L. & MacDonald. (1994). Regulation of NMDA receptors in cultured hippocampal neurons by protein phosphatases 1 and 2A. *Nature* Vol. 369, No. 6477, pp. 230-232, ISSN: 0028-0836.

Wang, W. & Giebisch, G. (1991b). Dual effect of adenosine triphosphate on the apical small conductance K^+ channel of the rat cortical collecting duct. *J. Gen. Physiol.* Vol. 98, No. 1, pp. 35-61, ISSN 0022-1295.

Wang, W. H. (1994). Two types of K^+ channel in thick ascending limb of rat kidney. *Am. J. Physiol. Renal Physiol.*, Vol. 267, No. 4, pp. F599-F605, ISSN: 0002-9513.

Wang. W.; Hebert, S. C. & Giebisch, G. (1997). Renal K^+ channels: Structure and function. *Annu. Rev. Physiol.*, Vol. 59, pp. 413-436, ISSN: 0066-4278.

Wang, W. H & Giebisch, G. (1991a). Dual modulation of renal ATP-sensitive K^+ channel by protein kinases A and C. *Proc. Natl. Acad. Sci. USA*, Vol. 88, No. 21, pp. 9722-9725, ISSN: 0027-8424.

Weinstein, A. M. (2000). Sodium and chloride transport: proximal nephron. In: *The kidney*, Seldin D.W., and Giebisch. G (Eds.), pp. 1286-1331, Lippincott Williams & Wilkins, ISBN: 0-397-58784-8, Philadelphia.

Welling, P.A. (1995). Cross-talk and the role of K_{ATP} channels in the proximal tubule. *Kidney Int.*, Vol. 48, No. 4, pp. 1017-23, ISSN: 0085-2538.

White, D. J. & Calne, R. Y. (1982). The use of cyclosporin A immunosuppression in organ grafting. *Immunol. Rev.*, Vol. 65, pp. 115-131. ISSN: 0105-2896.

Winters, C. J.; Reeves, W. B. & Andreoli, T. E. (1999). Cl^- channels in basolateral TAL membranes. XIV. Kinetic properties of a basolateral MTAL Cl^- channel. *Kidney Int.* Vol. 55, No. 4, pp. 1444-1449, ISSN 0085-2538.

Woda, C B.; Bragin, A.; Kleyman, T. R. & Satlin, L M. (2001). Flow-dependent K^+ secretion in the cortical collecting duct is mediated by a maxi-K channel. *Am J Physiol Renal Physiol.* Vol. 280, No.5, pp: F786-793, ISSN: 1931-857X.

Wu, H.; Kanatous, S. B.; Thurmond, F. A.; Gallardo. T.; Isotani, E.; Bassel-Duby, R. & Williams, R. S. (2002) Regulation of mitochondrial biogenesis in skeletal muscle by CaMK. *Science*, Vol. 296, No 5566, pp. 349-52, ISSN: 1095-9203.

Ye, B.; Liu, Y. & Zhang. Y. (2006). Properties of a potassium channel in the basolateral membrane of renal proximal convoluted tubule and the effect of cyclosporine on it. *Physiol. Res.*, Vol. 55, No. 6, pp. 617-622, ISSN: 0862-8408

6

BK Channels – Focus on Polyamines, Ethanol/Acetaldehyde and Hydrogen Sulfide (H_2S)

Anton Hermann[1], Guzel F. Sitdikova[2] and Thomas M. Weiger[1]

[1]University of Salzburg, Department of Cell Biology, Division of Cellular and Molecular Neurobiology, Salzburg

[2]Kazan Federal University, Department Physiology of Man and Animals, Kazan

[1]Austria

[2]Russia

1. Introduction

Calcium (Ca^{2+})-activated potassium (K^+) channels are activated by the synergistic action of voltage as well as by Ca^{2+} which links these channels to cell metabolism. Because of their high level of functional diversity the channels are widely expressed in a remarkable amount of different cells from bacteria to men and found in a great variety of tissues such as sensory, muscle, vascular or the brain. The channels are among the most frequently studied K^+ channels giving rise to an impressive amount of knowledge about their structure and function. The idea of a Ca^{2+}-activated conductance was born in 1958 during studies on erythrocytes by Gardos (1958) who showed that metabolically deprived cells in the presence of internal Ca^{2+} augment the permeability of the cell plasma membrane to K^+ ions. The finding was further elaborated by direct injection of Ca^{2+} ions into mollusc neurons (Meech & Standen 1975; Gorman & Hermann 1979) which supported the idea of a Ca^{2+}- and voltage dependent membrane K^+ conductance and showed that it is also present in excitable cells. Up to present Ca^{2+}-activated K^+ conductances were and still are studied in great detail concerning their biophysical, physiological, pathophysiological, pharmacological, structural and functional properties (for early and recent reviews see Meech 1978; Hermann & Hartung 1983; Latorre et al.. 1989; Kaczorowski et al. 1996; Gribkoff, et al. 2001; Jiang et al., 2001; Weiger et al. 2002; Calderone 2002; Jiang et al., 2002) Maher & Kuchel 2003; Salkoff et al. 2006; Pluznick & Sansom 2006; Cui et al. 2009; Wu et al. 2010; Lee & Cui 2010; Grimm & Sansom 2010; Hill et al. 2010; Berkefeld et al. 2010; Cui 2010). In the first sections of this chapter after we briefly describe techniques to record BK channels we review some properties of BK channels which appeared important in the context of our further presentations.

Ethanol is produced by the cell metabolism and is generally known as one of the most ancient and most ubiquitous psychoactive drugs consumed by humans. There are myriads of publications on the effects of alcohol on body functions, behavior, social interactions or cancer genesis. Research progressed rapidly in the field and scientists are vividly collecting data on the effects of alcohol and we experience growing understanding on the cellular level of some processes involved, however, many of its molecular mechanisms of action still remain elusive. We will review some aspects of the effects of ethanol as well as acetaldehyde - its first metabolite – on BK channels.

Polyamines (putrescine, spermidine and spermine), are simple molecules present in all eucaryotic cells. They have a wide array of functions from modulating ion channels, involvement in apoptosis and carcinogenicity and are required in cell proliferation and development. The Ca^{2+}-activated K^+ conductance was among the first to be reported being modulated by polyamines. We will briefly review the latest development in the field and cover the molecular mechanisms on polyamine interaction with BK channels.

Hydrogen sulfide (H_2S) is the third gasotransmitter discovered in brain next to nitric oxide and carbon monoxide. While H_2S is already well known to modulate ion channels, it was only recently discovered to also modulate BK channels. In the last section of this chapter we will briefly focus on this relatively new field in BK channel physiology.

2. Technical aspects of BK channel recordings

Due to their huge conductivity of 100 - 300 pikoSiemens (pS) BK channels are easily visible and discernible from other ion channels in single channel recordings. Since BK channels are well known to be asymmetric, i.e. drugs may act from the intracellular but not from the extracellular side, it is important to investigate BK channels in the inside out as well as in the outside out patch clamp mode. Choosing a model such as Chinese hamster ovarian (CHO) cells transfected with BK channels, inside out patches will allow to record macroscopic currents instead of single channels due to the huge number of channels expressed in a patch which add up to a macroscopic current. A good model for outside out single channel recordings are in our hands the GH3/GH4 cell lines from rat pituitary tumor cells. BK channels can be recorded in two different solution settings: a) a solution system which recalls the physiological situation with 3 - 5 milliMolar (mM) KCl in the extracellular bath and 100 - 145 mM KCl at the intracellular side, or b) in a more biophysical approach where a symmetric solution system with equal amounts of potassium (100 - 150 mM KCl) at either side of the membrane is used. The latter approach has been adopted by many researchers reported in the more recent literature. Since BK channels are Ca^{2+} sensitive a great deal of attention has to be paid to the Ca^{2+} concentration in the solution facing the intracellular side. Ca^{2+} has to be buffered and the resulting so called "free Ca^{2+} concentration" needs to be carefully adjusted according to the demands of the experiment. The Ca^{2+} concentration in a Ca^{2+} buffered solution reported as free Ca^{2+} contains only a fraction of the total Ca^{2+}. Depending on the buffer used the free Ca^{2+} concentration can be calculated using an online calculator (http://www.stanford.edu/~cpatton/webmaxcS.htm). Other Ca^{2+} buffering substances like magnesium or ATP have to be taken into account in these calculations. The best practice, however, is to finally measure the free Ca^{2+} concentration in the ready to use prepared solution with a Ca^{2+} sensitive electrode. At very low intracellular Ca^{2+} concentrations (below 1 microMolar (μM) free Ca^{2+}) and to remove potential other metal ion contaminants, solutions shall be passed over a Chelex 100 (BioRad) ion exchange column, prior to adding Ca^{2+} buffers and divalent ions (Erxleben et al. 2002). Low Ca^{2+} concentrations are in a range below 1 μM, while high Ca^{2+} concentrations for the BK channels are in a range of 10 - 100 μM free Ca^{2+}, depending on the type of BK Channel used.

The free Ca^{2+} concentration employed also determines the buffer to be used. BAPTA and EGTA are the best choice for low free Ca^{2+} concentration while HEDTA would be chosen for higher Ca^{2+} concentrations (Patton et al. 2004). It is good advice not to use these buffer systems at the edge of their buffer capacity since any additional Ca^{2+}, which may result for instance in whole cell recordings from Ca^{2+} influx by activation of Ca^{2+} channels, may not be buffered anymore and hence alter BK channel activity. Also higher concentrations of the Ca^{2+} buffer substance used like 10 mM are more favourable than low concentrations to make the system more stable. In addition small mistakes in balancing the salts for the solution or a sloppy adjustment of the pH can have serious consequences for the buffer range. Therefore great care has to be taken in the preparation of solutions and a freshly calibrated pH meter may help to adjust the free Ca^{2+} concentrations precisely. Ca^{2+} buffers can be of the fast type using BAPTA or of the slow type using EGTA. Fast buffers have the advantage that any input of additional Ca^{2+} from Ca^{2+} channels or a release of internal Ca^{2+} will not be sensed by the channel. Slow buffers like EGTA may be exceeded by the fast appearance of high amounts of Ca^{2+} but keep the overall Ca^{2+} concentration constant. For more information which Ca^{2+} buffer to use and how to calculate the free Ca^{2+} concentration see (Bers et al. 2010; Patton et al. 2004).

BK channels are located frequently in clusters in the cell membrane. This makes it sometimes almost impossible to obtain a patch with just a single channel. A way to work around this and to minimize the number of channels is to decrease the orifice of the tip of the patch electrode which increases the patch pipette resistance up to 5 - 6 MegaOhm. Indication that only one channel is in the patch, which is important for instance for kinetic analysis, can be obtained by increasing the Ca^{2+} concentration in the solution or by increasing the voltage to positive values and make sure that only one channel is observed. A good starting point to record single BK channels is to use a free Ca^{2+} concentration of 1 μM at a voltage of +30 mV. Submillimolar concentrations of tetraethylammonium (TEA) may be used as a low cost drug to block BK channels in initial experiments. To further specify the channel specific BK channel blockers such as iberiotoxin or paxilline shall be used.

3. Ca^{2+} activated K^+ channels

Ca^{2+}-activated K^+ channels channels are found in a great variety of excitable and non-excitable cells. The channels are broadly divided into three subfamilies mainly defined by their biophysical and pharmacological properties (Wei et al., 2005). In this chapter we will focus on the big (large or maxi conductance) K^+ channels (BK) which are also termed $K_{Ca}1.1$ or KCNM (gene name). The channels are also known as Slo1 channels - for "Slowpoke", a gene that was first cloned from the fruit fly *Drosophila* (Atkinson et al. 1991) and has since been cloned from a variety of organisms (Adelman et al. 1992; Salkoff et al., 2006). The channels are activated usually by both metal ions (Ca^{2+}/Mg^{2+}) and by membrane voltage synergistically, but can also be activated by either Ca^{2+}/Mg^{2+} or by voltage alone. In the absence of Ca^{2+} the channels require extremely large depolarization for activation (+100 to +200 mV). Some details of BK channels which bear relevance to the following section on ethanol/acetaldehyde, polyamines and H_2S are highlighted below.

3.1 BK channel properties

BK channels have a tetrameric structure with four independent alpha (α)-subunits containing the functional channel pore. The α-subunit subunit is a large protein of about

1,200 amino acids. Each BK channel α-subunit consists of a total of seven transmembrane segments with a unique S0 segment that precedes the usually six transmembrane segments (S1-S6). The total of seven segments (S0-S6) renders the N-terminus (amino terminal) at the extracellular side of the membrane (Meera et al., 1997). Multiple splice variants of the α-subunit have been identified resulting in a great variety of channel properties in various cell types (Fodor & Aldrich, 2009). The segments S1-S6 are conserved as in other voltage-dependent K^+ channels. BK channels consist of charged voltage sensing transmembrane segments (S1-S4) where charges appear to be functionally distributed (Ma et al. 2006; Aggarwal & MacKinnon 1996; Seoh et al., 1996). The S0 segment specific to BK channels appears to be involved in movements of the voltage sensor (Liu et al., 2008), and seems to be required for functional interaction of α-subunits and the accessory β-subunits as well as for insertion of the channels into the plasma membrane (Wallner et al. 1996; Morrow et al. 2006; Liu et al., 2008).

The pore forming segments (S5-S6) of each α-subunit have an amino acid sequence at the selectivity filter (glycine-tyrosine-glycine - GYG) which is also found in many other types of K^+ channels. The carboxyl (C) terminal tail comprises about two thirds of the α-subunit protein. In this region interactions take place with various channel modulating proteins including protein kinases and phosphatases (Wei et al., 1994; Schreiber & Salkoff 1997). It further includes a negatively charged Ca^{2+} binding region, the so called Ca^{2+} bowl (Wei et al., 1994; Schreiber & Salkoff 1997; Jiang et al. 2001) and a double negative charged region which is sensitive for Mg^{2+} as well as for Ca^{2+}, the so called RCK-domain (regulatory domain of K^+ conductance). In addition the biophysical functions of BK channels can be altered by interaction with auxiliary beta (β)-subunits. Tissue specificity is in part achieved by four different types of β-subunits (β1- β4) which associate with the α-subunit. β4 for instance is primarily expressed in the brain (Weiger et al., 2000) while the others are mainly found in the periphery (Torres et al. 2007). In addition to the β-subunits so called Slo binding proteins (Slob) have been identified which bind to and modulate Slo channels (Schopperle et al., 1998). Beside the complex pattern of channel gating by voltage, Ca^{2+} and β-subunits, other modulatory factors influence BK channel activity, like pH, the redox state or phosphoryation of the channel protein. Furthermore, gasotransmitters, like nitric oxide (NO) causing nitrosylation, carbon monoxide (CO) conveying carboxylation and H_2S imparting sulfuration may modulate channel activity (Wu & Wang 2005; Leffler et al., 2006; Kemp et al., 2009; Hou et al. 2009; Félétou 2009; Hu et al., 2011).

Through alternative splicing the pore forming α-subunit contains at its C-terminus a cysteine-rich 59-amino-acid insert between RCK1 and the Ca^{2+} bowl called stress-axis regulated exon (STREX). STREX exon expression is suppressed in hypophysectomized animals, whereas STREX exon expression is initiated by the stress-axis adrenocorticotropic hormone (Xie & McCobb 1998). Patch clamp recordings revealed that STREX causes BK channels to activate at more negative potentials and enhances activation and decreases deactivation which leads to increased repetitive firing of action potentials. STREX can be artificially induced by growing cells in phenol red which causes a significant increase in channel sensitivity to inhibition by oxidation but also to Ca^{2+} (Hall & Armstrong 2000). Coassembly of STREX/β1-subunits, however, could only be stimulated with a truncated N-terminus variation present which has physiological impact of channel regulation by Ca^{2+}, oxidation, and phosphorylation. β4-subunits together with the STREX insert alter BK channel biophysical properties in unexpected ways (Petrik & Brenner 2007). Individually β4 or the STREX insert promote channel opening by slowing deactivation at high Ca^{2+}.

BK channels have the largest single-channel conductance of all K^+ channels. The ideas why the conductance of these channels may be so large despite their high selectivity for K^+ can be summarized as followed: a) a negatively charged ring structure at the inner face of the channel which by electrostatic attraction of K^+ to the entrance approximately doubles the current amplitude (Brelidze et al. 2003; Nimigean et al. 2003; Zhang et al., 2006; Carvacho et al., 2008), b) a voluminous inner cavity with an excess of negatively charged amino acids near the selectivity filter which traps K^+ and facilitates their entrance into the selectivity filter (Brelidze & Magleby 2005; Li & Aldrich 2004), and c) a ring of four negative charges at the extracellular mouth of the channel (Haug et al., 2004), which pulls K^+ from the channel. The exact mechanism by which the high conductance of these channels is accomplished is still not fully understood in particular the contribution of the later two mechanisms to channel conductance have to be tested rigorously.

The dual modulation of BK channels by membrane voltage and by intracellular Ca^{2+} makes this channel to act as a molecular integrator of electrical events at the plasma membrane and intracellular signaling via Ca^{2+}. Since Ca^{2+} is involved in a multitude of cellular signaling processes this also provides a link to cell metabolism and gene activation. BK channels are widely distributed in brain and are often concentrated in neuronal cell bodies and nerve terminals (Knaus et al., 1996; Wanner et al., 1999). They facilitate membrane repolarization during action potential discharge and this way participate in the regulation of neurotransmitter release (Gho & Ganetzky 1992; Bielefeldt & Jackson 1994). BK channels play also a major role in relaxation of smooth muscles in the bladder, penis/clitoris or lung. The activity of BK channels therefore plays an essential role in controlling action potential discharge activity, hormone secretion or vasoconstriction (Weiger et al. 2002). The outward K^+ flux conducted by the BK channel moves the membrane potential in the hyperpolarizing direction suppressing activation of other voltage-dependent channels permeable to Ca^{2+}- or sodium. This provides a negative feedback for voltage-gated Ca^{2+} channels and hence prevents the accumulation of intracellular Ca^{2+}. Such a negative feedback system was already described for endogenous discharge activity in *Aplysia* pacemaker neurons (Gorman et al. 1981; Gorman et al. 1982).

There is a vast body of evidence to show that BK channels are also modulated by a antagonistic cycle of protein kinases/phosphatases as well as by G-proteins (Toro et al. 1990; Reinhart et al., 1991; Chung et al., 1991; Wei et al., 1994; Bielefeldt & Jackson 1994; Schreiber & Salkoff 1997; Schubert & Nelson 2001; Zhou et al., 2010; Tian et al., 2004; Xia et al., 1998). Channels remain functionally associated to kinase/phosphatase and G-proteins even after isolation and reconstitution into lipid bilayer membranes. Furthermore, BK channels are directly activated by internal GTP or GTPγS (a non-hydrolysable GTP analogue) in the presence of Mg^{2+}, characteristic for a G-protein mediated mechanism (Toro et al. 1990). Modulation of channels by kinases/phosphatases is involved in physiological processes such as transmitter release, hormone secretion or muscle contraction (Levitan 1994; Schubert & Nelson 2001; Newton & Messing 2006; Dai et al. 2009). In many cases BK channels and kinases/phosphatases are arranged in "nano-domains", and are constitutively attached to the channel proteins. The kinases themselves are regulated by substrate availability (ATP, GTP, phosphoinositoldiphosphat (PIP_2), by spatial factors (closeness of kinase to the channel within the membrane, association to the channel via specific binding sites) or by hydrolysis via phosphodiesterases.

The activity of BK channels is modulated by the redox state of critical cysteine sulfhydryl groups of the channel protein or an associated regulatory protein involving free thiols and disulfides (DiChiara & Reinhart 1997; Wang et al., 1997; Gong et al., 2000; Tang et al., 2001). Cysteine residues known for their responsibility of redox modulation are usually located at the cytoplasmic side of the channel. Under reducing conditions the channel activity is augmented as shown in different cell types (DiChiara & Reinhart 1997; Gong et al., 2000; Wang et al., 1997), whereas inclusion of the STREX insert makes the channels extremely sensitive to inhibition by oxidation (Erxleben et al., 2002).

BK channel activity is also influenced by their lipid surrounding. This has been studied by insertion of the channels into artificial lipid bilayer membranes. For example the probability of channel opening (Po) was significantly greater in phosphatidylethanolamine (PE) compared to phosphatidylserine (PS) at the same Ca^{2+} concentration and voltage (Moczydlowski et al., 1985). Also bilayer thickness and specific lipids such as sphingomyelin, which cluster in micro-domains have been identified as a critical factors that modulate BK channel conductance (reviewed in Yuan et al. 2004). Beside lipids cholesterol is a major component of cell membranes in animals. BK channels are generally inhibited by accessory cholesterol in native as well as in reconstituted cell membranes by shortening mean open and extending mean closed times. Depletion of membrane cholesterol results in an increase of channel open probability (Bolotina et al., 1989; Chang et al., 1995b; Crowley et al. 2003; Lin et al., 2006; Bukiya et al., 2008).

4. BK channels - and ethanol/acetaldehyde

Ethanol ($CH3-CH_2OH$) is a product of cell metabolism and can affect all living organisms from bacteria to men where it has a multitude of effects at the cellular level. For almost a century it was generally accepted that many of the pharmacological actions of ethanol result from nonspecific interactions with cellular membranes causing a „disordering" (fluidizing) effect. This was thought to alter membrane ionic conductances based on the „lipid theory of alcohol action" by Meyer and Overton (in Lynch 2008). Later, it was found that physiological concentrations of ethanol produced rather small disordering membrane effects and Franks & Lieb (1987) pointed out that a change in temperature of less than 1°C is sufficient to mimic the effects of anesthetics on lipid bilayers. During the last decades it became clear that ethanol directly acts on proteins such as receptors and ion channels located in the plasma membrane or at intracellular signalling molecules. Experimental evidence revealed that some effects of ethanol are due to specific actions including most ligand-gated ion channels, such as glutamate-, γ-aminobutyric acid- (GABA) (Lobo & Harris 2008), dopamine- (Di Chiara & Imperato 1986), 5-hydroxytryptamine-, or acetylcholine-, opioid-, (Di Chiara et al. 1996; Herz 1997; Gianoulakis 2009), adenosine-, ATP- (Asatryan et al., 2011; Ostrovskaya et al., 2011), or TRP receptors (Benedikt et al., 2007), as well as voltage-gated ion channels, such as K^+, Na^+, and in particular Ca^{2+} channels (Gonzales & Hoffman 1991; Crews et al., 1996b; Dopico et al. 1996; Jakab et al. 1997; Horishita & Harris 2008; Dopico & Lovinger 2009; Kerschbaum & Hermann 1997). Ethanol was also found to interact with signal-transduction mechanisms including G-proteins and protein kinases (Messing et al. 1991; Lahnsteiner & Hermann 1995; Newton & Ron 2007; Martin 2010; Kelm et al. 2011).

Ca^{2+} activated K^+ channels are among those channels being directly modulated by ethanol (in Dopico et al., 1999; Brodie et al., 2007; Mulholland et al., 2009; Dopico & Lovinger 2009; Treistman & Martin 2009; Martin et al., 2010). Activation of K^+ channels drives the membrane potential in hyperpolarizing direction which led to the speculation that these channels may be involved in the sedative action of ethanol (Nicoll & Madison, 1982). However, many of the early studies on the ethanol effects used very high ethanol concentrations far above the lethal dose in humans. For instance extracellular application of 500 - 2500 mM ethanol to cat trigeminal neurons caused a short burst of action potentials which was followed by hyperpolarization. This was interpreted as an ethanol-induced Ca^{2+} inward current that activated a Ca^{2+}-dependent electrogenic K^+-pump (Baranyi & Chase 1984). Studies at more relevant pharmacological concentrations showed that 20 mM ethanol (this equals the legal blood concentration in many countries) enhances the Ca^{2+}-dependent after-hyperpolarization, but not the Ca^{2+}-independent after-hyperpolarization in rat hippocampus CA1 cells (Carlen et al., 1982). Similar findings were reported in other studies for hippocampus CA3 neurons, granule cells and cerebellar Purkinje cells (Niesen et al., 1988). Initial evidence of an increase in a Ca^{2+} activated K^+ conductance by ethanol came from experiments on identified mollusc (*Helix*) neurons (Madsen & Edeson 1990). First studies showing the involvement of Ca^{2+} activated K^+ channels as a target of ethanol were presented in parallel by Dopico et al., (1996) and by Jakab et al., (1997). Ethanol augmented BK channel activity of isolated neuro-hypophyseal synaptic nerve terminals (Dopico et al. 1996) and increased BK channel open probability of rat pituitary tumor cells (Jakab et al. 1997). The increase in channel activity was considered as a result of modification of channel gating induced by ethanol acting on the channel protein or at some signalling mediator. The reduction of neuropeptide release (vasopressin, oxytocin) by ethanol from neuro-hypophyseal terminals was explained by inhibition of voltage-dependent Ca^{2+} channels (Wang et al., 1991) and it was speculated that the decrease in circulating vasopressin levels is involved in the generation of diuresis, a frequently observed phenomenon after alcohol ingestion.

4.1 Ethanol - BK channels – and cellular signalling

Ethanol/drugs and cellular signaling is covered extensively in several reviews (McIntire 2010; Ron & Messing 2011; Newton & Messing 2006; Harris et al. 2008; Chao & Nestler 2004; Newton & Ron 2007; Hoffman & Tabakoff 1990). In GH3 pituitary tumor cells the ethanol-induced potentiation of channel activity was prevented in the presence of PKC inhibitors and phosphatase inhibitors augmented the effect whereas blockade of phospholipase C was not able to prevent BK channel activation (Jakab et al. 1997). Taken together the experiments suggested a PKC-mediated phosphorylation and stimulation of the channels. PKC involvement in acute and chronic ethanol action has been summarized by Stubbs & Slater (1999) and Brodie et al., (2007). Using transgenic mice two PKC isoenzymes have been identified that mediate opposing behavioural effects of ethanol (Newton & Ron 2007). Deletion of PKCγ produced mice with high ethanol drinking phenotype requiring a high level of ethanol to reach intoxication - maybe similar to humans at risk to acquire alcoholism. On the other hand, deletion of PKCε produced animals with a low ethanol intake which were more sensitive to acute effects of ethanol - perhaps modelling humans with a low risk of developing alcoholism. The authors conclude that drugs interfering with different PKC isoforms may be beneficial in treating alcoholism. Ethanol has also been reported in cultured hippocampal neurons to transiently elevate intracellular Ca^{2+} by a Ca^{2+}-

induced Ca^{2+} release mechanism from internal stores by involvement of PKC activation (Mironov & Hermann 1996). Concomitant Ca^{2+} elevations in the cell soma as well as in dendrites were observed which appears important considering the effects of ethanol in the modulation of synaptic BK channels (Dopico et al. 1996). Ethanol activation of PKC was mimicked by application of the actin depolymerising drugs cytochalasin B and D suggesting that in intact cells cytoskeleton rearrangements may also contribute to Ca^{2+} liberation from internal pools (Mironov & Hermann 1996). This notion of an interaction of ion channels and the actin cytoskeleton is in concert with findings of BK channels in lipid rafts where they co-localize with the actin cytoskeleton (Brainard et al., 2005). Disruption or stabilization of actin increased or decreased BK channel activity, respectively. A similar finding of an ethanol increased elevation of intracellular Ca^{2+} was reported for GH4/C1 pituitary tumor cells which appeared to result from Ca^{2+} influx as well as liberation of Ca^{2+} from internal stores (Sato et al., 1990; Jakab et al., 2006). The ethanol initiated increase of internal Ca^{2+}, therefore, may be an additional factor to the activation of BK channels. Activation of BK channels is known to also derive from stretch activation of the cell membrane (Gasull et al., 2003; Kawakubo et al., 1999). Ethanol has been found to induce cell swelling even under isoosmotic conditions evoking transmitter and hormone secretion (Jakab et al., 2006). However, BK channels were reported to be stretch activated but insensitive to cell volume changes (Grunnet et al., 2002; Hammami et al., 2009) which makes it more likely that Ca^{2+} influx induced by ethanol activates BK channels but not cell swelling.

Experiments with cloned BK channels from mouse brain (*mslo* α-subunits) expressed in oocytes suggested that auxiliary subunits were not required for the action of ethanol (Dopico et al. 1998). Ethanol reversibly increased *mslo* activity in excised patches with a potency (EC_{50} = 24 mM) similar to native channels. Using this system it was concluded that the ethanol effect is unlikely to be mediated by second-messengers or G-proteins favouring a direct interaction of ethanol with the α-subunit of BK channels. Since BK channel activation by an increase of intracellular Ca^{2+} was reduced it was hypothesized that ethanol and intracellular Ca^{2+} act as agonists (Dopico et al. 1998). In further experiments BK channels were incorporated into artificial lipid bilayers to avoid complexities as from native cell membranes such as cytoplasmic constituents or complex membrane lipid composition. Even under these minimum conditions ethanol increased the activity of BK channels with a decrease of mean closed time or increase of mean open time, whereas channel conductance was not affected (Chu et al., 1998; Crowley et al.2005).

Recently the site of ethanol action at the BK channel protein has been targeted. A single mutation of threonine to valine (T107V) in the non-conserved S0-S1 linker loop has been identified to modify bovine BK channel (*bslo*) responses to acute ethanol exposure (Liu et al., 2006). Ethanol increased *bslo* T107V channel activity caused by augmenting frequency of channel openings. In addition, incremental phosphorylation at T107 by Ca^{2+}/calmodulin-dependent protein kinase II (CaMKII) progressively increased channel activity which depending on the state of phosphorylation was gradually inhibited by ethanol. Therefore, phosphorylation at T107 is considered as a "molecular dimmer switch" that via post-translational protein modification imposes tolerance to BK channels. It still remains to be seen where and how exactly ethanol impacts the channel structure to exert its effect and how tolerance is achieved. In intact cells the situation may be more complicated again since channels may be in different phosphorylated/dephorphorylated states and ethanol may also affect intracellular signalling systems. BK channels have been found to cluster into nano-domains including α-, β-subunits with Ca^{2+}/Mg^{2+}-binding sites and attachments of slob protein(s), as well as kinases and phosphatases. Isolation of channels and insertion into lipid bilayers therefore does not preclude the possibility that other constituents of the channel also respond to ethanol or to second messenger mediated interaction.

4.2 Ethanol – and membrane lipids

Although modern studies have produced a large amount of experimental evidence that ethanol directly affects proteins the lipid theory is not obviated by those findings. Indeed the lipid environment is an important modulator of channel properties. Ethanol action on channels is influenced by the composition of the native cell membrane which may differ in different cell types. The lipid composition and changes in the lipids environment by ethanol which interacts with lipids may modulate channel activity. Prolonged exposure to ethanol alters the lipid composition of membranes (Taraschi et al., 1991). Recent studies show that the lipid environment impacts BK channel function and is involved in causing acute tolerance to ethanol. BK channels reconstituted into lipid bilayers exhibit increased open probability by ethanol similar to native channels but the baseline characteristics of the channels differed depending on the lipid composition (Chu et al., 1998). BK channel activity induced by ethanol was dependent on the size and shape of the phospholipids independent of their charges (Crowley et al. 2005). Altering the thickness of the bilayer into which BK channels from HEK cells (human embryonic kidney cells) were inserted changed the ethanol response from potentiation in thin bilayers to inhibition in thick bilayers which correlated with mean closed time of the channels (Yuan et al., 2008). As mechanism for the biphasic channel modulation was proposed that forces of lateral stress within the lipid bilayer combine with hydrophobic mismatch to the channel gating spring structure (Yuan et al., 2007). It appears conceivable therefore that molecules such as cholesterol or alcohol inserted into the membrane bilayer may change its thickness and affect gating of BK channels. In fact elevation of membrane cholesterol decreased channel open probability (Bregestovski et al. 1989; Bolotina et al., 1989) and antagonized the potentiating effect of ethanol on BK channels (Crowley et al. 2003). Depletion of cholesterol resulted in activation of BK channels, an increase of BK current density and reduced firing of action potentials (Lam et al., 2004; Lin et al., 2006). Furthermore, the effect of ethanol as well as cholesterol was greatly reduced in the absence of phosphatidylserine in the bilayer membrane stressing the complexity of lipid impact on BK channel activity. This is of special interest since brain cholesterol in mice (Chin et al. 1978) or cerebellar granula cells is elevated after exposure to alcohol (Omodeo-Salé et al., 1995). Ethanol also reduced the asymmetric distribution of cholesterol between the cytofacial (higher cholesterol) and exofacial leaflet of the lipid bilayer (Wood et al., 1990). Cholesterol by itself concentration dependently moved BK channels into the closed state (Chang et al., 1995a) and hence appears to override the augmenting effect of ethanol. Furthermore, basal channel activity and its potentiation by ethanol in bilayers containing phosphatidylcholine are not as forceful as in those containing phosphatidylserine (PS). In natural membranes PS is abundant in the inner leaflet of the cell membrane and serves as an anchor for membrane-associated signalling molecules that regulate ion channel activity. PS is involved in Ca^{2+}-dependent PKC translocation to the cell membrane being a well-known modulator for both basal BK channel activity (Schubert & Nelson, 2001) as well as for ethanol potentiation of BK channels (Jakab et al., 1997). It is conceivable therefore that the presence of PS in cell membranes is specifically required for ethanol to modulate BK channel function given the links that exist between this phospholipid and signalling molecules (Crowley et al., 2005).

4.3 BK channels – ethanol and behaviour

BK channels play a pivotal role in behavioural responses to ethanol. Ethanol applied to the nematode *Caenorhabditis elegans* at human intoxicating concentrations dose-dependently and reversibly cause impairment of locomotion and egg-laying (Davies et al., 2003). Using BK channel knock outs the *slo-1* mutants were highly resistant to ethanol in behavioural assays. Behaviour of *slo-1* gain-of-function mutants again resembled those of ethanol-intoxicated animals as they show behavioural responses like in-coordination and a loss of social inhibition. Selective expression revealed that only *sol-1* in neurons but not in muscle rescued ethanol sensitivity. Investigation of excised BK channels showed that channel open probability was increased by ethanol as shown in previous single BK channel studies (Dopico et al. 1996; Jakab et al. 1997). In a molecular model for ethanol intoxication increased BK channel activity increases action potential repolarization and/or causes membrane hyperpolarization which shuts down Ca^{2+} channels and reduces transmitter release at synaptic terminals (Crowder, 2004). The experiments clearly demonstrate that mutation of a single gene affects ethanol sensitivity, although this is most probably not the only mechanism involved and it remains interesting to further monitor extensions of these findings to higher animals or to humans. Martin, et al. (2008) recently examined the generation of action potentials in brain spiny neurons using whole cell patch clamp recordings. They found that the number of action potentials evoked by current injection was increased in β4-subunit knockout mice compared to wild type under the influence of ethanol. However, the role of BK channels on the membrane resting potential was not investigated.

4.4 BK channels – and ethanol tolerance

Tolerance is generally defined as reduction or loss of response to a drug over time or after repeated exposure which may involve ion channels, receptors and/or gene expression (Chandler et al. 1998; Chao & Nestler 2004; Atkinson 2009; Treistman & Martin 2009). Tolerance in the nervous system is associated with down-regulation of excitatory receptors, such as NMDA-, nicotinic acetylcholine receptors or voltage dependent Ca^{2+} channels. It is also accompanied with up-regulation of inhibitory channels such as $GABA_A$, glycine or serotonin receptors (Harris et al.2008). Different types of tolerance may be categorized into: a) **acute tolerance** - which is a time-dependent type of tolerance that occurs during drug exposure in a time frame of seconds to minutes, b) **rapid tolerance** - occurs after a single usually high dosage drug experience, and c) **chronic tolerance**, which takes place after prolonged, repeated, identical, low dose drug exposures in a time frame of hours, days or weeks (Berger et al. 2004; Treistman & Martin 2009; McIntire 2010; Cowmeadow et al. 2005). Eventually drug tolerance may lead to increased consumption and addiction defined as compulsive drug-seeking and drug-taking behaviour (Chao & Nestler 2004).

In the early studies using excised single BK channel recordings from GH3 cells it was found that the potentiating effect after ethanol exposure rapidly declined. Within minutes both, mean open time and open probability of channels returned to control values (Jakab et al. 1997). In contrast, BK channel activity from synaptic terminals after application of ethanol remained elevated over minutes (Dopico et al. 1996). BK channels of rat hypothalamic-neurohypophysial terminals also become rapidly tolerant to ethanol including two components: decreased ethanol potentiation (short term within minutes) and decreased channel density (long term >24 hours) (Pietrzykowski et al., 2004). These two types of tolerance appear to reflect different mechanisms: a) decreased BK potentiation by ethanol and, b) down-regulation of BK channels and reduction of channel clustering associated with internalization of channels as suggested from immunolabeling. In the *Drosophila* nervous system a null mutation of the slowpoke gene completely eliminated rapid tolerance to ethanol (Cowmeadow et al. 2005). Ethanol increased slowpoke expression in the nervous system coincident with the induction of ethanol tolerance (Cowmeadow et al., 2006). Since an increase of slowpoke expression is also caused by cold, by CO_2 sedation (Ghezzi et al., 2010) or by heat-shock promoters (Cowmeadow et al., 2006) it was suggested that this is a more common mechanism for acquisition of tolerance. Interestingly the *Drosophila* slowpoke gene appears to contain a binding site for CREB (cyclic-AMP response element binding protein) which has been implicated in learning and memory and hence may also be involved in the ethanol response (Cowmeadow et al., 2006) and possibly in the memory deficits after excessive alcohol intake. Further experimentation into the molecular mechanism of tolerance using single channel recoding revealed that only after expression of the somatic BK α-subunit together with the brain specific β4-subunit ethanol dose-dependently increased the open probability of channels and decreased the duration of action potentials whereas BK α-subunit together with the ß1-subunit expressed in dendrites was insensitive to ethanol (Martin et al., 2004; Martin et al., 2008)

Human BK channels (*hslo*) are also potentiated by alcohol being dependent on the presence of auxiliary β-subunits (Feinberg-Zadek & Treistman 2007). BK channel activity containing only the α-subunit were substantially increased by ethanol, together with the β4-subunit the channel mean open time was also increased but to a lesser extent and channel activity was unaffected in the presence of β1-subunit. After prolonged ethanol exposure (24 h) down regulation of the BK current containing only *hslo* or *hslo*+β4 was observed - but not with β1 (Feinberg-Zadek, et al. 2008). Moreover, neuronal BK channels from wild-type mice expressing α- and β4-subunits show little tolerance whereas BK channels from β4 knockout (KO) mice also exhibit acute tolerance to ethanol. Studies at the behavioural level revealed that β4-KO mice drink more compared to wild-type companions (Martin et al., 2008). The authors point out that because subunit expression - in particular β4 - differs between many cells types, i.e. in neurons and even in neuronal compartments this could determine variations in individual alcohol responses such as tolerance which may lead to abuse and alcoholism.

Ethanol, via an epigenetic mechanism involving microRNA, induces alternative splicing and mediates rapid reorganization of BK α-isoforms (Pietrzykowski et al., 2008). This leads to destruction of a subset of BK α-subunits but persistence of ethanol-insensitive, mainly STREX BK channels. Acute molecular tolerance to ethanol was found to be a function of exposure time and once initiated tolerance persists in the absence of the drug (Velázquez-Marrero et al., 2011). During prolonged ethanol exposure (6 hours, but not at 1 or 3 hours) mRNA levels of the ethanol-insensitive STREX isoform were increased and transition to the biophysical properties of BK-STREX channels occurred.

Chronic tolerance to alcohol is observed in rats that have been maintained on an ethanol-containing diet for 3 to 4 weeks (Knott et al., 2002). On the cellular level it was found that long-term ethanol exposure leads to a compensatory change in the expression of two channels acting as functional dyads: L-type Ca^{2+} channels current density increased, whereas BK current decreased but BK channels also became less sensitive to ethanol.

Ethanol and other drugs such as benzyl alcohol, a common sedative, induces neural expression of the *slo* gene and the production of rapid tolerance (Cowmeadow et al. 2005; Ghezzi et al., 2004). The drugs increased expression of the *slo* gene, enhanced neuronal excitation by reducing the refractory period between action potentials and augmented seizure susceptibility (Ghezzi et al., 2010). Mutant BK channels exhibiting increased activity were found in humans to cause increased excitability due to rapid repolarization of action potentials (Du et al., 2005). This condition can lead to epilepsy and paroxysmal movement disorders and alcohol appears to be responsible for initiation of dyskinesia in these individuals. The molecular pathway that mediates the upregulation of *slo* transcription in *Drosophila* using benzyl alcohol has been linked to a CREB transcription factor. Down regulation of a CREB repressor isoform releases other CREB activator isoforms which after phosphorylation bind to CRE (cyclic AMP response element) within the *slo* promoter region and induces acetylation of histones (Wang et al., 2007). This eventually stimulates specific promoters to increase the expression of BK channels. Increased BK availability is suggested to enhance neural discharge activity by shorting action potentials. Reduced Ca^{2+} influx via voltage activated channels gives rise to sedation and development of rapid tolerance (Wang et al., 2009). If this mechanism also applies to ethanol remains to be investigated. Tolerance to alcohols may also include changes in membrane lipid composition (Yuan et al., 2007).

4.5 Ethanol – blocks BK channels

Although in most cases ethanol is found to increase BK channels activity it has also been reported to act as suppressant. Rat aortic myocyte BK channels expressed in *Xenopus* oocytes are in majority inhibited by 30 - 200 mM ethanol. Coexpression of the ß1-subunit together with the α-subunit in this tissue failed to influence ethanol action on *bslo* channels. The inhibition of BK channels in rat aortic myocytes may contribute to the direct contraction of aortic smooth muscle produced by acute alcohol exposure (Dopico, 2003). In supraoptic neuronal cell bodies ethanol failed to increase BK channel activity but increased nerve terminal BK channels (Dopico et al., 1999). Moreover, BK channels from vascular tissue are also blocked by ethanol (Walters et al. 2000; Liu et al., 2003). The reason for this difference is not clear but may include expression of different channel isoforms, different auxiliary proteins (β-subunits) or different lipid composition around the channels.

4.6 Ethanol – and transmitter/hormone secretion

Ethanol influences the duration of action potentials by facilitating their repolarization and their after-hyperpolarization (Gruss et al., 2001). This negative feedback on cell excitation closes Ca^{2+} channels, shortens the duration of Ca^{2+} entering the cells and decreases the Ca^{2+} triggered release of hormones or neurotransmitters (reviewed in Dopico et al., 1999).Ethanol also directly acts on Ca^{2+} channels. At low concentrations (10 mM - ca. 0.5 per mille) ethanol has been found to reduce vasopressin release from nerve terminals isolated from rat neurohypophysis by inhibition of the Ca^{2+} current which explains the reduction in plasma vasopression levels (Wang et al., 1991). In hippocampal CA1 neurons ethanol at extremely low concentration (0.01 per mille) enhanced, but at higher concentrations (5 per mille) decreased, synaptic transmission by activation of a G-protein/protein kinase C signalling pathway (Lahnsteiner & Hermann 1995). Voltage dependent Ca^{2+} currents were also suppressed by ethanol in invertebrate preparations (Camacho-Nasi & Treistman 1987; Oyama et al. 1986) by activation of a G-protein/protein kinase transduction pathway resulting in decreased action potential duration (Kerschbaum & Hermann 1997).

Despite the wealth of knowledge about alcohol interaction with receptors, ion channels, enzymes and signaling molecules questions about its main target(s) and its binding site(s) at these proteins still remain. It is thought that the most likely target sites of ethanol are amphipathic pockets in membrane proteins like K^+ channels of the inward rectifier type (Harris, et al. 2008; Howard, et al. 2011). Alcohol binding sites have been identified in the crystal structure of "alcohol dehydrogenase (ADH)" (Ramaswamy et al., 1996; Rosell et al., 2003) and for LUSH, an odorant binding protein from *Drosophila* (Kruse et al., 2003). This may help to develop further ideas on how the ethanol binding site may look like in other proteins. However, little is known if ethanol directly binds to these proteins or if accessory ethanol-binding proteins that target the functional protein are effective. Furthermore, it remains to be determined to which extent and how ethanol interferes with the lipid phase of the membrane or the lipid-protein interaction.

5. BK channels – and acetaldehyde

Acetaldehyde (ACA) is the primary metabolite of ethanol oxidation and in numerous studies a role for it in the action of ethanol on the brain has been proposed. Indeed evidence is accumulating that ACA is responsible for some of the effects that so far have been attributed to ethanol (reviewed in Hunt 1996; Quertemont et al. 2005; Correa et al., 2011). On basis that ACA has been generally considered as an aversive, treatment for alcoholics with disulfiram (Antabus, an inhibitor of ACA metabolism) has been established and used clinically. However, it was also noticed that ACA has central reinforcing effects (Melis et al., 2007; Quertemont & Tambour 2004; Rodd-Henricks et al., 2002; Quertemont & De Witte 2001). The metabolism and regulation of ACA particularly in blood or liver occurs via activities of alcohol dehydrogenase (ADH), cyctochrome P450, catalase and aldehyde dehydrogenase. The blood concentrations of ACA after ethanol consumption was found extremely low (<0.5 µM) (Eriksson & Fukunaga 1993; Eriksson 2007) and together with the activity of the blood-brain barrier it appeared unlikely to penetrate the brain in any pharmacological relevant amounts. However, ACA can be produced within the brain from ethanol through catalase and/or cytochrome P-4502E1 which makes it more likely that biologically significant concentrations at least in some brain areas can be achieved (Karahanian et al., 2011; Correa et al., 2011; Deng & Deitrich 2008; Quertemont et al. 2005). There is also evidence that ACA may mediate tolerance and dependence. Nevertheless, the actual ACA concentrations in the brain after ethanol consumption and its rapid oxidation remain to be determined. Most clear cut studies on the modulation of neurotransmission by acetaldehyde/alcohol have been performed on the dopaminergic system (reviewed in Correa et al., 2011). ACA appears to modulate dopaminergic function particularly in the mesolimbic pathway which indicates relevance to motivational behaviour. Studies of the action of ACA on the cellular level, on single channels or on electrical activity are scarce. In smooth muscle cells it was reported that ACA inhibits voltage-dependent Ca^{2+} currents (Morales et al., 1997). Furthermore, in vitro ACA was found to enhance firing of action potentials of dopaminergic neurons in the ventral tegmental area by reduction of the A-type K^+ current and activation of a hyperpolarization-activated inward current (Melis et al., 2007). The stimulating properties were prevented by blockade of local catalase.

In our laboratory we have investigated some of the effects of ACA on single BK channels from GH cells (Handlechner et al., 2008; Handlechner et al., 2011). Given the fact that the simultaneous presence of ACA and ethanol reflects the physiological situation in the brain after alcohol consumption we assumed that both molecules may either act synergistically or antagonistically. Hence we started to investigate the BK channel response to ethanol in the presence of ACA. Extracellular ethanol increased BK channel open probability as reported previously (Jakab et al. 1997). In the presence of intracellular ACA the ethanol related increment of BK channel activity was inhibited in a dose dependent manner. BK channel amplitudes were not affected but mean channel open time and open probability were significantly reduced. In contrast, extracellular ACA had no effect on ethanol induced channel activity. Our results reveal that ACA interferes with BK channel activity blunting the effect of ethanol. The action of ACA on the channel can be considered as direct and not through some metabolic product or adduct, activation of transmitters/hormones or gene expression since we use cell free recordings, ACA is always in excess and the effect is acute.

Our findings may have consequences for the pharmacological/toxicological effects of ACA/ethanol on the electrical activity of cells, on nervous function and animal behaviour. From our findings we may speculate that ACA counteracts the effect of ethanol and may potentiate tolerance to ethanol. In any case, in the context of ethanol actions ACA effects have to be considered carefully. Further investigation shall be concerned with the dependence of the ACA-mediated effect at variable concentrations of free internal Ca^{2+}, possible ACA interference with intracellular signaling cascades, i.e. the phosphorylation or redox state of the BK channels or interference with the brain specific β4 subunit in the action of EtOH/ACA on BK channel properties.

6. BK channels – modulation by hydrogen sulfide (H_2S)

H_2S is a colorless gas and well know because of its peculiar odor of rotten eggs. It also is an extremely toxic gas and inhaled in higher concentrations causes coma and eventually death (Reiffenstein et al. 1992; Beauchamp et al., 1984). H_2S is produced endogenously in many living cells from the amino acid L-cysteine. Three synthetic pathways in various organs have been described such as in vascular system, liver, kidneys and the brain (Shibuya et al., 2009; Ishigami et al., 2009; Stipanuk & Beck 1982; Łowicka & Bełtowski 2007). After its generation H_2S diffuses either immediately in the surrounding milieu or is bound to and stored in proteins until it is released by an adequate stimulus. H_2S - similar to the other gasotransmitters NO or CO - is water and lipid soluble and therefore also easily passes membranes. The physiology, pathophysiology, pharmacology of H_2S particularly in the vascular system and brain has been reviewed in an impressive amount of recent publications (Wang 2011; Kimura 2011; Hu et al., 2011; Bucci & Cirino 2011; Wang 2010; Tan et al. 2010; Gadalla & Snyder 2010; Mustafa et al., 2009; Mancardi et al., 2009; Qu et al., 2008; Li & Moore 2008; Li et al., 2011; Łowicka & Bełtowski 2007; Szabó 2007; Wallace 2007; Wallace 2010; Lloyd 2006; Wang 2002; Boehning and Snyder 2003; Caliendo et al., 2010).

Besides many other cellular targets H_2S also acts on ion channels. In neurons an increase of the cytosolic Ca^{2+}-concentration by H_2S appears to be caused by activation of Ca^{2+} entry through L-type Ca^{2+}-channels (García-Bereguiaín et al., 2008). Modulation of pain processing by H_2S appears to involve activation of T-type Ca^{2+} channels responsible for its pro-nociceptive effect, whereas analgesia is due to activation of K_{ATP} channels (Distrutti 2011). In peripheral tissue, however, H_2S reduces T-type Ca^{2+} channel activity leading to hyperalgesia (Kawabata et al., 2007). T-type calcium channels are also involved in pain processing of spinal nociceptive neurons (Maeda et al., 2009), in colon (Matsunami et al., 2009) and in pancreas (Nishimura et al., 2009). H_2S decreased the mechanical contraction of rat cardiomyocytes through inhibition of L-type calcium channels (Sun et al., 2008). One of the most well-known actions of H_2S is the activation of ATP-sensitive K^+ channels by which H_2S causes vasorelaxation (Zhao & Wang 2002; Tang et al., 2005; Zhao et al., 2001; Jiang et al., 2010; Liang et al., 2011; Liu et al., 2011), inhibits insulin secretion (Yang et al., 2005; Wu et al., 2009), or protects primary cortical neurons from oxidative stress (Kimura & Kimura 2004). However, the universal applicability of a K_{ATP} dependent action has been questioned (Kubo et al., 2007; Szabó 2007). In the gastrointestinal tract (human jejunum smooth muscle) H_2S activates sodium channels in a partially redox dependent manner (Strege et al., 2011). In contrast to other gasotransmittes H_2S appears not to act on the intracellular signaling pathway guanylyl cyclase (Garthwaite 2010). The interaction of H_2S with ion channels has been reviewed by Tang et al. (2010).

We choose GH3 cells since they are widely used as model cells to investigate BK channel activity in natural settings (Sitdikova et al. 2010). Sodium hydrosulfide (NaHS) was used as H_2S donor since it can be readily handled and quantified. Our experiments showed that H_2S dose-dependently increased single channel open probability (P_{open}) (Sitdikova et al. 2010). In our cell free, single channel recordings where Ca^{2+} is kept constant the increase of BK channel activity indicates that H_2S does not act via elevation of the Ca^{2+} concentration. The fast onset of the H_2S effect after application within seconds, but also the rapid decrease after washout of the drug, further suggests a direct effect at the channel protein. A half maximal effective concentration of 90 µM NaHS indicates that H_2S induces BK channel activation in a physiological relevant concentration range. To study the effect of H_2S on BK channel sensitivity to intracellular Ca^{2+} we used a range of Ca^{2+} concentrations at a constant membrane potential. The experiments show that there was no difference in H_2S effects on BK channel activity at different cytoplasmic Ca^{2+} concentrations. Hence H_2S appears not to interfere at the Ca^{2+}binding sites of the channel. Also ß4 subunits appear to be an unlikely target of our BK channels since iberiotoxin rapidly blocked the current indicating that BK channels in GH3 cells are not accompanied by ß4-subunits.

Redox modification is among the recognized mechanisms for cellular effects of H_2S including NMDA receptors (Kimura & Kimura 2004; Kabil & Banerjee 2010), K_{ATP} channels Zhao et al., 2001; Yang et al., 2005) or T-type Ca^{2+}-channels (Kawabata et al., 2007). We hypothesized that the increase of BK channel P_{open} may be mediated by redox modulation of cysteine residues. In our experiments the effect of NaHS was prevented when the reducing agent DTT was applied to the pipette solution accessing the cytoplasmic side of the channel. If channels were in the oxidized state by application of thimerosal, P_{open} was further increased by NaHS compared to the already increased thimerosal control.

In contrast to our findings a recent report indicates that BK channels expressed in HEK293 cells were inhibited by H_2S and activated by CO (Telezhkin et al., 2009; Telezhkin et al., 2010). In carotid body chemoreceptors, which are important to maintain oxygen homeostasis by regulating ventilation, H_2S caused an excitation of these cells by blocking BK channels which appear to play a crucial role in oxygen sensing (Li et al., 2010). In other preparations, however, H_2S causes dilatation and hyperpolarization of vascular smooth muscle (Jackson-Weaver et al., 2011) and activates BK channels in cultured endothelial cell (Zuidema et al., 2010). These differences in the response to H_2S are unclear but might be due to different tissues containing different BK channel splice variants or may be due to a different phosphorylation or redox state of the channels.

BK channels mediate or modulate many physiological functions as well as patho-physiological conditions. Future studies will have to show how H_2S or H_2S related substances may be involved and may contribute to those conditions. Techniques to determine H_2S even at low concentrations (in the micro- to nanomolar range) in biological preparations which are available now will help to facilitate the investigation of H_2S in biology and medicine (Doeller et al., 2005; Peng et al., 2011). In pharmacology the development of new drugs modulating H_2S signaling might be rewarding in the treatment of diseases like high blood pressure, pain therapy or erectile dysfunction.

7. BK channels – and polyamines

The polyamines putrescine, spermidine and spermine are hydrocarbon molecules with two, three or four positively charged amino groups under physiological conditions. Polyamines are metabolized from the decarboxylation products of ornithine and S-adenosyl-methionine in nearly all eukaryotic cells. They are multifunctional molecules which are inevitable for development or cell proliferation and modulate a number of cellular targets, like DNA, RNA or signaling proteins, but are also involved in pathological mechanisms, like cancer (Igarashi and Kashiwagi 2010; Bachrach, 2005). In addition to the above mentioned functions polyamines play a major role in modulating a number of ion channels. In the potassium channel family they act as modulators of the inward rectifiers K_{ir}, the BK, the TASK (two-pore-domain potassium channels), the KCNQ and the delayed rectifier channels (reviewed in Weiger & Hermann 2009). Furthermore, AMPA and NMDA receptors as well as Ca^{2+} and sodium channels are modulated by polyamines (Huang & Moczydlowski 2001; Williams, 1997). The ideas to test polyamines on ion channels was initially reported using mollusk neurons (Drouin & Hermann 1994; Drouin & Hermann 1990) and pituitary tumor GH cells (Weiger & Hermann 1994). Drouin & Hermann described a blocking action of polyamines on BK currents using whole cell two electrode voltage clamp experiments in *Aplysia californica* neurons on a K^+ channel which is pharmacologically similar to BK channels. They found spermine injected into the cell to have a dual action: immediately after injection the Ca^{2+} activated current was blocked, whereas after a prolonged time the current was increased. As explanation for these phenomena it was suggested that after prolonged Ca^{2+} injection the Ca^{2+} buffer capacity of the cells was exhausted or/and during the time course of the experiments the channels became more sensitive to Ca^{2+} which overcame the blocking effect caused by spermine. When they applied spermine in high concentrations up to 10 mM to the extracellular side of the cells they observed no or only a minor reduction (10%) of the current after prolonged application (10-15 min). The interpretation given for this result was that spermine possibly entered the cells by a polyamine transporter and acted at the intracellular face of the channels. To overcome the limitations of whole cells experiments Weiger & Hermann (1994) used a cell free patch system investigating single BK channel activity. They confirmed the blocking action of polyamines which acted in a voltage depended manner on the channel when applied to the intracellular face of the membrane but had no effect when applied extracellularly. The effect of polyamines on BK channel was dual: firstly, by a so called fast blocking mechanism the current amplitude was apparently reduced (caused by limitations of the recording system) and secondly, the open probability of the channel was decreased. The order of effectiveness of the various polyamines tested was: spermine > spermidine> putrescine. At high Ca^{2+} concentrations applied to the intracellular side polyamines were ineffective on single channel kinetics while the reduction of the amplitude remained. The stoichiometry of the channel block by spermine was 1:1, the reduction of the open probability had a 2:1 relationship. These data were in agreement with the whole cell recordings in *Aplysia* and suggested two interactions sites of BK channels with polyamines: namely the channel pore where the polyamine does not bind firmly but rather slips in and out at high frequency (flickery block, causing the reduced amplitude) as well as the Ca^{2+} sensor of the BK channel. The question why polyamines are not effective when applied to the outside the channel was probed with a series of diamines which differed in length up to 1,12 diaminododecane (Weiger et al. 1998). Diamine molecules are similar to polyamines in carrying a positively charged amino group at each end which is separated by a variable length CH-chain. Only 1,12-diaminododecane was found to act as a blocker from the extracellular face of the channel while diamines with a shorter chain length were ineffective. In silico molecular modeling revealed that 1, 12-diaminododecane and spermine although they have the same length the latter is more flexible and is completely hydrated. 1,12-diaminodocane has only small water caps at its ends, positioned over the charged amino groups separated by a long hydrophobic segment. It was hypothesized that spermine, putrescine or spermidine as well as the shorter diamines are not able to block the channel from the extracellular side due to energetic reasons which prevents to strip of the water shell in order to interact with the channel pore.

BK channels of rabbit pulmonary smooth muscle in contrast to other cells exhibit strong rectification (Snetkov et al., 1996). This was attributed to the presence of spermine and spermidine but not putrescine in the cytoplasm. Blocking polyamine synthesis with the ornithine decarboxylase inhibitor DFMO (difluoromethylornithine) released BK channel rectification supporting the notion of a rectifying action imposed by polyamines. Similar data were reported for BK channels in myocytes from the saphenous branch of the rat femoral artery (Catacuzzeno et al., 2000). These discoveries remind to the mechanism of current rectification caused by polyamines at inward rectifier channels (K_{ir}) (Fakler et al., 1995).

A more detailed molecular explanation of how polyamines block BK channels was presented by Zhang et al., 2006. They found the ring of 8 negative charges at the inner channel mouth to be responsible for the attraction of polyamines to the channel pore. Mutation of these charges to neutral amino acids reduced the blocking effect of polyamines 90-fold and reduced rectification. In another experiment they removed the polyamine block by a simple competition of positive charges at the negative ring at the channel entrance by applying 3 M KCl. Thus under physiological condition polyamines are attracted to channel by the ring of negative charges as well as the negative charges in the channel's pore driving them into the ion conduction pathway to block the channel when positive voltage is applied.

A study in humans suggests that BK channel block by polyamines may be a reason for the development of the overactive bladder syndrome (Li et al., 2009). In people with the syndrome high levels of polyamines were found in biopsies of the urothelium in parallel with a reduced or blocked BK channel activity. By preventing polyamine synthesis in these cells in vitro, BK channel activity could be restored to normal. This result opens a new window of opportunity for a possible future treatment of the disease.

While the majority of reports indicate a block of BK channels by polyamines, they were found to be ineffective in blocking the channel in retinal Müller glia cells (Biedermann et al., 1998). This result may be explained by the rather low concentration of polyamines used in these experiments or by a different, less sensitive splice variant of the channel being expressed in these cells. In summary polyamines appear to modulate BK channels by interacting with the channel pore from the inside of the cell membrane while they are not effective from the outside. They may either cause a block or rectification of the BK current.

8. Synopsis

BK channels are important integrators of cellular signals and hence are involved in a huge diversity of cellular actions and serve in initiating many cellular pathways. Here we summarized the action of ethanol/acetaldehyde, polyamines and hydrogen sulfide on BK channels - only a few of many modulators. Interestingly all these agents appear to interfere with quite different targets at the channel indicating its enormous plasticity. Although there is a vast array of input sites which modulate the channels its output is rather simple - once activated it hyperpolarizes the membrane potential. Since these channels use a combined mechanism of activation by voltage and intracellular Ca^{2+} concentration any of these signals and their minute manipulation by external factors is integrated by the channels imposing far-reaching effects for physiology, pathophysiology or pharmacology. These features makes BK channels so unique and warrants further interesting research in the future to discover even more interactions of this channel with its environment and its further modulatory action on the biology of cells.

9. Abbreviations

ACA = acetaldehyde; BK = maxi calcium-activated potassium channel; H_2S = hydrogen sulfide; EtOH = ethanol; STREX = stress-axis regulated exon; PS = phosphatiylserine; CREB = cyclic AMP response element-binding protein; P_{open} = open probability

10. References

Adelman, J.P.; Shen, K.Z.; Kavanaugh, M.P.; Warren, R.A.; Wu, Y.N.; Lagrutta, A.; Bond, C.T. & North, R.A. (1992) Calcium-activated potassium channels expressed from cloned complementary DNAs. *Neuron* Vol. 9, 209-216

Aggarwal, S.K. & MacKinnon, R. (1996) Contribution of the S4 segment to gating charge in the Shaker K+ channel. *Neuron* Vol. 16, 1169-1177

Asatryan, L.; Nam, H.W.; Lee, M.R.; Thakkar, M.M.; Saeed Dar, M.; Davies, D.L. & Choi, D.-S. (2011) Implication of the purinergic system in alcohol use disorders. *Alcoholism, Clinical and Experimental Research* Vol. 35, 584-594

Atkinson, N.S. (2009) Tolerance in Drosophila. *Journal of Neurogenetics* Vol. 23, 293-302

Atkinson, N.S.; Robertson, G.A. & Ganetzky, B. (1991) A component of calcium-activated potassium channels encoded by the Drosophila slo locus. *Science (New York, N.Y.)* Vol. 253, 551-555

Bachrach, U. (2005) Naturally occurring polyamines: interaction with macromolecules. *Current Protein & Peptide Science* Vol. 6, 559-566

Baranyi, A. & Chase, M.H. (1984) Ethanol-induced modulation of the membrane potential and synaptic activity of trigeminal motoneurons during sleep and wakefulness. *Brain Research* Vol. 307, 233-245

Beauchamp, R.O., Jr.; Bus, J.S.; Popp, J.A.; Boreiko, C.J. & Andjelkovich, D.A. (1984) A critical review of the literature on hydrogen sulfide toxicity. *Critical Reviews in Toxicology* Vol. 13, 25-97

Benedikt, J.; Teisinger, J.; Vyklicky, L. & Vlachova, V. (2007) Ethanol inhibits cold-menthol receptor TRPM8 by modulating its interaction with membrane phosphatidylinositol 4,5-bisphosphate. *Journal of Neurochemistry* Vol. 100, 211-224

Berger, K.H.; Heberlein, U. & Moore, M.S. (2004) Rapid and chronic: two distinct forms of ethanol tolerance in Drosophila. *Alcoholism, Clinical and Experimental Research* Vol. 28, 1469-1480

Berkefeld, H.; Fakler, B. & Schulte, U. (2010) Ca2+-activated K+ channels: from protein complexes to function. *Physiological Reviews* Vol. 90, 1437-1459

Bers, D.M.; Patton, C.W. & Nuccitelli, R. (2010) A practical guide to the preparation of Ca(2+) buffers. *Methods in Cell Biology* Vol. 99, 1-26

Biedermann, B.; Skatchkov, S.N.; Brunk, I.; Bringmann, A.; Pannicke, T.; Bernstein, H.G.; Faude, F.; Germer, A.; Veh, R. & Reichenbach, A. (1998) Spermine/spermidine is expressed by retinal glial (Müller) cells and controls distinct K+ channels of their membrane. *Glia* Vol. 23, 209-220

Bielefeldt, K. & Jackson, M.B. (1994) Phosphorylation and dephosphorylation modulate a Ca(2+)-activated K+ channel in rat peptidergic nerve terminals. *The Journal of Physiology* Vol. 475, 241-254

Boehning, D. & Snyder, S.H. (2003) Novel neural modulators. *Annual Review of Neuroscience* Vol. 26, 105-131

Bolotina, V.; Omelyanenko, V.; Heyes, B.; Ryan, U. & Bregestovski, P. (1989) Variations of membrane cholesterol alter the kinetics of Ca2(+)-dependent K+ channels and membrane fluidity in vascular smooth muscle cells. *Pflügers Archiv: European Journal of Physiology* Vol. 415, 262-268

Brainard, A.M.; Miller, A.J.; Martens, J.R. & England, S.K. (2005) Maxi-K channels localize to caveolae in human myometrium: a role for an actin-channel-caveolin complex in the regulation of myometrial smooth muscle K+ current. *American journal of physiology. Cell physiology* Vol. 289, C49-57

Bregestovski, P.D.; Bolotina, V.M. & Serebryakov, V.N. (1989) Fatty acid modifies Ca2+-dependent potassium channel activity in smooth muscle cells from the human aorta. *Proceedings of the Royal Society of London. Series B, Containing Papers of a Biological Character. Royal Society (Great Britain)* Vol. 237, 259-266

Brodie, M.S.; Scholz, A.; Weiger, T.M. & Dopico, A.M. (2007) Ethanol interactions with calcium-dependent potassium channels. *Alcoholism, Clinical and Experimental Research* Vol. 31, 1625-1632

Bucci, M. & Cirino, G. (2011) Hydrogen sulphide in heart and systemic circulation. *Inflammation & Allergy Drug Targets* Vol. 10, 103-108

Bukiya, A.N.; McMillan, J.; Parrill, A.L. & Dopico, A.M. (2008) Structural determinants of monohydroxylated bile acids to activate beta 1 subunit-containing BK channels. *Journal of Lipid Research* Vol. 49, 2441-2451

Calderone, V. (2002) Large-conductance, Ca(2+)-activated K(+) channels: function, pharmacology and drugs. *Current Medicinal Chemistry* Vol. 9, 1385-1395

Caliendo, G.; Cirino, G.; Santagada, V. & Wallace, J.L. (2010) Synthesis and biological effects of hydrogen sulfide (H_2S): development of H_2S-releasing drugs as pharmaceuticals. *Journal of Medicinal Chemistry* Vol. 53, 6275-6286

Camacho-Nasi, P. & Treistman, S.N. (1987) Ethanol-induced reduction of neuronal calcium currents in Aplysia: an examination of possible mechanisms. *Cellular and Molecular Neurobiology* Vol. 7, 191-207

Carlen, P.L.; Gurevich, N. & Durand, D. (1982) Ethanol in low doses augments calcium-mediated mechanisms measured intracellularly in hippocampal neurons. *Science (New York, N.Y.)* Vol. 215, 306-309

Carvacho, I.; Gonzalez, W.; Torres, Y.P.; Brauchi, S.; Alvarez, O.; Gonzalez-Nilo, F.D. & Latorre, R. (2008) Intrinsic electrostatic potential in the BK channel pore: role in determining single channel conductance and block. *The Journal of General Physiology* Vol. 131, 147-161

Catacuzzeno, L.; Pisconti, D.A.; Harper, A.A.; Petris, A. & Franciolini, F. (2000) Characterization of the large-conductance Ca-activated K channel in myocytes of rat saphenous artery. *Pflügers Archiv: European Journal of Physiology* Vol. 441, 208-218

Chandler, L.J.; Harris, R.A. & Crews, F.T. (1998) Ethanol tolerance and synaptic plasticity. *Trends in Pharmacological Sciences* Vol. 19, 491-495

Chang, H.M.; Reitstetter, R. & Gruener, R. (1995b) Lipid-ion channel interactions: increasing phospholipid headgroup size but not ordering acyl chains alters reconstituted channel behavior. *The Journal of Membrane Biology* Vol. 145, 13-19

Chang, H.M.; Reitstetter, R.; Mason, R.P. & Gruener, R. (1995a) Attenuation of channel kinetics and conductance by cholesterol: an interpretation using structural stress as a unifying concept. *The Journal of Membrane Biology* Vol. 143, 51-63

Chao, J. & Nestler, E.J. (2004) Molecular neurobiology of drug addiction. *Annual Review of Medicine* Vol. 55, 113-132

Chin, J.H.; Parsons, L.M. & Goldstein, D.B. (1978) Increased cholesterol content of erythrocyte and brain membranes in ethanol-tolerant mice. *Biochimica Et Biophysica Acta* Vol. 513, 358-363

Chu, B.; Dopico, A.M.; Lemos, J.R. & Treistman, S.N. (1998) Ethanol potentiation of calcium-activated potassium channels reconstituted into planar lipid bilayers. *Molecular Pharmacology* Vol. 54, 397-406

Chung, S.K.; Reinhart, P.H.; Martin, B.L.; Brautigan, D. & Levitan, I.B. (1991) Protein kinase activity closely associated with a reconstituted calcium-activated potassium channel. *Science (New York, N.Y.)* Vol. 253, 560-562

Correa, M.; Salamone, J.D.; Segovia, K.N.; Pardo, M.; Longoni, R.; Spina, L.; Peana, A.T.; Vinci, S. & Acquas, E. (2012) Piecing together the puzzle of acetaldehyde as a neuroactive agent. *Neuroscience and Biobehavioral Reviews* Vol. 35, 404-430

Cowmeadow, R.B.; Krishnan, H.R. & Atkinson, N.S. (2005) The slowpoke gene is necessary for rapid ethanol tolerance in Drosophila. *Alcoholism, Clinical and Experimental Research* Vol. 29, 1777-1786

Cowmeadow, R.B.; Krishnan, H.R.; Ghezzi, A.; Al'Hasan, Y.M.; Wang, Y.Z. & Atkinson, N.S. (2006) Ethanol tolerance caused by slowpoke induction in Drosophila. *Alcoholism, Clinical and Experimental Research* Vol. 30, 745-753

Crews, F.T.; Morrow, A.L.; Criswell, H. & Breese, G. (1996) Effects of ethanol on ion channels. *International Review of Neurobiology* Vol. 39, 283-367

Crowder, C.M. (2004) Ethanol targets: a BK channel cocktail in C. elegans. *Trends in Neurosciences* Vol. 27, 579-582

Crowley, J.J.; Treistman, S.N. & Dopico, A.M. (2003) Cholesterol antagonizes ethanol potentiation of human brain BKCa channels reconstituted into phospholipid bilayers. *Molecular Pharmacology* Vol. 64, 365-372

Crowley, J.J.; Treistman, S.N. & Dopico, A.M. (2005) Distinct Structural Features of Phospholipids Differentially Determine Ethanol Sensitivity and Basal Function of BK Channels. *Molecular Pharmacology* Vol. 68, 4-10

Cui, J. (2010) BK-type calcium-activated potassium channels: coupling of metal ions and voltage sensing. *The Journal of Physiology* Vol. 588, 4651-4658

Cui, J.; Yang, H. & Lee, U.S. (2009) Molecular mechanisms of BK channel activation. *Cellular and Molecular Life Sciences* Vol. 66, 852-875

Dai, S.; Hall, D.D. & Hell, J.W. (2009) Supramolecular assemblies and localized regulation of voltage-gated ion channels. *Physiological Reviews* Vol. 89, 411-452

Davies, A.G.; Pierce-Shimomura, J.T.; Kim, H.; VanHoven, M.K.; Thiele, T.R.; Bonci, A.; Bargmann, C.I. & McIntire, S.L. (2003) A central role of the BK potassium channel in behavioral responses to ethanol in C. elegans. *Cell* Vol. 115, 655-666

Deng, X.-s. & Deitrich, R.A. (2008) Putative role of brain acetaldehyde in ethanol addiction. *Current Drug Abuse Reviews* Vol. 1, 3-8

Di Chiara, G.; Acquas, E. & Tanda, G. (1996) Ethanol as a neurochemical surrogate of conventional reinforcers: the dopamine-opioid link. *Alcohol (Fayetteville, N.Y.)* Vol. 13, 13-17

Di Chiara, G. & Imperato, A. (1986) Preferential stimulation of dopamine release in the nucleus accumbens by opiates, alcohol, and barbiturates: studies with transcerebral dialysis in freely moving rats. *Annals of the New York Academy of Sciences* Vol. 473, 367-381

DiChiara, T.J. & Reinhart, P.H. (1997) Redox modulation of hslo Ca2+-activated K+ channels. *The Journal of Neuroscience: The Official Journal of the Society for Neuroscience* Vol. 17, 4942-4955

Distrutti, E. (2011) Hydrogen sulphide and pain. *Inflammation & Allergy Drug Targets* Vol. 10, 123-132

Doeller, J.E.; Isbell, T.S.; Benavides, G.; Koenitzer, J.; Patel, H.; Patel, R.P.; Lancaster, J.R., Jr.; Darley-Usmar, V.M. & Kraus, D.W. (2005) Polarographic measurement of hydrogen sulfide production and consumption by mammalian tissues. *Analytical Biochemistry* Vol. 341, 40-51

Dopico, A.M.; Anantharam, V. & Treistman, S.N. (1998) Ethanol increases the activity of Ca(++)-dependent K+ (mslo) channels: functional interaction with cytosolic Ca++. *The Journal of Pharmacology and Experimental Therapeutics* Vol. 284, 258-268

Dopico, A.M.; Chu, B.; Lemos, J.R. & Treistman, S.N. (1999) Alcohol modulation of calcium-activated potassium channels. *Neurochemistry International* Vol. 35, 103-106

Dopico, A.M.; Lemos, J.R. & Treistman, S.N. (1996) Ethanol increases the activity of large conductance, Ca(2+)-activated K+ channels in isolated neurohypophysial terminals. *Molecular Pharmacology* Vol. 49, 40-48

Dopico, A.M. (2003) Ethanol sensitivity of BK(Ca) channels from arterial smooth muscle does not require the presence of the beta 1-subunit. Am J Physiol Cell Physiol. Vol. 284(6), C1468-80.

Dopico, A.M. & Lovinger, D.M. (2009) Acute alcohol action and desensitization of ligand-gated ion channels. *Pharmacological Reviews* Vol. 61, 98-114

Dopico, A.M.; Widmer, H.; Wang, G.; Lemos, J.R. & Treistman, S.N. (1999) Rat supraoptic magnocellular neurones show distinct large conductance, Ca2+-activated K+ channel subtypes in cell bodies versus nerve endings. *The Journal of Physiology* Vol. 519 Pt 1, 101-114

Drouin, H. & Hermann, A. (1990) Intracellular spermine modifies neuronal electrical activity, in *Water and Ions in Biomolecular Systems; eds. VASILESCU, J.J., JAZ, J., PACKER, L. & PULLMAN, B* pp 213-220, Birkhäuser Verlag, Basel, Boston, Berlin.

Drouin, H. & Hermann, A. (1994) Intracellular action of spermine on neuronal Ca2+ and K+ currents. *The European Journal of Neuroscience* Vol. 6, 412-419

Du, W.; Bautista, J.F.; Yang, H.; Diez-Sampedro, A.; You, S.-A.; Wang, L.; Kotagal, P.; Lüders, H.O.; Shi, J.; Cui, J.; Richerson, G.B. & Wang, Q.K. (2005) Calcium-sensitive potassium channelopathy in human epilepsy and paroxysmal movement disorder. *Nature Genetics* Vol. 37, 733-738

Eriksson, C.J. & Fukunaga, T. (1993) Human blood acetaldehyde (update 1992). *Alcohol and Alcoholism (Oxford, Oxfordshire). Supplement* Vol. 2, 9-25

Eriksson, C.J.P. (2007) Measurement of acetaldehyde: what levels occur naturally and in response to alcohol? *Novartis Foundation symposium* Vol. 285, 247-255; discussion 256-260

Erxleben, C.; Everhart, A.L.; Romeo, C.; Florance, H.; Bauer, M.B.; Alcorta, D.A.; Rossie, S.; Shipston, M.J. & Armstrong, D.L. (2002) Interacting effects of N-terminal variation and strex exon splicing on slo potassium channel regulation by calcium, phosphorylation, and oxidation. *The Journal of Biological Chemistry* Vol. 277, 27045-27052

Esguerra, M.; Wang, J.; Foster, C.D.; Adelman, J.P.; North, R.A. & Levitan, I.B. (1994) Cloned Ca(2+)-dependent K+ channel modulated by a functionally associated protein kinase. *Nature* Vol. 369, 563-565

Fakler, B.; Brändle, U.; Glowatzki, E.; Weidemann, S.; Zenner, H.P. & Ruppersberg, J.P. (1995) Strong voltage-dependent inward rectification of inward rectifier K+ channels is caused by intracellular spermine. *Cell* Vol. 80, 149-154

Feinberg-Zadek, P.L.; Martin, G. & Treistman, S.N. (2008) BK channel subunit composition modulates molecular tolerance to ethanol. *Alcoholism, Clinical and Experimental Research* Vol. 32, 1207-1216

Félétou, M. (2009) Calcium-activated potassium channels and endothelial dysfunction: therapeutic options? *British Journal of Pharmacology* Vol. 156, 545-562

Fodor, A.A. & Aldrich, R.W. (2009) Convergent evolution of alternative splices at domain boundaries of the BK channel. *Annual Review of Physiology* Vol. 71, 19-36

Franks, N.P. & Lieb, W.R. (1987) Are the biological effects of ethanol due to primary interactions with lipids or with proteins? *Alcohol and Alcoholism (Oxford, Oxfordshire). Supplement* Vol. 1, 139-145

Gadalla, M.M. & Snyder, S.H. (2010) Hydrogen sulfide as a gasotransmitter. *Journal of Neurochemistry* Vol. 113, 14-26

Gardos, G. (1958) The function of calcium in the potassium permeability of human erythrocytes. *Biochimica Et Biophysica Acta* Vol. 30, 653-654

Garthwaite, J. (2010) New insight into the functioning of nitric oxide-receptive guanylyl cyclase: physiological and pharmacological implications. *Molecular and Cellular Biochemistry* Vol. 334, 221-232

Gasull, X.; Ferrer, E.; Llobet, A.; Castellano, A.; Nicolás, J.M.; Palés, J. & Gual, A. (2003) Cell membrane stretch modulates the high-conductance Ca2+-activated K+ channel in bovine trabecular meshwork cells. *Investigative Ophthalmology & Visual Science* Vol. 44, 706-714

Ghezzi, A.; Al-Hasan, Y.M.; Larios, L.E.; Bohm, R.A. & Atkinson, N.S. (2004) slo K(+) channel gene regulation mediates rapid drug tolerance. *Proceedings of the National Academy of Sciences of the United States of America* Vol. 101, 17276-17281

Ghezzi, A.; Pohl, J.B.; Wang, Y. & Atkinson, N.S. (2010) BK channels play a counter-adaptive role in drug tolerance and dependence. *Proceedings of the National Academy of Sciences of the United States of America* Vol. 107, 16360-16365

Gho, M. & Ganetzky, B. (1992) Analysis of repolarization of presynaptic motor terminals in Drosophila larvae using potassium-channel-blocking drugs and mutations. *The Journal of Experimental Biology* Vol. 170, 93-111

Gianoulakis, C. (2009) Endogenous opioids and addiction to alcohol and other drugs of abuse. *Current Topics in Medicinal Chemistry* Vol. 9, 999-1015

Gong, L.; Gao, T.M.; Huang, H. & Tong, Z. (2000) Redox modulation of large conductance calcium-activated potassium channels in CA1 pyramidal neurons from adult rat hippocampus. *Neuroscience Letters* Vol. 286, 191-194

Gonzales, R.A. & Hoffman, P.L. (1991) Receptor-gated ion channels may be selective CNS targets for ethanol. *Trends in Pharmacological Sciences* Vol. 12, 1-3

Goodwin, L.R.; Francom, D.; Dieken, F.P.; Taylor, J.D.; Warenycia, M.W.; Reiffenstein, R.J. & Dowling, G. (1989) Determination of sulfide in brain tissue by gas dialysis/ion chromatography: postmortem studies and two case reports. *Journal of Analytical Toxicology* Vol. 13, 105-109

Gorman, A.L. & Hermann, A. (1979) Internal effects of divalent cations on potassium permeability in molluscan neurones. *The Journal of Physiology* Vol. 296, 393-410

Gorman, A.L. & Hermann, A. (1982) Quantitative differences in the currents of bursting and beating molluscan pace-maker neurones. *The Journal of Physiology* Vol. 333, 681-699

Gorman, A.L.; Hermann, A. & Thomas, M.V. (1981) Intracellular calcium and the control of neuronal pacemaker activity. *Federation Proceedings* Vol. 40, 2233-2239

Gorman, A.L.; Hermann, A. & Thomas, M.V. (1982) Ionic requirements for membrane oscillations and their dependence on the calcium concentration in a molluscan pace-maker neurone. *The Journal of Physiology* Vol. 327, 185-217

Gribkoff, V.K.; Starrett, J.E., Jr. & Dworetzky, S.I. (2001) Maxi-K potassium channels: form, function, and modulation of a class of endogenous regulators of intracellular calcium. *The Neuroscientist: A Review Journal Bringing Neurobiology, Neurology and Psychiatry* Vol. 7, 166-177

Grimm, P.R. & Sansom, S.C. (2010) BK channels and a new form of hypertension. *Kidney International* Vol. 78, 956-962

Grunnet, M.; MacAulay, N.; Jorgensen, N.K.; Jensen, S.; Olesen, S.-P. & Klaerke, D.A. (2002) Regulation of cloned, Ca2+-activated K+ channels by cell volume changes. *Pflügers Archiv: European Journal of Physiology* Vol. 444, 167-177

Gruss, M.; Henrich, M.; König, P.; Hempelmann, G.; Vogel, W. & Scholz, A. (2001) Ethanol reduces excitability in a subgroup of primary sensory neurons by activation of BK(Ca) channels. *The European Journal of Neuroscience* Vol. 14, 1246-1256

Hall, S.K. & Armstrong, D.L. (2000) Conditional and unconditional inhibition of calcium-activated potassium channels by reversible protein phosphorylation. *The Journal of Biological Chemistry* Vol. 275, 3749-3754

Hammami, S.; Willumsen, N.J.; Olsen, H.L.; Morera, F.J.; Latorre, R. & Klaerke, D.A. (2009) Cell volume and membrane stretch independently control K+ channel activity. *The Journal of Physiology* Vol. 587, 2225-2231

Handlechner, A.; Weiger, T.M.; Kainz, V. & Hermann, A. (2011) Acetaldehye and ethanol interactions on calcium activated potassium (BK) channels in Pituitary (GH3/GH4) cells. *Alcohol and Alcoholism* Vol. 46, 3-3

Handlechner, A.G.; Weiger, T.M.; Kainz, V. & Hermann, A. (2008) Acetaldehyde blocks the augmenting action of ethanol on BK channels in pituitary (GH3) cells. *Alcohol – Clinical and Experimental Research* Vol. 32, 29A-29A

Harris, R.A.; Trudell, J.R. & Mihic, S.J. (2008) Ethanol's molecular targets. *Science Signaling* Vol. 1, re7

Haug, T.; Sigg, D.; Ciani, S.; Toro, L.; Stefani, E. & Olcese, R. (2004) Regulation of K+ flow by a ring of negative charges in the outer pore of BKCa channels. Part I: Aspartate 292 modulates K+ conduction by external surface charge effect. *The Journal of General Physiology* Vol. 124, 173-184

Hermann, A. & Hartung, K. (1983) Ca2+ activated K+ conductance in molluscan neurones. *Cell Calcium* Vol. 4, 387-405

Herz, A. (1997) Endogenous opioid systems and alcohol addiction. *Psychopharmacology* Vol. 129, 99-111

Hill, M.A.; Yang, Y.; Ella, S.R.; Davis, M.J. & Braun, A.P. (2010) Large conductance, Ca2+-activated K+ channels (BKCa) and arteriolar myogenic signaling. *FEBS Letters* Vol. 584, 2033-2042

Hoffman, P.L. & Tabakoff, B. (1990) Ethanol and guanine nucleotide binding proteins: a selective interaction. *The FASEB Journal: Official Publication of the Federation of American Societies for Experimental Biology* Vol. 4, 2612-2622

Horishita, T. & Harris, R.A. (2008) n-Alcohols inhibit voltage-gated Na+ channels expressed in Xenopus oocytes. *The Journal of Pharmacology and Experimental Therapeutics* Vol. 326, 270-277

Hou, S.; Heinemann, S.H. & Hoshi, T. (2009) Modulation of BKCa channel gating by endogenous signaling molecules. *Physiology (Bethesda, Md.)* Vol. 24, 26-35

Howard, R. J.; Slesinger, P.A.; Davies, D.L.; Das, J.; Trudell, J.R. & Harris, R.A. (2011) Alcohol-binding sites in distinct brain proteins: the quest for atomic level resolution. *Alcohol Clin Exp Res.* Vol. 35, 1561-1573

Hu, L.-F.; Lu, M.; Hon Wong, P.T. & Bian, J.-S. (2011) Hydrogen sulfide: neurophysiology and neuropathology. *Antioxidants & Redox Signaling* Vol. 15, 405-419

Huang, C.J. & Moczydlowski, E. (2001) Cytoplasmic polyamines as permeant blockers and modulators of the voltage-gated sodium channel. *Biophysical Journal* Vol. 80, 1262-1279

Hunt, W.A. (1996) Role of acetaldehyde in the actions of ethanol on the brain--a review. *Alcohol (Fayetteville, N.Y.)* Vol. 13, 147-151

Igarashi, K. & Kashiwagi, K. (2010) Modulation of cellular function by polyamines. *The International Journal of Biochemistry & Cell Biology* Vol. 42, 39-51

Ishigami, M.; Hiraki, K.; Umemura, K.; Ogasawara, Y.; Ishii, K. & Kimura, H. (2009) A source of hydrogen sulfide and a mechanism of its release in the brain. *Antioxid Redox Signal* Vol. 11, 205-14

Jackson-Weaver, O.; Paredes, D.A.; Gonzalez Bosc, L.V.; Walker, B.R. & Kanagy, N.L. (2011) Intermittent hypoxia in rats increases myogenic tone through loss of hydrogen sulfide activation of large-conductance Ca(2+)-activated potassium channels. *Circulation Research* Vol. 108, 1439-1447

Jakab, M.; Schmidt, S.; Grundbichler, M.; Paulmichl, M.; Hermann, A.; Weiger, T. & Ritter, M. (2006) Hypotonicity and ethanol modulate BK channel activity and chloride currents in GH4/C1 pituitary tumour cells. *Acta Physiologica (Oxford, England)* Vol. 187, 51-59

Jakab, M.; Weiger, T.M. & Hermann, A. (1997) Ethanol activates maxi Ca2+-activated K+ channels of clonal pituitary (GH3) cells. *The Journal of Membrane Biology* Vol. 157, 237-245

Jiang, B.; Tang, G.; Cao, K.; Wu, L. & Wang, R. (2010) Molecular mechanism for H(2)S-induced activation of K(ATP) channels. *Antioxidants & Redox Signaling* Vol. 12, 1167-1178

Jiang, Y.; Lee, A.; Chen, J.; Cadene, M.; Chait, B.T. & MacKinnon, R. (2002) Crystal structure and mechanism of a calcium-gated potassium channel. *Nature* Vol. 417, 515-522

Jiang, Y.; Pico, A.; Cadene, M.; Chait, B.T. & MacKinnon, R. (2001) Structure of the RCK domain from the E. coli K+ channel and demonstration of its presence in the human BK channel. *Neuron* Vol. 29, 593-601

Kabil, O. & Banerjee, R. (2010) Redox biochemistry of hydrogen sulfide. *The Journal of Biological Chemistry* Vol. 285, 21903-21907

Kaczorowski, G.J.; Knaus, H.G.; Leonard, R.J.; McManus, O.B. & Garcia, M.L. (1996) High-conductance calcium-activated potassium channels; structure, pharmacology, and function. *Journal of Bioenergetics and Biomembranes* Vol. 28, 255-267

Karahanian, E.; Quintanilla, M.E.; Tampier, L.; Rivera-Meza, M.; Bustamante, D.; Gonzalez-Lira, V.; Morales, P.; Herrera-Marschitz, M. & Israel, Y. (2011) Ethanol as a prodrug: brain metabolism of ethanol mediates its reinforcing effects. *Alcoholism, Clinical and Experimental Research* Vol. 35, 606-612

Kawabata, A.; Ishiki, T.; Nagasawa, K.; Yoshida, S.; Maeda, Y.; Takahashi, T.; Sekiguchi, F.; Wada, T.; Ichida, S. & Nishikawa, H. (2007) Hydrogen sulfide as a novel nociceptive messenger. *Pain* Vol. 132, 74-81

Kawakubo, T.; Naruse, K.; Matsubara, T.; Hotta, N. & Sokabe, M. (1999) Characterization of a newly found stretch-activated KCa,ATP channel in cultured chick ventricular myocytes. *The American Journal of Physiology* Vol. 276, H1827-1838

Kelm, M.K.; Criswell, H.E. & Breese, G.R. (2011) Ethanol-enhanced GABA release: a focus on G protein-coupled receptors. *Brain Research Reviews* Vol. 65, 113-123

Kemp, P.J.; Telezhkin, V.; Wilkinson, W.J.; Mears, R.; Hanmer, S.B.; Gadeberg, H.C.; Müller, C.T.; Riccardi, D. & Brazier, S.P. (2009) Enzyme-linked oxygen sensing by potassium channels. *Annals of the New York Academy of Sciences* Vol. 1177, 112-118

Kerschbaum, H.H. & Hermann, A. (1997) Ethanol suppresses neuronal Ca2+ currents by effects on intracellular signal transduction. *Brain Research* Vol. 765, 30-36

Kimura, H. (2011) Hydrogen sulfide: its production, release and functions. *Amino Acids* Vol. 41, 113-121

Kimura, Y. & Kimura, H. (2004) Hydrogen sulfide protects neurons from oxidative stress. *The FASEB Journal: Official Publication of the Federation of American Societies for Experimental Biology* Vol. 18, 1165-1167

Knaus, H.G.; Schwarzer, C.; Koch, R.O.; Eberhart, A.; Kaczorowski, G.J.; Glossmann, H.; Wunder, F.; Pongs, O.; Garcia, M.L. & Sperk, G. (1996) Distribution of high-conductance Ca(2+)-activated K+ channels in rat brain: targeting to axons and nerve terminals. *The Journal of Neuroscience: The Official Journal of the Society for Neuroscience* Vol. 16, 955-963

Knott, T.K.; Dopico, A.M.; Dayanithi, G.; Lemos, J. & Treistman, S.N. (2002) Integrated channel plasticity contributes to alcohol tolerance in neurohypophysial terminals. *Molecular Pharmacology* Vol. 62, 135-142

Kruse, S.W.; Zhao, R.; Smith, D.P. & Jones, D.N.M. (2003) Structure of a specific alcohol-binding site defined by the odorant binding protein LUSH from Drosophila melanogaster. *Nature Structural Biology* Vol. 10, 694-700

Kubo, S.; Doe, I.; Kurokawa, Y. & Kawabata, A. (2007) Hydrogen sulfide causes relaxation in mouse bronchial smooth muscle. *Journal of Pharmacological Sciences* Vol. 104, 392-396

Lahnsteiner, E. & Hermann, A. (1995) Acute action of ethanol on rat hippocampal CA1 neurons: effects on intracellular signaling. *Neuroscience Letters* Vol. 191, 153-156

Lam, R.S.; Shaw, A.R. & Duszyk, M. (2004) Membrane cholesterol content modulates activation of BK channels in colonic epithelia. *Biochimica Et Biophysica Acta* Vol. 1667, 241-248

Latorre, R.; Oberhauser, A.; Labarca, P. & Alvarez, O. (1989) Varieties of calcium-activated potassium channels. *Annual Review of Physiology* Vol. 51, 385-399

Lee, U.S. & Cui, J. (2010) BK channel activation: structural and functional insights. *Trends in Neurosciences* Vol. 33, 415-423

Leffler, C.W.; Parfenova, H.; Jaggar, J.H. & Wang, R. (2006) Carbon monoxide and hydrogen sulfide: gaseous messengers in cerebrovascular circulation. *Journal of Applied Physiology* Vol. 100, 1065-1076

Levitan, I.B. (1994) Modulation of ion channels by protein phosphorylation and dephosphorylation. *Annual Review of Physiology* Vol. 56, 193-212

Li, L. & Moore, P.K. (2008) Putative biological roles of hydrogen sulfide in health and disease: a breath of not so fresh air? *Trends in Pharmacological Sciences* Vol. 29, 84-90

Li, L.; Rose, P. & Moore, P.K. (2011) Hydrogen sulfide and cell signaling. *Annual Review of Pharmacology and Toxicology* Vol. 51, 169-187

Li, M.; Sun, Y.; Simard, J.M.; Wang, J.-Y. & Chai, T.C. (2009) Augmented bladder urothelial polyamine signaling and block of BK channel in the pathophysiology of overactive bladder syndrome. *Am J Physiol Cell Physiol* Vol. 297, C1445-C1451

Li, Q.; Sun, B.; Wang, X.; Jin, Z.; Zhou, Y.; Dong, L.; Jiang, L.-H. & Rong, W. (2010) A crucial role for hydrogen sulfide in oxygen sensing via modulating large conductance calcium-activated potassium channels. *Antioxidants & Redox Signaling* Vol. 12, 1179-1189

Li, W. & Aldrich, R.W. (2004) Unique inner pore properties of BK channels revealed by quaternary ammonium block. *The Journal of General Physiology* Vol. 124, 43-57

Liang, G.H.; Adebiyi, A.; Leo, M.D.; McNally, E.M.; Leffler, C.W. & Jaggar, J.H. (2011) Hydrogen sulfide dilates cerebral arterioles by activating smooth muscle cell plasma membrane KATP channels. *American Journal of Physiology. Heart and Circulatory Physiology* Vol. 300, H2088-2095

Lin, M.-W.; Wu, A.Z.; Ting, W.-H.; Li, C.-L.; Cheng, K.-S. & Wu, S.-N. (2006) Changes in membrane cholesterol of pituitary tumor (GH3) cells regulate the activity of large-conductance Ca2+-activated K+ channels. *The Chinese Journal of Physiology* Vol. 49, 1-13

Liu, G.; Zakharov, S.I.; Yang, L.; Deng, S.-X.; Landry, D.W.; Karlin, A. & Marx, S.O. (2008) Position and role of the BK channel alpha subunit S0 helix inferred from disulfide crosslinking. *The Journal of General Physiology* Vol. 131, 537-548

Liu, J.; Asuncion-Chin, M.; Liu, P. & Dopico, A.M. (2006) CaM kinase II phosphorylation of slo Thr107 regulates activity and ethanol responses of BK channels. *Nature Neuroscience* Vol. 9, 41-49

Liu, P.; Liu, J.; Huang, W.; Li, M.D. & Dopico, A.M. (2003) Distinct regions of the slo subunit determine differential BKCa channel responses to ethanol. *Alcoholism, Clinical and Experimental Research* Vol. 27, 1640-1644

Liu, W.Q.; Chai, C.; Li, X.Y.; Yuan, W.J.; Wang, W.Z. & Lu, Y. (2011) The cardiovascular effects of central hydrogen sulphide are related to K(ATP) channels activation. *Physiol Res* Vol. 60, 729-738

Lloyd, D. (2006) Hydrogen sulfide: clandestine microbial messenger? *Trends in Microbiology* Vol. 14, 456-462

Lobo, I.A. & Harris, R.A. (2008) GABA(A) receptors and alcohol. *Pharmacology, Biochemistry, and Behavior* Vol. 90, 90-94

Lorenz, S.; Heils, A.; Kasper, J.M. & Sander, T. (2007) Allelic association of a truncation mutation of the KCNMB3 gene with idiopathic generalized epilepsy. *American Journal of Medical Genetics. Part B, Neuropsychiatric Genetics: The Official Publication of the International Society of Psychiatric Genetics* Vol. 144B, 10-13

Łowicka, E. & Bełtowski, J. (2007) Hydrogen sulfide (H2S) - the third gas of interest for pharmacologists. *Pharmacological Reports: PR* Vol. 59, 4-24

Lu, R.; Alioua, A.; Kumar, Y.; Eghbali, M.; Stefani, E. & Toro, L. (2006) MaxiK channel partners: physiological impact. *The Journal of Physiology* Vol. 570, 65-72

Lynch, C., 3rd. (2008) Meyer and Overton revisited. *Anesthesia and Analgesia* Vol. 107, 864-867

Ma, Z.; Lou, X.J. & Horrigan, F.T. (2006) Role of charged residues in the S1-S4 voltage sensor of BK channels. *The Journal of General Physiology* Vol. 127, 309-328

Madsen, B.W. & Edeson, R.O. (1990) Ethanol enhancement of a calcium-activated potassium current in an identified molluscan neuron. *Brain Research* Vol. 528, 323-326

Maeda, Y.; Aoki, Y.; Sekiguchi, F.; Matsunami, M.; Takahashi, T.; Nishikawa, H. & Kawabata, A. (2009) Hyperalgesia induced by spinal and peripheral hydrogen sulfide: evidence for involvement of Cav3.2 T-type calcium channels. *Pain* Vol. 142, 127-32

Maher, A.D. & Kuchel, P.W. (2003) The Gárdos channel: a review of the Ca2+-activated K+ channel in human erythrocytes. *The International Journal of Biochemistry & Cell Biology* Vol. 35, 1182-1197

Mancardi, D.; Penna, C.; Merlino, A.; Del Soldato, P.; Wink, D.A. & Pagliaro, P. (2009) Physiological and pharmacological features of the novel gasotransmitter: hydrogen sulfide. *Biochimica Et Biophysica Acta* Vol. 1787, 864-872

Martin, G.; Puig, S.; Pietrzykowski, A.; Zadek, P.; Emery, P. & Treistman, S. (2004) Somatic localization of a specific large-conductance calcium-activated potassium channel subtype controls compartmentalized ethanol sensitivity in the nucleus accumbens. *The Journal of Neuroscience: The Official Journal of the Society for Neuroscience* Vol. 24, 6563-6572

Martin, G.E. (2010) BK channel and alcohol, a complicated affair. *International Review of Neurobiology* Vol. 91, 321-338

Martin, G.E.; Hendrickson, L.M.; Penta, K.L.; Friesen, R.M.; Pietrzykowski, A.Z.; Tapper, A.R. & Treistman, S.N. (2008) Identification of a BK channel auxiliary protein controlling molecular and behavioral tolerance to alcohol. *Proceedings of the National Academy of Sciences of the United States of America* Vol. 105, 17543-17548

Matsunami, M.; Tarui, T.; Mitani, K.; Nagasawa, K.; Fukushima, O.; Okubo, K.; Yoshida, S.; Takemura, M. & Kawabata, A. (2009) Luminal hydrogen sulfide plays a pronociceptive role in mouse colon. *Gut* Vol. 58, 751-61

McIntire, S.L. (2010) Ethanol. WormBook, ed. The C. elegans Research Community, WormBook, doi/10.1895/wormbook.1.40.1, http://www.wormbook.org.

Meech, R.W. (1978) Calcium-dependent potassium activation in nervous tissues. *Annual Review of Biophysics and Bioengineering* Vol. 7, 1-18

Meech, R.W. & Standen, N.B. (1975) Potassium activation in Helix aspersa neurones under voltage clamp: a component mediated by calcium influx. *The Journal of Physiology* Vol. 249, 211-239

Meera, P.; Wallner, M.; Song, M. & Toro, L. (1997) Large conductance voltage- and calcium-dependent K+ channel, a distinct member of voltage-dependent ion channels with seven N-terminal transmembrane segments (S0-S6), an extracellular N terminus, and an intracellular (S9-S10) C terminus. *Proceedings of the National Academy of Sciences of the United States of America* Vol. 94, 14066-14071

Meera, P.; Wallner, M. & Toro, L. (2000) A neuronal beta subunit (KCNMB4) makes the large conductance, voltage- and Ca2+-activated K+ channel resistant to charybdotoxin and iberiotoxin. *Proceedings of the National Academy of Sciences of the United States of America* Vol. 97, 5562-5567

Melis, M.; Enrico, P.; Peana, A.T. & Diana, M. (2007) Acetaldehyde mediates alcohol activation of the mesolimbic dopamine system. *The European Journal of Neuroscience* Vol. 26, 2824-2833

Messing, R.O.; Petersen, P.J. & Henrich, C.J. (1991) Chronic ethanol exposure increases levels of protein kinase C delta and epsilon and protein kinase C-mediated phosphorylation in cultured neural cells. *The Journal of Biological Chemistry* Vol. 266, 23428-23432

Mironov, S.L. & Hermann, A. (1996) Ethanol actions on the mechanisms of Ca2+ mobilization in rat hippocampal cells are mediated by protein kinase C. *Brain Research* Vol. 714, 27-37

Moczydlowski, E.; Alvarez, O.; Vergara, C. & Latorre, R. (1985) Effect of phospholipid surface charge on the conductance and gating of a Ca2+-activated K+ channel in planar lipid bilayers. *The Journal of Membrane Biology* Vol. 83, 273-282

Morales, J.A.; Ram, J.L.; Song, J. & Brown, R.A. (1997) Acetaldehyde inhibits current through voltage-dependent calcium channels. *Toxicology and Applied Pharmacology* Vol. 143, 70-74

Morrow, J.P.; Zakharov, S.I.; Liu, G.; Yang, L.; Sok, A.J. & Marx, S.O. (2006) Defining the BK channel domains required for beta1-subunit modulation. *Proceedings of the National Academy of Sciences of the United States of America* Vol. 103, 5096-5101

Mulholland, P.J.; Hopf, F.W.; Bukiya, A.N.; Martin, G.E.; Liu, J.; Dopico, A.M.; Bonci, A.; Treistman, S.N. & Chandler, L.J. (2009) Sizing up ethanol-induced plasticity: the role of small and large conductance calcium-activated potassium channels. *Alcoholism, Clinical and Experimental Research* Vol. 33, 1125-1135

Mustafa, A.K.; Gadalla, M.M.; Sen, N.; Kim, S.; Mu, W.; Gazi, S.K.; Barrow, R.K.; Yang, G.; Wang, R. & Snyder, S.H. (2009) H2S signals through protein S-sulfhydration. *Science Signaling* Vol. 2, ra72

Newton, P.M. & Messing, R.O. (2006) Intracellular signaling pathways that regulate behavioral responses to ethanol. *Pharmacology & Therapeutics* Vol. 109, 227-237

Newton, P.M. & Ron, D. (2007) Protein kinase C and alcohol addiction. *Pharmacological Research: The Official Journal of the Italian Pharmacological Society* Vol. 55, 570-577

Nicoll, R.A. & Madison, D.V. (1982) General anesthetics hyperpolarize neurons in the vertebrate central nervous system. *Science (New York, N.Y.)* Vol. 217, 1055-1057

Niesen, C.E.; Baskys, A. & Carlen, P.L. (1988) Reversed ethanol effects on potassium conductances in aged hippocampal dentate granule neurons. *Brain Research* Vol. 445, 137-141

Nimigean, C.M.; Chappie, J.S. & Miller, C. (2003) Electrostatic tuning of ion conductance in potassium channels. *Biochemistry* Vol. 42, 9263-9268

Nishimura, S.; Fukushima, O.; Ishikura, H.; Takahashi, T.; Matsunami, M.; Tsujiuchi, T.; Sekiguchi, F.; Naruse, M.; Kamanaka, Y. & Kawabata, A. (2009) Hydrogen sulfide as a novel mediator for pancreatic pain in rodents. *Gut* Vol. 58, 762-70

Omodeo-Salé, F.; Pitto, M.; Masserini, M. & Palestini, P. (1995) Effects of chronic ethanol exposure on cultured cerebellar granule cells. *Molecular and Chemical Neuropathology / Sponsored by the International Society for Neurochemistry and the World Federation of Neurology and Research Groups on Neurochemistry and Cerebrospinal Fluid* Vol. 26, 159-169

Ostrovskaya, O.; Asatryan, L.; Wyatt, L.; Popova, M.; Li, K.; Peoples, R.W.; Alkana, R.L. & Davies, D.L. (2011) Ethanol is a fast channel inhibitor of P2X4 receptors. *The Journal of Pharmacology and Experimental Therapeutics* Vol. 337, 171-179

Oyama, Y.; Akaike, N. & Nishi, K. (1986) Effects of n-alkanols on the calcium current of intracellularly perfused neurons of Helix aspersa. *Brain Research* Vol. 376, 280-284

Patton, C.; Thompson, S. & Epel, D. (2004) Some precautions in using chelators to buffer metals in biological solutions. *Cell Calcium* Vol. 35, 427-431

Peng, H.; Cheng, Y.; Dai, C.; King, A.L.; Predmore, B.L.; Lefer, D.J. & Wang, B. (2011) A Fluorescent Probe for Fast and Quantitative Detection of Hydrogen Sulfide in Blood. *Angewandte Chemie (International Ed. in English)* Vol. 50, 9672-9675

Petrik, D. & Brenner, R. (2007) Regulation of STREX exon large conductance, calcium-activated potassium channels by the beta4 accessory subunit. *Neuroscience* Vol. 149, 789-803

Pietrzykowski, A.Z.; Friesen, R.M.; Martin, G.E.; Puig, S.I.; Nowak, C.L.; Wynne, P.M.; Siegelmann, H.T. & Treistman, S.N. (2008) Posttranscriptional regulation of BK channel splice variant stability by miR-9 underlies neuroadaptation to alcohol. *Neuron* Vol. 59, 274-287

Pietrzykowski, A.Z.; Martin, G.E.; Puig, S.I.; Knott, T.K.; Lemos, J.R. & Treistman, S.N. (2004) Alcohol tolerance in large-conductance, calcium-activated potassium channels of CNS terminals is intrinsic and includes two components: decreased ethanol potentiation and decreased channel density. *The Journal of Neuroscience: The Official Journal of the Society for Neuroscience* Vol. 24, 8322-8332

Pluznick, J.L. & Sansom, S.C. (2006) BK channels in the kidney: role in K(+) secretion and localization of molecular components. *American Journal of Physiology. Renal Physiology* Vol. 291, F517-529

Qu, K.; Lee, S.W.; Bian, J.S.; Low, C.M. & Wong, P.T.H. (2008) Hydrogen sulfide: neurochemistry and neurobiology. *Neurochemistry International* Vol. 52, 155-165

Quertemont, E. & De Witte, P. (2001) Conditioned stimulus preference after acetaldehyde but not ethanol injections. *Pharmacology, Biochemistry, and Behavior* Vol. 68, 449-454

Quertemont, E. & Tambour, S. (2004) Is ethanol a pro-drug? The role of acetaldehyde in the central effects of ethanol. *Trends in Pharmacological Sciences* Vol. 25, 130-134

Quertemont, E.; Tambour, S. & Tirelli, E. (2005) The role of acetaldehyde in the neurobehavioral effects of ethanol: a comprehensive review of animal studies. *Progress in Neurobiology* Vol. 75, 247-274

Ramaswamy, S.; el Ahmad, M.; Danielsson, O.; Jörnvall, H. & Eklund, H. (1996) Crystal structure of cod liver class I alcohol dehydrogenase: substrate pocket and structurally variable segments. *Protein Science: A Publication of the Protein Society* Vol. 5, 663-671

Reiffenstein, R.J.; Hulbert, W.C. & Roth, S.H. (1992) Toxicology of hydrogen sulfide. *Annual Review of Pharmacology and Toxicology* Vol. 32, 109-134

Reinhart, P.H.; Chung, S.; Martin, B.L.; Brautigan, D.L. & Levitan, I.B. (1991) Modulation of calcium-activated potassium channels from rat brain by protein kinase A and phosphatase 2A. *The Journal of Neuroscience: The Official Journal of the Society for Neuroscience* Vol. 11, 1627-1635

Rodd-Henricks, Z.A.; Melendez, R.I.; Zaffaroni, A.; Goldstein, A.; McBride, W.J. & Li, T.-K. (2002) The reinforcing effects of acetaldehyde in the posterior ventral tegmental area of alcohol-preferring rats. *Pharmacology, Biochemistry, and Behavior* Vol. 72, 55-64

Ron, D. & Messing, R.O. (2011) Signaling Pathways Mediating Alcohol Effects. *Current Topics in Behavioral Neurosciences* Vol.

Rosell, A.; Valencia, E.; Parés, X.; Fita, I.; Farrés, J. & Ochoa, W.F. (2003) Crystal structure of the vertebrate NADP(H)-dependent alcohol dehydrogenase (ADH8). *Journal of Molecular Biology* Vol. 330, 75-85

Salkoff, L.; Butler, A.; Ferreira, G.; Santi, C. & Wei, A. (2006) High-conductance potassium channels of the SLO family. *Nature Reviews. Neuroscience* Vol. 7, 921-931

Sato, N.; Wang, X.B.; Greer, M.A.; Greer, S.E. & McAdams, S. (1990) Evidence that ethanol induces prolactin secretion in GH4C1 cells by producing cell swelling with resultant calcium influx. *Endocrinology* Vol. 127, 3079-3086

Schopperle, W.M.; Holmqvist, M.H.; Zhou, Y.; Wang, J.; Wang, Z.; Griffith, L.C.; Keselman, I.; Kusinitz, F.; Dagan, D. & Levitan, I.B. (1998) Slob, a novel protein that interacts with the Slowpoke calcium-dependent potassium channel. *Neuron* Vol. 20, 565-573

Schreiber, M. & Salkoff, L. (1997) A novel calcium-sensing domain in the BK channel. *Biophysical Journal* Vol. 73, 1355-1363

Schrofner, S.; Zsombok, A.; Hermann, A. & Kerschbaum, H.H. (2004) Nitric oxide decreases a calcium-activated potassium current via activation of phosphodiesterase 2 in Helix U-cells. *Brain Res* Vol. 999, 98-105

Schubert, R. & Nelson, M.T. (2001) Protein kinases: tuners of the BKCa channel in smooth muscle. *Trends in Pharmacological Sciences* Vol. 22, 505-512

Seoh, S.A.; Sigg, D.; Papazian, D.M. & Bezanilla, F. (1996) Voltage-sensing residues in the S2 and S4 segments of the Shaker K+ channel. *Neuron* Vol. 16, 1159-1167

Shibuya, N.; Mikami, Y.; Kimura, Y.; Nagahara, N. & Kimura, H. (2009) Vascular endothelium expresses 3-mercaptopyruvate sulfurtransferase and produces hydrogen sulfide. *Journal of Biochemistry* Vol. 146, 623-626

Sitdikova, G.F.; Weiger, T.M. & Hermann, A. (2010) Hydrogen sulfide increases calcium-activated potassium (BK) channel activity of rat pituitary tumor cells. *Pflügers Archiv: European Journal of Physiology* Vol. 459, 389-397

Snetkov, V.A.; Gurney, A.M.; Ward, J.P. & Osipenko, O.N. (1996) Inward rectification of the large conductance potassium channel in smooth muscle cells from rabbit pulmonary artery. *Experimental Physiology* Vol. 81, 743-753

Stipanuk, M.H. & Beck, P.W. (1982) Characterization of the enzymic capacity for cysteine desulphhydration in liver and kidney of the rat. *The Biochemical Journal* Vol. 206, 267-277

Strege, P.R.; Bernard, C.E.; Kraichely, R.E.; Mazzone, A.; Sha, L.; Beyder, A.; Gibbons, S.J.; Linden, D.R.; Kendrick, M.L.; Sarr, M.G.; Szurszewski, J.H. & Farrugia, G. (2011) Hydrogen sulfide is a partially redox-independent activator of the human jejunum Na+ channel, Nav1.5. *American Journal of Physiology. Gastrointestinal and Liver Physiology* Vol. 300, G1105-1114

Stubbs, C.D. & Slater, S.J. (1999) Ethanol and protein kinase C. *Alcoholism, Clinical and Experimental Research* Vol. 23, 1552-1560

Sun, Y.G.; Cao, Y.X.; Wang, W.W.; Ma, S.F.; Yao, T. & Zhu, Y.C. (2008) Hydrogen sulphide is an inhibitor of L-type calcium channels and mechanical contraction in rat cardiomyocytes. *Cardiovasc Res* Vol. 79, 632-41

Szabó, C. (2007) Hydrogen sulphide and its therapeutic potential. *Nature Reviews. Drug Discovery* Vol. 6, 917-935

Tan, B.H.; Wong, P.T.H. & Bian, J.-S. (2010) Hydrogen sulfide: a novel signaling molecule in the central nervous system. *Neurochemistry International* Vol. 56, 3-10

Tang, G.; Wu, L.; Liang, W. & Wang, R. (2005) Direct stimulation of K(ATP) channels by exogenous and endogenous hydrogen sulfide in vascular smooth muscle cells. *Mol Pharmacol* Vol. 68, 1757-64

Tang, G.; Wu, L. & Wang, R. (2010) Interaction of hydrogen sulfide with ion channels. *Clinical and Experimental Pharmacology & Physiology* Vol. 37, 753-763

Tang, X.D.; Daggett, H.; Hanner, M.; Garcia, M.L.; McManus, O.B.; Brot, N.; Weissbach, H.; Heinemann, S.H. & Hoshi, T. (2001) Oxidative regulation of large conductance calcium-activated potassium channels. *The Journal of General Physiology* Vol. 117, 253-274

Taraschi, T.F.; Ellingson, J.S.; Janes, N. & Rubin, E. (1991) The role of anionic phospholipids in membrane adaptation to ethanol. *Alcohol and Alcoholism (Oxford, Oxfordshire). Supplement* Vol. 1, 241-245

Telezhkin, V.; Brazier, S.P.; Cayzac, S.; Müller, C.T.; Riccardi, D. & Kemp, P.J. (2009) Hydrogen sulfide inhibits human BK(Ca) channels. *Advances in Experimental Medicine and Biology* Vol. 648, 65-72

Telezhkin, V.; Brazier, S.P.; Cayzac, S.H.; Wilkinson, W.J.; Riccardi, D. & Kemp, P.J. (2010) Mechanism of inhibition by hydrogen sulfide of native and recombinant BKCa channels. *Respiratory Physiology & Neurobiology* Vol. 172, 169-178

Tian, L.; Coghill, L.S.; McClafferty, H.; MacDonald, S.H.F.; Antoni, F.A.; Ruth, P.; Knaus, H.-G. & Shipston, M.J. (2004) Distinct stoichiometry of BKCa channel tetramer phosphorylation specifies channel activation and inhibition by cAMP-dependent protein kinase. *Proceedings of the National Academy of Sciences of the United States of America* Vol. 101, 11897-11902

Toro, L.; Ramos-Franco, J. & Stefani, E. (1990) GTP-dependent regulation of myometrial KCa channels incorporated into lipid bilayers. *The Journal of General Physiology* Vol. 96, 373-394

Torres, Y.P.; Morera, F.J.; Carvacho, I. & Latorre, R. (2007) A Marriage of Convenience: β-Subunits and Voltage-dependent K+ Channels. *Journal of Biological Chemistry* Vol. 282, 24485-24489

Treistman, S.N. & Martin, G.E. (2009) BK Channels: mediators and models for alcohol tolerance. *Trends in Neurosciences* Vol. 32, 629-637

Velázquez-Marrero, C.; Wynne, P.; Bernardo, A.; Palacio, S.; Martin, G. & Treistman, S.N. (2011) The relationship between duration of initial alcohol exposure and persistence of molecular tolerance is markedly nonlinear. *The Journal of Neuroscience: The Official Journal of the Society for Neuroscience* Vol. 31, 2436-2446

Wallace, J.L. (2007) Hydrogen sulfide-releasing anti-inflammatory drugs. *Trends in Pharmacological Sciences* Vol. 28, 501-505

Wallace, J.L. (2010) Physiological and pathophysiological roles of hydrogen sulfide in the gastrointestinal tract. *Antioxidants & Redox Signaling* Vol. 12, 1125-1133

Wallner, M.; Meera, P. & Toro, L. (1996) Determinant for beta-subunit regulation in high-conductance voltage-activated and Ca(2+)-sensitive K+ channels: an additional transmembrane region at the N terminus. *Proceedings of the National Academy of Sciences of the United States of America* Vol. 93, 14922-14927

Walters, F.S.; Covarrubias, M. & Ellingson, J.S. (2000) Potent inhibition of the aortic smooth muscle maxi-K channel by clinical doses of ethanol. *American journal of physiology. Cell physiology* Vol. 279, C1107-1115

Wang, R. (2002) Two's company, three's a crowd: can H2S be the third endogenous gaseous transmitter? *The FASEB Journal: Official Publication of the Federation of American Societies for Experimental Biology* Vol. 16, 1792-1798

Wang, R. (2010) Hydrogen sulfide: the third gasotransmitter in biology and medicine. *Antioxidants & Redox Signaling* Vol. 12, 1061-1064

Wang, R. (2011) Signaling pathways for the vascular effects of hydrogen sulfide. *Current Opinion in Nephrology and Hypertension* Vol. 20, 107-112

Wang, W.; Huang, H.; Hou, D.; Liu, P.; Wei, H.; Fu, X. & Niu, W. (2010) Mechanosensitivity of STREX-lacking BKCa channels in the colonic smooth muscle of the mouse. *American Journal of Physiology. Gastrointestinal and Liver Physiology* Vol. 299, G1231-1240

Wang, X.M.; Dayanithi, G.; Lemos, J.R.; Nordmann, J.J. & Treistman, S.N. (1991) Calcium currents and peptide release from neurohypophysial terminals are inhibited by ethanol. *The Journal of Pharmacology and Experimental Therapeutics* Vol. 259, 705-711

Wang, X.M.; Lemos, J.R.; Dayanithi, G.; Nordmann, J.J. & Treistman, S.N. (1991) Ethanol reduces vasopressin release by inhibiting calcium currents in nerve terminals. *Brain Research* Vol. 551, 338-341

Wang, Y.; Ghezzi, A.; Yin, J.C.P. & Atkinson, N.S. (2009) CREB regulation of BK channel gene expression underlies rapid drug tolerance. *Genes, Brain, and Behavior* Vol. 8, 369-376

Wang, Y.; Krishnan, H.R.; Ghezzi, A.; Yin, J.C.P. & Atkinson, N.S. (2007) Drug-induced epigenetic changes produce drug tolerance. *PLoS Biology* Vol. 5, e265-e265

Wang, Z.W.; Nara, M.; Wang, Y.X. & Kotlikoff, M.I. (1997) Redox regulation of large conductance Ca(2+)-activated K+ channels in smooth muscle cells. *The Journal of General Physiology* Vol. 110, 35-44

Wanner, S.G.; Koch, R.O.; Koschak, A.; Trieb, M.; Garcia, M.L.; Kaczorowski, G.J. & Knaus, H.G. (1999) High-conductance calcium-activated potassium channels in rat brain: pharmacology, distribution, and subunit composition. *Biochemistry* Vol. 38, 5392-5400

Wei, A.; Solaro, C.; Lingle, C. & Salkoff, L. (1994) Calcium sensitivity of BK-type KCa channels determined by a separable domain. *Neuron* Vol. 13, 671-681

Wei, A.D.; Gutman, G.A.; Aldrich, R.; Chandy, K.G.; Grissmer, S. & Wulff, H. (2005) International Union of Pharmacology. LII. Nomenclature and molecular relationships of calcium-activated potassium channels. *Pharmacological Reviews* Vol. 57, 463-472

Weiger, T. & Hermann, A. (1994) Polyamines block Ca(2+)-activated K+ channels in pituitary tumor cells (GH3). *The Journal of Membrane Biology* Vol. 140, 133-142

Weiger, T.M. & Hermann, A. (2009) Modulation of potassium channels by polyamines, in *Biological aspects of biogenic amines, polyamines and conjugates; Editor Dandrifosse G.* pp 185-199, Transworld Research Network, Kerala, India.

Weiger, T.M.; Hermann, A. & Levitan, I.B. (2002) Modulation of calcium-activated potassium channels. *Journal of Comparative Physiology. A, Neuroethology, Sensory, Neural, and Behavioral Physiology* Vol. 188, 79-87

Weiger, T.M.; Holmqvist, M.H.; Levitan, I.B.; Clark, F.T.; Sprague, S.; Huang, W.J.; Ge, P.; Wang, C.; Lawson, D.; Jurman, M.E.; Glucksmann, M.A.; Silos-Santiago, I.; DiStefano, P.S. & Curtis, R. (2000) A novel nervous system beta subunit that downregulates human large conductance calcium-dependent potassium channels. *The Journal of Neuroscience: The Official Journal of the Society for Neuroscience* Vol. 20, 3563-3570

Weiger, T.M.; Langer, T. & Hermann, A. (1998) External action of di- and polyamines on maxi calcium-activated potassium channels: an electrophysiological and molecular modeling study. *Biophysical Journal* Vol. 74, 722-730

Williams, K. (1997) Interactions of polyamines with ion channels. *The Biochemical Journal* Vol. 325 (Pt 2), 289-297

Wood, W.G.; Schroeder, F.; Hogy, L.; Rao, A.M. & Nemecz, G. (1990) Asymmetric distribution of a fluorescent sterol in synaptic plasma membranes: effects of chronic ethanol consumption. *Biochimica Et Biophysica Acta* Vol. 1025, 243-246

Wu, L. & Wang, R. (2005) Carbon monoxide: endogenous production, physiological functions, and pharmacological applications. *Pharmacological Reviews* Vol. 57, 585-630

Wu, L.; Yang, W.; Jia, X.; Yang, G.; Duridanova, D.; Cao, K. & Wang, R. (2009) Pancreatic islet overproduction of H2S and suppressed insulin release in Zucker diabetic rats. *Laboratory Investigation; a Journal of Technical Methods and Pathology* Vol. 89, 59-67

Wu, Y.; Yang, Y.; Ye, S. & Jiang, Y. (2010) Structure of the gating ring from the human large-conductance Ca(2+)-gated K(+) channel. *Nature* Vol. 466, 393-397

Xia, X.M.; Hirschberg, B.; Smolik, S.; Forte, M. & Adelman, J.P. (1998) dSLo interacting protein 1, a novel protein that interacts with large-conductance calcium-activated potassium channels. *The Journal of Neuroscience: The Official Journal of the Society for Neuroscience* Vol. 18, 2360-2369

Xie, J. & McCobb, D.P. (1998) Control of alternative splicing of potassium channels by stress hormones. *Science (New York, N.Y.)* Vol. 280, 443-446

Yang, W.; Yang, G.; Jia, X.; Wu, L. & Wang, R. (2005) Activation of KATP channels by H_2S in rat insulin-secreting cells and the underlying mechanisms. *The Journal of Physiology* Vol. 569, 519-531

Yuan, C.; O'Connell, R.J.; Feinberg-Zadek, P.L.; Johnston, L.J. & Treistman, S.N. (2004) Bilayer thickness modulates the conductance of the BK channel in model membranes. *Biophysical Journal* Vol. 86, 3620-3633

Yuan, C.; O'Connell, R.J.; Jacob, R.F.; Mason, R.P. & Treistman, S.N. (2007) Regulation of the gating of BKCa channel by lipid bilayer thickness. *The Journal of Biological Chemistry* Vol. 282, 7276-7286

Yuan, C.; O'Connell, R.J.; Wilson, A.; Pietrzykowski, A.Z. & Treistman, S.N. (2008) Acute alcohol tolerance is intrinsic to the BKCa protein, but is modulated by the lipid environment. *The Journal of Biological Chemistry* Vol. 283, 5090-5098

Zhang, Y.; Niu, X.; Brelidze, T.I. & Magleby, K.L. (2006) Ring of negative charge in BK channels facilitates block by intracellular Mg2+ and polyamines through electrostatics. *The Journal of General Physiology* Vol. 128, 185-202

Zhao, H. & Sokabe, M. (2008) Tuning the mechanosensitivity of a BK channel by changing the linker length. *Cell Research* Vol. 18, 871-878

Zhao, W. & Wang, R. (2002) H(2)S-induced vasorelaxation and underlying cellular and molecular mechanisms. *American Journal of Physiology. Heart and Circulatory Physiology* Vol. 283, H474-480

Zhao, W.; Zhang, J.; Lu, Y. & Wang, R. (2001) The vasorelaxant effect of H(2)S as a novel endogenous gaseous K(ATP) channel opener. *The EMBO Journal* Vol. 20, 6008-6016

Zhou, X.-B.; Wulfsen, I.; Utku, E.; Sausbier, U.; Sausbier, M.; Wieland, T.; Ruth, P. & Korth, M. (2010) Dual role of protein kinase C on BK channel regulation. *Proc Natl Acad Sci U S A* Vol. 107, 8005-8010

Zuidema, M.Y.; Yang, Y.; Wang, M.; Kalogeris, T.; Liu, Y.; Meininger, C.J.; Hill, M.A.; Davis, M.J. & Korthuis, R.J. (2010) Antecedent hydrogen sulfide elicits an anti-inflammatory phenotype in postischemic murine small intestine: role of BK channels. *American Journal of Physiology. Heart and Circulatory Physiology* Vol. 299, H1554-1567

7

From Action Potential-Clamp to "Onion-Peeling" Technique – Recording of Ionic Currents Under Physiological Conditions

Ye Chen-Izu[1], Leighton T. Izu[1], Peter P. Nanasi[2] and Tamas Banyasz[2]
[1]*University of California Davis*
[2]*University of Debrecen*
[1]*USA*
[2]*Hungary*

1. Introduction

Upon stimulation, excitable cells generate a transient change in the membrane potential called Action Potential (AP). The AP is governed by numerous ionic currents that flow in or out of the cell membrane. The goal of cellular electrophysiology is to understand the role of individual ionic currents and the interplay between currents in determining the profile and the time course of AP. A critically important question of the field is how different ionic currents behave individually and interact collectively during the AP cycle in an excitable cell. To answer this question we need to know the dynamic behavior of ionic currents during AP and how these currents work in concert to determine the cell's membrane potential at every moment.

Ionic currents are studied with voltage-clamp technique. Since the introduction of this method, intensive research has been conducted to characterize the kinetic properties of ionic currents. Various versions of the method were used to determine the charge carrier, voltage gating, ligand gating, activation, inactivation, recovery, etc. of individual ionic currents. One variation of the voltage-clamp is the Action Potential-clamp (AP-clamp) which can record the ionic currents during the AP cycle. In this chapter we will review the principles and variations of the AP-clamp technique and discuss the advantages and limitations of the technique. We will discuss and demonstrate how AP-clamp can help us to understand the ionic mechanisms underlying AP by using the experimental data obtained from cardiac cells where these techniques are extensively used to study the fundamental role of the ionic currents and AP dynamics in governing the cardiac function and heart diseases.

2. The principles of the AP-clamp technique

In order to determine the profile of ionic currents during AP, extensive efforts have been made in two different approaches: one is to use mathematical models based on the traditional voltage-clamp data to simulate the current profile during AP; another is to experimentally measure the ionic currents under AP-clamp.

The modeling approach uses the method pioneered by Hodgkin-Huxley (Hodgkin & Huxley, 1952) to describe the currents using ordinary differential equations and parameters derived from the voltage-clamp data. The standard voltage-clamp experiments were designed to isolate each individual ionic current, and to investigate its kinetic properties including activation, inactivation, recovery etc. Later, more detailed single channel kinetic state models (Mahajan et al., 2008) and molecular structure models (Silva et al., 2009) were also used to describe some channels. However, standard voltage-clamp experiments often used non-physiological conditions. The ion concentrations in the internal and external solutions were usually different from the physiological ionic milieu in order to isolate a particular current. Rectangular voltage pulses, instead of the AP waveform, were used to characterize the biophysical properties of the channel/transporter. In most cases, the intracellular Ca^{2+} was buffered with chelants and the currents were measured without the Ca^{2+} transients. These artificial conditions can bring inaccuracies into the data and cause the resultant models to deviate from the physiological reality, as evidenced by the differences between model simulations and experimental measurements. (Decker et al., 2009; Mahajan et al., 2008)

The experimental approach uses the AP-clamp technique to directly record the current profile during AP. The AP-clamp technique is quite simple in principle, although can be challenging in practice. (The technical aspects are discussed below.) The experimental protocol mainly involves the following steps (Figure-1) with some variations. (1) Record the steady state AP of the cell under current-clamp mode (I=0). (2) Apply this AP waveform as the voltage command onto the same cell under voltage-clamp. After reaching steady state, the net current output, I_{Ref} should be zero. (3) Isolate the current of interest by using its specific blocker to remove it from the net current output, seen as a compensation current from the amplifier, I_{Comp}. (4) The current of interest is then obtained as the difference current: $I_{Diff} = I_{Ref} - I_{Comp}$. What happens behind the scene is that the cell's AP, generated by all the membrane currents working in concert, is recorded under current-clamp with I=0. When this AP is applied as the command voltage onto the same cell under voltage-clamp, the net current output (seen as I_{Ref}) should be zero (Doerr et al., 1989; M.E. Starzak & R. J. Starzak, 1978), and the amplifier does not need to inject any compensation current to maintain the AP as long as the cell condition remains stable. When a particular current is blocked by its specific inhibitor, then the amplifier has to inject a compensation current (seen as I_{Comp}) in place of the blocked current in order to maintain the AP under voltage-clamp. This compensation current is the mirror image (negative) of the particular current that had been blocked. Hence, by subtracting I_{Comp} from I_{Ref} we can obtain the current that was originally flowing during the AP prior to the specific inhibitor application.

A fundamental difference between the AP-clamp technique and the conventional voltage-clamp technique is in the way of separating the current of interest from the other currents. In conventional voltage-clamp experiments, the current of interest is recorded under conditions that suppress all other currents by using custom-made voltage protocols, simplified ionic solutions, and sometimes blocking other 'contaminating' currents. In contrast, the AP-clamp experiments take an opposite approach to record the 'absence of the current' by blocking the current of interest using its specific inhibitor while allowing all other currents to flow during AP. By subtracting the net current output before and after the blocker application, all other currents (unaltered by the blocker) are cancelled out, and the current of interest is thereby obtained. The major advantage of the AP-clamp technique is that it enables us to record the ionic currents under *in situ* conditions (i.e. during AP, with Ca^{2+} cycling, in a physiological milieu, and undergoing contraction).

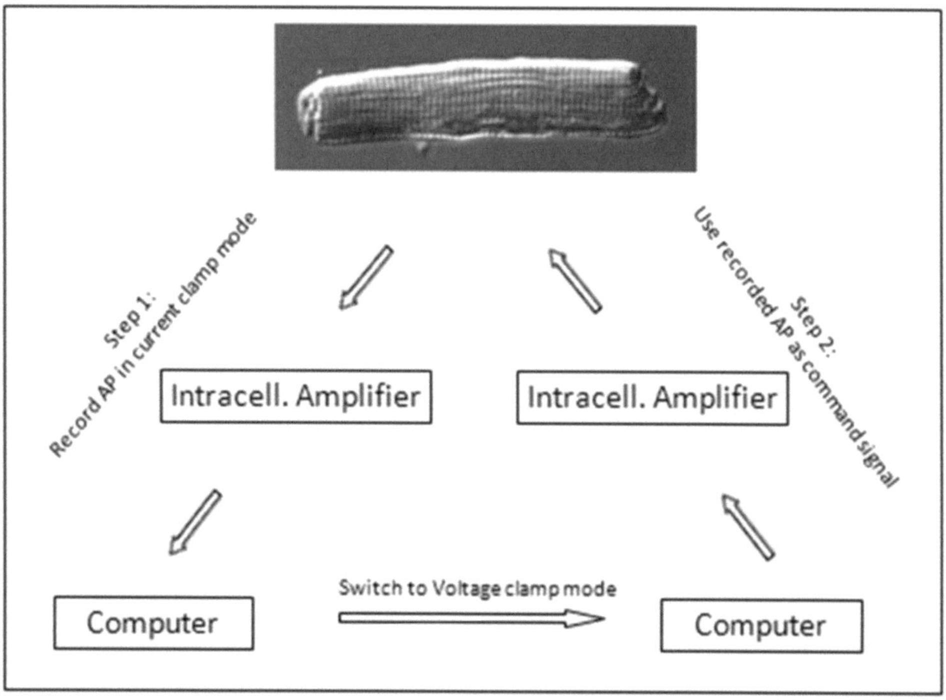

Fig. 1. The Flow chart of the Action Potential-clamp technique.

3. A short historical review

Attempts to determine the transmembrane currents during AP-clamp can be dated back to 1970s. One approach used in those pioneering experiments was to switch the amplifier from current-clamp mode to voltage-clamp mode and record the instantaneous current at different phase of AP. The cell was stimulated by a brief current pulse and the action potential was allowed to develop in a free running mode until the amplifier was switched to voltage-clamp mode and the voltage was frozen to the value at that moment for measuring the instantaneous current. (Bezanilla et al., 1970)

At the beginning, the idea to record the membrane current using a pre-recorded AP as the voltage command was used for various applications. Bastian & Nakajima studied the T-tubule function in skeletal muscle fiber with double sucrose gap method using pre-recorded AP as the voltage command. (Bastian & Nakajima, 1974; Nakajima & Bastian, 1974). In other experiments AP-clamp was used simply to test the effectiveness of the space clamp in axon (M.E. Starzak & R. J. Starzak, 1978; M.E. Starzak & Needle, 1983). Though the goals of those early studies were very different from the later studies in which AP-clamp was used to study the contributions of the membrane currents to shaping the AP, we can identify the essential elements of the AP-clamp technique in those experiments. The AP was recorded under current-clamp mode and then stored and used as the command voltage waveform in voltage-clamp mode. Most importantly the concept of 'zero current' is already established in those publications: "To produce the most accurate reproduction of this action potential, the voltage clamp currents must include no contributions due to ineffective space clamp" (M.E. Starzak & R. J. Starzak, 1978).

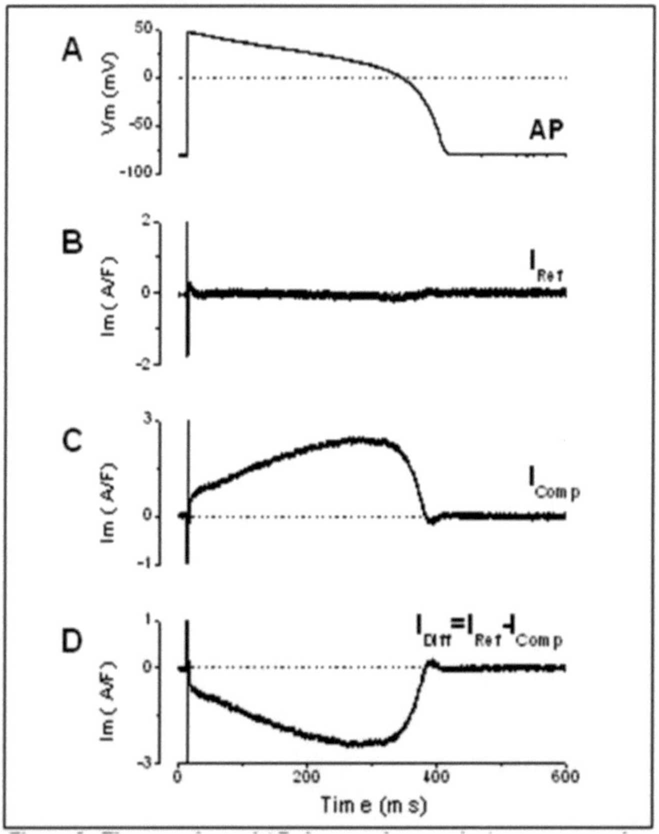

Fig. 2. The procedure of AP-clamp technique. **A**: Action potential is recorded in current clamp mode. **B**: Zero current or Reference current (I_{ref}) is recorded in voltage clamp mode. **C**: Using a specific channel inhibitor compensation current (I_{comp}) is recorded. **D**: The inhibitor sensitive current is determined as the difference current

Another technique to measure the membrane conductance during AP (without using pre-recorded AP) utilizes two electrodes (Fischmeister et al., 1984; Mazzanti & DeFelice, 1987). The first electrode had access to the cell's interior for recording the membrane potential under current-clamp mode; the second electrode was sealed to a patch of membrane in cell-attached configuration and clamped to the desired voltage (under voltage-clamp mode. The fundamental difference between this technique and the AP-clamp method are twofold. First, the two electrode technique recording of the current is limited only to the membrane patch inside the second pipette; hence only the current of limited number of channels and sometimes only single channel recording were performed. Second, the AP was not controlled with this method; hence variability of the APs would result in variability of the currents and sometimes this method was used to record membrane currents during spontaneous AP (Jackson et al., 1982).

The breakthrough took place during 1990s following three important publications. Trautwein and his colleagues used digitized AP from spontaneously beating rabbit sinoatrial node cell (Doerr et al., 1989) to stimulate the guinea pig ventricular myocyte (Doerr et al., 1990) and to record membrane currents during AP. In both cases, specific blockers (D-600, Ni^{2+}) and current subtraction were used to dissect the ionic currents during AP. These two papers were the first to describe the profile of individual ionic currents (L and T-type calcium currents) directly recorded during AP. By using specific blockers and current subtraction, they established the basic principles of the AP-clamp method. The third paper published AP clamp data obtained in nerve fiber also using the specific blocker and the current subtraction method to visualize sodium and potassium currents during AP (de Haas & Vogel, 1989). AP-clamp technique became a popular tool during 1990s and was used for mapping the key membrane currents that shape AP in several cell types including cardiac myocytes (Bouchard et al., 1995), neurocytes (Barra, 1996), as well as plant cells (Thiel, 1994). Combined with epifluorescent Ca^{2+} measurement, AP-clamp technique became a powerful tool for studying the Ca^{2+} dynamics in cardiac myocytes (Arreola et al., 1991; Grantham & Cannell, 1996; Puglisi et al., 1999).

During 2000s, a unique variation of the AP-clamp, called Dynamic Clamp technique, was developed in which an isolated cell (or a mathematical model of the cell) is coupled electrically to another cell. The first cell or the model provides the AP that is used as the voltage command onto the second cell (Bereczki et al., 2005; Wilders, 2006). The greatest advantage of the Dynamic Clamp is that the AP obtained from the first cell or model can be manipulated by changing the conditions (ionic milieu, stimulation parameters…etc.) or model parameters. This allows the experimenter to study the effects of changing AP on the dynamics of the currents.

A further extension of AP clamp technique is the sequential dissection of membrane currents or the "Onion-Peeling" (O-P) method. Previously, traditional voltage-clamp and AP-clamp technique was used to record only one current in any one cell. The O-P method uses a series of channel blockers to sequentially dissect out the currents in a single cell under AP-clamp. (Banyasz et al., 2011; Banyasz et al., 2012) The ability to measure many currents in a single cell enables study of the Individual Cell Electrophysiology of excitable cells.

4. Variations of the AP-clamp technique

Since the AP-clamp technique was introduced, several variations of the technique have been implemented. These variants use different modifications to circumvent technical limitations and provide new information on the properties of ionic currents. It is not our intension to discuss all possible modifications; here we list the characteristic features of several most frequently used variants.

4.1 Using 'standardized' or 'typical' AP

Individual cells display distinctive APs with some degree of cell-to-cell variations. To eliminate the individual variations of AP, a "characteristic or standardized" AP can be used instead of the cell's own AP as the command voltage in AP-clamp (Arreola et al., 1991). This AP could be obtained from a 'typical' cell, tissue or generated by a mathematical model. The consequence of using such standardized AP instead of the cell's own AP is that the reference current is no longer flat ($I_{Ref} \neq 0$). Development of any kind of deflection from the zero level or appearance of a hump on the reference (zero) current may be an indicator of rundown or some other type of instability. This advantage, i.e. the possibility of monitoring the stability of our preparations is lost when a foreign AP is used as a voltage command. The pharmacological subtraction is also a crucial and necessary step here; the difference current gives the drug-sensitive current.

4.2 Using 'modified' or 'reconstructed' AP

Modified or reconstructed AP was used in some AP-clamp experiments for various reasons. Some were designed to tease out certain properties of the currents; some were to circumvent technical difficulties. For example, such a situation is generated when we need to record a delayed potassium current with small amplitude in the range of 0.1-0.5 nA. This current can be recorded with good resolution if the amplifier is set to ±1-2 nA input range. Nevertheless, this setting cannot reliably hold the voltage clamp during the upstroke of the AP where the voltage dependent Na^+ channels generate a large current with 100-150 nA peak amplitude. We have two equally poor options here. If we keep the amplifier Gain high to maintain the high resolution, the voltage clamp at the beginning of the AP would be lost. If we lower the Gain, the fidelity of the voltage clamp would be kept but the resolution of the current recording would be poor. To prevent loosing the voltage clamp, we can modify the AP by adding a short depolarizing step (i.e. 10 ms, to -30 mV) prior to the upstroke of the AP. (Varro et al., 2000) This depolarizing step can inactivate the voltage dependent Na^+ channels to allow the amplifier to hold the voltage clamp, and thereby circumvent this technical problem. Modification of other parameters of the AP (duration, plateau height, diastolic interval etc.) can also be used in AP-clamp experiments to study the ionic mechanisms that shape the AP. (Rocchetti et al., 2001)

4.3 Dynamic Clamp

A special variant of the AP-clamp method is the Dynamic Clamp (Bereczki et al., 2005; Wilders, 2006). In this case the AP voltage command comes from a current-clamped cell (Cell -1) or, alternatively, from a mathematical model. This AP waveform is used as the command voltage to voltage-clamp another cell (Cell-2). The current recorded from the Cell-2 is then fed back to the Cell-1 or to the mathematical model so it can modify the morphology of the AP accordingly. In this configuration, the two systems are in dynamic connection and in real time coupling.

Originally developed to study interactions between neural cells, the Dynamic Clamp technique provides a powerful tool for studying the dynamic interaction of currents and AP. For example, Weiss et al (Mahajan et al., 2008) used Dynamic Clamp method to investigate the Ca^{2+} modulation of ionic currents during AP in the cardiac myocyte. They first eliminated the intracellular Ca^{2+} cycling (by depleting the SR load), and then used mathematical model ('Cell-1') to generate the Ca^{2+} transient and feed the data into an AP-clamped cell (Cell-2) to record the L-type Ca^{2+} current. The data provide valuable information on how the L-type Ca^{2+} channel are modulated by the Ca^{2+} transient during AP cycle.

4.4 The "Onion-Peeling" (O-P) technique

Recently, a new version of the AP-clamp method was developed, called "Onion-Peeling" (O-P) technique, by making two significant modifications (Banyasz et al., 2011; Banyasz et al., 2012): (1) a triad of conditions is used to directly record ionic currents during AP, with Ca^{2+} cycling, in a physiological milieu; (2) multiple ionic currents are recorded from the same cell by sequentially applying the channel blockers one-by-one to dissect out each of the ionic currents. Figuer-2 shows the recording of four different currents (I_{Ks}: chromanol, I_{Kr}: E4031, I_{K1}: Ba^{2+}, and I_{NISO}: nisoldipine sensitive current) in a single guinea pig ventricular myocyte using the O-P technique. All currents are recorded after reaching steady state. To directly record the ionic currents during AP with Ca^{2+} cycling in a physiological milieu allows us to construct accurate and realistic models. The unprecedented ability to measure multiple currents in the same cell enables us to study the Individual Cell Electrophysiology (see below).

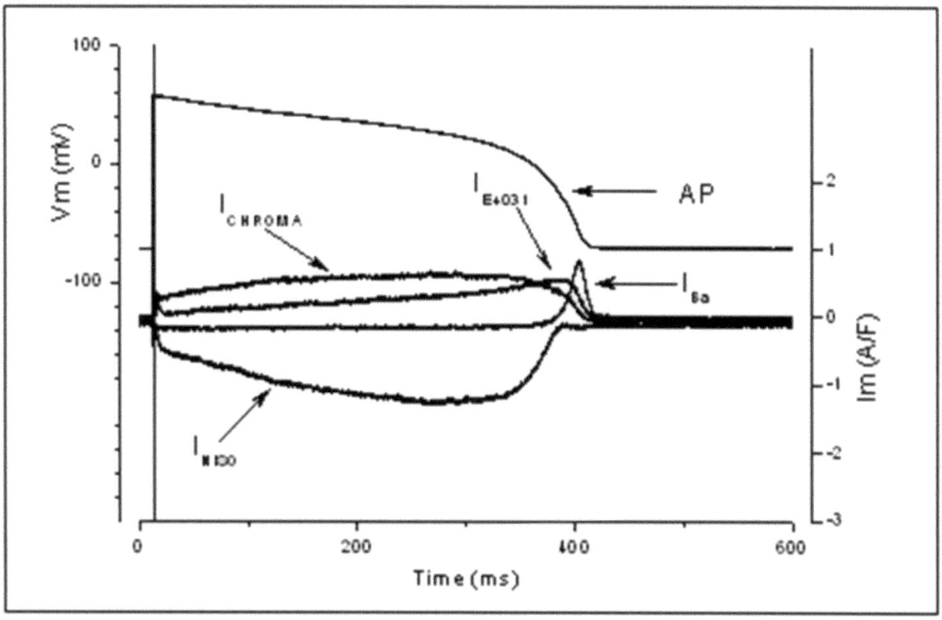

Fig. 3. Onion Peeling recording of multiple currents in guinea pig ventricular myocytate.
AP: Action Potential.
I_{CHROMA}, I_{Ba}, I_{NISO}, I_{E4031} detones Chromanol-293B, Ba^{2+}, Nisoldipine and E4031 sensitive currents respectively.

5. Applications of the AP-clamp and Onion-Peeling (O-P) methods

5.1 Study of the Individual Cell Electrophysiology (ICE)

The cell's AP is a finely choreographed dance involving many ion channels and transporters interacting with each other via the membrane potential and intracellular Ca^{2+}. How do we determine the role of individual ion channel/transporter in shaping the AP and which channels/transporters are altered by disease, stress, drug etc.? Traditional voltage-clamp studies record one current from any one cell, and different currents from different cells. The data averaged from many cells are then used to construct a canonical AP model. However, the averaged canonical AP model may not reflect the behavior of single cells due to cell-to-cell variability. (Marder & Taylor, 2011) It has been known that discrepancies exist between the model simulations and the AP-clamp measured current (Banyasz et al., 2011), and model simulations still fail to reproduce some important AP dynamic behaviors (i.e. EAD, adaptation, restitution) (Decker et al., 2009; Faber et al., 2007; Mahajanaw et al., 2008; Pasek et al., 2008).

The significant contribution of the O-P technique is that it provides a new and different type of data from all previous voltage-clamp experiments; the O-P technique allows recording of *multiple* ionic currents flowing through different channels/transporters during AP in the same single cell (see Figure 3). The ability to measure multiple ionic currents in the same cell provide experimental data to directly study how inward and outward currents work concertedly to shape the AP in individual cells, and hence the Individual Cell Electrophysiology (ICE).

The cell's AP is shaped by all of the currents flowing across the cell membrane via ion channels, exchangers, transporters etc. In order to gain in-depth quantitative understanding of the ionic mechanisms that control the AP and arrhythmias, we need to study how different currents interact and integrate at the single cell level. Due to cell-to-cell variability in the protein expression and modulation, the density of certain ionic currents may vary in different cells. Nevertheless if the inward and the outward currents can counterbalance one another, different combinations of the currents can generate similar APs. (Banyasz et al., 2011) Under pathological conditions, disproportional changes in some currents may upset the balance to cause long or short AP duration, early or delayed afterdepolarizations that lead to arrhythmias. Using O-P to measure multiple currents in the same cell provides necessary data for determining how the proportion of the currents is altered to upset the balance and cause arrhythmogenic APs. Furthermore, disease conditions may widen the heterogeneity among cells to cause more pronounced cell-to-cell variability. Using O-P to conduct the ICE study will be critically important for understanding the impact of cellular variability (inherent variance, transmural gradient, regional difference, disease associated variance, etc.) on the population behavior at tissue and organ levels.

5.2 Mapping the regional and transmural differences

It has long been known that the AP morphology has characteristic epicardial versus endo-cardial differences: a spike-and-dome AP profile was seen in the epi-, but not in the endo-cardial ventricular myocytes from canine and human hearts. Early studies show that different K^+ current density in different layers of the myocardium may contribute to this heterogeneity (Litovsky & Antzelevitch, 1988; Liu et al., 1993; Liu & Antzelevitch, 1995). When L-type Ca^{2+} currents ($I_{Ca,L}$) of epi- and endocardial ventricular myocytes were compared using traditional voltage-clamp technique, the voltage dependence, activation and inactivation kinetics of the currents were found to be identical. Interestingly, when the profile of $I_{Ca,L}$ was studied as the nisoldipine-sensitive current using the AP-clamp technique, marked differences were observed between epi- and endocardial cells. Although $I_{Ca,L}$ showed a sharp spike followed by a rapid decay in both endo- and epicardial cells following the upstroke of the AP, a hump developed in the $I_{Ca,L}$ record in the epi- but not in the endocardial cells. (Banyasz et al., 2003) Similar transmural differences were also obtained in ventricular myocytes isolated from human hearts (Fulop et al., 2004). This double-peaked morphology of $I_{Ca,L}$ in epi-cardial cells and the relationship between the profile of AP and $I_{Ca,L}$ was revealed only under AP-clamp but not with traditional voltage-clamp method. The AP-clamp technique provides a direct method for studying the relationship between currents and membrane potential during AP.

5.3 Studying the complex effects of pathological conditions

Cardiac arrhythmias are a leading cause of morbidity and mortality in the developed world. Yet, pharmacological approaches to treat arrhythmias have met with only limited success, in large part due to incomplete understanding of the cellular and molecular mechanisms of arrhythmias. Pathological conditions wrought by disease, stress, drug etc. usually cause complex changes in multiple ion channels/transporters and Ca^{2+} handling molecules. The changes in proteins involved in determining electrical currents of the cardiac myocytes can cause abnormalities in the profile and the time course of APs, leading to arrhythmias. To elucidate the pathologic mechanisms behind abnormal APs one needs to understand the changes in the kinetics of ionic currents. For example, in a rabbit model of pressure & volume overload induced non-ischemic heart failure, researchers have found changes in several key channel/transporter that determine the cardiac AP and arrhythmias (Pogwizd, 1995; Pogwizd et al., 2001): I_{K1}, an important K^+ current that controls the repolarization of AP, was reduced; the Na^+/Ca^{2+} exchanger, a key regulator of the Ca^{2+} signaling and AP profile, was up-regulated; the expression of ß-adrenergic receptors, which profoundly influence the AP and the Ca^{2+} signaling was also altered. The reduced repolarization power (weakened I_{K1}) combined with augmented depolarizing drive (up-regulated Na^+/Ca^{2+} exchanger current), and probably more, changes at molecular level manifest at cellular level to cause delayed afterdepolarizations which can trigger premature APs. The abnormal APs at the cellular level further manifest at the tissue and organ level to cause cardiac arrhythmias. These changes in the rabbit heart failure model resemble some of the changes found in human congestive heart failure (Nattel et al., 2007).

Traditionally, the dynamic behavior of ionic currents during AP was studied using voltage-clamp experiments in conjunction with mathematical modeling. The standard voltage-clamp technique uses custom-made voltage protocols to determine the voltage dependency and the kinetic properties of each individual current. As a basic rule, only one single current can be measured from any one cell because different experimental conditions (ionic milieu, voltage protocol, inhibitors to block contaminating currents etc.) are required to record different current. For example, it is impossible to record I_{K1} and Na^+/Ca^{2+} exchanger current from the same cell using standard voltage-clamp technique. The novelty of the O-P technique is that it enables recording of multiple currents from the same single cell, by sequentially applying specific inhibitors of different currents under AP-clamp. The O-P technique gives us unprecedented ability to conduct the ICE study of the complex effects of pathological conditions on altering multiple currents and to study their collective effects on shaping APs in the single cell. The O-P recording of multiple currents enables us to study how the proportions of different currents are altered in diseased myocardium compared to healthy hearts (see the previous chapter on individual cell electrophysiology).

5.4 Characterization of drug effects

The AP-clamp technique can be used economically for studying drugs with multiple actions. As shown previously, each ionic current has its own characteristic profile during AP, called fingerprint. Knowing the fingerprint of each individual current, we can identify the suspected drug-sensitive currents (perhaps several of them) in a single AP-clamp experiment (Szabo et al., 2007; Szabo et al., 2008). Hence, the profile of the drug-sensitive current recorded under AP-clamp can give clues towards the identity of currents that might be affected by the drug. This initial screening with AP-clamp prior to the systematic pharmacological study can save significant time and resource in studying the drug effects.

Effects of receptor agonists or antagonists (hormones, receptor modulators or drugs) can also be readily studied by AP-clamp. Even in the case when agonist/antagonist affects only one population of receptors, signal transduction pathways often couple to more than one ionic current; AP-clamp is uniquely suited for studying the drug concentration-dependent effects on multiple currents. As example, the frequency-dependent effects of isoproterenol on $I_{Ca,L}$, I_{Ks}, and I_{Cl} in guinea pig ventricular cells (Rocchetti et al., 2006) and the effects of acetylcholine and adenosine on I_{K1} in ferret cardiac myocytes (Dobrzynski et al., 2002) were clarified using the AP-clamp technique.

6. Technical aspects

To successfully perform AP-clamp experiments requires high quality cells and instrumentation. In this chapter we will review the essential technical requirements for the AP-clamp technique.

6.1 Cell quality and the solutions

Robust cells with physiological resting membrane potential and stable APs are necessary requirements for using the AP-clamp technique. The experiments are performed in isolated single cells. Preparation of the cells is basically the same as in other voltage-clamp experiments. The only difference is that the composition of the bath solution should mimic the physiologic extracellular milieu during experiments. If the cell preparation requires using non-physiological media, try to minimize the time the cells spend in that media. The pH and the osmolarity of the solutions must be tightly controlled, and membrane permeable buffers (bicarbonate) seem to improve robustness of the cells. We found that the cells worked better if stored for 1-2 hours in the same extracellular medium as used in the AP-clamp experiments. Continuous perfusion of the bath solution is needed to maintain the cell quality throughout the experiments.

To work with a contracting muscle cell, the whole-cell seal usually last longer if we lift the cell up from the bottom of perfusion chamber after establishing the seal (no need to do this for other non-contracting cell types). However, sometimes cells tend to stick to the bottom of the perfusion chamber. Adding a small drop of the albumen solution into the chamber before the cells are placed can reduce the stickiness of the cells without reducing the success rate for seal formation.

The pipette solution used for the AP-clamp experiment should mimic the intracellular ionic milieu of the cell. Again, tight control of the pH and the osmolarity of the solutions are crucial. Traditionally, 2-10 mmol/L EGTA is added to the pipette solution in voltage-clamp experiments to buffer cytosolic Ca^{2+} and eliminate cell contraction when working with muscle cells. Buffering the cytosolic Ca^{2+} can lengthen the lifetime of the seal significantly, and the previous voltage-clamp and AP-clamp experiments used Ca^{2+} buffers in most cases. However, our recent studies using the O-P technique (Banyasz et al., 2012) demonstrate that the current profiles during AP with Ca^{2+} transient (Ca^{2+} cycling preserved by not using exogenous Ca^{2+} buffer in the pipette solution) can be significantly different from the currents recorded with buffered Ca^{2+} (2-10 mM EGTA or BAPTA were used). Therefore, in order to record the ionic currents under physiological condition in situ, we suggest to minimize the use of exogenous Ca^{2+} buffer in the pipette solution and to measure the Ca^{2+} transient simultaneously with the O-P recording of the currents during AP.

Before recording the steady state AP to use as the voltage command for AP-clamp, we need to first test stability of the AP parameters. Keep in mind that the AP duration is never strictly constant, as a living system usually has some fluctuation in parameters. If the beat-to-beat fluctuation in the APs is large, the cell cannot be used for the AP-clamp experiment. We usually use the cells with no more than 4-5% fluctuation in the AP duration. At the same time, we use very stringent requirements for the voltage parameters. The cell must be discarded if any of the voltage parameters (resting membrane potential, peak amplitude, plateau height… etc.) displays more than 1 mV instability. Depending on the input impedance of the cell, ≤1 mV of instability in the voltage could translate into several hundred pA of fluctuation in the current measurements. We found it useful to stimulate the cell for 15-20 minutes at a constant pacing rate until it truly reaches a steady state, and then record the AP under current-clamp mode. Also, depending on the geometry of the cell it might take 15-20 minutes of pacing for the cell to reach equilibrium between the cytoplasm and the pipette solution after establishing the whole-cell ruptured seal configuration.

6.2 The patch pipette and the whole-cell seal configuration

The requirements for patch pipette used for the AP-clamp technique are the same as for those used in traditional voltage-clamp experiments. We use borosilicate glass pipette with tip resistance of 1.5-2.5 MΩ. The technical details of pipette fabrication follow the direction from the pipette puller manufacturer.

After compensating for the junction potential between the pipette solution and the bath solution, the pipette tip is placed on the cell membrane and a gentle suction is used to make a GΩ seal. Then the membrane patch inside the pipette is ruptured by suction to establish the whole-cell seal configuration. As a critical requirement for the AP-clamp technique, the access resistance must be kept low (< 5 MΩ) and constant throughout the entire experiment, because a change in the access resistance exerts significant impact on the magnitude and the dynamics of the compensation current. We suggest recording the access resistance at the beginning and at the end of the AP-clamp experiment and checking it several times during the experiment. It is important that the access resistance should remain close to the initial value throughout; if it changes significantly, the cell should be discarded from data collection.

The cell capacitance cancellation and the series resistance compensation are necessary for speeding up the capacitive transient and maintaining high fidelity of the actual voltage command across the cell membrane. When both are engaged, it is somewhat difficult to adjust the degrees of compensation without provoking 'ringing' (oscillations in the compensation circuitry), especially when dealing with large size cardiac myocytes (150-300 pF). Hence it is practically impossible to achieve 100% compensation. In our experience, about 80% compensation with 20 µs Lag is feasible for a typical cardiac myocyte without risking the 'ringing'. Higher percentage of compensation should be achievable for other cell types with smaller whole cell capacitance.

After breaking the membrane patch to establish the whole-cell seal, a junction potential would build up between the pipette solution and the cytosol. The magnitude of this junction potential can be as high as 10-15 mV depending on the composition of the pipette solution, and seen as an apparent depolarization of the membrane. It is essential not to compensate this junction potential to avoid imposing a voltage shift into the measurement. It can be somewhat deceptive when the reading of the resting membrane potential appears 10-15 mV higher than the known physiologic value and the temptation might be very strong to eliminate this apparent voltage shift. But doing so would destroy the AP-clamp measurements. In the design of the AP-clamp technique, this junction potential is automatically taken care of (dropped out from the equation) during the current subtraction. Nevertheless, if we want to analyze the voltage dependence of the currents, this junction potential must be corrected in the voltage values and taken into consideration during the analysis.

6.3 Instrumentation

The instruments (amplifier, A/D converter, computer etc.) used in AP-clamp experiments are basically the same as for voltage-clamp experiments. The specifications are determined by the cell type and current parameters we want to measure. However, there are some special considerations in deciding what particular instruments to use for the AP-clamp experiment.

- External stimulator: it is very useful (albeit not essential) to have an external stimulator to generate the electric pulses for evoking the AP. Although most electrophysiology software offers the option to program square pulse protocol to stimulate the cell in current-clamp mode, modification/adjustment of stimulation parameters (changing the amplitude or pacing rate) is more complicated and time consuming (if we have to overwrite protocol parameters) than simply turning the control knobs on the stimulator (analogue stimulator, for example). If we decide to use a square pulse generator to stimulate the cell it must be a DC stabile device, and the zero level should be truly zero with extremely low noise. Note that the manufacturer specification usually uses the maximum output voltage as reference to calculate the noise level, but we use no more than 5-10 V voltage range for the AP-clamp experiment, so the noise of the instrument needs to be extremely low.
- Amplifier: If we decide to use an external stimulator, the amplifier should have at least one command potential input compatible with the stimulator. It is also necessary to calibrate the amplifier frequently, because the AP-clamp experiment is more sensitive to voltage drift than traditional voltage-clamp experiments.

6.4 Channel inhibitors

The quality of AP-clamp data is determined primarily by the selectivity of the channel blocker used in pharmacological dissection of the current. In ideal situation, we have a highly specific drug which can selectively block the current under study without affecting any other currents. In such cases, using maximal concentration of the drug is suggested. Unfortunately, some drugs are known to have side effects on other channels besides their primary target. Using these non-specific drugs in the AP-clamp experiment would yield a composite current contaminated with other current(s).

Ion Channel & Transporter	Inhibitor	Dosage	Reference
I_{Na}	TTX	1-10 µm	Chorvatova et al 2004; Yuill et al 2000
$I_{Na\text{-}Late}$	Ranolazine	10 µm	Rajamani et al 2009
$I_{Ca\text{-}L}$	Nifedipine	1 µm	Horiba et al 2008
$I_{Ca\text{-}L}$	Nisoldipine	0.1 µm	Banyasz et al 2003
I_{NCX}	SEA0400	3 µm	Birinyi et al 2005; Ozdemir et al 2008
$I_{Cl\text{-}Ca}$	N-(p-amylcinnamoyl) anthranilic acid	5 µm	Gwanyanya et al 2010
I_{to}	4-aminopyridine	1 mM	Banyasz et al 2007; Patel & Campbell 2005
I_{kr}	E4031	1 µM	Banyasz et al 2007; Varro et al 2000
I_{ks}	HMR-1556	30 nM	Gogelein et al 2000; Thomas et al 2003
I_{ks}	Chromanol-293B	1-10 µM	Yamada et al 2008
I_{k1}	$BaCl_2$	50 µM	Banyasz et al 2008; Banyasz et al 2007
$I_{K\text{-}Ca}$	Apmin	100 pM, 1 nM	Xu et al 2003; Özgen et al 2007
$I_{Cl\text{-}Ca}$	Niflumic acid	50 µM	Greenwood & Leblanc 2007; Saleh et al 2007
$I_{Cl\ \text{-small}}$	Chlorotoxin		Borg et al 2007
$I_{Cl\text{-vol}}$	Tamoxifen		Borg et al 2007
$I_{Cl,ligand}$	Picrotoxin		Etter et al 1999
$I_{Ca\text{-}T}$	NNC 55-0396		Huang 2004
$I_{Ca\text{-}T}$	R(-)efodipine		Tanaka 2004

Table 1. Specific inhibitors for the major ionic currents in cardiac cells.

How to make sure the O-P current dissected out by the specific channel blocker is pure?

First and foremost, we should use the most specific channel blocker known in literature. Several things can be done to check the specificity of drugs. (***A***) We do drug dose-response to check that the profile of current scales proportionally. (***B***) We use low dosage to dissect out a proportion of the target current and minimize non-specific effects. The full magnitude of the current can be recovered using the drug dose-response curve determined by using the traditional voltage-clamp experiments where other (contaminating) currents are eliminated. (***C***) We scramble the sequence of drug application to make sure there is no crosstalk (Banyasz et al., 2011). If the drug is found to have non-specific effects even at very low dosage, we will search for a more specific blocker (i.e. experimental drugs, venoms, toxins, peptide inhibitors, antibody-targeted inhibitors, etc.) The drug list in Table-1 is our first pass, but can and will be modified as more specific drugs become available.

Special consideration is needed for recording Ca^{2+}-sensitive currents using the O-P technique. As a charge carrier, Ca^{2+} current influences the membrane potential. As a ubiquitous second messenger, Ca^{2+} also influences a number of Ca^{2+}-sensitive currents. The intracellular Ca^{2+} concentration is kept very low (100 nM) in cells at resting state (diastole in cardiac myocytes). Upon excitation, the Ca^{2+} inflow during AP drastically elevate the intracellular Ca^{2+} concentration which triggers many biochemical events and modulate numerous Ca^{2+}-sensitive currents. For example, to record the L-type Ca^{2+} current using the O-P technique, we use nifedipine to block the L-type Ca^{2+} channel. However, blocking the L-type Ca^{2+} current during AP also abolishes the Ca^{2+} transient, which, in turn, affects other Ca^{2+}-sensitive currents. Thus the nifedipine-sensitive current recorded using O-P technique would consist not only the L-type Ca^{2+} current but also other Ca^{2+}-sensitive currents including the Na^+/Ca^{2+} exchanger, Ca^{2+}-activated K^+ currents, Ca^{2+}-sensitive Cl^- currents and so on. One way to avoid this problem is to first record Ca^{2+}-sensitive currents while Ca^{2+} is normally cycling, and subsequently to add nifedipine to block the L-type Ca^{2+} current.

7. Conclusion

The AP-clamp methods provide powerful tools for studying the ionic currents under physiological conditions, and the complex effects of disease, stress, drug, mutation etc. on the ion channels and transporters. The O-P technique is uniquely suited for studying the Individual Cell Electrophysiology which is necessary for investigating the cell-to-cell variability. The AP-clamp and the O-P techniques can be used in muscle cells, neural cells, and any excitable cells where ionic currents and membrane potential play important roles in cell function.

8. References

Arreola, J.; Dirksen, R.T.; Shieh, R.C.; Williford, D.J. & Sheu, SS. (1991). Ca^{2+} current and Ca^{2+} transients under action potential clamp in guinea pig ventricular myocytes. *American Journal of Physiology,* Vol.261, pp. C393-C397, ISSN: 00029513

Banyasz, T.; Fulop, L.; Magyar, J.; Szentandrassy, N.; Varro, A. & Nanasi PP. (2003). Endocardial versus epicardial differences in L-type calcium current in canine ventricular myocytes studied by action potential voltage clamp. *Cardiovascular Research,* Vol.58, pp. 66-75, ISSN 00086363

Banyasz, T.; Magyar, J.; Szentandrassy, N.; Horvath, B.; Birinyi, P.; Szentmiklosi, J. & Nanasi, PP. (2007). Action potential clamp fingerprints of K^+ currents in canine cardiomyocytes: their role in ventricular repolarization. *Acta Physiologica (Oxford),* Vol.190, pp. 189-198, ISSN 17481708

Banyasz, T.; Lozinskiy, I.; Payne, C.E.; Edelmann, S.; Norton, B.; Chen, B.; Chen-Izu, Y.; Izu, L.T. & Balke, C.W. (2008). Transformation of adult rat cardiac myocytes in primary culture. *Experimental Physiology,* Vol.93, pp. 370-382, ISSN 09580670

Banyasz, T.; Horvath, B.; Jiang, Z.; Izu, L.T. & Chen-Izu, Y. (2011). Sequential dissection of multiple ionic currents in single cardiac myocytes under action potential-clamp. *Journal of Molecular and Cellular Cardiology,* Vol.50, pp. 578-581, ISSN 00222828

Banyasz, T.; Horvath, B.; Jiang, Z.; Izu, L.T. & Chen-Izu, Y. (2012). Profile of L-type Ca^{2+} current and Na^+/Ca^{2+} exchange current during cardiac action potential in ventricular myocytes. *Heart Rhythm,* Accepted for publication, ISSN 15475271

Barra, P.F.A. (1996). Ionic currents during the action potential in the molluscan neurone with the self-clamp technique. *Comparative Biochemistry and Physiology Part A: Physiology,* Vol.113, pp. 185-194, ISSN 03009629

Bastian, J. & Nakajima, S. (1974). Action Potential in the Transverse Tubules and Its Role in the Activation of Skeletal Muscle. *The Journal of General Physiology,* Vol.63, pp. 257-278, ISSN 15407748

Bereczki, G.; Zegers, J.G.; Verkerk, A.O.; Bhuiyan, Z.A.; de Jonge, B.; Veldkamp, M.W.; Wilders, R. & van Ginneken A.C.G. (2005). HERG Channel (Dys)function Revealed by Dynamic Action Potential Clamp Technique. *Biophysical Journal,* Vol.88, pp. 566-578, ISSN 00063495

Bezanilla, F.; Rojas E. & Taylor, RE. (1970). Sodium and potassium conductance changes during a membrane action potential. *Journal of Physiology,* Vol.211, pp. 729-751, ISSN 00223751

Birinyi, P.; Acsai, K.; Banyasz, T.; Toth, A.; Horvath, B.; Virag, L.; Szentandrassy, N.; Magyar, J.; Varro, A.; Fulop, F. & Nanasi PP. (2005). Effects of SEA0400 and KB-R7943 on Na^+/Ca^{2+} exchange current and L-type Ca^{2+} current in canine ventricular cardiomyocytes. *Naunyn Schmiedebergs Archives of Pharmacology,* Vol.372, pp. 63-70, ISSN 00281298

Borg, J.J.; Hancox, J.C.; Zhang, H.; Spencer, C.I.; Li, H. & Kozlowski, R.Z. (2007). Differential pharmacology of the cardiac anionic background current I(AB). *European Journal of Pharmacology,* Vol.569, pp. 163-170, ISSN 00142999

Bouchard, R.A.; Clark, R.B. & Giles, W.R. (1995). Effects of Action Potential Duration on Excitation-Contraction Coupling in Rat Ventricular Myocytes Action Potential Voltage-Clamp Measurements. *Circulation Research,* Vol.76, pp. 790-801, ISSN 00097330

Chorvatova, A.; Snowdon, R.; Hart, G. & Hussain, M. (2004). Effects of pressure overload-induced hypertrophy on TTX-sensitive inward currents in guinea pig left ventricle. *Molecular and Cellular Biochemistry,* Vol.261, pp. 217-226, ISSN 03008177

Decker, K.F.; Heijman, J.; Silva, J.R.; Hund, T.J. & Rudy, Y. (2009). Properties and ionic mechanisms of action potential adaptation, restitution, and accommodation in canine epicardium. *Am J Physiol Heart Circ Physiol.,* Vol.296, pp. H1017-26, ISSN 03636135

Dobrzynski, H.; Janvier, N.C.; Leach, R.; Findlay, J.B.C. & Boyett, M.R. (2002). Effects of ACh and adenosine mediated by Kir3.1 and Kir3.4 on ferret ventricular cells. *American Journal of Physiology - Heart and Circulatory Physiology,* Vol.283, pp. H615-H630, ISSN 00029513

Doerr, Th.; Denger, R. & Trautwein W. (1989). Calcium currents in single SA nodal cells of the rabbit heart studied with action potential clamp. *Pflügers Archiv European Journal of Physiology,* Vol.413, pp. 599-603, ISSN 00316768

Doerr, Th.; Denger, R.; Doerr, A. & Trautwein W. (1990). Ionic currents contributing to the action potential in single ventricular myocytes of the guinea pig studied with action potential clamp. . *Pflügers Archiv European Journal of Physiology,* Vol.416, pp. 230-237, ISSN 00316768

Etter, A.; Cully, D.F.; Liu, K.K.; Reiss, B.; Vassilatis, D.K.; Schaeffer, J.M. & Arena JP. (1999). Picrotoxin blockade of invertebrate glutamate-gated chloride channels: subunit dependence and evidence for binding within the pore. *Journal of Neurochemistry,* Vol.72, pp. 318-326, ISSN 00223042

Faber, G.M.; Silva, J.; Livshitz, L. & Rudy Y. (2007). Kinetic properties of the cardiac L-type Ca2+ channel and its role in myocyte electrophysiology: a theoretical investigation. *Biophys. J.,* Vol.92, pp. 1522-43, ISSN 15420086

Fischmeister, R.; DeFelice, L.J.; Ayer, R.K.; Jr., Levi, R. & DeHaan, R.L. (1984). Channel Currents During Spontaneous Action Potentials in Embryonic Chick Heart Cells – The Action Potential Patch Clamp. *Biophysical Journal,* Vol.46, pp. 267-272, ISSN 00063495

Fulop, L.; Banyasz, T.; Magyar, J.; Szentandrassy, N.; Varro A & Nanasi P.P. (2004). Reopening of L-type calcium channels in human ventricular myocytes during applied epicardial action potentials *Acta Physiologica Scandinavica,* Vol.180, pp. 39-47, ISSN 00016772

Gogelein, H.; Bruggemann, A.; Gerlach, U.; Brendel, J. & Busch, A.E. (2000). Inhibition of I_{Ks} channels by HMR 1556. *Naunyn Schmiedebergs Archives of Pharmacology,* Vol.362, pp. 480-488, ISSN 00281298

Grantham, C.J. & Cannell, M.B. (1996). Ca^{2+} Influx During the Cardiac Action Potential in Guinea Pig Ventricular Myocytes. *Circulation Research,* Vol.79, pp. 194-200, ISSN 00097330

Greenwood, I.A. & Leblanc, N. (2007). Overlapping pharmacology of Ca^{2+}-activated Cl^- and K^+ channels. *Trends in Pharmacological Sciences,* Vol.28, pp. 1-5, ISSN 01656147

Gwanyanya, A.; Macianskiene, R.; Bito, V.; Sipido, K.R.; Vereecke, J. & Mubagwa, K. (2010). Inhibition of the calcium-activated chloride current in cardiac ventricular myocytes by N-(p-amylcinnamoyl)anthranilic acid (ACA). *Biochemical and Biophysical Research Communications,* Vol.402, pp. 531-536, ISSN 0006291X

de Haas, V. & Vogel, W. (1989). Sodium and Potassium Currents Recorded During an Action Potential. *European Biophysics Journal,* Vol.17, pp. 49-51, ISSN 01757571

Hodgkin, A. L. & Huxley, A. F. (1952). A quantitative description of membrane current and its application to conduction and excitation in nerve. *Journal of Physiology,* Vol.117, pp. 500-544. ISSN 00223751

Horiba, M.; Muto, T.; Ueda, N.; Opthof, T.; Miwa, K.; Hojo, M.; Lee, J.K.; Kamiya, K.; Kodama, I. & Yasui, K. (2008). T-type Ca^{2+} channel blockers prevent cardiac cell hypertrophy through an inhibition of calcineurin-NFAT3 activation as well as L-type Ca^{2+} channel blockers. *Life Sciences,* Vol.82, pp. 554-560, ISSN 00243205

Huang, L.; Keyser, B.M.; Tagmose, T.M.; Hansen, J.B.; Taylor, J.T.; Zhuang, H.; Zhang, M.; Ragsdale, D.S. & Li M. (2004). NNC 55-0396 [(1S,2S)-2-(2-(N-[(3-benzimidazol-2-yl)propyl]-N-methylamino)ethyl)-6-fluo ro-1,2,3,4-tetrahydro-1-isopropyl-2-naphtyl cyclopropanecarboxylate dihydrochloride]: a new selective inhibitor of T-type calcium channels. *Journal of Pharmacology and Experimental Therapeutics,* Vol.309, pp. 193-199, ISSN: 00223565

Jackson, M.B.; Lecar, H.; Brenneman, D.E.; Fitzgerald, S. & Nelson P.G. (1982). Electrical Development in Spinal Cord Cell Culture. *The Journal of Neuroscience,* Vol.2, pp. 1052-1061, ISSN 15292401

Litovsky, S.H. & Antzelevitch, C. (1988). Transient outward current prominent in canine ventricular epicardium but not endocardium. *Circulation Research,* Vol.62, pp. 116-126, ISSN 00097330

Liu, D.W.; Gintant, G.A. & Antzelevitch, C. (1993). Ionic bases for electrophysiological distinctions among epicardial, midmyocardial, and endocardial myocytes from the free wall of the canine left ventricle. *Circulation Research,* Vol.72, pp. 671-687, ISSN 00097330

Liu, D.W. & Antzelevitch, C. (1995). Characteristics of the delayed rectifier current (I_{Kr} and I_{Ks}) in canine ventricular epicardial, midmyocardial and endocardial myocytes. A weaker I_{Ks} contributes to the longer action potential of the M cell. *Circulation Research,* Vol.76, pp. 351-365, ISSN 00097330

Mahajan, A.; Shiferaw, Y.; Sato, D.; Baher, A.; Olcese, R.; Xie, L-H.; Yang, M.; Chen, P-S.; Restrepo, J. G.; Karma, A.; Garfinkel, A.; Qu, Z. & Weiss, J. N. (2008). A rabbit ventricular action potential model replicating cardiac dynamics at rapid heart rates. *Biophys. J.,* Vol.94, pp. 392-410, ISSN 00063495

Marder, E. & Taylor, A. L. (2011). Multiple models to capture the variability in biological neurons and networks. *Nat. Neurosci.* Vol.14, pp. 133-138, ISSN 1097-6256

Mazzanti, M. & DeFelice L.J. (1987). Na Channel Kinetics During the Spontaneous Heart Beat in Embryonic Chick Ventricular Cells. *Biophysical Journal,* Vol.52, pp. 95-100, ISSN 00063495

Nakajima, S. & Bastian, J (1974). Double Sucrose-Gap Method Applied to Single Muscle Fiber of Xenopus laevis. *The Journal of General Physiology,* Vol.63, pp. 235-256, ISSN 15407748

Nattel, S.; Maguy, A.; Le Bouter, S. & Yeh, Y-H. (2007). Arrhythmogenic Ion-Channel Remodeling in the Heart: Heart Failure, Myocardial Infarction, and Atrial Fibrillation. *Physiological Reviews*, Vol.87, pp. 425-456, ISSN 00319333

Ozdemir, S.; Bito, V.; Holemans, P.; Vinet, L.; Mercadier, J.J.; Varro, A. & Sipido, K.R. (2008) Pharmacological inhibition of na/ca exchange results in increased cellular Ca^{2+} load attributable to the predominance of forward mode block. *Circulation Research,* Vol.102, pp. 1398-1405, ISSN 00097330

Özgen, N.; Dun, W.; Sosunov, E.A.; Anyukhovsky, E.P.; Hirose, M.; Duffy, H.S.; Boyden, P.A. & Rosen, M.R. (2007). Early electrical remodeling in rabbit pulmonary vein results from trafficking of intracellular SK2 channels to membrane sites. *Cardiovascular Research,* Vol.75, pp. 758-769, ISSN 00086363

Pasek, M.; Simurda, J.; Orchard, C.H. & Christe, G. (2008). A model of the guinea-pig ventricular cardiac myocyte incorporating a transverse-axial tubular system. *Progress in Biophysics and Molecular Biology,* Vol.96, pp. 258-80, ISSN 00796107

Patel, S.P. & Campbell, D.L. (2005). Transient outward potassium current, 'Ito', phenotypes in the mammalian left ventricle: underlying molecular, cellular and biophysical mechanisms. *Journal of Physiology,* Vol.569, pp. 7-39, ISSN 00223751

Pogwizd, S. M. (1995). Nonreentrant Mechanisms Underlying Spontaneous Ventricular Arrhythmias in a Model of Nonischemic Heart Failure in Rabbits. *Circulation* Vol.92, pp. 1034-1048, ISSN 00097322

Pogwizd, S.M.; Schlotthauer, K.; Li, L.; Yuan, W. & Bers, D.M. (2001). Arrhythmogenesis and Contractile Dysfunction in Heart Failure : Roles of Sodium-Calcium Exchange, Inward Rectifier Potassium Current, and Residual beta-Adrenergic Responsiveness. *Circulation Research* Vol.88, pp. 1159-1167, ISSN 00097330

Puglisi, J.L.; Yuan, W.; Bassani, W.M & Bers, D.M. (1999). Ca^{2+} Influx Through Ca^{2+} Channels in Rabbit Ventricular Myocytes During Action Potential Clamp: Influence of Temperature. *Circulation Research,* Vol.85, pp. e7-e16, ISSN 00097330

Rajamani, S.; El-Bizri, N.; Shryock, J.C.; Makielski, J.C. & Belardinelli, L. (2009). Use-dependent block of cardiac late Na^{+} current by ranolazine. *Heart Rhythm,* Vol.6, pp. 1625-1631, ISSN 15475271

Rocchetti, M.; Besana, A.; Gurrola, G.B.; Possani, L.D. & Zaza A. (2001). Rate dependency of delayed rectifier currents during the guinea-pig ventricular action potential. *Journal of Physiology,* Vol.534, pp. 721–732, ISSN 00223751

Rocchetti, M.; Freli, V.; Perego, V.; Altomare, C.; Mostacciuolo, G. & Zaza, A. (2006). Rate dependency of β-adrenergic modulation of repolarizing currents in the guinea-pig ventricle. *Journal of Physiology,* Vol.574, pp. 183-193, ISSN 00223751

Saleh, S.N.; Angermann, J.E.; Sones, W.R.; Leblanc, N.; Greenwood I.A. (2007). Stimulation of Ca^{2+}-gated Cl^{-} currents by the calcium-dependent K^{+} channel modulators NS1619 [1,3-dihydro-1-[2-hydroxy-5-(trifluoromethyl)phenyl]-5-(trifluoromethyl)-2 H-benzimidazol-2-one] and isopimaric acid. *Journal of Pharmacology and Experimental Therapeutics,* Vol.321, pp. 1075-1084, ISSN: 00223565

Starzak, M. E. & Starzak R. J. (1978). An Action Potential Clamp to Probe the Effectiveness of Space Clamp in Axons. *IEEE Transactions On Biomedical Engineering,* Vol.BME-25, pp. 201-204, ISSN 00189294

Starzak, M. E. & Needle M. (1983). The Action Potential Clamp as a Test of Space-Clamp Effectiveness - The Lettwin Analog Axon. *IEEE Transactions On Biomedical Engineering,* Vol.BME-30, pp. 139-140, ISSN 00189294

Silva, J.R.; Pan, H.; Wu, D.; Nekouzadeh, A.; Decker, K.F.; Cui, J.; Baker, N.A.; Sept, D. & Rudy, Y. (2009). A multiscale model linking ion-channel molecular dynamics and electrostatics to the cardiac action potential. Proc Natl Acad Sci USA, Vol.106, pp. 11102-6, ISSN 1091-6490

Szabó, A.; Szentandrássy, N.; Birinyi, P.; Horváth, B.; Szabó, G.; Bányász, T.; Márton, I.; Nánási, P.P. & Magyar, J. (2007). Effects of articaine on action potential characteristics and the underlying ion currents in canine ventricular myocytes. *British Journal of Anaesthesia,* Vol.99, pp. 726-733, ISSN 00070912

Szabó, A.; Szentandrássy, N.; Birinyi, P.; Horváth, B.; Szabó, G.; Bányász, T.; Márton, I.; Magyar, J. & Nánási, P.P. (2008). Effects of ropivacaine on action potential configuration and ion currents in isolated canine ventricular cardiomyocytes. *Anesthesiology,* Vol.108, pp. 693-702, ISSN 00033022

Tanaka, H.; Komikado, C.; Shimada, H.; Takeda, K.; Namekata, I.; Kawanishi, T. & Shigenobu K. (2004). The R(-)-enantiomer of efonidipine blocks T-type but not L-type calcium current in guinea pig ventricular myocardium. *Journal of Pharmacological Sciences,* Vol.96, pp. 499-501 ISSN 13478613

Thiel, G. (1995). Dynamics of chloride and potassium currents during the action potential in Chara studied with action potential clamp. *European Biophysics Journal,* Vol.24, pp. 85-92, ISSN 01757571

Thomas, G.P.; Gerlach, U. & Antzelevitch, C. (2003). HMR 1556, a potent and selective blocker of slowly activating delayed rectifier potassium current. *Journal of Cardiovascular Pharmacology,* Vol.41, pp. 140-147, ISSN: 01602446

Varro, A.; Balati, B.; Iost, N.; Takacs, J.; Virag, L.; Lathrop, D.A.; Csaba, L.; Talosi, L. & Papp, J.G. (2000). The role of the delayed rectifier component IKs in dog ventricular muscle and Purkinje fibre repolarization. *Journal of Physiology,* Vol.523, pp. 67-81, ISSN 00223751

Wilders, R. (2006). Dynamic clamp: a powerful tool in cardiac electrophysiology. *Journal of Physiology,* Vol.576, pp. 349–359, ISSN 00223751

Xu, Y.; Tuteja, D.; Zhang, Z.; Xu, D.; Zhang, Y.; Rodriguez, J.; Nie, L.; Tuxson, H.R.; Young, J.N.; Glatter, K.A.; VÃ¡zquez, A.E.; Yamoah, E.N. & Chiamvimonvat, N. (2003). Molecular Identification and Functional Roles of a Ca^{2+}-activated K^+ Channel in Human and Mouse Hearts. *Journal of Biological Chemistry,* Vol.278, pp. 49085-49094, ISSN 00219258

Yamada, M.; Ohta, K.; Niwa, A.; Tsujino, N.; Nakada, T. & Hirose, M. (2008). Contribution of L-type Ca^{2+} channels to early afterdepolarizations induced by I_{Kr} and I_{Ks} channel suppression in guinea pig ventricular myocytes. *Journal of Membrane Biology,* Vol.222, pp. 151-166, ISSN 00222631

Yuill, K.H.; Convery, M.K.; Dooley, P.C.; Doggrell, S.A. & Hancox, J.C. (2000). Effects of BDF 9198 on action potentials and ionic currents from guinea-pig isolated ventricular myocytes. *British Journal of Pharmacology,* Vol.130, pp.1753-1766, ISSN 00071188.

8

Electrophysiological Techniques for Mitochondrial Channels

Rainer Schindl[1] and Julian Weghuber[2,*]
[1]Institute of Biophysics, Johannes Kepler University, Linz
[2]University of Applied Sciences Upper Austria, Wels
Austria

1. Introduction

The patch-clamp technique has revolutionized studies of ion channels in a wide variety of cells. Its development in the late 1970s and early 1980s made it possible to analyze the currents of single ion channels. However, electrophysiological recordings of these proteins were limited to the plasma-membrane of living cells: the small size and the double-membrane organization of some cell organelles and the reduced stability of intracellular membranes precluded an implementation of this technique to investigate the properties of channels located in mitochondria, chloroplasts, the endoplasmic reticulum or lysosomes. For the last 25 years great effort has been made to implement the patch-clamp technique besides the planar lipid bilayer approach on these organelles. In this mini-review we focus on the developments and applications of the patch-clamp technique on mitochondrial and other cytosolic membranes and discuss the challenges of different, innovative approaches.

2. First attempts

The patch-clamp technique developed by Neher and Sakmann in 1976 as an effective refinement of the voltage-clamp-assay proved the function of single ion channels in fundamental cellular processes (Neher and Sakmann, 1976). The effectiveness of this technique to study plasma-membrane localized channels soon became of interest to characterize putative ion channels in intra-cellular membranes. Special attention was paid to mitochondria and their main function to provide the cell with energy in the form of ATP. Still in 1976, members from the lab of Alan Finkelstein inserted mitochondrial membranes into planar lipid bilayers using a Triton X-100 extract of rat liver mitochondria (Schein *et al.*, 1976). Single channels were directly detected by elevated ion permeability through the bilayer. This innovative work and further studies in the following years (Roos *et al.*, 1982) provided a good deal of information on the probable behavior of the voltage-dependent anion channel (VDAC, "mitochondrial porin"). However, the permeability properties of the intact outer mitochondrial membrane (OMM) remained hidden. In 1987, Tedeschi and colleagues managed to fill gaps in information about the electrical characterization of the OMM (Tedeschi *et al.*, 1987): they successfully introduced the patch-clamp technique on

* Corrresponding Author

outer membrane vesicles isolated from mitochondria of the fungus *Neurospora crassa* and giant mitochondria isolated from cuprizone-fed mice. The cupper chelator Cuprizone (biscyclohexanone oxaldihydrazone) had been shown to induce an increase of the size of mitochondria more than twofold (Flatmark *et al.*, 1980). Their results confirmed the presence of the VDAC, which facilitates the exchange of ions and molecules up to 5 kDa. Kathleen W. Kinnally, a pioneer in the field of mitochondrial electrophysiology, and co-workers continued their work on channels of the OMM. Among other findings they presented first studies on the kinetics of conductance changes induced by negative voltages and speculated on the multimerization of the high-conductance channel in the OMM (Kinnally *et al.*, 1989b). It took more than 30 years after its initial discovery, to unravel the structure of the VDAC channel by NMR spectroscopy and X-ray crystallographic techniques (Bayrhuber *et al.*, 2008;Hiller *et al.*, 2008;Ujwal *et al.*, 2008).

Apart from the progress that had been made on channels of the OMM, the importance of putative channels in the inner mitochondrial membrane (IMM) was neglected. Until the late eighties it was widely believed that the IMM was unlikely to contain any channels. The small size and the double-membrane structure of mitochondria had prohibited researchers from using the patch-clamp technique on the IMM. This prevailing opinion was finally disproved by the group of Walter Stühmer in 1987 by patch-clamp recordings of the inner mitochondrial membranes (IMM) (Sorgato *et al.*, 1987). They circumvented the restriction in size and the double-membraned organization by removing the OMM using a swelling-shrinking-sonication procedure from mitochondria isolated from cuprizone-fed mice. This led to the formation of vesicles termed mitoplasts with diameters of 3-6 μm. Measurements in whole mitoplast configuration as well as single channel recordings resulted in the identification of a voltage-dependent, anion permeable channel. The properties of this 107 pS channel, which was later found to correspond to the Inner Membrane Anion channel (IMAC) (Borecky *et al.*, 1997;Schonfeld *et al.*, 2004), were reported to be remarkably different from those of the VDAC of the OMM with a calculated mean conductance of 480 pS (Roos *et al.*, 1982).

3. Pros and cons of electrophysiological methods

In order to characterize the electrophysiological properties of single channels, two methods are commonly used: a voltage-clamp assay using planar lipid bilayers and the glass-pipette based patch-clamp method. The former one is based on the incorporation of purified membrane vesicles into the bilayer. The method allows studying channels, which are not easily accessible for the patch-clamp technique. Thus, the IMM but also other intracellular organelles including endosomes, lysosomes or the Golgi-apparatus may be analyzed. Furthermore this method is capable of detecting changes in the channel properties upon modulation by other proteins or lipids, i.e. variations in the lipid or protein composition of the bilayer. The major advantage of the patch-clamp method is the suitability to study channels in their natural environment, as the isolation process for the bilayer method may be harmful for the channel. Exemplarily, the recorded VDAC currents through single channels were 350 pS in patch-clamp recordings of whole mitochondria (Sorgato *et al.*, 1987), while a single channel conductance of 480 pS was found for the same channel after reconstitution in lipid bilayers (Roos *et al.*, 1982). The differences in the conductance might originate from a distinct membrane environment. However, the purity of the vesicle preparation is crucial due to the different fusion probability of vesicles from various membranes. This problem can be partly solved by using vesicles lacking any other contaminating membranes. Disadvantages of the patch-clamp technique should be mentioned at this point as well: first, mitochondria and possibly other cell organelles are in close contact with membranes of the endoplasmic reticulum, forming so-called "mitochondria associated membranes" (Garcia-Perez *et al.*, 2008). It might be difficult to completely remove these membranes. Second, some membranes, e.g. mitochondrial Cristae tubules, might not be accessible for the glass pipette (Zick *et al.*, 2009). Thus, the patch clamp technique might fail to detect some important ion channel currents.

4. Mitochondrial electrophysiology in the nineties

A report of the group of T. Higuti in 1991 can be considered another milestone in mitochondrial electrophysiology: they characterized a highly K^+ selective, low conductance channel in the IMM, using mitochondria isolated from fresh liver of rats (Inoue *et al.*, 1991). To date the activity identified as corresponding to that of the ATP-sensitive K^+ channel (mito KATP or mtK$_{ATP}$) remains a matter of controversy. The molecular composition of the mito KATP has not been unraveled and some reports challenge the presence of this channel in the IMM (Brustovetsky *et al.*, 2005;Das *et al.*, 2003), while most studies support its existence (Dahlem *et al.*, 2004;Zhang *et al.*, 2001). However, the chosen patch-clamp strategy for the IMM appeared promising: To circumvent the double-membrane structure and small size of mitochondria, they removed the OMM by digitonin treatment and fused the generated giant mitoplasts using a low-pH solution containing Ca^{2+}. This method does not require additional lipids necessary for the vesicle fusion process that could influence the channel in its native environment. In addition the achieved size of the fused mitoplasts of up to 15 µm allows for a robust giga-seal formation with the patch-pipette tip.

A first report on the mitochondrial multiconductance channel by Kinnally and co-workers in 1989 (Kinnally *et al.*, 1989a) opened up a new chapter in mitochondrial electrophysiology, which has not lost its relevance. The MCC, also termed mitochondrial megachannel, was later described as the mitochondrial permeability transition pore (MPT pore or MPTP), which is formed under certain pathological conditions (Lemasters *et al.*, 2009). At that time, first amphiphilic drugs were found to affect the mitochondrial permeability (Beavis and Powers, 1989;Beavis, 1989), which was regarded as a prerequisite for the study of channels in the IMM (Antonenko *et al.*, 1991;Gunter and Pfeiffer, 1990). Electrophysiological data on mitoplasts created in the lab of Mario Zoratti proposed the MCC may comprise VDAC molecules (Szabo *et al.*, 1993). Further biochemical investigations by Andrew Halestrap et al. and electrophysiology studies performed in the Kinnally-lab identified the adenine nucleotide translocase (ANT) as the central unit forming the MPTP (Halestrap *et al.*, 1998;Lohret *et al.*, 1996). Although these findings have been challenged by studies published only a few years ago (Baines *et al.*, 2007;Kokoszka *et al.*, 2004), the efforts of these few pioneers during the nineties accelerated the progress for a better understanding of the complex behavior of mitochondrial channels.

In contrast to the work of Inoue et al, 1991 (Inoue *et al.*, 1991) on the mito KATP channel, most researchers did not work on fused mitoplasts, but continued to directly patch-clamp whole mitoplasts prepared from mammalian cell lines or tissue. These included a glioma cell line (Siemen *et al.*, 1999), an osteosarcoma cell line (Murphy *et al.*, 1998), rat liver tissue (Szabo *et al.*, 1992) or brown adipose tissue from hamsters (Borecky *et al.*, 1997). However, only mitochondria isolated from a limited number of mammalian and other eukaryotic cells formed mitoplasts with an adequate size. Thus, most IMM channel activities were recorded after reconstitution into proteoliposomes or giant liposomes (Lohret *et al.*, 1996;Lohret *et al.*, 1997;Paliwal *et al.*, 1992). This holds especially true for studies of yeast mitochondrial channels. The small size of mitoplasts from this organism (0.3-1.5 µm), either isolated by osmotic swelling or using the French press method, makes the patch clamp technique very difficult. The advantages of the model-system yeast, e.g. fast growth, easy handling and especially the possibility of a fast gene knock-out, justify the technically challenging and laborious reconstitution process. Still, the purity of the used membrane fractions and a potential influence of externally added lipids for the generation of giant liposomes must be taken into account.

5. Mitochondrial electrophysiology over the past 10 years

We may start with an outstanding finding published in 2004 (Kirichok *et al.*, 2004). Yuriy Kirichok *et al.* from the lab of David Clapham described the properties of a channel consistent with the predicted mitochondrial Calcium uniporter (miCa). The experiments were accomplished by patch-clamp recordings of whole mitoplasts without any additional fusion or reconstitution processes. An interesting feature of the chosen COS-7 cell-line is that intact mitoplasts of 2-5 µm size could be prepared. For that purpose mitochondria were isolated by differential centrifugation and mitoplasts were formed by osmotic swelling. Interestingly, no additional steps to remove the OMM had been performed, which enhances the risk that remnants of the OMM or the tightly attached ER-membrane falsify the results. According to the authors such artifacts had been excluded as these structures appeared as black spots on the transparent vesicles. Taken together extreme care has to be taken and numerous technical problems have to be solved when performing such experiments. Very recently, the molecular identity of the miCa has been clarified: a channel protein termed MCU and a regulating protein named MICU1 have been reported to fulfill the requirements for the long searched mitochondrial Calcium uniporter (De *et al.*, 2011;Perocchi *et al.*, 2010).

Inspired by the work of these pioneers our group started electrophysiological studies on mitochondrial Magnesium transport. In 2003-2006 we had shown, using Magnesium selective fluorescent dyes, that the IMM localized protein Mrs2 is a major player for the influx of Magnesium into yeast mitochondria (Kolisek *et al.*, 2003;Weghuber *et al.*, 2006). At that time we could not exclude that Mrs2p forms a transporter rather than a high-conductance channel. To answer this question we generated mitoplasts using osmotic swelling or the French press method. However, consistent with the work performed in the Kinnally lab in the 90ies, the small size of yeast < 1 µm did not allow for effective patch-clamp recordings. Thus we decided to fuse IMM with lipid vesicles: submitochondrial particles (SMPs) (Froschauer *et al.*, 2009) were mixed with asolectin vesicles. According to the report of Criado and Keller (Criado and Keller, 1987) in 1987 we finally succeeded in generating giant vesicles (5-50 µm in size) with reconstituted IMM fractions from yeast overexpressing or having *MRS2* disrupted. The used method is characterized by dehydration-rehydration cycles and has also been successfully used by other groups working on yeast mitochondrial channels (Lohret *et al.*, 1996). During our studies the crystal structure of the bacterial homologue CorA had been reported (Lunin *et al.*, 2006;Eshaghi *et al.*, 2006;Payandeh and Pai, 2006), but the mode of ion transport remained obscure. We could finally prove that Mrs2p forms a Magnesium selective channel in the IMM with a calculated conductance of 155 pS (Schindl *et al.*, 2007). We furthermore showed that co-expression of a homologue of Mrs2p termed Lpe10p (or Mfm1p) yielded a unique, reduced conductance in comparison to the one of Mrs2p channels (Sponder *et al.*, 2010).

6. Outlook

The progress on the electrophysiological characterization of mitochondrial channels was rather limited in the nineties with only a few labs being able to deal with the emerging technical challenges. This situation has definitely changed in the last decade: while there are not more than 100 reports on patch-clamp recordings on mitochondria indexed in pubmed in the nineties, this number has changed to more than 400 in the last decade. However, there is still a lot of work to be done. The molecular identities of most channel proteins forming the MPTP, the 107 pS channel or the mito KATP have not been defined. From that point of view the studies on the yeast Magnesium channel proteins Mrs2p and its homolog Mfm1p are of even greater relevance.

The experiences in the field of protein- and membrane reconstitution and in mitochondrial electrophysiology rendered these techniques also interesting for the characterization of other intracellular membranes. First studies by a pioneer in this field Igor Pottosin in the early nineties reported on the investigation of ion channels located in chloroplast envelope membranes (Pottosin, 1993; Heiber *et al.*, 1995) or thylakoid membranes (Pottosin and Schonknecht, 1995). Further reports described the action of a voltage-gated ion channel in the outer envelope (Hinnah *et al.*, 1997), focused on the inactivation of a 50 pS envelope anion channel necessary for import of proteins into the chloroplast (van den Wijngaard *et al.*, 1999) or characterized a fast-activating channel regulating various fluxes in the native envelope of chloroplasts (Pottosin *et al.*, 2005). The methods of choice to cope with the severe technical challenges (e.g. the maintenance of the organelle integrity) were either the reconstitution of isolated membranes into lipid vesicles - mainly based on asolectin - or measurements on whole chloroplasts. In both cases great experience as well as patience appears indispensable. That is probably the main reason for the limited number of studies in this field.

To efficiently investigate intracellular ion channels, new methods will be necessary. Assays like the one presented by Schieder and co-workers (Schieder *et al.*, 2010) using solid-matrix planar glass chips, which even enables patch-clamp recordings of lysosomes, are a step in the right direction.

7. Acknowlegdements

This work was supported by the Austrian Science Foundation (FWF): project P22747 to R.S. We thank Jeffrey Nesbitt (Haverford High School, PA) for critically reading the manuscript.

8. References

Antonenko YN, Kinnally KW, Perini S, Tedeschi H, 1991. Selective effect of inhibitors on inner mitochondrial membrane channels. FEBS Lett 285:89-93.

Baines CP, Kaiser RA, Sheiko T, Craigen WJ, Molkentin JD, 2007. Voltage-dependent anion channels are dispensable for mitochondrial-dependent cell death. Nat Cell Biol 9:550-555.

Bayrhuber M, Meins T, Habeck M, Becker S, Giller K, Villinger S, Vonrhein C, Griesinger C, Zweckstetter M, Zeth K, 2008. Structure of the human voltage-dependent anion channel. Proc Natl Acad Sci U S A 105:15370-15375.

Beavis AD, 1989. On the inhibition of the mitochondrial inner membrane anion uniporter by cationic amphiphiles and other drugs. J Biol Chem 264:1508-1515.

Beavis AD, Powers MF, 1989. On the regulation of the mitochondrial inner membrane anion channel by magnesium and protons. J Biol Chem 264:17148-17155.

Borecky J, Jezek P, Siemen D, 1997. 108-pS channel in brown fat mitochondria might Be identical to the inner membrane anion channel. J Biol Chem 272:19282-19289.

Brustovetsky T, Shalbuyeva N, Brustovetsky N, 2005. Lack of manifestations of diazoxide/5-hydroxydecanoate-sensitive KATP channel in rat brain nonsynaptosomal mitochondria. J Physiol 568:47-59.

Criado M, Keller BU, 1987. A membrane fusion strategy for single-channel recordings of membranes usually non-accessible to patch-clamp pipette electrodes. FEBS Lett 224:172-176.

Dahlem YA, Horn TF, Buntinas L, Gonoi T, Wolf G, Siemen D, 2004. The human mitochondrial KATP channel is modulated by calcium and nitric oxide: a patch-clamp approach. Biochim Biophys Acta 1656:46-56.

Das M, Parker JE, Halestrap AP, 2003. Matrix volume measurements challenge the existence of diazoxide/glibencamide-sensitive KATP channels in rat mitochondria. J Physiol 547:893-902.

De SD, Raffaello A, Teardo E, Szabo I, Rizzuto R, 2011. A forty-kilodalton protein of the inner membrane is the mitochondrial calcium uniporter. Nature 476:336-340.

Eshaghi S, Niegowski D, Kohl A, Martinez MD, Lesley SA, Nordlund P, 2006. Crystal structure of a divalent metal ion transporter CorA at 2.9 angstrom resolution. Science 313:354-357.

Flatmark T, Kryvi H, Tangeras A, 1980. Induction of megamitochondria by cuprizone(biscyclohexanone oxaldihydrazone). Evidence for an inhibition of the mitochondrial division process. Eur J Cell Biol 23:141-148.

Froschauer EM, Schweyen RJ, Wiesenberger G, 2009. The yeast mitochondrial carrier proteins Mrs3p/Mrs4p mediate iron transport across the inner mitochondrial membrane. Biochim Biophys Acta 1788:1044-1050.

Garcia-Perez C, Hajnoczky G, Csordas G, 2008. Physical coupling supports the local Ca2+ transfer between sarcoplasmic reticulum subdomains and the mitochondria in heart muscle. J Biol Chem 283:32771-32780.

Gunter TE, Pfeiffer DR, 1990. Mechanisms by which mitochondria transport calcium. Am J Physiol 258:C755-C786.

Halestrap AP, Kerr PM, Javadov S, Woodfield KY, 1998. Elucidating the molecular mechanism of the permeability transition pore and its role in reperfusion injury of the heart. Biochim Biophys Acta 1366:79-94.

Heiber T, Steinkamp T, Hinnah S, Schwarz M, Flugge Ui, Weber A, Wagner R, 1995. Ion channels in the chloroplast envelope membrane. Biochemistry 34:15906-15917.

Hiller S, Garces RG, Malia TJ, Orekhov VY, Colombini M, Wagner G, 2008. Solution structure of the integral human membrane protein VDAC-1 in detergent micelles. Science 321:1206-1210.

Hinnah SC, Hill K, Wagner R, Schlicher T, Soll J, 1997. Reconstitution of a chloroplast protein import channel. EMBO J 16:7351-7360.

Inoue I, Nagase H, Kishi K, Higuti T, 1991. ATP-sensitive K+ channel in the mitochondrial inner membrane. Nature 352:244-247.

Kinnally KW, Campo ML, Tedeschi H, 1989a. Mitochondrial channel activity studied by patch-clamping mitoplasts. J Bioenerg Biomembr 21:497-506.

Kinnally KW, Tedeschi H, Mannella CA, FRISCH HL, 1989b. Kinetics of voltage-induced conductance increases in the outer mitochondrial membrane. Biophys J 55:1205-1213.

Kirichok Y, Krapivinsky G, Clapham DE, 2004. The mitochondrial calcium uniporter is a highly selective ion channel. Nature 427:360-364.

Kokoszka JE, Waymire KG, Levy SE, Sligh JE, Cai J, Jones DP, Macgregor GR, Wallace DC, 2004. The ADP/ATP translocator is not essential for the mitochondrial permeability transition pore. Nature 427:461-465.

Kolisek M, Zsurka G, Samaj J, Weghuber J, Schweyen RJ, Schweigel M, 2003. Mrs2p is an essential component of the major electrophoretic Mg2+ influx system in mitochondria. EMBO J 22:1235-1244.

Lemasters JJ, Theruvath TP, Zhong Z, Nieminen AL, 2009. Mitochondrial calcium and the permeability transition in cell death. Biochim Biophys Acta 1787:1395-1401.

Lohret TA, Jensen RE, Kinnally KW, 1997. Tim23, a protein import component of the mitochondrial inner membrane, is required for normal activity of the multiple conductance channel, MCC. J Cell Biol 137:377-386.

Lohret TA, Murphy RC, Drgon T, Kinnally KW, 1996. Activity of the mitochondrial multiple conductance channel is independent of the adenine nucleotide translocator. J Biol Chem 271:4846-4849.

Lunin VV, Dobrovetsky E, Khutoreskaya G, Zhang R, Joachimiak A, Doyle Da, Bochkarev A, Maguire Me, Edwards AM, Koth CM, 2006. Crystal structure of the CorA Mg2+ transporter. Nature 440:833-837.

Murphy RC, Diwan JJ, King M, Kinnally KW, 1998. Two high conductance channels of the mitochondrial inner membrane are independent of the human mitochondrial genome. FEBS Lett 425:259-262.

Neher E, Sakmann B, 1976. Single-channel currents recorded from membrane of denervated frog muscle fibres. Nature 260:799-802.

Paliwal R, Costa G, Diwan JJ, 1992. Purification and patch clamp analysis of a 40-pS channel from rat liver mitochondria. Biochemistry 31:2223-2229.

Payandeh J, Pai EF, 2006. Crystallization and preliminary X-ray diffraction analysis of the magnesium transporter CorA. Acta Crystallogr Sect F Struct Biol Cryst Commun 62:148-152.

Perocchi F, Gohil Vm, Girgis Hs, Bao Xr, Mccombs Je, Palmer Ae, Mootha VK, 2010. MICU1 encodes a mitochondrial EF hand protein required for Ca(2+) uptake. Nature 467:291-296.

Pottosin II, 1993. One of the chloroplast envelope ion channels is probably related to the mitochondrial VDAC. FEBS Lett 330:211-214.

Pottosin Ii, Muniz J, Shabala S, 2005. fast-activating channel controls cation fluxes across the native chloroplast envelope. J Membr Biol 204:145-156.

Pottosin Ii, Schonknecht G, 1995. Patch clamp study of the voltage-dependent anion channel in the thylakoid membrane. J Membr Biol 148:143-156.

Roos N, Benz R, Brdiczka D, 1982. Identification and characterization of the pore-forming protein in the outer membrane of rat liver mitochondria. Biochim Biophys Acta 686:204-214.

Schein Sj, Colombini M, Finkelstein A, 1976. Reconstitution in planar lipid bilayers of a voltage-dependent anion-selective channel obtained from paramecium mitochondria. J Membr Biol 30:99-120.

Schieder M, Rotzer K, Bruggemann A, Biel M, Wahl-Schott C, 2010. Planar patch clamp approach to characterize ionic currents from intact lysosomes. Sci Signal 3:13.

Schindl R, Weghuber J, Romanin C, Schweyen RJ, 2007. Mrs2p forms a high conductance Mg2+ selective channel in mitochondria. Biophys J 93:3872-3883.

Schonfeld P, Sayeed I, Bohnensack R, Siemen D, 2004. Fatty acids induce chloride permeation in rat liver mitochondria by activation of the inner membrane anion channel (IMAC). J Bioenerg Biomembr 36:241-248.

Siemen D, Loupatatzis C, Borecky J, Gulbins E, Lang F, 1999. Ca2+-activated K channel of the BK-type in the inner mitochondrial membrane of a human glioma cell line. Biochem Biophys Res Commun 257:549-554.

Sorgato MC, Keller BU, Stuhmer W, 1987. Patch-clamping of the inner mitochondrial membrane reveals a voltage-dependent ion channel. Nature 330:498-500.

Sponder G, Svidova S, Schindl R, Wieser S, Schweyen RJ, Romanin C, Froschauer EM, Weghuber J, 2010. Lpe10p modulates the activity of the Mrs2p-based yeast mitochondrial Mg2+ channel. FEBS J 277:3514-3525.

Szabo I, Bernardi P, Zoratti M, 1992. Modulation of the mitochondrial megachannel by divalent cations and protons. J Biol Chem 267:2940-2946.

Szabo I, De P, V, Zoratti M, 1993. The mitochondrial permeability transition pore may comprise VDAC molecules. II. The electrophysiological properties of VDAC are compatible with those of the mitochondrial megachannel. FEBS Lett 330:206-210.

Tedeschi H, Mannella CA, Bowman CL, 1987. Patch clamping the outer mitochondrial membrane. J Membr Biol 97:21-29.

Ujwal R, Cascio D, Colletier JP, Faham S, Zhang J, Toro L, Ping P, Abramson J, 2008. The crystal structure of mouse VDAC1 at 2.3 A resolution reveals mechanistic insights into metabolite gating. Proc Natl Acad Sci U S A 105:17742-17747.

Van den Wijngaard PW, Dabney-Smith C, Bruce BD, Vredenberg WJ, 1999. The mechanism of inactivation of a 50-pS envelope anion channel during chloroplast protein import. Biophys J 77:3156-3162.

Weghuber J, Dieterich F, Froschauer EM, Svidova S, Schweyen RJ, 2006. Mutational analysis of functional domains in Mrs2p, the mitochondrial Mg2+ channel protein of Saccharomyces cerevisiae. FEBS J 273:1198-1209.

Zhang DX, Chen YF, Campbell WB, Zou AP, Gross GJ, Li PL, 2001. Characteristics and superoxide-induced activation of reconstituted myocardial mitochondrial ATP-sensitive potassium channels. Circ Res 89:1177-1183.

Zick M, Rabl R, Reichert AS, 2009. Cristae formation-linking ultrastructure and function of mitochondria. Biochim Biophys Acta 1793:5-19.

9

Patch-Clamp Analysis of Membrane Transport in Erythrocytes

Guillaume Bouyer, Serge Thomas and Stéphane Egée
Centre National de la Recherche Scientifique, Université Pierre et Marie Curie Paris6, Station Biologique, Roscoff
France

1. Introduction

Among all the models used to study membrane transport, erythrocytes (Red Blood Cells, RBCs) have probably been the most utilised cell type. Radioisotopes fluxes, isosmotic haemolysis, ion content analysis (e.g. flame photometry), or fluorescence techniques have been widely used to characterise the various transporters present in the RBCs membrane. These techniques have allowed the description of several types of transporters such as pumps, specific solute transporters, symporters or antiporters, and even ion channels. However, the physiology of RBCs and their maintenance of homeostasis remains incompletely understood, and electrophysiology has proven, since the first single-channel recording on a human erythrocyte membrane thirty years ago, to be a very useful tool to understand more deeply RBC membrane transport. Why does one use these techniques on a small, non-excitable cell that has long been considered no more than an empty bag of haemoglobin? The diversity of transporters in the RBC membrane, including ion channels, shows that these cells are much more complex than expected. Indeed, ion channels now described in the RBC membrane (from Mammals to other Vertebrates) are implicated in important phenomena and functions throughout the cells lifespan (gas transports, cell volume regulation, differentiation and death). In this chapter, we will describe the main properties of the erythrocyte's membrane transport system, how electrophysiological techniques can be applied, and how they have contributed to the comprehension of erythrocyte physiology with the description of the various ion channels that can be found in RBC membrane.

2. Red blood cell membrane description

2.1 Why one studies red blood cell membrane transport properties?

RBCs are highly specialised cells, present in all vertebrates (except some cold/ice-water fish (Ruud, 1954)). Their main role is the transport of respiratory gases, between tissues and lungs or gills. Encapsulation of the respiratory pigment haemoglobin in a cell in vertebrates has hugely increased the gas transport capacity of blood, and is a key point throughout evolution of the animal kingdom. Erythrocytes are produced in the bone marrow, differentiating from pluripotent cells during erythropoïesis. A human RBC has a lifespan of around 120 days, before being removed from the circulation by macrophages, essentially in the spleen.

RBCs have always occupied a primordial place in the investigations on membrane transport. First of all, even if their major role in respiration processes of vertebrates has been known for long time, deciphering the precise role of the different membrane transporters involved has been a long story. Indeed, membrane transport and especially ion permeability are inseparable from the description of gas transport by erythrocytes. Oxygen diffuses freely across RBC membranes, but its affinity to haemoglobin is highly dependant on cell homeostasis and thus to transport regulation across cell membrane. Moreover, the high carbon dioxide transport capacity of blood is essentially supported by the Jacobs-Stewart cycle between red cells and plasma, relying on the existence of specialised ion transporters in the erythrocyte membrane (Jacobs & Stewart, 1942).

Moreover, the cellular structure and particularly that of mammalian of RBCs, has made them an ideal model for studying membrane transport. Mature mammalian erythrocytes are devoid of intracellular organites, and this means that they consist of a single compartment, simplifying many approaches for transport studies. During the end of cellular differentiation, the nucleus is extruded from the normoblats and engulfed by surrounding macrophages (Yoshida et al., 2005), and the other organelles are removed during the maturation of reticulocytes into erythrocytes, probably mainly via autophagy (Kundu et al., 2008; Mortensen et al., 2010). This makes mature erythrocytes from mammals a very easy-to-use model for plasma membrane transport studies: intracellular constant measurements (ion or metabolite concentrations, pH) and flux experiments are easier than in any other type of cell containing multiple compartments.

Finally, another reason for studying red blood cell membrane transporters is the nature of blood, as a non-fibrous connective tissue : the fact that these cells are naturally in suspension and thus do not need any mechanical, enzymatical or chemical treatment before use in any kind of experiment also makes them easy to handle. Moreover, apart from ethical questions, it is always technically easy to draw blood from animals or humans, and purification of RBCs from blood only requires few centrifugation steps.

Thus, many scientists have used RBCs throughout history to describe the diversity of transporters in the plasma membrane, and to understand their role in the maintenance of homeostasis. Among all these studies, several have revealed essential characteristics of cell membrane permeability. In particular, as early as 1960 work on RBCs allowed Tosteson and Hoffman to complete the description of the "pump and leak" steady-state concept using sheep RBCs (Tosteson & Hoffman, 1960). In 1966, Schatzmann described for the first time an ATP-fuelled Ca^{2+} pump and this discovery was made using human erythrocytes (Schatzmann, 1966). Furthermore, as early as 1981 these cells were among the first using the patch-clamp technique that provided direct electrophysiological evidence for the presence of ionic channels in the plasma membrane (first recordings of a K_{Ca} channel by Owen Hamill) (Hamill, 1981). They were also the cells in which aquaporins were first described and for which Peter Agre won the Nobel Prize (Agre et al., 1993). Nowadays, many studies on the properties of RBC membrane transport are made either in physiological or pathophysiological situations, and the patch clamp technique has become an essential tool in their characterisation and comprehension.

2.2 The basis of red cell membrane permeability

The transport of the respiratory gases within the blood is highly dependent on electrolytes and the acid-base status of RBCs and they are strongly correlated with the permeability properties of the membrane. Indeed, the erythrocyte membrane is endowed with a variety of membrane transporters, whose role is absolutely vital to maintain cell homeostasis. In this section, we will first present the different types and roles of membrane ion transporters that have been described in human RBCs, as they are the most studied among vertebrates.

The main characteristics of human RBC membrane ion permeability are linked to the unusual composition of these cells. The encapsulation of ~ 5 mmol of impermeable haemoglobin per litre of intracellular water in a cell moving in a plasma environment, that has a much lower protein concentration, creates a huge osmotic pressure. As explicitly formulated in the 'pump-leak' concept (Tosteson & Hoffman, 1960), the risk of colloidosmotic swelling and bursting is prevented by a very low membrane permeability to cations, allowing the pumps Na^+/K^+-ATPase and Ca^{2+}-ATPase to extrude the residual Na^+ and Ca^{2+} leaks at minimal metabolic cost. The red cell Ca^{2+}-ATPase is so powerful that it maintains intracellular concentration below micromolar concentrations (Lew et al., 1982; Schatzmann, 1983). The Na^+/K^+-ATPase maintains gradients for Na^+ and K^+, fuelling the secondary transporters present in the membrane. Indeed, a potassium/chloride cotransporter (KCC, identified as KCC1 (C.M. Pellegrino et al., 1998)), a potassium/sodium/two chloride cotransporter (NKCC) (Haas, 1989) and a sodium/proton (Na^+/H^+) exchanger (Semplicini et al., 1989) have been described in the RBC membrane.

By contrast, the RBC membrane is characterised by a huge anion permeability that is essentially linked to the respiratory function: a million copies per cell of electroneutral Cl^-/HCO_3^- exchanger (called Band 3) permit 85% of the CO_2 produced in the tissues to be transported in the blood as HCO_3^- ions, *via* the Jacobs/Stewart cycle (Figure 1). This protein was identified in 1972 by Cabantchik and Rothstein (Cabantchik & Rothstein, 1972), even if RBC anion permeability had been studied for long time.

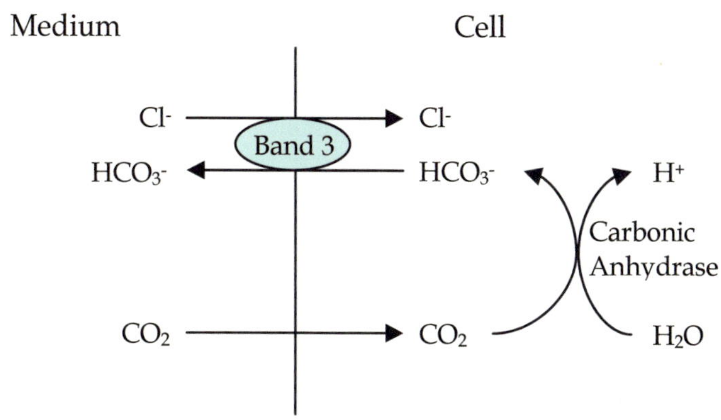

Fig. 1. The Jacobs/Stewart cycle in tissues. In lungs, cycle goes the other way.

It was known for long that RBCs anion permeability could be divided into two components: a large exchange component fundamental for the CO_2-carrying capacity of the blood (Gunn et al., 1973), and a much smaller electrogenic component that normally determines the RBC resting potential (Hunter, 1977; Lassen et al., 1978). This conductive part of chloride permeability ensures a dissipation of chloride gradient across red cell membrane: the membrane potential is clamped at the equilibrium potential for chloride (-12mV) ensuring that Band 3 never has to fight against a chloride gradient to transport bicarbonate ions across red cell membrane. Before the use of patch clamp, both components were frequently attributed to Band 3 protein activity, the electrogenic part resulting from either slippage in the exchange mechanism (Kaplan et al., 1983; Knauf et al., 1977), or tunneling (Frohlich, 1984; Knauf et al., 1983).

Besides these cotransports, a single conductive pathway had been described in RBCs before the appearance of patch clamp techniques: Gardos had shown in the 50's that an electrogenic, calcium-dependent potassium pathway was present in the human red cell membrane. This has since become known as the Gardos effect, linked to the activity of a calcium-activated potassium channel (Gardos, 1956, 1958).

For a long time, this list of membrane transporters was assumed to be complete: Band 3 mediated the very high anion permeability of red cells (*via* anion exchange and a much smaller electrogenic gating), and the tiny cationic permeability was supported by various powerful pumps and cation transporters carrying out homeostasis maintenance. The emergence of patch clamp studies on red cell membranes (during the 80's and 90's for single channel, 2000's for whole cell) first confirmed the presence of a calcium-activated potassium channel. It was described by single channel recordings by Hamill in 1981 (Hamill, 1981), and identified later as a member of the K_{Ca} channel family (hSK4, now called Gardos channel) (Hoffman et al., 2003). But these patch clamp studies, first using single channel (cell-attached and excised inside-out) and then whole-cell configurations, also brought evidence of a more complex situation than expected. The groups in the fields showed the existence of various anion and cation channels in the red cell membrane from different vertebrate species, and even if their role in physiological situations still remains poorly understood, their implication in several red cell pathologies is unambiguous and considerable.

It has taken much effort to first apply and then adapt patch clamp techniques to these tiny, non excitable cells; especially with the previous models of red cell membrane permeability that did not predicted the presence of various ion channels. Indeed, description of these different channel types in RBCs, using patch clamp techniques, has raised many questions: how do they interfere with the regulation of intracellular homeostasis, cell differentiation or death? But their presence also fits with a new vision of red cells, being much more than an empty bag of haemoglobin, notably regarding the recently discovered role of red cells in vascular tone regulation (Sprague et al., 2007).

3. Technical specificities of red blood cells regarding the patch-clamp technique

The technical aspects described here will focus on human red blood cells, as they are more described and studied than erythrocytes from any other species.

3.1 Red blood cell specificities

3.1.1 Size and deformability

The main problem when attempting to perform patch clamp on RBCs lies in their very small size. The smallest RBCs are encountered in mammals and are enucleated (mean diameter: human 8µm, mammalian 2.1-9.4µm). They are nucleated and slightly bigger in other vertebrates: amphibian: 16-70µm; Birds: 9.7-15.4µm, fish: 6.5-44.6µm. Nevertheless, deformability of the membrane remains a problem that could impair seal formation. This small size (notably for human RBCs) imposes working with high quality microscopes and objectives (at least X20) and a correct micromanipulation set up.

The pipettes must be really thin, with an average opening below 1µm. The main problem is to avoid entry of the cells into the pipette. RBCs are highly deformable, a property linked to their unique membrane composition and cytoskeleton structure. This allows correct blood circulation *via* all the narrow capillaries that are essential for correct transport of respiratory gases to each cell. In the human circulation, when passing through the spleen, RBCs have to go through tiny slits whose mean size has been recently measured: 1.89µm length and 0.65µm wide (Deplaine et al., 2011). Thus, considerable efforts have been necessary to adapt pipette shape and size for patch clamping RBCs. Their characteristics will be given in section 3.2.

3.1.2 Preparation

Blood is a non-fibrous connective tissue and collection and isolation of RBCs is therefore, a rather easy process. This is a great advantage, meaning that to be patched these cells do not require any mechanic or enzymatic dissociation, and that they do not adhere to solid surface *in vitro*.

Blood must be drawn from human or animal with an anticoagulant (such as heparin or EDTA), and can be stored at 4°C for several days. To isolate RBCs, simple centrifugation steps (usually four successive washings) at a speed of 3000 to 4000 rpm (around 2000g) are needed. RBCs are the densest elements of blood, thus supernatant (plasma) and buffy coat (white blood cells and platelets) can be discarded after each step. RBCs are usually washed with an appropriate saline isosmotic buffered solution.

3.1.3 Cell ion composition

The ion composition of human RBCs is given in Table 1. Plasma concentrations are also given in order to calculate ion gradients occurring in a physiological situation across RBCs membrane.

	Erythrocyte	Plasma
Cations		
Na^+	10 - 20	145
K^+	140	4.5
Ca^{2+}	<0.001	2
Mg^{2+}	1.5 - 2.4	1
Anions		
Cl^-	80	105
HCO_3^-		24
HPO_4^{2-}		4
protein	340g/l	60-80g/l

Table 1. Ion composition of human erythrocytes and plasma (in mM, except for proteins).

3.2 Patch clamping technique on red cells

3.2.1 Remarks on the use of patch clamp on small cells

Erythrocytes are small cells when compared to most commonly studied cells using patch clamp techniques. This induces two different types of problem that had to be considered. The first one was to adapt the shape and size of the pipette, so that a fragment of the membrane could seal to the tip of the pipette without this highly deformable cell entering totally the pipette. The second was that consequent to this adapted shape, electrical issues needed to be considered, especially for the whole cell configuration. Figure 2 describes the electric model for cell-attached configuration and whole-cell configuration.

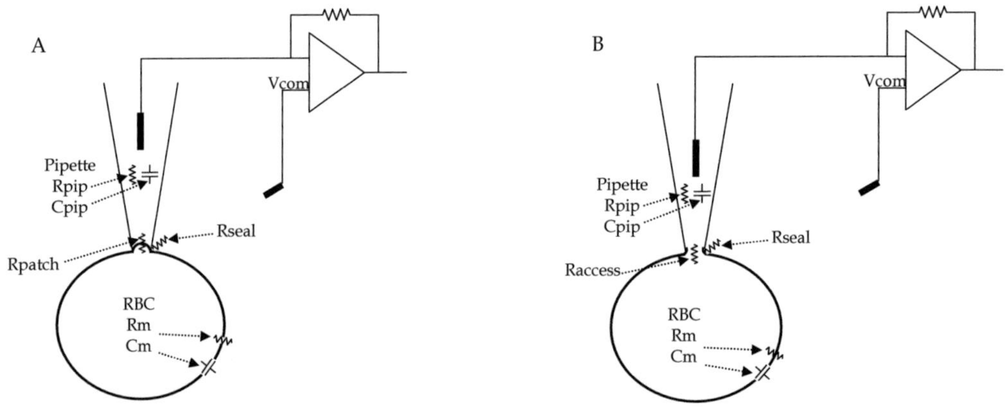

Fig. 2. Electrical model of patch clamp for cell-attached (A) and whole-cell (B) configuration

Single channel studies

When narrowing the tip of the pipette, its resistance R_{pip} is increased. Work on an adapted shape of the pipette for RBCs (especially human RBCs) has led to the use of pipette with an R_{pip} between 10 and 15MΩ. Up to 80% of seal attempts are successful with such pipettes, depending on the solutions used. These values are not a problem for single channel studies, as they remain well below the patch resistance R_{patch} through which single channel currents are recorded, thus these currents can be easily recorded. However, as suggested by Barry and Lynch (Barry & Lynch, 1991), distortion of the potential really applied to the membrane can happen in small cells, in relation with the global membrane resistance of the cell R_m. This resistance (see Figure 2) in small cells can be of the order of magnitude of several gigaOhms, not much lower than R_{patch}. Then variations in the pipette potential can induce changes in the global membrane potential, and this would result in large errors in the estimation of the single channel conductance and of the reversal potential. Barry and Lynch conclude their work (Barry & Lynch, 1991) by the equation (1) revealing the distortion between the apparent and real conductance of the single channel conductance :

$$\gamma_c=\gamma_{app}(1+R_m/R_{patch}) \quad (1)$$

where γ_c and γ_{app} are the real and apparent conductance of the channel, respectively. This clearly highlights the possible distortion of conductance estimation with cells showing a high membrane resistance.

Whole-cell studies

For whole-cell experiments, two resistances are in parallel in the electrical model: global membrane resistance R_m and seal resistance R_{seal} (see Figure 2). Once again, the complexity when using this configuration with small cells is linked to the possible high value of membrane resistance. Seal resistance is barely higher than 100GΩ, and small cell global resistance can reach this order of magnitude. Mammalian T-lymphocytes for example are reported to have a membrane resistance of 100GΩ (Cahalan et al., 1985). Then the contribution of the seal leak to current recorded in this configuration is not negligible: the current measured between the two electrodes can be attributed either to membrane current or seal leak, and the voltage applied to the pipette might be different of the voltage really occurring at the membrane level. This can induce an underestimation of the channel conductance and a discrepancy between real and apparent reversal potentials in cell-attached experiments, and wrong global conductance and shift in I/V curves in whole-cell experiments.

Do these limitations apply to RBCs?

From suspension experiments, membrane resistance R_m of human RBCs was estimated in the range $1.10^6\Omega.cm^2$, with chloride resistance R_{Cl} ranging between 10^5 and 10^6 $\Omega.cm^2$ (Hoffman et al., 1980). Given a mean surface of $135\mu m^2$, the whole-cell resistance can be estimated at 10 to 40 GOhms. This is in range of the seal values obtained on RBCs when using the patch clamp techniques. This explains why whole-cell experiments were not used on red blood cells: the conductance was estimated to be too low to discriminate properly between channel activity and seal leakage. However the membrane resistance can be considerably lowered when the ion channels present in the red cell membrane are activated: anion conductance rises to several nS in PKA-activated, oxidized or malaria-infected RBCs, and cation conductance can be increased when cation channels are activated (with low external chloride concentrations for example). Thus, for whole cell experiments, R_m remains much lower than R_{seal} and the current and voltage do not suffer high distortion.

Another parameter to take into account is the pipette access resistance R_{access}. Since the pipette tip must be rather narrow when patching RBCs, precautions must be taken. R_{access} value is determined by pipette geometry, solution composition and possibly by the presence of cell debris generated by membrane rupture. The use of pipette immediately after pulling, their adapted shape, with rapidly tapering geometry, and the filtering of all pipette solutions with 0.2µm filters can maintain low R_{access} values.

3.2.2 Electrodes, pipettes and seal

In our group, we use Ag/AgCl electrodes. Ag wires are regularly anodised to maintain uniform oxidized coating. The reference electrode is connected to the bath solution *via* an Agar/KCl 3M bridge. When perfusion of bath solution is performed, it modifies junction potentials. Liquid junction potentials must be corrected *a posteriori*. For that purpose, the JPCalc software developed by Peter Barry is very convenient (Barry, 1994).

Pipettes must be pulled by good pullers and have a tip diameter well below 1µm. We use thin borosilicate pipette with filament (this helps liquid filling) (GC150TF-10, Clark Electromedical Instruments), and pull them with a horizontal DMZ puller (Zeitz instruments, Germany). The pulling is concluded by automatic heat-polishing step ensuring highly reproducible geometries and electrical properties. The pipette resistance in physiological saline solutions is usually around 10-15MΩ. The pipette capacitance can be measured and compensated, it is usually around 3pF.

As mentioned previously, RBCs are non-adhesive cells; thus pipettes can approach the bottom of the plate, using a 10X and then a 40X objective (Figure 3a-b). When the pipette is in contact with a RBC a small, calibrated depression can be applied (around 20mbars) and the cell can be lifted above the supporting plate (Figure 3c). The seal formation is helped by this negative pressure, and by the presence in the pipette solution of divalent cations (Mg^{2+} and/or Ca^{2+}). A pulse of 10mV is imposed on the pipette to monitor the resistance of the contact between membrane and pipette (seal resistance), from a holding potential around -30mV (figure 3d). When the seal value is higher than 10GΩ, the system is ready for cell-attached recordings.

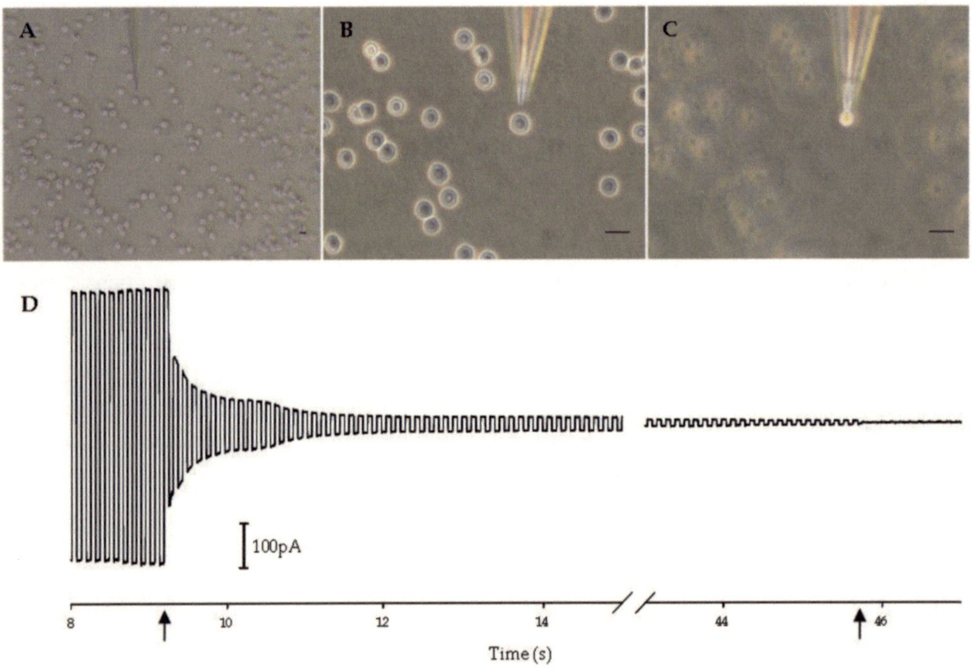

Fig. 3. Human RBC patch clamp - approach, seal and electric monitoring of seal formation. A, pipette approach with 10X objective. B, pipette approach with 40x objective. C, Patched RBC. D, monitoring of cell/pipette contact resistance (seal formation). Cell is touched by the pipette after 9s (arrow) with calibrated depression, and seal (around 10GΩ) is complete after 45s (arrow). Scale bar for A, B and C : 10μm.

3.2.3 Single channel recordings

The ion channels present in the RBC membrane can have a small unitary conductance, thus the most important task is to track and eliminate noise. Efficient Faraday cages and link of all metallic elements (microscope, anti-vibration table, and micromanipulator) to the ground using low resistance cables are of primary importance. Noise is also dependent on filtering settings. In our studies, recordings are filtered with a low pass 3 kHz filter. This allows correct transition detection of events up to 300μs, and elimination of noise faster events. When needed, these records can be refiltered digitally during analysis, with lower low-pass cut-off. Before digitalisation, signal is displayed on an oscilloscope to monitor analogic signal live and continuously. We use a 1401 or micro 1401 acquisition interface (Cambridge Electronic Design, Cambridge, UK) for analogic/digital conversion. Sampling rate is set at 10 kHz, above the Nyquist-Shannon sampling theorem value. Recordings for each voltage should last at least 1min, to obtain a sufficient number of events that will allow correct kinetic analysis (dwell time, open probability, burst duration). From cell-attached configuration, excised inside-out configuration of the patch clamp can be obtained by exposing very briefly the patched cell to air (this will rupture the membrane around the pipette tip). Then perfusion can modify the intracellular side solution and help characterising the channel activity.

3.2.4 Whole-cell recordings

After obtaining seal, records can be made for single channel studies. For whole-cell studies, the fragment of membrane inside the tip of the pipette must be ruptured. This is achieved via imposition in the pipette of a brief electrical pulse (200ms, 500mV). This rarely provokes damages to the seal. Successful whole-cell configuration can be checked via the sudden appearance of membrane capacitance transient currents. Mammalian erythrocyte capacitance is estimated at 0.8 $\mu F.cm^{-2}$ (Fettiplace et al., 1971), this gives a membrane capacitance around 1-1.3 pF for red blood cells. A measure of membrane capacitance for human RBC gives a value of 1pF (Rodighiero et al., 2004), and this can be compensated on the amplifier. Seal leak contribution can be estimated. If seal resistance is around 10GΩ, current leaking at +100mV can be calculated at 10pA. We usually have recordings showing current values well above 300 to 1000pA; then leak always remains below 1-3% of total current.

The nystatin-perforated patch clamp has also been used on human RBCs, especially to study cation channel activity. This technique avoids the dialysis of important substances from the RBCs cytoplasm. In these studies, nystatin was used at a final concentration of 150μg/mL in the pipette (Rodighiero et al., 2004; Vandorpe et al., 2010).

4. The contributions of patch clamp techniques in describing human red blood cell membrane properties

In this part of the chapter, we will list the various ion channels that have been described or suggested in the human RBC membrane, and list their main properties as well as recording conditions. Their possible physiological role will be evoked, before a description of their implication in various pathological situations.

4.1 Cation channels

4.1.1 Gardos channel

As mentioned previously, the Gardos channel was the first channel described in RBCs membrane. The pioneering experiments were performed by Georges Gardos who showed that metabolic poisoning of human RBCs was able to elicit K^+ leak via an ion channel depending on Ca^{2+} (Gardos, 1956, 1958). The phenomenon was termed Gardos effect; and later on this was particularly studied, since it was shown that the Gardos channel plays a role in cell dehydration during sickling at low oxygen tension in sickle cell disease (see §4.3.1). Then in the early years of patch clamp in 1981, RBCs were used to set up the technique, and Owen Hamill demonstrated using cell-attached recordings that the Gardos effect was truly supported by the activity of a Ca^{2+}-activated K^+ channel (Hamill, 1981). The channel remained known as the Gardos channel. It was later identified as the SK4 channel, belonging to the Small Conductance Calcium-Activated Potassium Channel family (SK channels). It is encoded by the gene kcnn4 (Hoffman et al., 2003). In an attempt to reconcile fluxes and electrophysiological data, Grygorczick and Schwarz estimated the number of channels per cell to be between 10 (Grygorczyk & Schwarz, 1983) and 55 (Grygorczyk et al., 1984). However, using inside-out vesicles from RBCs and $^{86}Rb^+$ fluxes, Alvarez and Garcia-Sancho estimated this number to be around 150 (Alvarez & Garcia-Sancho, 1987). Today this value seems the closer to reality. Until now, the Gardos channel remains the best characterised channel in human RBC membrane.

Examples of Gardos channel activity are given in Figure 4, showing cell-attached recordings. It was recently shown that the deformation of the membrane during seal formation was able to induce transient activity of the Gardos channel, mainly due to the entry of Ca^{2+} via a finite Ca^{2+} permeability (Dyrda et al., 2010). This induces a quick loss of cellular potassium via Gardos channel activation in all the membrane. Then when using KCl in the pipette solution, the current remains inwardly directed.

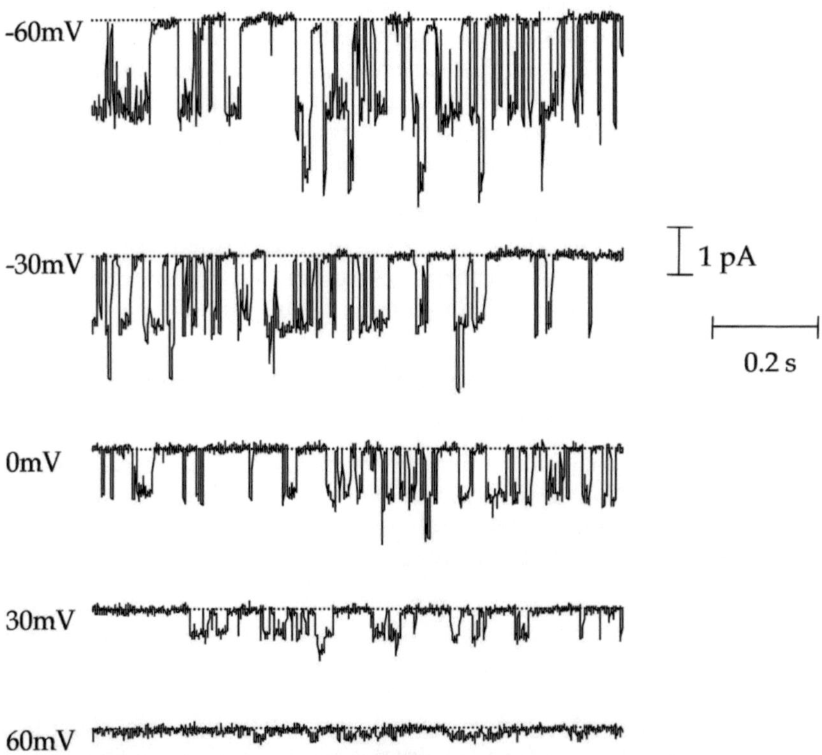

Fig. 4. Example of Gardos channel activity
Bath solution (mM): 115 NaCl, 5 KCl, 10 $MgCl_2$, 1.4 $CaCl_2$, 10 Hepes, 10 glucose. Pipette solution (mM): 120 KCl, 10 $MgCl_2$, 1.4 $CaCl_2$, 10 Hepes, 10 glucose. Voltages indicated refer to -Vp.

When recorded with a K^+ concentration of 100mM in the bath and pipette, the Gardos channel displays a conductance of 18pS (Hamill, 1981), or 25pS with 140mM K^+ (Grygorczyk et al., 1984). The channel shows slight inward rectification at high voltages (Grygorczyk et al., 1984). The open probability does not appear to be voltage-dependant. The channel has a clear selectivity for K^+, with $pK^+/pNa^+ \sim 15$ (Grygorczyk & Schwarz, 1983). However, Christophersen gave a more precise permeability sequence for the Gardos channel, from bi-ionic reversal potential (Christophersen, 1991): $K^+ \geq Rb^+ > NH_4^+ >> Li^+$; Na^+; Cs^+. The channel shows a susceptibility to temperature, with increased conductance and reduced open probability with increasing temperature (Grygorczyk, 1987).

Ca^{2+} at the intracellular face of the channel is necessary for channel activation. Open probability raises from 0.1 to 0.9 with an increase in free $[Ca^{2+}]_i$ from 500nM to 60μM. It is admitted that 2-3 μM of free $[Ca^{2+}]_i$ is required to activate the channel (Yingst & Hoffman, 1984). Calcium acts via binding to calmodulin, which is constitutively associated with the Gardos channel (Del Carlo et al., 2002). Protein kinase A induces a dramatic increase in Gardos channel activity, an effect that might be linked to enhancement of Ca^{2+} sensitivity (M. Pellegrino & Pellegrini, 1998). Lead is also a known activator of the Gardos channel, and it was shown that Pb^{2+} ions act independently and on the same site as Ca^{2+} ions (Shields et al., 1985). NS309 (6,7-dichloro-1H-indole-2,3-dione 3-oxime) is also a powerful agonist that decreases $K_{1/2}$ (Ca^{2+}) for IK and SK channels (Strobaek et al., 2004), and can be used to activate Gardos channel at subphysiological extracellular Ca^{2+} concentrations (Baunbaek & Bennekou, 2008). Extracellularly applied clotrimazole (Brugnara et al., 1993b) and charybdotoxin (Brugnara et al., 1995; Brugnara et al., 1993a) inhibit Gardos channels with IC50s of 50 nM and 5 nM, respectively. On the intracellular side, when the inside-out configuration is used, Ba^{2+} and tetraethylammonium exert a voltage-dependent channel inhibition by binding to the cytoplasmic domains of the channel with Kds of 150nM and 20mM, respectively (Dunn, 1998). In addition, the channel activity needs extracellular potassium, and incubation of RBCs ghost in a K^+-free solution irreversibly inhibits the activation of the channel by Ca^{2+} (Grygorczyk et al., 1984).

The resting free $[Ca^{2+}]_i$ in an unstimulated cell is about 100nM, consequently Gardos channel activation in a physiological situation seems rare. Thus, the physiological role of the channel still remains unclear. However, since deformation tends to activate the Gardos channel (at least indirectly) (Dyrda et al., 2010), the presence of the Gardos channel could make sense during the normal aging of RBCs. Density of RBCs is known to increase during the 120 days of their lifespan, and the channel has been involved in this process, with a decrease in maximal activity of the channel with cell age (Tiffert et al., 2007). A recent study using Gardos channel knock-out mice also showed modulation of various RBCs parameters. Mouse Gardos-deficient RBCs are mildly macrocytic, their susceptibility to osmotic lysis is increased and their filterability is impaired (Grgic et al., 2009), suggesting a role for Gardos channel in RBCs volume maintenance. Besides, the Gardos channel is also involved in various pathophysiological situations, notably in sickle cell anaemia where part of the cell dehydration leading to sickling is *via* the Gardos channel (Lew et al., 2002).

4.1.2 Non-selective cation channel

A voltage-activated cation channel in the human RBC membrane was originally proposed by Halperin (Halperin et al., 1989), and it was later described electrophysiologically using single channel recordings (Christophersen & Bennekou, 1991). This channel was further described by Bennekou (Bennekou, 1993), and by Kaestner (Kaestner et al., 1999). The channel is coupled to an acetylcholine receptor of nicotinic type (Bennekou, 1993), and can be activated by prostaglandin E2 (Kaestner & Bernhardt, 2002) and clotrimazole and its analogues (Barksmann et al., 2004). The channel shows a conductance around 20pS (Christophersen & Bennekou, 1991), and is permeant to mono- and di-valent cations, including Ca^{2+}, Ba^{2+} and Mg^{2+}. It shows a hysteresis voltage dependence (Kaestner et al., 2000), that was shown using patch clamp as well as cell suspension potential measurements (Bennekou et al., 2004a). Indeed, the half-maximal voltage of activation curve is 25mV higher than for the deactivation curve. The voltage-activated non-selective cation channel was also described using whole-cell and nystatin-perforated patch-clamp recordings, with a half-maximal conductance reached at 42mV (Rodighiero et al., 2004). It is inhibited by Ruthenium Red, N-ethylmaleimide (NEM), La^{3+}, or iodoacetamide (Bennekou et al., 2004b).

During the last decade the group of Florian Lang has also described a voltage-independent cation channel in the human RBC membrane. This channel can be activated by oxidative stress (Duranton et al., 2002), hyperosmotic shrinkage (Huber et al., 2001), and replacement of extracellular chloride by gluconate, NO_3^-, Br^- or SCN^- (Duranton et al., 2002; K.S. Lang et al., 2003). Monovalent cations are barely discriminated by this channel, with a selectivity $Cs^+ > K^+ > Na^+ = Li^+ >>> NMDG^+$ (Duranton et al., 2002); the channel is also permeable to Ca^{2+}. This voltage-independent channel is inhibited by amiloride, gadolinium or EIPA (ethylisopropylamiloride) (Duranton et al., 2002).

One cation channel seems to be identified as the Transient Receptor Potential Cation channel 6 (TRPC6), which fits with properties of the non-selective voltage-dependent cation channel. Expression has been detected in erythroid progenitor cells, and the protein has been detected in mature erythrocyte membrane by western blot analysis (Foller et al., 2008b).

The non-selective cation channels described so far in RBCs membrane are permeant to Ca^{2+}, leading to the hypothesis that the conductive pathway for Ca^{2+} entry is such a route. Nevertheless, a few studies have given evidence that true Ca^{2+} channels might also be present in the RBCs membrane. Even if the data regarding the exact nature of the Ca^{2+} pathway are confusing, the presence of both L-type and R-type Ca^{2+} channels subunits in RBCs membrane has been detected using western blots and tracers fluxes with adapted inhibitors (Romero et al., 2006). The group of Pedro Romero has also shown that the channel types appear to be different according to the RBC age. The pharmacology experiments suggest activating effects of vanadate (Romero & Romero, 2003) and caffeine (Cordero & Romero, 2002) on these Ca^{2+} channels. However, until now these channels have never been characterised electrophysiologically, but only with tracer fluxes using $^{45}Ca^{2+}$. The work of Pinet *et al* (Pinet et al., 2002) also suggests that the Ca^{2+} pump in human RBCs may behave as a Ca^{2+} channel sharing many similarities with the B-type Ca^{2+} channel. It can be seen when adapted voltage protocols are used on RBCs or when the Ca^{2+} pump is inserted into liposomes.

Thus, it seems that at least two distinct cationic pathways coexist in the human RBC membrane. Their physiological roles are probably mainly linked to regulation of free $[Ca^{2+}]_i$. It is known that prostaglandin E2 increases free $[Ca^{2+}]_i$ leading to Gardos channel activation

(Li et al., 1996), and decreases in erythrocyte deformability and filterability with a maximal effect at 0.1 nM (Allen & Rasmussen, 1971). Beyond Gardos channel activation and cell shrinkage, an increase in free $[Ca^{2+}]_i$ could also lead to scramblase activation and phosphatidylserine exposure on the outer membrane leaflet. Cation channels are thus implicated in the cascade of events leading to RBCs death, a phenomenon called eryptosis (as an erythrocytic-specific apoptosis) and described by the group of Florian Lang (Foller et al., 2008a; F. Lang et al., 2006; K.S. Lang et al., 2005).

4.2 Anion channels

As described in the first part of this chapter, conductive anion permeability of the RBC membrane has long been exclusively attributed to band 3 via slippage, or tunneling mechanisms. Nevertheless, various anion channel activities have been described essentially in the last decade, giving unambiguous evidence of the presence of anion channel in the human RBC membrane. Knowledge has come partly from studies on *Plasmodium*-infected RBCs, showing spontaneous anion channel activity. It is now known that these channels are endogenous proteins, upregulated in parasitized cells. Nonetheless, anion channel inhibitors are generally poorly specific, and in cells lacking expression machinery, and where membrane majoritary proteins complicate proteomics studies, precise identification of these anion channels has been and still remains difficult. So far two functional anion channels have been identified in the human RBC membrane, plus the CFTR channel that probably acts more as a regulator.

4.2.1 Cystic Fibrosis Transmembrane conductance Regulator (CFTR)

The presence of CFTR protein in the human RBC membrane has long been debated. CFTR transcripts in human RBCs progenitors and CFTR protein were not found by Hoffman et al. (Hoffman et al., 2004). On the contrary, other groups using various techniques suggested the presence of this channel in the RBC membrane. Indeed, CFTR activity seems necessary for deformation-induced ATP release and the protein was detected using western blots (Abraham et al., 2001; Sprague et al., 1998). Moreover, using atomic force microscopy combined with quantum-dot-labelled anti-CFTR antibodies, Lange et al. could estimate the number of CFTR per cell at around 700 and 170 in RBCs from healthy donors and Cystic Fibrosis patients (homozygous ΔF508 mutation), respectively (Lange et al., 2006).

However, no recording of CFTR activity in the RBC membrane at the single channel level has ever been given. Comparisons between RBCs from healthy donors and CF patients using whole-cell configuration allowed the description of a tiny current attributed to CFTR activity; but the role of CFTR remains unclear. Indeed, its regulatory properties seem more important. Plasmolysis, or *Plasmodium*-infection induced a chloride current (different from CFTR activity) in normal RBCs, but not in RBCs from CF patients (Verloo et al., 2004). This was confirmed at the single channel level in non-infected human RBCs, with behaviour similar to a small anionic channels (around 10pS) in these two types of cells: its gating and kinetics were different, but the properties did not correspond to CFTR activity (Decherf et al., 2007).

This seems to confirm the presence of a CFTR protein in the human RBC membrane acting rather as a regulator of other channels/transporters.

4.2.2 Small conductance chloride channel / ClC-2

A small anion channel has been described in the human RBC membrane, and was designed as a ClC-2 channel. Western-blots showed its presence in the human RBC membrane, in keeping with swell-activation and Zn^{2+} inhibition of a whole-cell chloride conductance (Huber et al., 2004; Shumilina & Huber, 2011).

We also described a small conductance chloride channel, spontaneously active using cell-attached configuration in malaria-infected human RBCs, showing a conductance of 4-5pS (Bouyer et al., 2006). Relying on the number of apparitions of active channel, we estimated the density of this channel to be approximately 60-80 copies per cell. In a subsequent study, we showed that this channel corresponds to the ClC-2 channel already described, thus being an endogenous channel activated upon infection (Bouyer et al., 2007). Using the whole-cell configuration it was shown that this channel is activated by oxidation using 1mM *tert*-butylhydroperoxide (t-BHP) (Huber et al., 2004; Huber et al., 2002). Zinc shows an inhibitory effect, with an IC50 around 100μM, whereas NPPB or furosemide are relatively ineffective (Shumilina & Huber, 2011).

This channel might contribute to the basal anionic conductance of human RBCs, clamping the cell potential to the chloride equilibrium potential. Since ClC-2 channels are activated by cell swelling in many cell types (Roman et al., 2001; Strange et al., 1996) including RBCs (Huber et al., 2004), a role for this channel in cell volume regulation seems more than probable.

4.2.3 Voltage-dependent anion channel / peripheral benzodiazepine receptor

Two recent works, using both cell-attached and whole-cell configurations have provided evidence for the presence of another type of anion channel in human RBC membrane. The first study showed that a maxi-anion channel is present in the human RBC membrane (Glogowska et al., 2010). This channel shows multiple conductance substates that are dependent on the presence of serum in the bath solution. Activity was classified in various patterns, with a basal level in the absence of serum showing a conductance of 8-12pS. Multiple substates were described in the presence of serum in the bath, with a maximal conductance of 600pS and a more stable conductance of 300pS. Figure 5 gives examples of this maxi-anion channel activity. The channel showed a preference for SCN^- ions, since some of these substates could be seen in serum-free conditions when cells were bathed in solutions where 10mM of Cl^- ions was replaced by SCN^- ions.

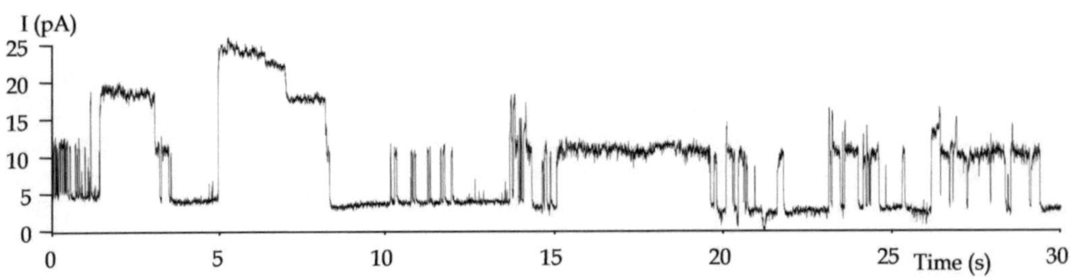

Fig. 5. Example of VDAC recordings with multiple conductance substates.
Bath solution (mM): 115 NaCl, 5 KCl, 10 $MgCl_2$, 1.4 $CaCl_2$, 10 Hepes, 10 glucose plus 0.5% human serum
Pipette solution: 115 NaCl, 5 KCl, 10 $MgCl_2$, 1.4 $CaCl_2$, 10 Hepes, 10 glucose
-Vp=50mV

The identity of this channel remained undetermined until a second study by our group showed that a Voltage Dependent Anion Channel (VDAC) was present in the human RBC membrane (Bouyer et al., 2011a). VDAC are components of the Peripheral Benzodiazepine Receptor (PBR), composed of at least three components: VDAC, an Adenine Nucleotide Transporter (ANT) and a translocator protein (TSPO) that probably modulates VDAC activity (Veenman et al., 2008). Presence of binding sites for specific ligands of this receptor in the RBC membrane had already been suggested (Canat et al., 1993; Olson et al., 1988), and in this study evidence for presence of transcripts in differentiating erythrocytes and of proteins in mature erythrocytes was given (Bouyer et al., 2011a).

VDAC conductance and selectivity are voltage-dependent and shows multiple substates with permeability for small and large anions (SCN^-, Cl^-, glutamate, ATP …), it also displays low conductance substates with cation permeability (Na^+, K^+, Ca^{2+}) (Bathori et al., 2006; Benz et al., 1990; Gincel et al., 2000; Rostovtseva & Colombini, 1997). The channel activity can be activated, or modulated by multiple ways including phosphorylation by PKA, oxidation, Ca^{2+} ions, or serum components (Banerjee & Ghosh, 2004; Bera et al., 1995; Madesh & Hajnoczky, 2001; Shoshan-Barmatz et al., 2010). The channel is sensitive to DIDS, La^{3+} or ruthenium red (Shoshan-Barmatz & Gincel, 2003). In the PBR complex, three specific ligands are generally used as modulators of ion transport activity: PK11195, Ro5-4864 and diazepam. The TSPO component is considered to be primarily responsible for binding to PK 11195, while Ro5-4864 and other benzodiazepines may bind to all components of the PBR complex (Le Fur et al., 1983; McEnery et al., 1992).

The presence of such proteins in the RBC membrane raises many questions regarding their possible physiological role, but according to the properties of its components we can predict a major role in membrane transport, volume and redox status regulation, as well as cell differentiation and senescence. The multiple behaviours of VDAC allow us to propose a unifying hypothesis where VDAC could explain parts of both anionic and cationic channel activity already described in human RBCs, as suggested in a recent review (Thomas et al., 2011). VDAC could then be a common pathway for Ca^{2+} entry (leading to Gardos channel modulation), Cl^- equilibrium regulation or ATP release.

4.3 Implication of RBC ion channels in pathophysiological situations

Although the physiological role of ion channels in the RBC membrane is still under investigation, their implication in various pathologies has been known for decades. In particular, cation channel activity is a key factor of sickle cell disease pathology (Lew & Bookchin, 2005), as is anion channel activity in *Plasmodium*-infected human RBCs (Ginsburg et al., 1983; Kirk et al., 1994). Implication of the channels in senescence of RBCs will also be evoked.

4.3.1 Sickle cell disease

In the sickle cell disease polymerisation of HbS haemoglobin under deoxygenated conditions leads to cell dehydration via efflux of potassium, chloride and osmotically obliged water. Pathways for these effluxes are the Gardos channel and the potassium-chloride cotransporter KCC. Both are activated during sickling consequent to the entry of calcium in the red cell, consecutive of HbS polymerisation, via an abnormal cation permeability pathway termed P_{sickle}, known and described from a functional point of view, but never characterised at the molecular level. P_{sickle} is a poorly selective permeability pathway for small inorganic mono- and divalent cations (Joiner et al., 1993). Ion movements are non-saturable, voltage-dependent and not obligatory coupled, characteristics of a conductive pathway (Joiner, 1993). However, studies run in two laboratories in Oxford and Cambridge have also reported that under certain circumstances, P_{sickle} might also be permeable to non-electrolytes (Ellory et al., 2007; Ellory et al., 2008), including sugars (notably sucrose or lactose) and other molecules such as taurine or glutamine. This pathway has been characterised mainly using fluxes or haemolysis experiments, yet patch-clamp has also been recently used. Whole-cell cationic conductance can be induced by deoxygenation of HbSS cells. The channel shows equal permeability for Na^+, K^+ or Ca^{2+}, and is inhibited by DIDS (100μM), Zn^{2+} (100μM) and Gd^{3+} (2μM) (Browning et al., 2007). Nystatin-perforated whole-cell recordings and single channel recordings gave a unitary conductance of 27pS, and a sensitivity to dipyridamole (100μM) and GsMTx (*Grammastola spatulata* mechanotoxin IV) (1μM) (Vandorpe et al., 2010).

P_{sickle} is an obvious target of primary interest in the pharmacology of sickle cell disease, because of its initial role in the cascade of events leading to cell dehydration. Future research should focus on its identification. The Gardos channel is also a molecular target in the pharmacology of sickle cell disease, and research is in progress using of clotrimazole and various analogues (Brugnara, 2003). Finally, RBC dehydration in sickle cell disease is also mediated by anion channels and by loss of Cl^- accompanying K^+ loss via the Gardos channel. Thus, these anion channels are another target of interest, an hypothesis successfully tested in seminal work, where specific RBCs anion conductance inhibitors were used to prevent cell dehydration either *in vitro* or *in vivo* on the SAD mouse model for sickle cell disease (Bennekou, 1999; Bennekou et al., 2001; Bennekou et al., 2000).

4.3.2 Malaria and *Plasmodium*-infection

Another pathophysiological situation, where anion channels display a critical role is malaria, caused by infection by parasites of the genus *Plasmodium*. The malaria parasite invades and multiplies within RBCs in about 48h. To accomplish this cycle and ensure the supply of nutrients and the release of waste products, it relies on large, poorly selective anion channels in the host RBC membrane. Because of a dramatic gap in the original knowledge on RBC native anion channel, this transport properties have been termed New Permeability Pathways (NPPs) (Kirk, 2001) and their molecular nature has long been debated. This pathway was characterised using fluxes and haemolysis experiments during the 80's and 90's. Patch-clamp techniques were first applied to infected RBCs by the group of Sanjay Desai who demonstrated an inwardly rectified anion current in infected RBCs (Figure 6) (Desai et al., 2000). Electrophysiological description of the spontaneously active ion channels in infected RBCs membrane has been highly controversial, owing to the multiplicity of experimental conditions used by the different groups in the field. Indeed, a negative holding potential imposed between ramps in whole-cell experiments inactivates inward currents, and the presence of serum in the bath solution activates outward and inward current in a different manner (see Figure 6) (Staines et al., 2003). It was also shown that supraphysiological ionic concentrations used in bath and pipette solutions modify anion channel activity: saturation of conductance and inhibition by lower open probability appeared beyond 0.6M of Cl^- in solutions (Bouyer et al., 2007).

Several studies have since showed that this channel activity was supported by endogenous channels activated upon infection and that the channels were sensitive to PKA phosphorylation, oxidation or serum presence (Bouyer et al., 2007; Bouyer et al., 2011b; Egee et al., 2002; Huber et al., 2005; Huber et al., 2002; Merckx et al., 2008; Staines et al., 2003). Examples of whole-cell currents are given in Figure 6. One channel involved is the ClC-2 channel, but its activation is rather a side effect, since growth of *P. berghei* is not affected in ClC-2 knock-out mice (Huber et al., 2004).

We recently showed that the main component of NPPs was the PBR complex including the VDAC channel, since the specific PBR ligands could prevent both parasite growth *in vitro*, NPP-mediated sorbitol permeability and whole cell anion currents (Figure 6) of infected cells (Bouyer et al., 2011a).

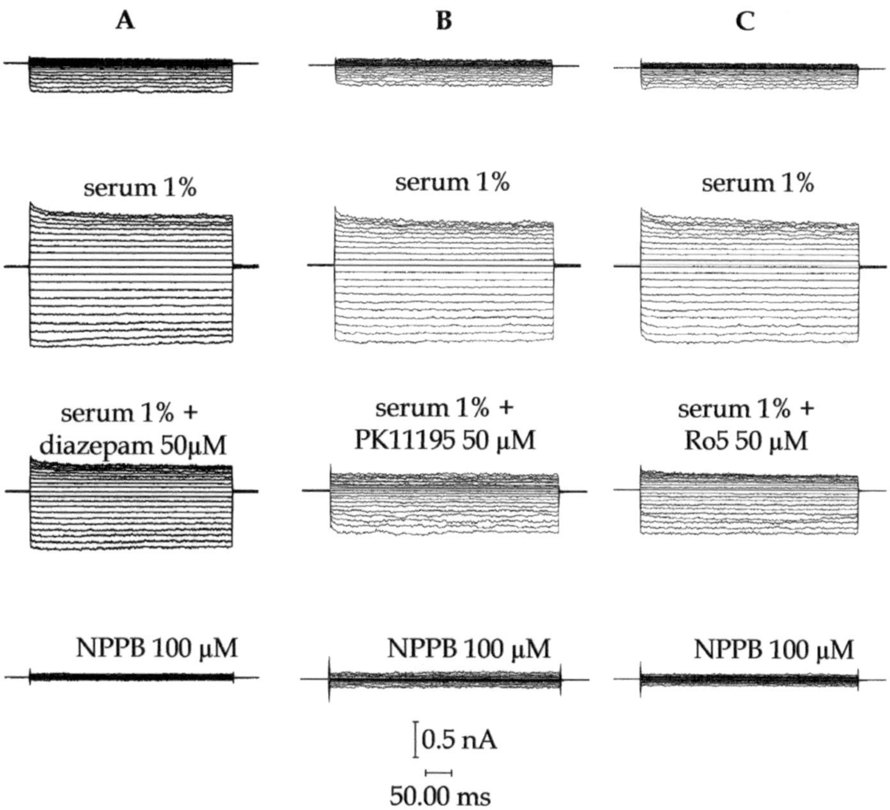

Fig. 6. Anion channel activity in *Plasmodium falciparum*-infected human RBCs
3 cells A, B and C were studied using the whole-cell configuration, and serial perfusion were performed showing the typical serum effect and the inhibitory effect of PBR ligands and NPPB, as described in (Bouyer et al., 2011a). Stimulation was made with 500ms ramps between +100mV and -100mV, with -10mV increments

The identification of PBR/VDAC activity in the *Plasmodium*-infected RBC membrane is an important step in the description of the pathophysiology of malaria. This suggests that the pharmacopoeia of benzodiazepine, as well as benzodiazepine scaffolds for the production of new inhibitors could become a novel strategic approach for future antimalarial chemotherapies.

The discrepancies between results obtained by patch-clamp studies on infected RBCs in the last decade proved how important the specific experimental conditions are. Indeed, presence of various substances in the bath, holding potential, ionic concentrations, or duration of recordings are able to alter ion channel activity and extrapolation of results to physiological situation is then difficult (Staines et al., 2007). Thus, extra precaution should be taken when designing experiments and analysing results and detailed methodology should be clearly stated in communications.

4.3.3 Senescence

During their circulatory life, RBCs tend to become progressively denser. This correlates with a decline in Ca^{2+} pump activity that leads to KCl loss (via Gardos and anion channels) overcompensated by NaCl gain (Lew et al., 2007). Activity of a non-selective cation channel has been linked to this phenomenon that tends to dissipate Na^{+} and K^{+} gradients late in the lifespan of RBCs.

A cascade of events leading to programmed cell death of RBCs has been described by the group of Florian Lang. This follows the same path: an external signal (oxidative stress, for example) triggers a rise in intracellular Ca^{2+}, probably by activation of a non-selective cation channel by prostaglandin E2. This provokes cell shrinkage, scramblase and calpain activation resulting in phosphatidylserine exposure and degradation of the cytoskeleton (reviewed in (Foller et al., 2008a; F. Lang et al., 2004; F. Lang et al., 2006)).

Thus, the non-selective cation channel Gardos channel together with anion channels play a central role in RBCs aging process and cell death.

5. Comparative physiology: Use of patch-clamp in a evolutionary approach on vertebrates red cells

Comparative physiology of red blood cells membrane have been evident for many years regarding respiratory function and many studies highlighted the role of ion transporters in resting and challenging situations. Although, the use of the patch-clamp technique on models other than mammals are relatively limited. Nonetheless, major groups have been studied and Table 2 summarizes current knowledge on vertebrates' erythrocytes ion channels.

As stated in the table below, most of the studies performed so far were conducted in the framework of cell volume regulation. Volume regulation is of importance to cells exposed to anisotonic extracellular media and to cells where solute transport could change intracellular osmolality. Exposure of vertebrate cells to a hypotonic solution results in an initial increase in cell volume due to the relatively rapid influx of water. During continuous hypotonic stress increases in cell volume are followed by a slower, spontaneous recovery towards the pre-shock level, a process known as regulatory volume decrease (RVD). This recovery is accomplished by selectively increasing the permeability of the plasma membrane during cell swelling to allow for efflux of specific intracellular osmolytes, thereby generating a driving force for water efflux. Most vertebrate cells lose K^{+} and Cl^{-} during RVD (Hoffmann et al., 2009). This may occur by electroneutral ion transport pathways, or by the separate activation of K^{+} and Cl^{-} channels. Loss of organic anions and osmolytes also may occur during RVD.

Species	Ion channels	Configu-ration	Unitary conductance (pS)	selectivity	inhibitors	modulators	references
Lamprey *(Lampetra fluviatilis)*	K^+	CA, IO, WC	25	K^+>>Na^+	Ba^{2+}	swelling	(Virkki & Nikinmaa, 1996; Virkki & Nikinmaa, 1998)
	K^+		65	K^+>>Na^+		Mg^{2+} block	
Lamprey *(Petromyzon marinus)*	K^+	CA, IO, WC	80 inward 35 outward	K^+ (1) >Rb^+(2.0) >Cs^+ (4.6) >Li^+ (17.2) >Na^+ (22.4)	Ba^{2+}, ATP, glibenclamide, lidocaine		(Lapaix et al., 2002)
	K^+		25 inward 10 outward	K^+ (1)>Li^+ (2.3)>Rb^+ (2.6) >Cs^+ (6.5) >Na^+ (10.4).	TEA, Ba^{2+}, apamin	swelling	
Dogfish *(Scyliorhinus canicula)*	NSC	CA, IO	18	K^+=Na^+	Ba^{2+}		Unpublished data
Trout *(Oncorhy-nchus mykiss)*	NSC	WC, CA, IO	15	K^+~Na^+~Ca^{2+} >>NMDG	Ba^{2+}, quinine, Gd^{3+}	swelling	(Egee et al., 1997; Egee et al., 2000; Egee et al., 1998)
	SCC		6		NPPB, DIDS glibenclamide		
	ORCC		80		NPPB, DIDS	swelling	
Frog *(Rana sp.)*	K^+	IO	56		ATP, Ba^{2+}	Ca^{2+} increase	(Shindo et al., 2000)
Mudpuppy *(Necturus maculosus)*	K^+	WC			Ba^{2+}	Cell swelling	(Bergeron et al., 1996; Light et al., 2003; Light et al., 2001; Light et al., 1997)
Chicken *(Gallus gallus)*	NSC	WC, IO	24	K^+~Na^+≥Ca^{2+} >>NMDG		stretch	(Lapaix et al., 2008; Thomas et al., 2001)
	NSC		62	K^+~Na^+>>Ca^{2+}>>NMDG		cAMP	
	Cl^-		255			Swelling ?	

Table 2. Ion channels described in vertebrates RBCs.
NSC: Non Selective Cation Channel. SCC: Small Chloride Channel. ORCC: Outwardly Rectified Chloride Channel. CA: Cell-Attached. IO: Inside-Out. WC: Whole-Cell.

A series of articles on *Necturus* erythrocytes showed the common schema of RVD in erythrocytes during hypotonic stress. It was demonstrated that it depends on a quinine-inhibitable K^+ conductance that is regulated during cell swelling by a calmodulin-dependent mechanism (Bergeron et al., 1996), and by a 5-lipoxygenase metabolite of arachidonic acid (Light et al., 1997), as well as by extracellular ATP activation of P2 receptors (Light et al., 2001). Eventually the triggering factor was shown to be the initial increase of intracellular Ca^{2+} concentration (Light et al., 2003). If all other studies do not reach such complete signalling pathway for RVD, it is intriguing and tempting to think that a common schema is conserved throughout evolution of vertebrates RBCs with only minor differences.

Moreover, when RBCs are replaced in their context of respiratory function, and in the light of recent advances on human RBCs regarding anionic transporters, one may think that CO_2/HCO_3^- within the blood occurs originally through anion channels. Indeed, Agnathans (jawless vertebrates) are devoid of Cl^-/HCO_3^- exchangers, but possess like other vertebrates a powerful anion conductance with low selectivity and which presents similar electrophysiological characteristics, as VDAC/PBR found in human RBCs (unpublished data). These types of ionic channel activities have been also reported in trout, chicken as well as in amphibian (table 2). Thus, future comparative studies should go further in deciphering the role played by ion conductance in the success of intracellular O_2/CO_2 transport thought vertebrates evolution. The molecular control in transporters expression and activity and their integrative physiology represent future areas of interest.

6. Conclusion

Though having been explorated by various techniques during decades, the membrane permeability of RBCs is still not fully understood. The use of patch-clamp techniques has proven to be very useful shedding light on a much more complicated situation than expected. Indeed, RBCs are equipped with multiple transporters including various ion channels. These allow precise and fast regulation of volume, acid-base and electrolyte status and they are essential for adequate respiratory functions of RBCs. The external environment of RBCs is constantly changing: with a mean cardiac output of 5L/min and a blood volume of 5L an erythrocyte of healthy human adult has circulate every minute. This means travelling through narrow capillaries (smaller than RBCs own diameter), or bigger vessels in multiple tissues, where acid-base conditions, partial gas pressure, temperature or nutrient concentrations is highly variable. Thus, RBCs are far more than an empty bag of haemoglobin.

There is much evidence that RBCs plays a more complex role than simple oxygen supplier to tissues. Indeed, excessive tissue demand (*i.e.* low oxygen tension) activates RBC signalling pathways resulting in the release of ATP acting in a paracrine fashion to increase vascular calibre (Sprague et al., 2007). RBCs now become vital sensors in matching microvascular oxygen delivery with local tissue oxygen demand (Ellsworth et al., 2009). The pathways involved in this process are not yet fully described, but the presence of calcium pathways plus the Gardos channel and PBR/VDAC provide adequate machinery that could explain these functions of RBCs.

Finally, the characterisation of the various ion channels in diverse vertebrate species is of high interest and constitutes an important field for future research. Indeed if as postulated above, anion channels originally mediated bicarbonate exchange across RBCs membrane at the root of vertebrates (and Agnathans are a very good model), an evolution process has led to the apparition of Band 3; while anion and cation channels have been maintained in the RBCs membrane throughout this process. Then, the various ion channels in vertebrate RBCs could help describe a phylogeny of respiratory mechanisms throughout evolution, and lead to a better understanding of the role of ion channels in the human RBC membrane.

7. Acknowledgment

The authors thank Dr Gordon Langsley for reading and improving the manuscript.

8. References

Abraham, E. H., Sterling, K. M., Kim, R. J., Salikhova, A. Y., Huffman, H. B., Crockett, M. A., Johnston, N., Parker, H. W., Boyle, W. E., Jr., Hartov, A., Demidenko, E., Efird, J., Kahn, J., Grubman, S. A., Jefferson, D. M., Robson, S. C., Thakar, J. H., Lorico, A., Rappa, G., Sartorelli, A. C. & Okunieff, P. (2001). Erythrocyte membrane ATP binding cassette (ABC) proteins: MRP1 and CFTR as well as CD39 (ecto-apyrase) involved in RBC ATP transport and elevated blood plasma ATP of cystic fibrosis. *Blood Cells, Molecules and Diseases*, Vol. 27, No. 1, (Jan-Feb), pp. 165-180, ISSN 1079-9796

Agre, P., Preston, G. M., Smith, B. L., Jung, J. S., Raina, S., Moon, C., Guggino, W. B. & Nielsen, S. (1993). Aquaporin CHIP: the archetypal molecular water channel. *The American Journal of Physiology*, Vol. 265, No. 4 Pt 2, (Oct), pp. F463-476, ISSN 0002-9513

Allen, J. E. & Rasmussen, H. (1971). Human red blood cells: prostaglandin E2, epinephrine, and isoproterenol alter deformability. *Science*, Vol. 174, No. 8, (Oct 29), pp. 512-514, ISSN 0036-8075

Alvarez, J. & Garcia-Sancho, J. (1987). An estimate of the number of Ca2+-dependent K+ channels in the human red cell. *Biochimica et Biophysica Acta*, Vol. 903, No. 3, pp. 543-546, ISSN 0006-3002

Banerjee, J. & Ghosh, S. (2004). Interaction of mitochondrial voltage-dependent anion channel from rat brain with plasminogen protein leads to partial closure of the channel. *Biochimica et Biophysica Acta*, Vol. 1663, No. 1-2, (May 27), pp. 6-8, ISSN 0006-3002

Barksmann, T. L., Kristensen, B. I., Christophersen, P. & Bennekou, P. (2004). Pharmacology of the human red cell voltage-dependent cation channel; Part I. Activation by clotrimazole and analogues. *Blood Cells, Molecules and Diseases*, Vol. 32, No. 3, (May-Jun), pp. 384-388, ISSN 1079-9796

Barry, P. H. & Lynch, J. W. (1991). Liquid junction potentials and small cell effects in patch-clamp analysis. *The Journal of Membrane Biology*, Vol. 121, No. 2, (Apr), pp. 101-117, ISSN 0022-2631

Barry, P. H. (1994). JPCalc, a software package for calculating liquid junction potential corrections in patch-clamp, intracellular, epithelial and bilayer measurements and for correcting junction potential measurements. *Journal of Neuroscience Methods*, Vol. 51, No. 1, (Jan), pp. 107-116, ISSN 0165-0270

Bathori, G., Csordas, G., Garcia-Perez, C., Davies, E. & Hajnoczky, G. (2006). Ca2+-dependent control of the permeability properties of the mitochondrial outer membrane and voltage-dependent anion-selective channel (VDAC). *The Journal of Biological Chemistry*, Vol. 281, No. 25, (Jun 23), pp. 17347-17358, ISSN 0021-9258

Baunbaek, M. & Bennekou, P. (2008). Evidence for a random entry of Ca2+ into human red cells. *Bioelectrochemistry*, Vol. 73, No. 2, (Aug), pp. 145-150, ISSN 1567-5394

Bennekou, P. (1993). The voltage-gated non-selective cation channel from human red cells is sensitive to acetylcholine. *Biochimica et Biophysica Acta*, Vol. 1147, No. 1, pp. 165-167, ISSN 0006-3002

Bennekou, P. (1999). The feasibility of pharmacological volume control of sickle cells is dependent on the quantization of the transport pathways. A model study. *Journal of Theoretical Biology*, Vol. 196, No. 1, pp. 129-137, ISSN 0022-5193

Bennekou, P., Pedersen, O., Moller, A. & Christophersen, P. (2000). Volume control in sickle cells is facilitated by the novel anion conductance inhibitor NS1652. *Blood*, Vol. 95, No. 5, pp. 1842-1848, ISSN 0006-4971

Bennekou, P., de Franceschi, L., Pedersen, O., Lian, L., Asakura, T., Evans, G., Brugnara, C. & Christophersen, P. (2001). Treatment with NS3623, a novel Cl-conductance blocker, ameliorates erythrocyte dehydration in transgenic SAD mice: a possible new therapeutic approach for sickle cell disease. *Blood*, Vol. 97, No. 5, pp. 1451-1457, ISSN 0006-4971

Bennekou, P., Barksmann, T. L., Jensen, L. R., Kristensen, B. I. & Christophersen, P. (2004a). Voltage activation and hysteresis of the non-selective voltage-dependent channel in the intact human red cell. *Bioelectrochemistry*, Vol. 62, No. 2, (May), pp. 181-185, ISSN 1567-5394

Bennekou, P., Barksmann, T. L., Kristensen, B. I., Jensen, L. R. & Christophersen, P. (2004b). Pharmacology of the human red cell voltage-dependent cation channel. Part II: inactivation and blocking. *Blood Cells, Molecules and Diseases*, Vol. 33, No. 3, (Nov-Dec), pp. 356-361, ISSN 1079-9796

Benz, R., Kottke, M. & Brdiczka, D. (1990). The cationically selective state of the mitochondrial outer membrane pore: a study with intact mitochondria and reconstituted mitochondrial porin. *Biochimica et Biophysica Acta*, Vol. 1022, No. 3, (Mar), pp. 311-318, ISSN 0006-3002

Bera, A. K., Ghosh, S. & Das, S. (1995). Mitochondrial VDAC can be phosphorylated by cyclic AMP-dependent protein kinase. *Biochemical and Biophysical Research Communications*, Vol. 209, No. 1, (Apr 6), pp. 213-217, ISSN 0006-291X

Bergeron, L. J., Stever, A. J. & Light, D. B. (1996). Potassium conductance activated during regulatory volume decrease by mudpuppy red blood cells. *The American Journal of Physiology*, Vol. 270, No. 4 Pt 2, pp. R801-810, ISSN 0002-9513

Bouyer, G., Egee, S. & Thomas, S. L. (2006). Three types of spontaneously active anionic channels in malaria-infected human red blood cells. *Blood Cells, Molecules and Diseases*, Vol. 36, No. 2, (Mar-Apr), pp. 248-254, ISSN 1079-9796

Bouyer, G., Egee, S. & Thomas, S. L. (2007). Toward a unifying model of malaria-induced channel activity. *Proceedings of the National Academy of Sciences of the United States of America*, Vol. 104, No. 26, (Jun 26), pp. 11044-11049, ISSN 0027-8424

Bouyer, G., Cueff, A., Egee, S., Kmiecik, J., Maksimova, Y., Glogowska, E., Gallagher, P. G. & Thomas, S. L. (2011a). Erythrocyte peripheral type benzodiazepine receptor/voltage-dependent anion channels are upregulated by Plasmodium falciparum. *Blood*, Vol. 118, No. 8, (Aug 25), pp. 2305-2312, ISSN 0006-4971

Bouyer, G., Thomas, S. L. Y. & Egee, S. (2011b). Protein Kinase-Regulated Inwardly Rectifying Anion and Organic Osmolyte Channels in Malaria-Infected Erythrocytes. *The Open Biology Journal*, Vol. 4, No. pp. 10-17, ISSN 1874-1967

Browning, J. A., Staines, H. M., Robinson, H. C., Powell, T., Ellory, J. C. & Gibson, J. S. (2007). The effect of deoxygenation on whole-cell conductance of red blood cells from healthy individuals and patients with sickle cell disease. *Blood*, Vol. 109, No. 6, (Mar 15), pp. 2622-2629, ISSN 0006-4971

Brugnara, C., De Franceschi, L. & Alper, S. L. (1993a). Ca(2+)-activated K+ transport in erythrocytes. Comparison of binding and transport inhibition by scorpion toxins. *The Journal of Biological Chemistry*, Vol. 268, No. 12, pp. 8760-8768, ISSN 0021-9258

Brugnara, C., de Franceschi, L. & Alper, S. L. (1993b). Inhibition of Ca(2+)-dependent K+ transport and cell dehydration in sickle erythrocytes by clotrimazole and other imidazole derivatives. *Journal of Clinical Investigation*, Vol. 92, No. 1, pp. 520-526., ISSN 0021-9738

Brugnara, C., Armsby, C. C., De Franceschi, L., Crest, M., Euclaire, M. F. & Alper, S. L. (1995). Ca(2+)-activated K+ channels of human and rabbit erythrocytes display distinctive patterns of inhibition by venom peptide toxins. *The Journal of Membrane Biology*, Vol. 147, No. 1, pp. 71-82, ISSN 0022-2631

Brugnara, C. (2003). Sickle cell disease: from membrane pathophysiology to novel therapies for prevention of erythrocyte dehydration. *Journal of Pediatric Hematology/Oncology*, Vol. 25, No. 12, (Dec), pp. 927-933, ISSN 1077-4114

Cabantchik, Z. I. & Rothstein, A. (1972). The nature of the membrane sites controlling anion permeability of human red blood cells as determined by studies with disulfonic stilbene derivatives. *The Journal of Membrane Biology*, Vol. 10, No. 3, (Dec 29), pp. 311-330, ISSN 0022-2631

Cahalan, M. D., Chandy, K. G., DeCoursey, T. E. & Gupta, S. (1985). A voltage-gated potassium channel in human T lymphocytes. *The Journal of Physiology*, Vol. 358, No. (Jan), pp. 197-237, ISSN 0022-3751

Canat, X., Carayon, P., Bouaboula, M., Cahard, D., Shire, D., Roque, C., Le Fur, G. & Casellas, P. (1993). Distribution profile and properties of peripheral-type benzodiazepine receptors on human hemopoietic cells. *Life Sciences*, Vol. 52, No. 1, pp. 107-118, ISSN 0024-3205

Christophersen, P. (1991). Ca2(+)-activated K+ channel from human erythrocyte membranes: single channel rectification and selectivity. *The Journal of Membrane Biology*, Vol. 119, No. 1, pp. 75-83, ISSN 0022-2631

Christophersen, P. & Bennekou, P. (1991). Evidence for a voltage-gated, non-selective cation channel in the human red cell membrane. *Biochimica et Biophysica Acta*, Vol. 1065, No. 1, pp. 103-106, ISSN 0006-3002

Cordero, J. F. & Romero, P. J. (2002). Caffeine activates a mechanosensitive Ca(2+) channel in human red cells. *Cell Calcium*, Vol. 31, No. 5, pp. 189-200, ISSN 0143-4160

Decherf, G., Bouyer, G., Egee, S. & Thomas, S. L. (2007). Chloride channels in normal and cystic fibrosis human erythrocyte membrane. *Blood Cells, Molecules and Diseases*, Vol. 39, No. 1, (Jul-Aug), pp. 24-34, ISSN 1079-9796

Del Carlo, B., Pellegrini, M. & Pellegrino, M. (2002). Calmodulin antagonists do not inhibit IK(Ca) channels of human erythrocytes. *Biochimica et Biophysica Acta*, Vol. 1558, No. 2, pp. 133-141, ISSN 0006-3002

Deplaine, G., Safeukui, I., Jeddi, F., Lacoste, F., Brousse, V., Perrot, S., Biligui, S., Guillotte, M., Guitton, C., Dokmak, S., Aussilhou, B., Sauvanet, A., Cazals Hatem, D., Paye, F., Thellier, M., Mazier, D., Milon, G., Mohandas, N., Mercereau-Puijalon, O., David, P. H. & Buffet, P. A. (2011). The sensing of poorly deformable red blood cells by the human spleen can be mimicked in vitro. *Blood*, Vol. 117, No. 8, (Feb 24), pp. e88-95, ISSN 0006-4971

Desai, S. A., Bezrukov, S. M. & Zimmerberg, J. (2000). A voltage-dependent channel involved in nutrient uptake by red blood cells infected with the malaria parasite. *Nature*, Vol. 406, No. 6799, pp. 1001-1005., ISSN 0028-0836

Dunn, P. M. (1998). The action of blocking agents applied to the inner face of Ca(2+)-activated K+ channels from human erythrocytes. *The Journal of Membrane Biology*, Vol. 165, No. 2, pp. 133-143, ISSN 0022-2631

Duranton, C., Huber, S. M. & Lang, F. (2002). Oxidation induces a Cl(-)-dependent cation conductance in human red blood cells. *The Journal of Physiology*, Vol. 539, No. Pt 3, pp. 847-855, ISSN 0022-3751

Dyrda, A., Cytlak, U., Ciuraszkiewicz, A., Lipinska, A., Cueff, A., Bouyer, G., Egee, S., Bennekou, P., Lew, V. L. & Thomas, S. L. (2010). Local membrane deformations activate Ca2+-dependent K+ and anionic currents in intact human red blood cells. *PLoS ONE*, Vol. 5, No. 2, pp. e9447, ISSN 1932-6203

Egee, S., Harvey, B. J. & Thomas, S. (1997). Volume-activated DIDS-sensitive whole-cell chloride currents in trout red blood cells. *The Journal of Physiology*, Vol. 504, No. Pt 1, pp. 57-63, ISSN 0022-3751

Egee, S., Mignen, O., Harvey, B. J. & Thomas, S. (1998). Chloride and non-selective cation channels in unstimulated trout red blood cells. *The Journal of Physiology*, Vol. 511, No. Pt 1, pp. 213-224, ISSN 0022-3751

Egee, S., Lapaix, F., Cossins, A. R. & Thomas, S. L. (2000). The role of anion and cation channels in volume regulatory responses in trout red blood cells. *Bioelectrochemistry*, Vol. 52, No. 2, pp. 133-149, ISSN 1567-5394

Egee, S., Lapaix, F., Decherf, G., Staines, H. M., Ellory, J. C., Doerig, C. & Thomas, S. L. (2002). A stretch-activated anion channel is up-regulated by the malaria parasite Plasmodium falciparum. *The Journal of Physiology*, Vol. 542, No. Pt 3, pp. 795-801, ISSN 0022-3751

Ellory, J. C., Robinson, H. C., Browning, J. A., Stewart, G. W., Gehl, K. A. & Gibson, J. S. (2007). Abnormal permeability pathways in human red blood cells. *Blood Cells, Molecules and Diseases*, Vol. 39, No. 1, (Jul-Aug), pp. 1-6, ISSN 1079-9796

Ellory, J. C., Sequeira, R., Constantine, A., Wilkins, R. J. & Gibson, J. S. (2008). Non-electrolyte permeability of deoxygenated sickle cells compared. *Blood Cells, Molecules and Diseases*, Vol. 41, No. 1, (Jul-Aug), pp. 44-49, ISSN 1079-9796

Ellsworth, M. L., Ellis, C. G., Goldman, D., Stephenson, A. H., Dietrich, H. H. & Sprague, R. S. (2009). Erythrocytes: oxygen sensors and modulators of vascular tone. *Physiology (Bethesda)*, Vol. 24, No. (Apr), pp. 107-116, ISSN 1548-9221

Fettiplace, R., Andrews, D. M. & Haydon, D. A. (1971). The thickness, composition and structure of some lipid bilayers and natural membranes. *The Journal of Membrane Biology*, Vol. 5, No. 3, pp. 277-296, ISSN 0022-2631

Foller, M., Huber, S. M. & Lang, F. (2008a). Erythrocyte programmed cell death. *IUBMB Life*, Vol. 60, No. 10, (Oct), pp. 661-668, ISSN 1521-6543

Foller, M., Kasinathan, R. S., Koka, S., Lang, C., Shumilina, E., Birnbaumer, L., Lang, F. & Huber, S. M. (2008b). TRPC6 contributes to the Ca(2+) leak of human erythrocytes. *Cellular Physiology and Biochemistry*, Vol. 21, No. 1-3, pp. 183-192, ISSN 1015-8987

Frohlich, O. (1984). Relative contributions of the slippage and tunneling mechanisms to anion net efflux from human erythrocytes. *The Journal of General Physiology*, Vol. 84, No. 6, (Dec), pp. 877-893, ISSN 0022-1295

Gardos, G. (1956). The permeability of human erythrocytes to potassium. *Acta Physiologica*, Vol. 4, No. pp. 185-189, ISSN 0231-424X

Gardos, G. (1958). The function of calcium in the potassium permeability of human erythrocytes. *Biochimica et Biophysica Acta*, Vol. 30, No. 3, (Dec), pp. 653-654, ISSN 0006-3002

Gincel, D., Silberberg, S. D. & Shoshan-Barmatz, V. (2000). Modulation of the Voltage-Dependent Anion Channel (VDAC) by Glutamate1. *Journal of Bioenergetics and Biomembranes*, Vol. 32, No. 6, (Dec), pp. 571-583, ISSN 0145-479X

Ginsburg, H., Krugliak, M., Eidelman, O. & Cabantchik, Z. I. (1983). New permeability pathways induced in membranes of Plasmodium falciparum infected erythrocytes. *Molecular and Biochemical Parasitology*, Vol. 8, No. 2, (Jun), pp. 177-190, ISSN 0166-6851

Glogowska, E., Dyrda, A., Cueff, A., Bouyer, G., Egee, S., Bennekou, P. & Thomas, S. L. (2010). Anion conductance of the human red cell is carried by a maxi-anion channel. *Blood Cells, Molecules and Diseases*, Vol. 44, No. 4, (Apr 15), pp. 243-251, ISSN 1079-9796

Grgic, I., Kaistha, B. P., Paschen, S., Kaistha, A., Busch, C., Si, H., Kohler, K., Elsasser, H. P., Hoyer, J. & Kohler, R. (2009). Disruption of the Gardos channel (KCa3.1) in mice causes subtle erythrocyte macrocytosis and progressive splenomegaly. *Pflügers Archiv - European Journal of Physiology*, Vol. 458, No. 2, (Jun), pp. 291-302, ISSN 0031-6768

Grygorczyk, R. & Schwarz, W. (1983). Properties of the CA2+-activated K+ conductance of human red cells as revealed by the patch-clamp technique. *Cell Calcium*, Vol. 4, No. 5-6, pp. 499-510,ISSN 0143-4160

Grygorczyk, R., Schwarz, W. & Passow, H. (1984). Ca2+-activated K+ channels in human red cells. Comparison of single- channel currents with ion fluxes. *Biophysical Journal*, Vol. 45, No. 4, pp. 693-698, ISSN 0006-3495

Grygorczyk, R. (1987). Temperature dependence of Ca2+-activated K+ currents in the membrane of human erythrocytes. *Biochimica et Biophysica Acta*, Vol. 902, No. 2, pp. 159-168, ISSN 0006-3002

Gunn, R. B., Dalmark, M., Tosteson, D. C. & Wieth, J. O. (1973). Characteristics of chloride transport in human red blood cells. *The Journal of General Physiology*, Vol. 61, No. 2, pp. 185-206, ISSN 0022-1295

Haas, M. (1989). Properties and diversity of (Na-K-Cl) cotransporters. *Annual Review of Physiology*, Vol. 51, No. pp. 443-457, ISSN 0066-4278

Halperin, J. A., Brugnara, C., Tosteson, M. T., Van Ha, T. & Tosteson, D. C. (1989). Voltage-activated cation transport in human erythrocytes. *The American Journal of Physiology*, Vol. 257, No. 5 Pt 1, (Nov), pp. C986-996, ISSN 0002-9513

Hamill, O. P. (1981). Potassium channel currents in human red blood cells. *The Journal of Physiology*, Vol. 319 (suppl), No. pp. 97P-98P, ISSN 0022-3751

Hoffman, J. F., Kaplan, J. H., Callahan, T. J. & Freedman, J. C. (1980). Electrical resistance of the red cell membrane and the relation between net anion transport and the anion exchange mechanism. *Ann N Y Acad Sci*, Vol. 341, No. pp. 357-360, ISSN 0077-8923

Hoffman, J. F., Joiner, W., Nehrke, K., Potapova, O., Foye, K. & Wickrema, A. (2003). The hSK4 (KCNN4) isoform is the Ca2+-activated K+ channel (Gardos channel) in human red blood cells. *Proceedings of the National Academy of Sciences of the United States of America*, Vol. 100, No. 12, (Jun 10), pp. 7366-7371, ISSN 0027-8424

Hoffman, J. F., Dodson, A., Wickrema, A. & Dib-Hajj, S. D. (2004). Tetrodotoxin-sensitive Na+ channels and muscarinic and purinergic receptors identified in human erythroid progenitor cells and red blood cell ghosts. *Proceedings of the National Academy of Sciences of the United States of America*, Vol. 101, No. 33, (Aug 17), pp. 12370-12374, ISSN 0027-8424

Hoffmann, E. K., Lambert, I. H. & Pedersen, S. F. (2009). Physiology of cell volume regulation in vertebrates. *Physiological Reviews*, Vol. 89, No. 1, (Jan), pp. 193-277, ISSN 0031-9333

Huber, S. M., Gamper, N. & Lang, F. (2001). Chloride conductance and volume-regulatory nonselective cation conductance in human red blood cell ghosts. *Pflügers Archiv - European Journal of Physiology*, Vol. 441, No. 4, pp. 551-558, ISSN 0031-6768

Huber, S. M., Uhlemann, A. C., Gamper, N. L., Duranton, C., Kremsner, P. G. & Lang, F. (2002). Plasmodium falciparum activates endogenous Cl(-) channels of human erythrocytes by membrane oxidation. *The Embo Journal*, Vol. 21, No. 1_2, pp. 22-30, ISSN 0261-4189

Huber, S. M., Duranton, C., Henke, G., Van De Sand, C., Heussler, V., Shumilina, E., Sandu, C. D., Tanneur, V., Brand, V., Kasinathan, R. S., Lang, K. S., Kremsner, P. G., Hubner, C. A., Rust, M. B., Dedek, K., Jentsch, T. J. & Lang, F. (2004). Plasmodium induces swelling-activated ClC-2 anion channels in the host erythrocyte. *The Journal of Biological Chemistry*, Vol. 279, No. 40, (Oct 1), pp. 41444-41452, ISSN 0021-9258

Huber, S. M., Duranton, C. & Lang, F. (2005). Patch-clamp analysis of the "new permeability pathways" in malaria-infected erythrocytes. *Int Rev Cytol*, Vol. 246, No. pp. 59-134, ISSN 0074-7696

Hunter, M. J. (1977). Human erythrocyte anion permeabilities measured under conditions of net charge transfer. *The Journal of Physiology*, Vol. 268, No. 1, (Jun), pp. 35-49, ISSN 0022-3751

Jacobs, M. H. & Stewart, D. (1942). The role of carbonic anhydrase in certain ionic exchanges involving the erythrocyte. *The Journal of General Physiology*, Vol. 25, No. pp. 539-552, ISSN 0022-1295

Joiner, C. H. (1993). Cation transport and volume regulation in sickle red blood cells. *The American Journal of Physiology*, Vol. 264, No. 2 Pt 1, pp. C251-270, ISSN 0002-9513

Joiner, C. H., Morris, C. L. & Cooper, E. S. (1993). Deoxygenation-induced cation fluxes in sickle cells. III. Cation selectivity and response to pH and membrane potential. *The American Journal of Physiology*, Vol. 264, No. 3 Pt 1, (Mar), pp. C734-744, ISSN 0002-9513

Kaestner, L., Bollensdorff, C. & Bernhardt, I. (1999). Non-selective voltage-activated cation channel in the human red blood cell membrane. *Biochimica et Biophysica Acta*, Vol. 1417, No. 1, pp. 9-15, ISSN 0006-3002

Kaestner, L., Christophersen, P., Bernhardt, I. & Bennekou, P. (2000). The non-selective voltage-activated cation channel in the human red blood cell membrane: reconciliation between two conflicting reports and further characterisation. *Bioelectrochemistry*, Vol. 52, No. 2, pp. 117-125, ISSN 1567-5394

Kaestner, L. & Bernhardt, I. (2002). Ion channels in the human red blood cell membrane: their further investigation and physiological relevance. *Bioelectrochemistry*, Vol. 55, No. 1-2, pp. 71-74, ISSN 1567-5394

Kaplan, J. H., Pring, M. & Passow, H. (1983). Band-3 protein-mediated anion conductance of the red cell membrane. Slippage vs ionic diffusion. *FEBS Letters*, Vol. 156, No. 1, (May 30), pp. 175-179, ISSN 0014-5793

Kirk, K., Horner, H. A., Elford, B. C., Ellory, J. C. & Newbold, C. I. (1994). Transport of diverse substrates into malaria-infected erythrocytes via a pathway showing functional characteristics of a chloride channel. *The Journal of Biological Chemistry*, Vol. 269, No. 5, pp. 3339-3347, ISSN 0021-9258

Kirk, K. (2001). Membrane transport in the malaria-infected erythrocyte. *Physiological Reviews*, Vol. 81, No. 2, pp. 495-537, ISSN 0031-9333

Knauf, P. A., Fuhrmann, G. F., Rothstein, S. & Rothstein, A. (1977). The relationship between anion exchange and net anion flow across the human red blood cell membrane. *The Journal of General Physiology*, Vol. 69, No. 3, (Mar), pp. 363-386, ISSN 0022-1295

Knauf, P. A., Law, F. Y. & Marchant, P. J. (1983). Relationship of net chloride flow across the human erythrocyte membrane to the anion exchange mechanism. *The Journal of General Physiology*, Vol. 81, No. 1, (Jan), pp. 95-126, ISSN 0022-1295

Kundu, M., Lindsten, T., Yang, C. Y., Wu, J., Zhao, F., Zhang, J., Selak, M. A., Ney, P. A. & Thompson, C. B. (2008). Ulk1 plays a critical role in the autophagic clearance of mitochondria and ribosomes during reticulocyte maturation. *Blood*, Vol. 112, No. 4, (Aug 15), pp. 1493-1502, ISSN 0006-4971

Lang, F., Birka, C., Myssina, S., Lang, K. S., Lang, P. A., Tanneur, V., Duranton, C., Wieder, T. & Huber, S. M. (2004). Erythrocyte ion channels in regulation of apoptosis. *Advances in Experimental Medicine and Biology*, Vol. 559, No. pp. 211-217, ISSN 0065-2598

Lang, F., Lang, K. S., Lang, P. A., Huber, S. M. & Wieder, T. (2006). Mechanisms and significance of eryptosis. *Antioxidants and Redox Signaling*, Vol. 8, No. 7-8, (Jul-Aug), pp. 1183-1192, ISSN 1523-0864

Lang, K. S., Myssina, S., Tanneur, V., Wieder, T., Huber, S. M., Lang, F. & Duranton, C. (2003). Inhibition of erythrocyte cation channels and apoptosis by ethylisopropylamiloride. *Naunyn-Schmiedeberg's Archives of Pharmacology*, Vol. 367, No. 4, pp. 391-396, ISSN 0028-1298

Lang, K. S., Lang, P. A., Bauer, C., Duranton, C., Wieder, T., Huber, S. M. & Lang, F. (2005). Mechanisms of suicidal erythrocyte death. *Cellular Physiology and Biochemistry*, Vol. 15, No. 5, pp. 195-202, ISSN 1015-8987

Lange, T., Jungmann, P., Haberle, J., Falk, S., Duebbers, A., Bruns, R., Ebner, A., Hinterdorfer, P., Oberleithner, H. & Schillers, H. (2006). Reduced number of CFTR molecules in erythrocyte plasma membrane of cystic fibrosis patients. *Molecular Membrane Biology*, Vol. 23, No. 4, (Jul-Aug), pp. 317-323, ISSN 0968-7688

Lapaix, F., Egee, S., Gibert, L., Decherf, G. & Thomas, S. L. (2002). ATP-sensitive K(+) and Ca(2+)-activated K(+) channels in lamprey (Petromyzon marinus) red blood cell membrane. *Pflügers Archiv - European Journal of Physiology*, Vol. 445, No. 1, pp. 152-160, ISSN 0031-6768

Lapaix, F., Bouyer, G., Thomas, S. & Egee, S. (2008). Further characterization of cation channels present in the chicken red blood cell membrane. *Bioelectrochemistry*, Vol. 73, No. 2, (Aug), pp. 129-136, ISSN 1567-5394

Lassen, U. V., Pape, L. & Vestergaard-Bogind, B. (1978). Chloride conductance of the amphiuma red cell membrane. *The Journal of Membrane Biology*, Vol. 39, No. 1, (Feb 6), pp. 27-48, ISSN 0022-2631

Le Fur, G., Vaucher, N., Perrier, M. L., Flamier, A., Benavides, J., Renault, C., Dubroeucq, M. C., Gueremy, C. & Uzan, A. (1983). Differentiation between two ligands for peripheral benzodiazepine binding sites, [3H]RO5-4864 and [3H]PK 11195, by thermodynamic studies. *Life Sciences*, Vol. 33, No. 5, (Aug 1), pp. 449-457, ISSN 0024-3205

Lew, V. L., Tsien, R. Y., Miner, C. & Bookchin, R. M. (1982). Physiological [Ca2+]i level and pump-leak turnover in intact red cells measured using an incorporated Ca chelator. *Nature*, Vol. 298, No. 5873, (Jul 29), pp. 478-481, ISSN 0028-0836

Lew, V. L., Etzion, Z. & Bookchin, R. M. (2002). Dehydration response of sickle cells to sickling-induced Ca(++) permeabilization. *Blood*, Vol. 99, No. 7, pp. 2578-2585, ISSN 0006-4971

Lew, V. L. & Bookchin, R. M. (2005). Ion transport pathology in the mechanism of sickle cell dehydration. *Physiological Reviews*, Vol. 85, No. 1, (Jan), pp. 179-200, ISSN 0031-9333

Lew, V. L., Daw, N., Etzion, Z., Tiffert, T., Muoma, A., Vanagas, L. & Bookchin, R. M. (2007). Effects of age-dependent membrane transport changes on the homeostasis of senescent human red blood cells. *Blood*, Vol. 110, No. 4, (Aug 15), pp. 1334-1342, ISSN 0006-4971

Li, Q., Jungmann, V., Kiyatkin, A. & Low, P. S. (1996). Prostaglandin E2 stimulates a Ca2+-dependent K+ channel in human erythrocytes and alters cell volume and filterability. *The Journal of Biological Chemistry*, Vol. 271, No. 31, pp. 18651-18656, ISSN 0021-9258

Light, D. B., Mertins, T. M., Belongia, J. A. & Witt, C. A. (1997). 5-Lipoxygenase metabolites of arachidonic acid regulate volume decrease by mudpuppy red blood cells. *The Journal of Membrane Biology*, Vol. 158, No. 3, pp. 229-239, ISSN 0022-2631

Light, D. B., Dahlstrom, P. K., Gronau, R. T. & Baumann, N. L. (2001). Extracellular atp activates a p2 receptor in necturus erythrocytes during hypotonic swelling. *The Journal of Membrane Biology*, Vol. 182, No. 3, pp. 193-202, ISSN 0022-2631

Light, D. B., Attwood, A. J., Siegel, C. & Baumann, N. L. (2003). Cell swelling increases intracellular calcium in Necturus erythrocytes. *Journal of Cell Science*, Vol. 116, No. Pt 1, (Jan 1), pp. 101-109, ISSN 0021-9533

Madesh, M. & Hajnoczky, G. (2001). VDAC-dependent permeabilization of the outer mitochondrial membrane by superoxide induces rapid and massive cytochrome c release. *Journal of Cell Biology*, Vol. 155, No. 6, (Dec 10), pp. 1003-1015, ISSN 0021-9525

McEnery, M. W., Snowman, A. M., Trifiletti, R. R. & Snyder, S. H. (1992). Isolation of the mitochondrial benzodiazepine receptor: association with the voltage-dependent anion channel and the adenine nucleotide carrier. *Proceedings of the National Academy of Sciences of the United States of America*, Vol. 89, No. 8, (Apr 15), pp. 3170-3174, ISSN 0027-8424

Merckx, A., Nivez, M. P., Bouyer, G., Alano, P., Langsley, G., Deitsch, K., Thomas, S., Doerig, C. & Egee, S. (2008). Plasmodium falciparum regulatory subunit of cAMP-dependent PKA and anion channel conductance. *PLoS Pathogens*, Vol. 4, No. 2, (Feb 8), pp. e19, ISSN 1553-7366

Mortensen, M., Ferguson, D. J. & Simon, A. K. (2010). Mitochondrial clearance by autophagy in developing erythrocytes: clearly important, but just how much so? *Cell Cycle*, Vol. 9, No. 10, (May 15), pp. 1901-1906, ISSN 1551-4005

Olson, J. M., Ciliax, B. J., Mancini, W. R. & Young, A. B. (1988). Presence of peripheral-type benzodiazepine binding sites on human erythrocyte membranes. *European Journal of Pharmacology*, Vol. 152, No. 1-2, (Jul 26), pp. 47-53, ISSN 0014-2999

Pellegrino, C. M., Rybicki, A. C., Musto, S., Nagel, R. L. & Schwartz, R. S. (1998). Molecular identification and expression of erythroid K:Cl cotransporter in human and mouse erythroleukemic cells. *Blood Cells, Molecules and Diseases*, Vol. 24, No. 1, (Mar), pp. 31-40, ISSN 1079-9796

Pellegrino, M. & Pellegrini, M. (1998). Modulation of Ca2+-activated K+ channels of human erythrocytes by endogenous cAMP-dependent protein kinase. *Pflügers Archiv - European Journal of Physiology*, Vol. 436, No. 5, pp. 749-756, ISSN 0031-6768

Pinet, C., Antoine, S., Filoteo, A. G., Penniston, J. T. & Coulombe, A. (2002). Reincorporated Plasma Membrane Ca2+-ATPase can Mediate B-Type Ca2+ Channels Observed in Native Membrane of Human Red Blood Cells. *The Journal of Membrane Biology*, Vol. 187, No. 3, pp. 185-201, ISSN 0022-2631

Rodighiero, S., De Simoni, A. & Formenti, A. (2004). The voltage-dependent nonselective cation current in human red blood cells studied by means of whole-cell and nystatin-perforated patch-clamp techniques. *Biochimica et Biophysica Acta*, Vol. 1660, No. 1-2, (Jan 28), pp. 164-170, ISSN 0006-3002

Roman, R. M., Smith, R. L., Feranchak, A. P., Clayton, G. H., Doctor, R. B. & Fitz, J. G. (2001). ClC-2 chloride channels contribute to HTC cell volume homeostasis. *The*

American Journal of Physiology Gastrointest Liver Physiol, Vol. 280, No. 3, pp. G344-353, ISSN 0002-9513

Romero, P. J. & Romero, E. A. (2003). New vanadate-induced Ca2+ pathway in human red cells. *Cell Biology International*, Vol. 27, No. 11, pp. 903-912, ISSN 1065-6995

Romero, P. J., Romero, E. A., Mateu, D., Hernandez, C. & Fernandez, I. (2006). Voltage-dependent calcium channels in young and old human red cells. *Cell Biochemistry and Biophysics*, Vol. 46, No. 3, pp. 265-276, ISSN 1085-9195

Rostovtseva, T. & Colombini, M. (1997). VDAC channels mediate and gate the flow of ATP: implications for the regulation of mitochondrial function. *Biophysical Journal*, Vol. 72, No. 5, (May), pp. 1954-1962, ISSN 0006-3495

Ruud, J. T. (1954). Vertebrates without erythrocytes and blood pigment. *Nature*, Vol. 173, No. 4410, (May 8), pp. 848-850, ISSN 0028-0836

Schatzmann, H. J. (1966). ATP-dependent Ca++-extrusion from human red cells. *Experientia*, Vol. 22, No. 6, (Jun 15), pp. 364-365, ISSN 0014-4754

Schatzmann, H. J. (1983). The red cell calcium pump. *Annual Review of Physiology*, Vol. 45, No. pp. 303-312, ISSN 0066-4278

Semplicini, A., Spalvins, A. & Canessa, M. (1989). Kinetics and stoichiometry of the human red cell Na+/H+ exchanger. *The Journal of Membrane Biology*, Vol. 107, No. 3, (Mar), pp. 219-228, ISSN 0022-2631

Shields, M., Grygorczyk, R., Fuhrmann, G. F., Schwarz, W. & Passow, H. (1985). Lead-induced activation and inhibition of potassium-selective channels in the human red blood cell. *Biochimica et Biophysica Acta*, Vol. 815, No. 2, pp. 223-232, ISSN 0006-3002

Shindo, M., Imai, Y. & Sohma, Y. (2000). A novel type of ATP block on a Ca(2+)-activated K(+) channel from bullfrog erythrocytes. *Biophysical Journal*, Vol. 79, No. 1, pp. 287-297, ISSN 0006-3495

Shoshan-Barmatz, V. & Gincel, D. (2003). The voltage-dependent anion channel: characterization, modulation, and role in mitochondrial function in cell life and death. *Cell Biochemistry and Biophysics*, Vol. 39, No. 3, pp. 279-292, ISSN 1085-9195

Shoshan-Barmatz, V., De Pinto, V., Zweckstetter, M., Raviv, Z., Keinan, N. & Arbel, N. (2010). VDAC, a multi-functional mitochondrial protein regulating cell life and death. *Molecular Aspects of Medicine*, Vol. 31, No. 3, (Jun), pp. 227-285, ISSN 0098-2997

Shumilina, E. & Huber, S. (2011). ClC-2 Channels in Erythrocytes. *The Open Biology Journal*, Vol. 4, No. pp. 18-26, ISSN 1874-1967

Sprague, R. S., Ellsworth, M. L., Stephenson, A. H., Kleinhenz, M. E. & Lonigro, A. J. (1998). Deformation-induced ATP release from red blood cells requires CFTR activity. *The American Journal of Physiology*, Vol. 275, No. 5 Pt 2, pp. H1726-1732., ISSN 0002-9513

Sprague, R. S., Stephenson, A. H. & Ellsworth, M. L. (2007). Red not dead: signaling in and from erythrocytes. *Trends in Endocrinology & Metabolism*, Vol. 18, No. 9, (Nov), pp. 350-355, ISSN 1043-2760

Staines, H. M., Powell, T., Ellory, J. C., Egee, S., Lapaix, F., Decherf, G., Thomas, S. L., Duranton, C., Lang, F. & Huber, S. M. (2003). Modulation of whole-cell currents in Plasmodium falciparum-infected human red blood cells by holding potential and serum. *The Journal of Physiology*, Vol. 552, No. Pt 1, (Oct 1), pp. 177-183, ISSN 0022-3751

Staines, H. M., Alkhalil, A., Allen, R. J., De Jonge, H. R., Derbyshire, E., Egee, S., Ginsburg, H., Hill, D. A., Huber, S. M., Kirk, K., Lang, F., Lisk, G., Oteng, E., Pillai, A. D., Rayavara, K., Rouhani, S., Saliba, K. J., Shen, C., Solomon, T., Thomas, S. L., Verloo, P. & Desai, S. A. (2007). Electrophysiological studies of malaria parasite-infected erythrocytes: current status. *International Journal for Parasitology*, Vol. 37, No. 5, (Apr), pp. 475-482, ISSN 0020-7519

Strange, K., Emma, F. & Jackson, P. S. (1996). Cellular and molecular physiology of volume-sensitive anion channels. *The American Journal of Physiology*, Vol. 270, No. 3 Pt 1, pp. C711-730., ISSN 0002-9513

Strobaek, D., Teuber, L., Jorgensen, T. D., Ahring, P. K., Kjaer, K., Hansen, R. S., Olesen, S. P., Christophersen, P. & Skaaning-Jensen, B. (2004). Activation of human IK and SK Ca2+ -activated K+ channels by NS309 (6,7-dichloro-1H-indole-2,3-dione 3-oxime). *Biochimica et Biophysica Acta*, Vol. 1665, No. 1-2, (Oct 11), pp. 1-5, ISSN 0006-3002

Thomas, S. L., Egee, S., Lapaix, F., Kaestner, L., Staines, H. M. & Ellory, J. C. (2001). Malaria parasite Plasmodium gallinaceum up-regulates host red blood cell channels. *FEBS Letters*, Vol. 500, No. 1-2, pp. 45-51, ISSN 0014-5793

Thomas, S. L., Bouyer, G., Cueff, A., Egee, S., Glogowska, E. & Ollivaux, C. (2011). Ion channels in human red blood cell membrane: actors or relics? *Blood Cells, Molecules and Diseases*, Vol. 46, No. 4, (Apr 15), pp. 261-265, ISSN 1079-9796

Tiffert, T., Daw, N., Etzion, Z., Bookchin, R. M. & Lew, V. L. (2007). Age decline in the activity of the Ca2+-sensitive K+ channel of human red blood cells. *The Journal of General Physiology*, Vol. 129, No. 5, (May), pp. 429-436, ISSN 0022-1295

Tosteson, D. C. & Hoffman, J. F. (1960). Regulation of cell volume by active cation transport in high and low potassium sheep red cells. *The Journal of General Physiology*, Vol. 44, No. (Sep), pp. 169-194, ISSN 0022-1295

Vandorpe, D. H., Xu, C., Shmukler, B. E., Otterbein, L. E., Trudel, M., Sachs, F., Gottlieb, P. A., Brugnara, C. & Alper, S. L. (2010). Hypoxia activates a Ca2+-permeable cation conductance sensitive to carbon monoxide and to GsMTx-4 in human and mouse sickle erythrocytes. *PLoS ONE*, Vol. 5, No. 1, pp. e8732, ISSN 1932-6203

Veenman, L., Shandalov, Y. & Gavish, M. (2008). VDAC activation by the 18 kDa translocator protein (TSPO), implications for apoptosis. *Journal of Bioenergetics and Biomembranes*, Vol. 40, No. 3, (Jun), pp. 199-205, ISSN 0145-479X

Verloo, P., Kocken, C. H., Van der Wel, A., Tilly, B. C., Hogema, B. M., Sinaasappel, M., Thomas, A. W. & De Jonge, H. R. (2004). Plasmodium falciparum-activated chloride channels are defective in erythrocytes from cystic fibrosis patients. *The Journal of Biological Chemistry*, Vol. 279, No. 11, (Mar 12), pp. 10316-10322, ISSN 0021-9258

Virkki, L. V. & Nikinmaa, M. (1996). Conductive Ion Transport across the Erythrocyte Membrane of Lamprey (Lampetra fluviatilis) in Isotonic Conditions is Mainly via an Inwardly Rectifying K+ Channel. *Comparative Biochemistry and Physiology Part A: Physiology*, Vol. 115, No. 2, (1996/10), pp. 169-176, ISSN 1095-6433

Virkki, L. V. & Nikinmaa, M. (1998). Two distinct K+ channels in lamprey (Lampetra fluviatilis) erythrocyte membrane characterized by single channel patch clamp. *The Journal of Membrane Biology*, Vol. 163, No. 1, pp. 47-53, ISSN 0022-2631

Yingst, D. R. & Hoffman, J. F. (1984). Ca-induced K transport in human red blood cell ghosts containing arsenazo III. Transmembrane interactions of Na, K, and Ca and the relationship to the functioning Na-K pump. *The Journal of General Physiology*, Vol. 83, No. 1, (Jan), pp. 19-45, ISSN 0022-1295

Yoshida, H., Kawane, K., Koike, M., Mori, Y., Uchiyama, Y. & Nagata, S. (2005). Phosphatidylserine-dependent engulfment by macrophages of nuclei from erythroid precursor cells. *Nature*, Vol. 437, No. 7059, (Sep 29), pp. 754-758, ISSN 0028-0836

10

Electrical Membrane Properties in the Model *Leishmania*-Macrophage

Marcela Camacho
Universidad Nacional de Colombia, Sede Bogotá, and Centro Internacional de Física
Colombia

1. Introduction

Leishmaniasis, a disease caused by parasites of the genus *Leishmania*, constitutes a worldwide health problem. Since 1993 the disease has spread over wider areas of the world, and the situation has further deteriorated due to the AIDS pandemic. No vaccine is available to control the disease, and current therapies have problems of resistance, therapeutic failure, and cost (González et al., 2009). In Colombia, the recent incidence of most parasitic diseases has tended to stabilize, but in contrast the last decade has seen a doubling of the incidence of cutaneous Leishmaniasis (INS, 2009). Several reasons explain this, among which climate changes, deforestation, migration, and vector changes are in common with other areas of the world, but local circumstances such as illegal cultivars, the internal armed conflict, and recent health reforms are also important.

The parasite *Leishmania* transits between different environments in its life cycle, from the mosquito gut to the salivary glands, and for a short period in the vertebrate skin, before entering a compartment known as the parasitophorous vacuole (PV) inside macrophages and dendritic cells. The transformation between the two major parasite stages, from promastigote to amastigote, is the result of some of the changes that occur between the mosquito gut and the PV, among which temperature and pH have been implicated (Zilberstein & Shapira, 2004). On the other hand, osmolarity and ionic concentration are known to affect the ability of any cell, presumably including *Leishmania*, to control membrane potential, ion and nutrient transport, osmolarity, volume, and pH, though the precise effects are yet poorly understood. Our major interest is based on the assumption that *Leishmania* survival in the macrophage is also the result of the integrated function of the three concentric membranes found by the intracellular (amastigote) form of this parasite: the host cell plasma membrane (i.e., the macrophage plasma membrane), the parasitophorous vacuole membrane (PVM), and the *Leishmania* plasma membrane.

2. Macrophage plasma membrane

In the vertebrate host, *Leishmania* is an obligatory intracellular parasite that infects cells of macrophagic and dendritic lineage. In other intracellular parasite-host cell relationships, in particular *Plasmodium*, the demands of the replicating parasite are met by incorporating parasite membrane channels and transporters, or by modulating those of the host cell

(Ginsburg & Stein, 2005; Staines et al., 2007; Martin et al., 2009). Altered calcium homeostasis in erythrocytes and muscle cells has been reported in *Plasmodium* and *Trypanosoma* infection (Tanabe, 1990; Olivier, 1996; Tardieux et al., 1994) as well as in macrophages infected with *Leishmania* (Eilam, 1985; Olivier, 1996). Though there is no evidence of *Leishmania* induced alterations of macrophage nutrient transport, changes in macrophage membrane permeability, particularly in the electrical membrane properties, may alter its ability to activate and present antigen, therefore affecting the immune response.

Several functional activation stages have been proposed for the macrophage (Gordon, 2003; Mosser & Edwards, 2008). For example, macrophages stimulated *in vitro* with lipopolysaccharide (LPS) and interferon gamma (INF-γ) activate in a way that has been designated as classical. This functional stage is characterized by morphological changes, increased macrophage surface area, nitric oxide (NO) production, upregulation of tumour necrosis alpha (TNF-α) secretion, and induction of interleukin 12, among others (reviewed by Gordon, 2003). There is so far no coherent view of macrophage electrical membrane properties and their functional significance and association with the different activation stages, but some associations are now becoming apparent.

2.1 Macrophage passive membrane properties

Macrophage passive electrical membrane properties have been recorded in primary macrophages of mouse (Buisman et al., 1988; Vicente et al., 2003), man (Gallin & Gallin, 1977; Holevinsky & Nelson, 1998), various mouse (Randriamampita & Trautmann, 1987; Buisman et al., 1988; Holevinsky & Nelson, 1998; Gallin, 1991; Forero et al., 1999; Camacho et al., 2008; Quintana et al., 2010; Villalonga et al., 2010) and human (McCann et al., 1987) cell lines. Differences between some measurements of electrophysiological parameters may be related to cell culture conditions, though some authors argue that they represent genuine innate characteristics of currents expressed by macrophages (Randriamampita & Trautmann, 1987). Several studies support this hypothesis (Randriamampita & Trautmann, 1987; McKinney & Gallin, 1990; Forero et al., 1999; Camacho et al., 2008). In this view, variations in characterization are explained by differences in the ionic solutions used and the corresponding change to the Nernst equilibrium and the electrochemical force imposed on each ion, and of course by differences in the ion channel populations expressed (Vicente et al., 2006; Villalonga et al., 2007; Vicente et al., 2008).

In our experience, adherence onto glass, a step used to differentiate monocytes to macrophages, critically affects the electrical properties of these cells. The passive electrical membrane properties of J774.A1, a mouse macrophage-like cell line, vary with time after adherence onto glass, where increased membrane capacitance (Cm) and hyperpolarization have been observed (Gallin & Sheehy, 1985; McKinney & Gallin, 1990; Camacho, unpublished data). Active electrical properties also vary, with large outward currents (I_{OUT}) recorded during the first 8 hours post-adherence with minimal inward rectifying current (I_{KIR}) (Gallin & Sheehy, 1985; Randriamampita & Trautmann, 1987). I_{OUT} and I_{KIR} have similar amplitudes after 24 hours post-adherence (Gallin & Sheehy, 1985), but after 48 hours the situation reverses, with predominant I_{KIR} and negligible I_{OUT} (McKinney & Gallin, 1990). We have therefore controlled for time of adherence onto glass in some of our experiments to guarantee similar amplitudes for I_{OUT} and I_{KIR} (Forero et al., 1999; Camacho et al., 2008; Quintana et al., 2010).

Membrane capacitance is an electrical parameter that is proportional to surface area because the ability to store charge of a capacitor depends on this geometry. The lipid bilayer of the cell membrane can be modelled as a capacitor of two parallel conducting plates separated by a dielectric, so changes in Cm reflect direct changes in membrane area. Cm varies with post-adherence time (McKinney & Gallin, 1990) and *Leishmania amazonensis* post-infection time (Forero et al., 1999; Camacho et al., 2008; Quintana et al., 2010). J774.A1 cells exposed to cytochrome C to induce apoptosis are smaller than control cells, but have a similar Cm (Clavijo et al., 2009) suggesting loss of volume without decrease in surface area.

Macrophages are professional phagocytes, and phagocytosis implies the remodelling of the plasma membrane by the cytoskeleton and the incorporation of the phagocytic load into an intracellular compartment, the phagosome (PG). Entry of any load results in plasma membrane donation and should impact macrophage Cm. Phagocytosis of immune complexes or inert particles is associated with a reduction of Cm proportional to the load. After phagocytosis of 3 μm particles, Cm dropped around 7-10% (Holevinsky & Nelson, 1998; Quintana et al., 2010). *L. amazonensis* infected macrophage-like cells lose nearly a third of their Cm by 3 hours post-entry (Quintana et al., 2010), corresponding well to an average parasite load of two to three promastigotes (Hoyos et al., 2009). Cm recovers in this model, and by 24 hours post-infection the values are above control levels (Forero et al., 1999). Membrane donation by the macrophage plasma membrane upon phagocytosis requires recruitment of intracellular membranes from endosomes and lysosomes (Pitt et al., 1992; Desjardins et al., 1994a, Desjardins et al., 1994b, 1995; Beron et al., 1995; Idone et al., 2008). The fusion of these membranes fulfils two purposes: extra membrane for the nascent PG, and early release of immune effectors as shown for TNF-α (Murray et al., 2005, 2005b). Recruitment of endoplasmic reticulum (ER) membrane (Gagnon et al., 2002; Becker et al., 2005) may be important for PG formation when phagocytosing large loads, though this proposed role is controversial. After *L. amazonensis* entry we observe positive labelling with anti-LAMP antibodies, a lysosomal marker, as well as labelling of PV membranes with anti-IP_3 receptor (Perez, 2008), a marker of ER, suggesting membrane donation from this organelle.

Other processes in which the macrophage recruits more membrane to the macrophage plasma membrane should polarize more charge and increase Cm. During classical activation, J774.A1 becomes larger and adds membrane (Camacho et al., 2008), but LPS alone surprisingly reduces this parameter in RAW 264.7 (Villalonga et al., 2010), suggesting reduction of surface area. In the *Leishmania*-macrophage model, evidence of reduction of endosomal and lysosomal compartments (Barbiéri et al., 1990) has been interpreted as a way to concentrate lysosomal activity into the PV to control the parasite. This phenomenon may also indicate alterations in the ratio of membrane fusion/fission in the exocytic/endocytic rate. Assuming that the dielectric properties of J774.A1 plasma membrane do not change during *Leishmania* infection, the augmentation of macrophage Cm (Forero et al., 1999) may constitute a defect in macrophage plasma membrane recycling.

In most animal cells, the resting Vm is set by the activity of leak channels selective for K^+, along with the contribution of the electrogenic activity of ion pumps. Thus the resting potential of many animal cells is close to E_{K+}. Most data suggest a Vm for the macrophage close to E_{K+} that varies with changes in $[K^+]_o$ (Gallin & Sheehy, 1985; Judge et al., 1994). The slope was -49 mV/ 10 fold $[K^+]_o$, indicating higher permeability to K^+ under resting conditions (Gallin & Sheehy, 1985), but suggesting the contribution of other ions. A contribution of the electrogenic activity of a Na^+/K^+ ATPase pump (Gallin & Livengood, 1983) of about 6 mV to the macrophage Vm could explain the deviation found. The depolarization obtained after inhibition of I_{KIR} by Ba^{2+} has implicated this current in the macrophage Vm value (Randriamampita & Trautmann, 1987), particularly after long periods of adherence (Gallin & Sheehy, 1985). More recently it has been proposed that Kv1.3 voltage gated channels establish macrophage Vm (Mackenzie et al., 2003) as shown in T lymphocytes (Panyi et al., 2004).

Early measurements of Vm were misinterpreted as action potentials (McCann et al., 1983). Values of around -75 mV have been recorded (McCann et al., 1987; Gallin & Sheehy, 1985; Buisman et al., 1988), and it has been suggested that time of adherence is not important because rapid hyperpolarization was observed after only 30 minutes post-adhesion (Gallin & Sheehy, 1985). Values of -60 to -70 mV were recorded in J774 and in peritoneal macrophages (Randriamampita & Trautmann, 1987). In whole cell configuration we have measured Vm when the membrane current is zero during the first minute after attaining this configuration, and have recorded potentials between -40 and -50 mV (Forero et al., 1999; Camacho et al., 2008; Quintana et al., 2010) at 24 or less hours post-adherence onto glass. Similar values (-42 to -58 mV) were reported in the same cell line by McKinney & Gallin, (1990). In T lymphocytes, Vm is close to the gating potential of Kv1.3 channels and this argument has been used to suggest that one of the functions of this channel is to establish Vm (Panyi et al., 2004). After 24 hours post-phagocytosis the average Vm found was depolarized compared to control macrophages (Camacho et al., 2008). Similar Vm values and depolarizations were found in macrophages stimulated with LPS and INF-γ (Camacho et al., 2008) at 24 hours post-treatment or exposed to cytochrome C to induce apoptosis 2 hours post-treatment (Clavijo et al., 2009).

Macrophage membrane resistance (Rm) has been measured. We have found membrane resistance of 2 GΩ in macrophage-like cells not altered by *L. amazonensis* infection (Forero et al., 1999). After reaching the whole cell configuration, a 10 mV step pulse was applied to the macrophage from a holding potential of -60 mV and the recorded current averaged to calculate Rm from Ohm's law. ATP^{4-} exposure reduced Rm by a factor of 10 in a related macrophage cell line (J774.2), where membrane depolarization was also observed (Buisman et al., 1988). Functional evidence of a poorly selective conductance was recorded, based on the reversal potential measurement (0 mV; Buisman et al., 1988). This is compatible with the permeation effect achieved after the ATP exposure used to load macrophages with anionic fluorescent probes (Steinberg et al., 1987a), explaining the reduction in Rm.

2.2 Macrophage ion currents

With respect to active electrical properties of the macrophage plasma membrane, Cl^- currents (I_{Cl}), K^+ outward (K^+ I_{OUT}) and inward currents (I_{KIR}) have been described. Large conductance chloride channels were recorded in patches of J774 and peritoneal macrophages (Randriamampita & Trautmann, 1987). Thioglycolate-recruited mouse peritoneal cells dialyzed in K^+-glutamate containing $InsP_3$ showed a rapidly activating outward current that depolarized the cells, suggesting that part of the current was carried out by Cl^- (Judge et al., 1994). In human monocyte-derived macrophages, rises of $[Ca^{2+}]_i$ induced an I_{Cl} current of properties similar to those described previously, and sensitive to

DIDS (Holevinsky et al., 1994). A third of the total current recorded in our model was inhibited by DIDS involving an anion, most probably Cl^- (Camacho et al., 2008).

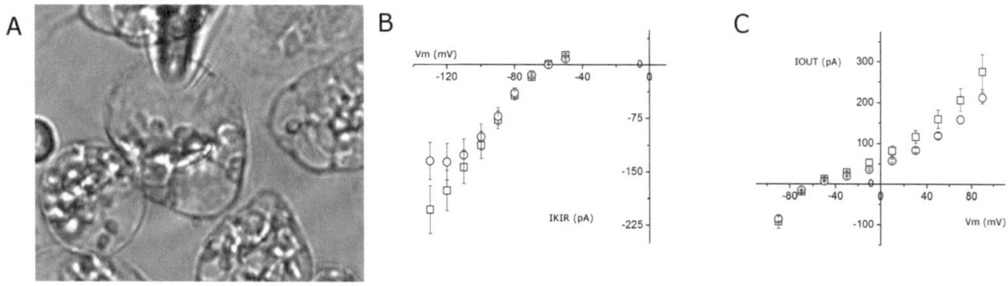

Fig. 1. Currents in J774.A1 macrophage-like cells. A. Light microscopy image of J774.A1 macrophage-like cells infected with *Leishmania amazonensis*. Note the tip of the recording pipette and the intracellular parasites in the parasitophorous vacuoles. B. I/V curve of inward rectifying potassium currents. C. I/V curve of outward currents. (□) Mean I peak at the beginning of the trace and after the capacitive transient. (O) Mean I of the steady state current in the last 20 s of the recording. Typical recordings of currents elicited by applying 9-10 pulses of 1 s (B) or 100 ms (C) at 10-20 mV intervals, from -50 to -130 mV (B) or -90 to 90 mV (C) from a holding potential of -60 mV. Data represent the mean values ± SE with n = 10.

I_{KIR} is induced by hyperpolarizing voltages and is inactivating (Gallin & Sheehy, 1985; Randriamampita & Trautmann, 1987; McKinney & Gallin, 1988; Judge et al., 1994; Forero et al., 1999). The voltage range of activation changes to less negative potentials with rises in $[K^+]_o$, following a square root relation between the peak G_{KIR} and $[K^+]_o$ (Gallin and Sheehy, 1985; Randriamampita & Trautmann, 1987). I_{KIR} inactivation is accelerated by voltage (Gallin & Sheehy, 1985; Randriamampita & Trautmann, 1987; Judge et al., 1994; Forero et al., 1999; Figure 1B), and its amplitude inhibited by extracellular Ba^{2+} (Gallin & Sheehy, 1985; Randriamampita & Trautmann, 1987; Forero et al., 1999) and Cs^+ (Gallin & Sheehy, 1985; Randriamampita & Trautmann, 1987; Judge et al., 1994). I_{KIR} goes to zero over time during recording (Gallin and Sheehy, 1985; Randriamampita and Trautmann, 1987), but the current is preserved in time with the addition of ATP and GTP to the pipette solution (McKinney & Gallin, 1990; Judge et al., 1994), suggesting dependence on metabolism (Judge et al., 1994). I_{KIR} was reduced by H_2O_2 production (Judge et al., 1994) although the authors did not indicate this. Colony stimulating factor was reported to have no effect on this current (Judge et al., 1994) but more recent data shows induction of this current by macrophage colony stimulating factor (M-CSF; Vicente et al., 2003). Kir2.1 channels are held responsible for this current (Vicente et al., 2003).

K^+ I_{OUT} appears to be the result of at least three different types of ion channel populations: inactivating, non-inactivating and calcium dependent K^+ current(Ca^{2+} K^+ I_{OUT}; Gallin, 1991). Total K^+ I_{OUT} is rapidly activated with depolarization (Figure 1C) and inactivates over time. The selectivity was studied in tail currents with different $[K^+]_o$; the reversal potential was also shifted to lower potentials with rises in $[K^+]_o$, implicating this ion in these currents (Gallin & Sheehy, 1985). K^+ I_{OUT} is inhibited by 4-aminopyridine (4-AP) and tetraethyl ammonium (TEA; Gallin & Sheehy, 1985; Camacho et al., 2008;), two classical potassium channel blockers, by margatoxin, a specific inhibitor of Kv1.3, and by charybdotoxin, a specific inhibitor of calcium dependent potassium channels (Camacho et al., 2008). Ca^{2+} K^+ I_{OUT} attains a large amplitude with $[Ca^{2+}]_i$ of 1 μM and is inhibited by quinine (Randriamampita & Trautmann, 1987). The relative contributions of these currents may however vary with cell type, culture conditions, and cell status (Randriamampita & Trautmann, 1987; Gallin, 1991; DeCoursey et al., 1996; Eder et al., 1997; Camacho et al., 2008). More recently it has been shown that transcription of specific genes and expression of Kv1.3, Kv1.5 are associated with K^+ I_{OUT} (Mackenzie et al., 2003; Vicente et al., 2006; Vicente et al., 2006; Park et al., 2006).

Despite several reports that outward currents are altered when macrophages are activated (Nelson et al., 1992; Ichinose et al., 1992; McKinney & Gallin, 1992), a coherent view of macrophage functional status and electrical membrane properties is just emerging. Using primary cultured cells and a macrophage cell line, Felipe's group initially showed that the majority of the K^+ I_{OUT} in mouse bone marrow macrophages is generated almost exclusively by Kv1.3 voltage gated potassium ion channels (Vicente et al., 2003). In an elegant set of experiments they characterised the electrophysiological response of these cells and showed K^+ I_{OUT} and I_{KIR}. The currents were associated with transcription of mRNA, protein expression, and localization to the macrophage plasma membrane of Kv1.3 and Kir2.1. They showed the expression pattern of these ion channels after stimulation with M-CSF, LPS, and TNF-α. Over-expression of Kv1.3 and Kir2.1 translated functionally to approximately three times more K^+ I_{OUT} and around four times more I_{KIR} during macrophage proliferation (M-CSF), contrasting with nearly 15 times more K^+ I_{OUT} and a reduction to half in I_{KIR} after macrophage activation with LPS (classical activation; Vicente et al., 2003). The differences in the timing of mRNA transcription resulted in electrophysiological differences. Furthermore, they recorded evidence of the impact of differential expression of Kvβ subunits in macrophages during proliferation and differential activation between LPS and TNF-α (Vicente et al., 2005). The expression of Kv1.5 resulted in associations with Kv1.3 to conform hetero-dimers to distinct biophysical properties (Vicente et al., 2006; Vicente et al., 2007). They also observed modulation by Kv1.5 and LPS of intracellular traffic of Kv1.3 and Kv1.3/Kv1.5 channels (Vicente et al., 2008). Their suggestion is that expression of Kv1.5 does not generate homodimers, but rather the heterodimers Kv1.3/Kv1.5 with different stoichiometries that alter macrophage electrical properties and trafficking to the membrane. In humans, hKv1.5 is transcribed and expressed and a drop is observed in its typical outward current after exposure of antisense oligonucleotides against this channel; functional consequences followed, with a 50% reduction in migration of human alveolar macrophages (Park et al., 2006). Using the same conditions reported by Vicente et al. (2003), we have amplified products the expected weight for Kir2.1 and Kv1.3 in J774.A1 cells from mRNA (Figure 2).

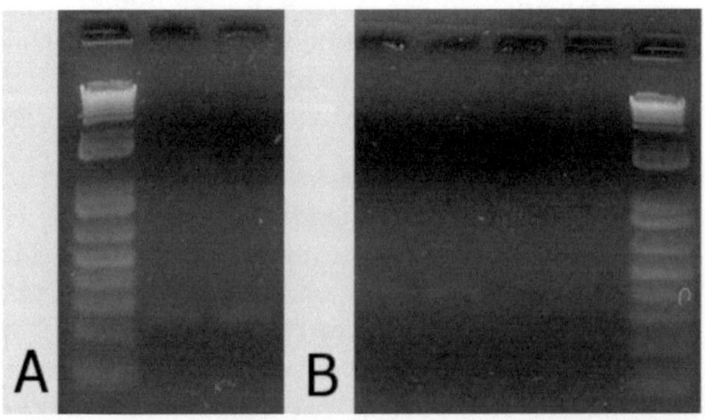

Fig. 2. Transcripts of ion channels in J774.A1. A. Transcripts of the inward rectifier potassium channel Kir2.1. B. Transcripts of the voltage gated potassium ion channel Kv1.3. Primers and PCR conditions are as reported by Vicente et al., (2003).

TNF-α stimulation alters expression of different Kvβ, inducing more inactivation of K^+ I_{OUT} but less expression to the plasma membrane of Kv1.3 after 24 hours compared to LPS stimulation (Vicente et al., 2005). In the mouse macrophage cell line RAW 264.7, there is evidence of transcription, protein expression, protein localization, and functional expression of Kv1.3 and Kv1.5 (Vicente et al., 2006; Villalonga et al., 2007). RAW 264.7 has K^+ I_{OUT} of smaller amplitude, expresses more Kv1.5 and therefore has less sensitivity to margatoxin (Vicente et al., 2006). Stimulation with LPS of this line resulted in higher transcription and protein expression and increases in Kv1.3 K^+ I_{OUT}, in contrast to the downregulation in transcription and protein expression and lower K^+ I_{OUT} inactivation after exposure to the immunosuppressor agent dexamethasone (Villalonga et al., 2010). However, it has been suggested that, by inducing Kv1.3 and repressing Kir2.1 after activation, macrophages reduce Ca^{2+} driving force and intracellular K^+ concentration (Vicente et al., 2003). In the proposed model of the T lymphocyte, the influx of Ca^{2+} after activation depolarizes the membrane and reduces the electrical driving force in the absence of a counterbalancing cation flux, suggesting that Kv1.3 is involved in maintenance of Ca^{2+} entry (Panyi et al., 2004). A biphasic intracellular increase of calcium has been observed in mouse macrophages. The initial transient is attributed to release of Ca^{2+} from endoplasmic reticulum, and the second transient, with less amplitude but longer time course, to Ca^{2+} influx through the plasma membrane (Randriamampita et al., 1991). In TRPV2 (Transient Receptor Potential Vanilloid 2) knockout mice, this cation channel generates a biphasic entry of Ca^{2+}, with a rapid initial transient followed by a sustained increase. The lack of this channel alters macrophage phagocytosis (Link et al., 2010). We have seen evidence of higher Ca^{2+} concentration in activated macrophage-like cells (Camacho, unpublished data) in which the Vm measured was less than in control cells, suggesting depolarization (Holevinsky & Nelson, 1995; Camacho et al., 2008). The currents show increased K^+ I_{OUT} and lowered I_{KIR} (Camacho et al., 2008), similar to the description by Vicente et al., 2003. Thus far unavailable simultaneous measurements of calcium and other potential currents could bolster support for this hypothesis, but a proposal in macrophages similar to that of T lymphocytes is more coherent. Thus, some of the functional stages of the macrophage (phagocytosis, classical activation) generate an increase in intracellular Ca^{2+}. The entry of this divalent cation depolarizes the membrane, inducing gating of Kv1.3 channels and counterbalancing Ca^{2+} entry, thereby maintaining the driving force for a sustained response.

We have studied the impact of *Leishmania* infection on the electrical properties of infected macrophage-like cells using the classical whole cell configuration of the patch clamp technique (Hamill et al., 1981). The model initially chosen was infection of J774.A1 by *L. amazonensis* because this parasite induces a large PV that is easily recognised in a light microscope (Figure 1A). The pipette solution included ATP and GTP to maintain stable currents, particularly I_{KIR}, and glutamate was the predominant anion (Forero et al., 1999; Camacho et al., 2008; Quintana et al., 2010). The solution flow during pharmacological testing was carefully controlled to avoid rises of the total I_{OUT} (Randriamampita & Trautmann, 1987). The recordings were easier at earlier times but seal stability was compromised with time post-infection.

Leishmania infection alters macrophage plasma membrane electrical properties and K^+ ion currents. However, there are differences between the state induced during the first hours post-infection and the state of an established infection of more than 24 hours. Macrophages were seen to depolarize during the first 12 hours post-infection. A decrease in Vm was associated with a change in conductance and lower amplitude of I_{KIR} density, and a rise of I_{OUT} density (Quintana et al., 2009). Reductions of Cm and depolarization, with similar ion current density, were found in macrophages after phagocytosis of latex beads (Quintana et al., 2009), suggesting that the changes observed during early *Leishmania* infection are associated with phagocytosis. In contrast, we have shown that an established infection with *L. amazonensis* of non-activated J774A.1 is associated with increased I_{KIR} density (Forero et al., 1999) and a rise in K^+ I_{OUT} (Camacho et al., 2008), which would tend to counter activation (Vicente et al., 2003) and would be consistent with observed suppression of activation by *Leishmania* infection (Liew et al., 1998). There is evidence that depolarization is associated with less I_{KIR} and a rise in K^+ I_{OUT}, leading to hyperpolarization (Randriamampita & Trautmann, 1987). The depolarization during early infection (Quintana et al., 2010) can be explained by lower I_{KIR}, and the hyperpolarization observed in established infection is explained by higher K^+ I_{OUT} (Camacho et al., 2008).

Scott et al., 2003, working with macrophages stimulated with LPS and INF-γ, have shown that *Leishmania major* infection, or treatment with a wide range of K^+ channel blockers, suppresses NO production, consistent with a role for K^+ currents in the deactivation effect of *Leishmania*. We found no differences between non-activated control and infected macrophages in either the K^+ I_{OUT} density, its Iss/Ip ratio over the period studied, or susceptibility to 4-AP inhibition. However, the K^+ I_{OUT} time to peak and sensitivity to TEA were altered by infection, suggesting greater contribution of some potassium channels to the I_{OUT} (Camacho et al., 2008). This is significant because alteration of K^+ currents may compromise the ability of the macrophage to phagocytose, to be activated (Randriamampita & Trautmann, 1989; Buchmüller-Rouiller & Mauël, 1991; Fischer & Eder, 1995; Holevinsky & Nelson, 1995; Vicente et al., 2003; Villalonga et al., 2010), to present antigen (McKinney & Gallin, 1992), and to do transmigration (Gendelman et al., 2009). In particular, inhibition of K^+ I_{OUT} has been associated with reduced phagocytic ability (Hara et al., 1990) and less NO production (Scott et al., 2003; Vicente et al., 2003), suggesting that these currents may be critical in control of infections in macrophages.

Plasma membrane hyperpolarization has been associated with calcium replenishment, production of oxygen radicals (Gamaley et al., 1998; Hattori et al., 2003; Hanley et al., 2004), myoblast membrane fusion (Liu et al., 1998), and protection against apoptosis (Dallaporta et al., 1999; Liu et al., 2005) as well as *Leishmania* infection (Forero et al., 1999; Fajardo et al., 2007). *Leishmania* infection raises macrophage intracellular calcium (Eilam et al., 1985; Olivier, 1996), and promotes production of oxygen radicals (Sousa-Franco et al., 2005). Fusion to form giant cells is a physiological response in macrophages (Vignery, 2005; McNally & Anderson, 2005; Cui et al., 2006) and proceeds via mechanisms similar to those in phagocytosis and with the participation of ER membranes (McNally & Anderson 2005). There is evidence of protection against apoptosis in macrophages infected with *Leishmania* (Aga et al., 2002; Lisi et al., 2005). We found that 1 μM staurosporin induces apoptosis of J774A.1, but that infection with *Leishmania braziliensis* protects these cells against this agent. Macrophage apoptosis was accompanied by depolarization of the mitochondrial and macrophage plasma membrane (Clavijo et al., 2009). *L. amazonensis* and *L. braziliensis* infection induce macrophage plasma membrane hyperpolarization (Forero et al., 1999; Fajardo et al., 2005). The significance of this hyperpolarization is not clear, though it may contribute to macrophage deactivation. Interestingly, we have observed that the changes in volume of the PV are associated with this membrane potential change (Fajardo et al., 2005).

We intend to continue with the characterisation of macrophage membrane properties, to understand them in relation to the cell's repertoire of functional stages.

3. Parasitophorous vacuole membrane

Leishmania is an intracellular parasitic protozoon of macrophages and dendritic cells, confined in an endolysosomal compartment, the parasitophorous vacuole (PV). Assuming that the parasitophorous vacuole membrane (PVM) permits ion and nutrient exchange between the lumen of the PV and the infected cell cytoplasm, the PVM will be involved in parasite survival and replication. The majority of studies on the biogenesis and membrane composition of phagolysosomes have been made on the model of phagocytosis of latex beads by macrophages (Desjardins & Griffiths, 2003). These studies have shown highly regulated sequential acquisition into the phagosome of proteins from the plasma, endosomal and lysosomal membranes of the macrophage (Pitt et al., 1992; Desjardins et al., 1994a, 1994b; Desjardins, 1995; Andrews, 1995; Beron et al., 1995; Idone et al., 2008). The compartment is thereby acidified, and with concentrated hydrolytic activity becomes a microbicidal environment. Moreover, once the phagosome (PG) matures to a phagolysosome, enzymatic activity allows protein degradation of intracellular pathogens and coupling of small peptides for antigen presentation in the MCH-II context (Harding et al., 1995; Germain, 1995). The correct presentation is vital to orchestrate the immune response. *Leishmania* delays PG maturation and phagolysosomal formation (Burchmore & Barrett, 2001). This delay is associated with lipophosphoglycan (LPG), a component of the complex parasite glycocalyx. It has been proposed that LPG disorganises PVM lipid domains and interferes with recruitment of synaptotagmin V, a regulator of the exocytic pathway, delaying the arrival of cathepsin F and V-Type H^+ pumps to the PVM (Vinet et al., 2009). The consequence is altered PV enzymatic activity and acidification. Furthermore, *Leishmania* sequesters in the interface between the PVM and the parasite MCH-II molecules, altering antigen presentation (Antoine et al., 1998).

The permeability of the PVM depends on whether it interacts with the endocytic pathway of the host cell, which varies with different intracellular pathogens (Meirelles & De Souza, 1983). *Toxoplasma gondii* PV appears to be excluded from the phagosome maturation process within the endocytic pathway (Lingelbach & Joiner, 1998). Its PVM comes from host-cell membranes, plasma membrane, ER and mitochondria, in addition to membranes from specialized parasite organelles that appear to interfere with fusion of the PVM with endocytic components (Lingelbach & Joiner, 1998; Marti et al., 2007). Schwab et al., (1994) have documented bidirectional movement of charged and non-charged molecules of less than 1900 Da between the lumen of the *Toxoplasma* PV and the host-cell cytoplasm. It has been suggested that this permeability could be involved in nutrient transport, as proposed in *Plasmodium* where ion channel activity of small cationic and anionic molecules was recorded (Desai et al., 1993; Desai & Rosenberg, 1997). However, *Chlamydia trachomatis* PVM, located to the same level in the endocytic pathway as that of *Toxoplasma*, is able to exclude anionic molecules of low molecular weight (Heinzen & Hackstadt, 1997), suggesting differences in the transporter repertoire. The molecules described in the PVM of *T. gondii* and *Plasmodium* are poorly selective, similar to porin molecules. In the case of *Trypanosoma cruzi* such molecules are implicated in the lysis of the PVM that sets the parasite free into the host cell cytoplasm (Andrews, 1990). Porin-like molecules that induce ion currents (Noronha et al., 2000) have been implicated in *Leishmania* exit from the macrophage (Horta, 1997; Aleida-Campos & Horta, 2000).

The *Leishmania* PV matures in the endosomal pathway to a late endo-lysosomal compartment and has features such as acidic pH, acid phosphatase and hydrolytic activity, and late endosomal and lysosomal membrane markers (Shepherd et al., 1983; Rabinovitch et al., 1985; Prina et al., 1990; Russell et al., 1992; Russell, 1995; Antoine et al., 1998). Despite this apparent hostility, the amastigote is adapted to the pH, and the PV could also be seen as a land of milk and honey since this compartment is rich in products from the degradation of sugars, proteins and nucleic acids (Prina et al., 1990; reviewed by Burchmore & Barrett, 2001).

The acidification of the PV is assumed to be the result of the V-Type H^+ pumps in charge of endosomal acidification (Sturgill-Koszycki et al., 1994; Lamb et al., 2009), with a contribution from the parasite metabolism. The pump function may be associated with cationic as well as anionic shunt currents (Harvey & Wieczorek, 1997; Grabe & Oster, 2001; Haggie et al., 2007; Carraro-Lacroix et al., 2009; Wienert et al., 2010; Steinberg et al., 2010; Dong et al., 2010) for proper function. In macrophages the presence of the V-Type H^+ pump is also vital for efficient phagocytosis (Gagnon et al., 2002) and macrophage fusion into giant cells (McNally & Anderson, 2005).

Leishmania PVM permeability depends on transporters present in the donating membranes: the macrophage plasma membrane (Antoine et al., 1998, Quintana et al., 2009), the endosomal membranes (Veras et al., 1992, 1994, 1995, 1996; Russell et al., 1992; Collins et al., 1997; Schaible et al., 1999; Cortázar et al., 2006), the parasite membrane (Henriques et al., 2003), membrane transporters from the endoplasmic reticulum (Gagnon et al., 2002; Becker et al., 2005; McNally & Anderson, 2005), and transporters from compartments that exchange membrane with the PVM, endosomes and autophagic vacuoles (Schaible et al., 1999). Thus, the possible membrane transporters present on the PVM include macrophage plasma membrane ion pumps and channels where K^+ channels (Randriamampita & Trautmann, 1987; Gallin, 1991; Nelson et al., 1992; Ichinose et al., 1992; McKinney & Gallin, 1992; Holevinsky & Nelson, 1995; DeCoursey et al., 1996; Eder et al., 1997; Forero et al., 1999; Scott et al., 2003; Vicente et al., 2003; Mackenzie et al., 2005; Vicente et al., 2006; Park et al., 2006; Hanley et al., 2004; Scheel et al., 2005; Camacho et al., 2008; Villalonga et al., 2010) and transporters of the ABC family (Di et al., 2006) are important. Transporters present in the macrophage intracellular membranes that could contribute to the PV are V-Type H^+ pumps (Lamb et al., 2009), ionic channels (Dong et al., 2010), H^+ channels (Grabe & Hoster, 2001), iron Nramp transporters (Hackam, 1998; Jabado, 2000; Gomez et al., 2007), voltage gated chloride ClC channel/H^+ transporters (Jentsch, 2007; Steinberg et al., 2010; Wienert et al., 2010), and ABC transporters (Russell et al., 1992; Cortázar et al., 2006; Di et al., 2006), not to mention the repertoire of ion channels and pumps present on the ER membrane. The conditions on the PVM in terms of membrane potential, pH and electrochemical driving force are different for transporters coming from the macrophage membrane, but relatively similar for those coming from intracellular membranes, suggesting that many of the latter are functional.

Several transporters have been found in the *Leishmania* PVM. The acidification of the compartment is associated with the activity of a V-Type H^+ pump (Sturgill-Koszycki et al., 1994). The distribution of anionic fluorescent probes of different sizes into *in situ* and isolated PVs suggests the expression of transporters of the ABC superfamily (Steinberg et al., 1987; Russell et al., 1992; Cortázar et al., 2006), which could be involved in nutrient transport or could function as anionic shunts. There are also Narmp1 iron transporters that in some mouse models compromise *Leishmania* survival (Huynh & Andrews, 2008).

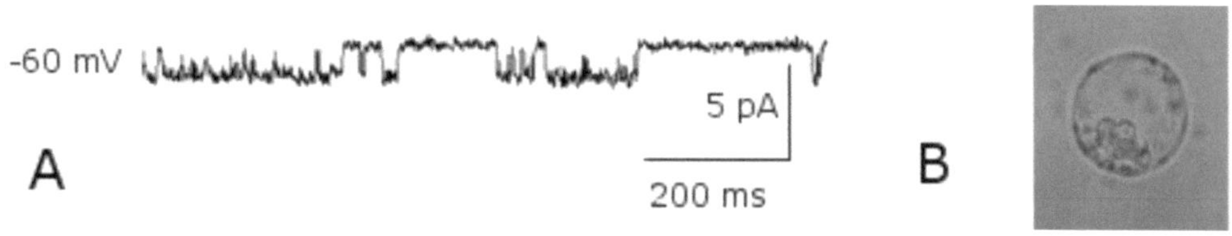

Fig. 3. Patch clamp recording on the parasitophorous vacuole membrane of *Leishmania amazonensis*. A. Inside-out single channel recording in symmetric solutions consisting of 150 mM KCl, 100 μM $CaCl_2$ and 20 mM HEPES, pH 7.2 in the presence of 50 μM DIDS. B. Light microscopy of an isolated PV. Note the parasites polarized to the membrane (Perez et al., 2009).

We have recorded ion currents through the PVM (Cortázar et al., 2006; Perez et al., 2009) in the giant PV induced by *L. amazonensis* in J774.A1 using the PV-attached configuration of the patch clamp technique (Hamill et al., 1981). Isolation of the PV can be achieved by rupturing macrophages with hypotonic solutions and mechanical force (Chakraborty et al., 1994). We substituted a hypotonic solution for an isotonic one in the presence of protein inhibitors and mechanical force (Cortázar et al., 2006), because acidic pH and probenecid-sensitive Lucifer yellow load were better maintained. Though PV purification can be achieved with either sucrose (Chakraborty et al., 1994) or Percoll gradients (Cortázar et al., 2006), we chose the latter method as in our experience the PV deteriorates in presence of sucrose gradients, most probably due to high osmotic pressure. The purified PV maintains low pH and ABC transport activity (Cortázar et al., 2006), though washout of excess Percoll from the PVM has proved difficult, particularly in patch clamp experiments in which seal formation is compromised. We explored the use of differential centrifugation and more recently Split Flow Lateral Transport Thin Cell Fractionation (SPLITT) techniques to enrich the PV fraction, avoiding the problem of further centrifugation and wash (Perez et al., 2007) and preserving the PVM for electrophysiological recording. We used bath solutions with high Na^+ as well as high K^+, with relatively high calcium, and with high Na^+ in the pipette, and recorded a current with 46 pS conductance in the PV-attached configuration (Cortázar et al., 2006). In high symmetrical K^+ in the PV-attached configuration there are bursts of anion currents induced by hyperpolarized potentials that are sensitive to broad range anion channel inhibitors but not to probenecid (Perez et al., 2009), ruling out the presence of some ABC transporters. In inside-out patches bathed in high K^+ and DIDS 50 μM, cationic single channel activity was recorded that was sensitive to broad range potassium channel blockers (Perez et al., 2009; Figure 3). We are now refining the electrophysiological characterization of these currents before determining the molecular nature of the transporters responsible for them.

4. *Leishmania* plasma membrane

Leishmania adaptation and proliferation during its life cycle demands appropriate regulatory mechanisms for parasite survival. During its life cycle the parasite encounters environmental differences in ion concentrations, osmolarity, pH, temperature and nutrient availability, among others. Adaptation to some of these changes requires the expression of ion channels and transporters, and it has been suggested that the parasite will therefore have higher demands for energy (ter Kulle, 1993).

Leishmania relies on purine transport by specific transporters (Ogbunude & Dzimidi, 1993), has a biopterin transporter capable of also moving folate (Ouellette et al., 2002, Dridi et al., 2010), imports Fe by specific transporters (Huynh et al., 2006; Jacques et al., 2010) to warranty its replication within the PV, and is able to extrude drugs (Ouellette et al., 2001). Of particular importance in *Leishmania* is the presence of a homologue of the mammalian AQP9 aquaglyceroprotein (Figarella et al., 2007) associated with parasite resistance to one of the first line therapies (Gourbal, 2004; Figarella et al., 2007; Maharjan et al., 2008). LmAQP1 has been shown to contribute to parasite water and glycerol transport (Figarella et al., 2007). More recently a family of aquaporins was described in *Leishmania donovani*. They localise intracellularly, have differential expression, and can potentially contribute to parasite osmolarity regulation by transporting water because once transfected into yeast they alter the osmolarity properties of these cells (Biyani et al., 2011).

Molecules involved in nutrient uptake are found in *Trypanosomatidae* (Glaser & Mukkada, 1992; ter Kulle, 1993; Vieira et al., 1996; Tetaud et al., 1994). Transport of D-glucose and D-proline is mediated by H^+ symporters in *Leishmania* (Zilberstein & Dwyer, 1985; Zilberstein, 1993). The transport of 2-deoxy-D-glucose was used as a measure of D-glucose and it was shown that *Leishmania donovani* secondarily transports glucose. Glucose can be concentrated to a level of about 80 times the extracellular concentration by an electrochemical gradient of protons (Zilberstein & Dwyer, 1985). Sucrose is similarly concentrated by a sucrose/H^+ symporter in promastigotes of *L. donovani* (Singh & Mandal, 2011) and forms a potential source of nutrients relevant for this form living in the mosquito gut. Arginine is an essential amino acid for *Leishmania. Leishmania donovani* LdAAP3 permease has highly specific arginine transport activity in *Saccharomyces cerevisiae,* optimal at low pH, and when coupled to GFP localises to the *Leishmania* plasma membrane (Shaked-Mishan et al., 2006). Lysine is an essential amino acid of many eukaryotes including *Leishmania.* Unlike mammalian cells, *Leishmania* has independent permeases for arginine and lysine. LdAAP7 expressed in *S. cerevisiae* transports lysine, but when over-expressed in promastigotes has no impact on the uptake of this amino acid (Inbar et al., 2010). LdAA7 is essential for parasite survival and localises to the promastigote plasma membrane and flagella (Inbar et al., 2010).

Leishmania as other eukaryotes is capable of maintaining lower levels of cytoplasmic Ca^{2+}. The parasite stores this ion in the usual compartments, the ER and mitochondria (Philosoph & Zilberstein, 1989), but also in specialized vesicles, the acidocalcisomes (Lu et al., 1997), which are important during parasite invasion, differentiation and replication (Moreno & Docampo, 2003). Evidence suggests that calmodulin (Moreno & Docampo, 2003) and calcium ATPases are present in the ER membrane (Philosoph & Zilberstein, 1989) and in the *Leishmania* plasma membrane (Mandal et al., 1997; Corte-Real et al., 1995), where they will pump calcium out of the parasite cytoplasm. In addition to controlling calcium, *Leishmania* controls volume and osmolarity. The parasite transits from the insect gut, where fluctuations in osmolarity are expected, to the PV, which is believed to be hypo-osmotic compared to the macrophage cytoplasm (LeFurgey et al., 2005). In hypotonic solutions, water enters *Leishmania,* increasing its volume, which is later restored to normal levels. The recovery is associated with extrusion of ions, saccharides and amino acids (Darling et al., 1990;

LeFurgey et al., 2005), as well as with minor changes of K^+ and efflux of alanine through a DIDS sensitive transporter that alters Vm (Vieira et al., 1996). This latter transporter possesses ion channel kinetics (Vieria et al., 1996) and is regulated by protein kinases A and C and arachidonic acid (Vieira et al., 1997). In addition to acidocalcisomes, other intracellular organelles may be involved, by moving Na^+ and Cl^- (LeFurgey et al., 2001). *Leishmania* also maintains a large pool of free intracellular amino acids that appears to help the parasite to respond to osmolarity changes (Shaked-Mishan et al., 2006).

Ion and nutrient transport are fundamental in all cells for function and survival. Cells build up ion gradients by investing energy, generating an electrochemical force that when dissipated allows secondary transport of ions, sugars and amino acids and indirect regulation of volume, pH and osmolarity. In eukaryotic cells the pumps that acidify in the endocytic pathway are V-Type (Lamb et al., 2009). H^+ concentration induces a proton motive force that opposes further acidification. Cationic and anionic shunt currents present in the endocytic membranes counteract the effect of this force on pump function (Harvey & Wieczorek, 1997; Graben & Oster, 2001; Scheel et al., 2005; Carraro-Lacroix et al., 2009; Wienert et al., 2010; Steinberg et al., 2010; Dong et al., 2010). Recent evidence suggests that anion entry may not be required for maintenance of the electromotive force for pump functioning and, surprisingly, that the H^+ gradient may serve to concentrate Cl^- that might be important in the progression in the endocytic pathway (Novarino et al., 2010). The maintenance of intracellular pH within the physiological range is important in this parasite, particularly when adapting to the low pH of the PV (Antoine et al., 1990). The acidification of the PV results from V-Type H^+ pump activity (Sturgill-Koszycki et al., 1994) coupled to anionic or cationic currents. In macrophages, the currents generated by chloride channels CFRT (Di et al., 2006) and ClC/H^+ transporters (Steinberg et al., 2010) are the candidate anion current shunts.

Leishmania maintains an intracellular pH that ranges between 6.4 and 6.8 (Vieira et al., 1994; Marchesini & Docampo, 2002) independent of the environmental pH. Amastigotes show tighter pH regulation compared to promastigotes (Marchesini & Docampo, 2002). In promastigotes, the insect form, a proton electrochemical gradient drives secondary transport of D-glucose and D-proline (Zilberstein & Dwyer, 1985); this gradient is built up by the activity of P-type K^+/H^+ pumps that are Mg^{2+} dependent and orthovanadate sensitive (Jiang et al., 1994; Mukherjee et al., 2001; Burchmore & Barrett, 2001). This pump indirectly effects pH regulation by keeping the chemiosmotic energy constant (Zilberstein, 1993). Two P-Type H^+ pumps have been cloned with differential expression between amastigotes and promastigotes (Meade et al., 1987), one of which is expressed on the parasite plasma membrane (Anderson & Mukkada, 1994; Marchesini & Docampo, 2002). Functional complementation of *S. cerevisiae* with the putative P-Type H^+ pumps LDH1A and LDH1B induces an electrochemical gradient and allows survival of this yeast at low pH (Grigore & Meade, 2006).

Other pumps and transporters are also found in *Leishmania* membranes. A different H^+ pump has been suggested in *Leishmania* pH regulation. The use of DCCD has implicated H^+ pump activity in parasite pH control. It has been suggested that, in *Leishmania major* promastigotes, a H^+ pump and a Cl^- channel contribute to acid secretion and to the maintenance of a hyperpolarized membrane potential (Vieira et al., 1994; Vieira et al., 1995). These authors refer to the inhibition of a H^+ pump by the use of DCCD, which is more compatible with the V-PP-H^+ pump (vacuolar proton-pumping pyrophosphatase) described in plants and protozoa. The V-PP-H^+ pump activity of digitonin-permeabilized *L. donovani* promastigotes is K^+ dependent, and inhibited by Na^+, DCCD and N-ethylmaleimide, but not by vanadate (a P-Type H^+ pump inhibitor) or bafilomycin-A (a V-Type H^+ pump inhibitor). The pump localises to subfractions of electron-dense organelles with contents similar to acidocalcisomes, and it is suggested that it acts to degrade cytosolic pyrophosphatase (Rodrigues et al., 1999). Docampos´ group later described, in amastigotes of *L. amazonensis*, a H^+ pump sensitive to DCCD and N-ethylmaleimide but not to bafilomycin-A; based on its apparent K^+ independence, they concluded it has P-Type H^+ pump activity. The positive labelling of the parasite plasma membrane with antibodies against the TcHA2 of *Trypanosoma*, that recognises LDH1 in promastigotes and amastigotes, suggests that the H^+ pump seen by them is the P-Type H^+ pump described by Zilberstein´s group and cloned by Meade's group, but the activity reported corresponds to the V-PP-H^+ pump described by Rodrigues et al., 1999. Besides the two type pumps, a Na^+/K^+ pump has been suggested on the *Leishmania mexicana* plasma membrane based on ouabain sensitivity (Felibertt et al., 1995). Also P-type K^+ and a Na^+ pumps (Stiles et al., 2003), Na^+/H^+ exchanger (Vercesi et al., 2000) and Mg^{2+}/H^+ pump activities were found in *L. donovani.* The antiporter accumulates H^+ in everted vesicles in the presence of ATP and Mg^{2+}, releases it when exposed to FCCF, and is unable to transport Rb^+, therefore implicating Mg^{2+} as the co-transported ion (Mukherjee et al., 2001).

Leishmania Vm has been measured in promastigotes and amastigotes. Very hyperpolarized Vm values are found (-113 mV; Vieira et al., 1995; Marchesini & Docampo, 2002) after bisoxonol distribution in promastigotes, and less hyperpolarized Vm are found in amastigotes (-75 mV; Marchesini & Docampo, 2002). The activity of DCCD on the electrochemical gradient created by H^+ pumps coupled to DIDS sensitive anion transporters have been implicated in the promastigote Vm (Vieira et al., 1995; Marchesini & Docampo 2002) and Vieira et al., (1995), concluded that there is no contribution from K^+ or Na^+ to the *Leishmania* Vm, despite previous data where distribution of tetraphenylphosphonium bromide points to K^+ (Glaser et al., 1992) as in other eukaryotes. Again the presumed DCCD-sensitive H^+ pump activity was shown in intracellular compartments of *Leishmania* promastigotes, but the proposed model couples its function to DIDS sensitive transporters on the parasite plasma membrane.

Direct measure of an anion conductance was recorded in plasma membranes of *Leishmania mexicana* reconstituted in lipid planar bilayers (DiFranco et al., 1995). These authors described two anionic currents but suggested the presence of other channels. Their finding is consistent with *Leishmania* anion fluxes involved in parasite survival (Ponte-Sucre et al., 1998) and electron transport by redox enzymes (Bera et al., 2005). We have been interested in the study of ion channels of *Leishmania,* with a particular emphasis on chloride channels, assuming that these molecules are relevant in parasite survival and are an adaptation to the acidic pH of the PV. Direct recordings on the parasite have not been possible due to its small size, shape, movement, and the presence of a complex glycocalyx on its surface that interferes with seal formation. To overcome this difficulty we have used amphibian oocytes to express total mRNA from *Leishmania.* In combination with the model of *Xenopus* oocytes (Stühmer, 1992) we have shown that other species are capable of expressing ion channels efficiently (*Bufo marinus*). However, we have encountered some difficulties in both models in particular low oocyte production and viability compared to other reports (Chaves et al.,

2003). We attribute these to reduced partial oxygen pressure due to the altitude of Bogotá (2600 m above sea level) where the laboratory is located, because control of temperature and hormone replacement did not improve production or oocyte viability (Arroyo & Camacho, 2006). Despite these difficulties it is possible to record the expected potassium currents after injection of cRNA of the ion channel Kv1.1 (Vargas et al., 2004).

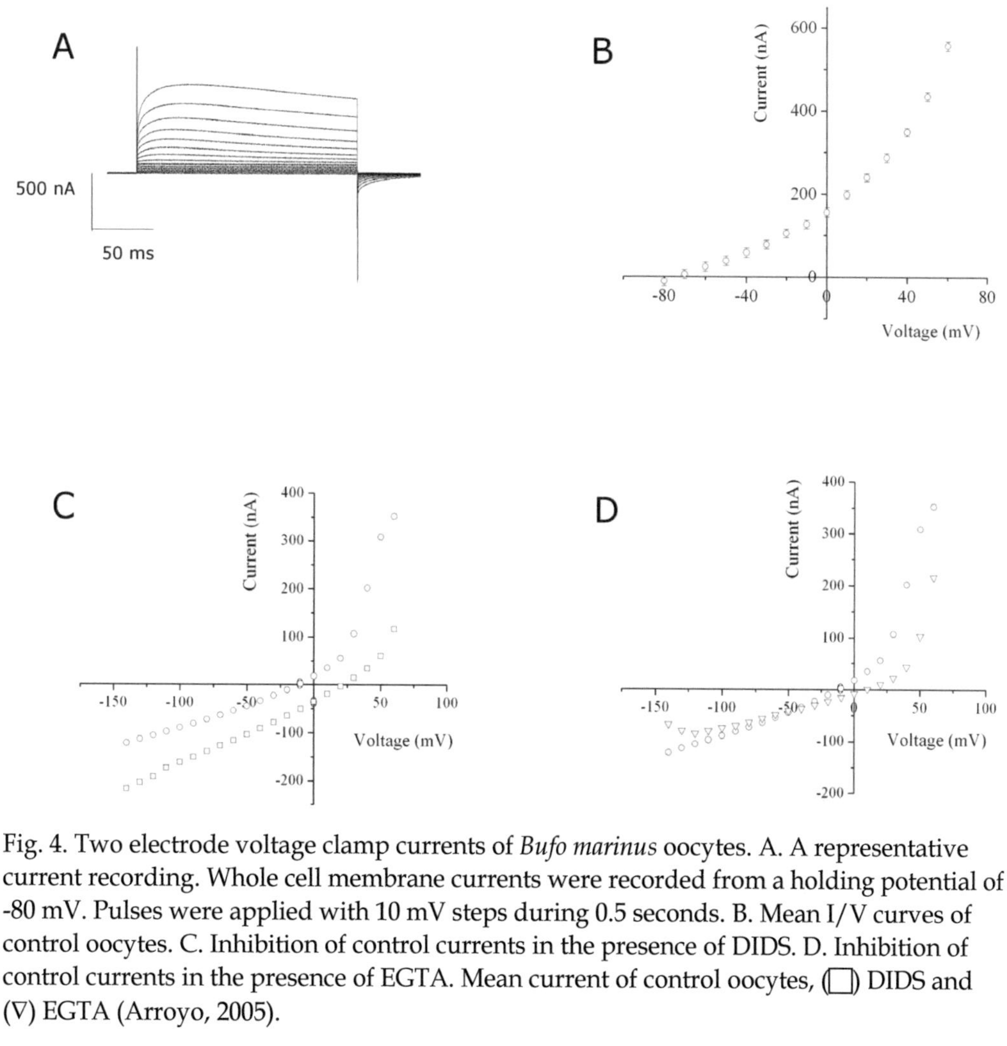

Fig. 4. Two electrode voltage clamp currents of *Bufo marinus* oocytes. A. A representative current recording. Whole cell membrane currents were recorded from a holding potential of -80 mV. Pulses were applied with 10 mV steps during 0.5 seconds. B. Mean I/V curves of control oocytes. C. Inhibition of control currents in the presence of DIDS. D. Inhibition of control currents in the presence of EGTA. Mean current of control oocytes, (□) DIDS and (∇) EGTA (Arroyo, 2005).

B. marinus can be kept in captivity at intermediate altitudes (e.g. in Bogotá) in controlled temperature environments and 12 hour light cycles. They can be fed with insects or earthworms, and oocyte production can be induced with gonadotropic hormone (Arroyo & Camacho, 2006). The basal oocyte ion currents of *B. marinus* recorded with two-electrode voltage clamp are mainly due to chloride. The reversal potential of the currents is close to the chloride equilibrium potential, which becomes more negative in chloride-free medium, suggesting that potassium permeability is also important (Figure 4). The currents are sensitive to broad anion inhibitors of these channels such as DIDS, which caused the inhibition (Figure 4C). These currents are also calcium dependent because recording in the presence of the chelator EGTA reduces their current amplitude (Figure 4D). The electrophysiological properties found resemble those of the currents described for *X. laevis* oocytes (Miledi et al., 1989).

Using *X. laevis* and *B. marinus* oocytes we have studied the ion currents induced after injection of polyA mRNA of *L. amazonensis* (Arroyo, 2005; Lagos et al., 2007) and *Leishmania braziliensis* (Garzon et al., 2009). We recorded three types of chloride currents, one of which (Type 3) appears to be specific to parasite mRNA injection. Type 3 currents were characterized by a slow activation to a stationary level, without any inactivation. Reduction of extracellular Cl^- resulted in an altered current, indicating that the current was mediated by chloride ions. Substitution of extracellular Cl^- by other halogens indicated a relative anion permeability sequence $I^- > Br^- > Cl^-$. Negative results with the broad spectrum K^+ channel inhibitors TEA and 4-AP eliminated the possibility of contribution of K^+ currents. Conversely, these currents showed sensitivity to the broad spectrum Cl^- channel blockers DIDS and niflumic acid; one-third of the current was blocked by these two inhibitors. Reduction of extracellular divalent cations using a nominally Ca^{2+}-free solution with 1 mM EGTA resulted in only a 10% decrease of the amplitude of this current (Lagos et al., 2007). There are several classes of chloride channels and three molecularly distinct chloride channel families (ClC, CFTR and ligand-gated GABA and glycine receptors; Jentsch et al., 2005). In animal cells, the activity of V-Type H^+ pumps in endosomal compartments is associated with ClC chloride channel activity.

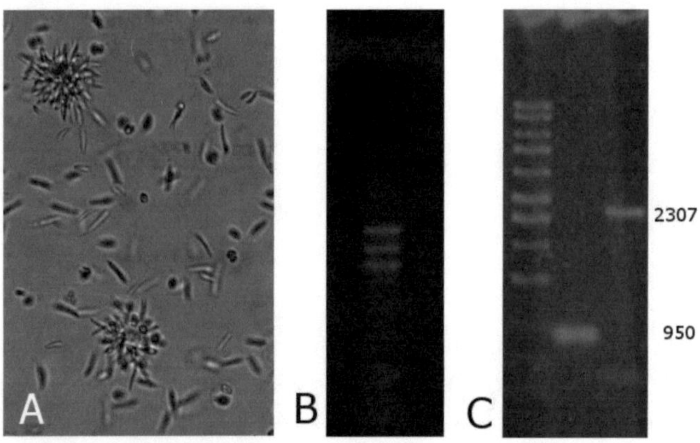

Fig. 5. Transcription of the putative chloride channel LbrM01 V2.0210. A. Promastigotes of *Leishmania braziliensis*. Note typical rosettes. B. Total RNA of *Leishmania*. A triple band of ribosomal RNA is found and is typical of this parasite. Agarose gel of a partial and total product of the putative ClC channel (Lozano et al., 2009).

The *Leishmania* amastigote inside the PV is constantly flooded by protons, and ClC channel/H^+ transporter expression will contribute to efflux of H^+, alleviating pH changes, and may concentrate Cl^- or other relevant anions. With the information available on the *L. braziliensis* genome we have identified four potential putative genes for chloride channels, one of which is classified as ClC. We have recently cloned the putative ClC channel of *L. braziliensis* (Lozano et al., 2009; Figure 5) which is homologous to intracellular mammalian ClC channels, some of which are co-transporters of Cl^-/H^+. We have inserted this putative channel into an expression vector and recorded HEK293 cells with patch clamp in the whole-cell configuration after transfection of LbrM01 V2.0210. A Cl^- current 10 times the amplitude of control currents was recorded (Lozano et al., 2011). We are further characterising this channel and hope to understand its importance for *Leishmania* physiology.

5. Acknowledgement

The author thanks the Departamento de Biología, Facultad de Ciencias, Universidad Nacional de Colombia, Sede Bogotá, and the Laboratorio de Biofísica, Centro Internacional de Física; the Colombian agency Colciencias (projects 222852129302, 222851928951 and 2228519289191); and the División de Investigación de Bogotá, DIB, Universidad Nacional de Colombia, Sede Bogotá. The author wants to thank Michel Delay, INDEC Systems, Inc., who kindly edited this manuscript.

6. References

Aga E, Katschinski DM, van Zandbergen G, Laufs H, Hansen B, Müller K, Solbach W, Laskay T. Inhibition of the spontaneous apoptosis of neutrophil granulocytes by the intracellular parasite Leishmania major. J Immunol. 2002;169(2):898-905.

Almeida-Campos FR, Horta MF. Proteolytic activation of leishporin: evidence that Leishmania amazonensis and Leishmania guyanensis have distinct inactive forms. Mol Biochem Parasitol. 2000;111(2):363-375.

Anderson SA, Mukkada AJ. Biochemical and immunochemical characterization of a P-type ATPase from Leishmania donovani promastigote plasma membrane. Biochim Biophys Acta. 1994 ;1195(1):71-80.

Andrews NW. The acid-active hemolysin of Trypanosoma cruzi. Exp Parasitol. 1990;71(2):241-4.

Andrews NW. Lysosome recruitment during host cell invasion by *Trypanosoma cruzi*. Trends Cell Biol. 1995;5:133-137

Antoine JC, Prina E, Jouanne C, Bongrand P. Parasitophorous vacuoles of Leishmania amazonensis-infected macrophages maintain an acidic pH. Infect Immun. 1990;58(3):779-787.

Antoine JC, Prina E, Lang T, Courret N. The biogenesis and properties of the parasitophorous vacuoles that harbour Leishmania in murine macrophages. Trends Microbiol. 1998;6(10):392-401.

Arroyo R. Expresión heteróloga de ARNm poli(A)+ de *Leishmania* (*Trypanosomatidae*) en ovocitos de anfibio (*Bufonidae*), Departamento de Biología, Facultad de Ciencias, Universidad Nacional de Colombia; 2005

Arroyo R, Camacho M. Generación de una colonia de *Chaunus marinus* como fuente de ovocitos para estudios electrofisiológicos. Tema Zoocría. II Congreso colombiano de zoología; 2006:557.

Barbiéri CL, Doine AI, Freymuller E. Lysosomal depletion in macrophages from spleen and foot lesions of Leishmania-infected hamster. Exp Parasitol. 1990;71(2):218-28.

Becker T, Volchuk A, Rothman JE. Differential use of endoplasmic reticulum membrane for phagocytosis in J774 macrophages. Proc Natl Acad Sci U S A. 2005;102(11):4022-6.

Bera T, Lakshman K, Ghanteswari D, Pal S, Sudhahar D, Islam MN, Bhuyan NR, Das P. Characterization of the redox components of transplasma membrane electron transport system from Leishmania donovani promastigotes. Biochim Biophys Acta. 2005;1725(3):314-326.

Beron W, Alvarez-Dominguez C, Mayorga L, Stahl PD. Membrane trafficking along the phagocytic pathway. Trends Cell Biol. 1995;5:100-104

Biyani N, Mandal S, Seth C, Saint M, Natarajan K, Ghosh I, Madhubala R. Characterization of Leishmania donovani Aquaporins Shows Presence of Subcellular Aquaporins Similar to Tonoplast Intrinsic Proteins of Plants. PLoS One. 2011;6(9):e24820.

Buisman HP, Steinberg TH, Fischbarg J, Silverstein SC, Vogelzang SA, Ince C, Ypey DL, Leijh PC. Extracellular ATP induces a large nonselective conductance in macrophage plasma membranes. Proc Natl Acad Sci U S A. 1988;85(21):7988-92.

Burchmore RJ, Barrett MP. Life in vacuoles--nutrient acquisition by Leishmania amastigotes. Int J Parasitol. 2001;31(12):1311-20.

Buchmüller-Rouiller Y, Mauël J. Macrophage activation for intracellular killing as induced by calcium ionophore. Correlation with biologic and biochemical events. J Immunol. 1991;146(1):217-223.

Chakraborty P, Sturgill-Koszycki S, Russell DG. Isolation and characterization of pathogen-containing phagosomes. Methods Cell Biol. 1994;45:261-276.

Chaves H, Villalba C, Lagos L, Vargas R, Martínez-Wittinghan F, Clavijo C, Camacho M. Expresión de canales de potasio voltaje dependientes en ovocitos de Xenopus laevis (amphibia) Acta biol Colomb. 2003;8(1):59-67.

Clavijo LM, Forero ME, Matta NE, Camacho M. Impacto de la infección de *L. (V.) braziliensis* sobre apoptosis y propiedades eléctricas de macrófagos. Biomedica. 2009;29(S)351.

Camacho M, Forero ME, Fajardo C, Niño A, Morales P, Campos H. Leishmania amazonensis infection may affect the ability of the host macrophage to be activated by altering their outward potassium currents. Exp Parasitol. 2008;120(1):50-6.

Carraro-Lacroix LR, Lessa LM, Fernandez R, Malnic G. Physiological implications of the regulation of vacuolar H+-ATPase by chloride ions. Braz J Med Biol Res. 2009;42(2):155-163.

Collins HL, Schaible UE, Ernst JD, Russell DG. Transfer of phagocytosed particles to the parasitophorous vacuole of Leishmania mexicana is a transient phenomenon preceding the acquisition of annexin I by the phagosome. J Cell Sci. 1997;110 (Pt 2):191-200.

Cortázar TM, Hernández J, Echeverry MC, Camacho M. Role of the parasitophorous vacuole of murine macrophages infected with Leishmania amazonensis in molecule acquisition. Biomedica. 2006;26;S1:26-37.

Corte-Real S, Santos CB, Meirelles MN. Differential expression of the plasma membrane Mg2+ ATPase and Ca2+ ATPase activity during adhesion and interiorization of Leishmania amazonensis in fibroblasts in vitro. J Submicrosc Cytol Pathol. 1995;27(3):359-366.

Cui W, Ke JZ, Zhang Q, Ke HZ, Chalouni C, Vignery A.The intracellular domain of CD44 promotes the fusion of macrophages. Blood. 2006;107(2):796-805.

Dallaporta B, Marchetti P, de Pablo MA, Maisse C, Duc HT, Métivier D, Zamzami N, Geuskens M, Kroemer G. Plasma membrane potential in thymocyte apoptosis. J Immunol. 1999;162(11):6534-65G42.

DeCoursey TE, Kim SY, Silver MR, Quandt FN. Ion channel expression in PMA-differentiated human THP-1 macrophages. J Membr Biol. 1996;152(2):141-57.

Darling T N, Burrows CM, Blum JJ. Rapid shape change and release of ninhydrin-positive substances by Leishmania major promastigotes in response to hypo-osmotic stress. J Protozool. 1990;37:493-499.

Desai SA, Krogstad DJ, McCleskey EW. A nutrient-permeable channel on the intraerythrocytic malaria parasite. Nature. 1993;362(6421):643-646.

Desai SA, Rosenberg RL. Pore size of the malaria parasite's nutrient channel. Proc Natl Acad Sci U S A. 1997;94(5):2045-2049.

Desjardins M, Celis JE, van Meer G, Dieplinger H, Jahraus A, Griffiths G, Huber LA. Molecular characterization of phagosomes. J Biol Chem. 1994a;269(51):32194-32200.

Desjardins M, Huber LA, Parton RG, Griffiths G. Biogenesis of phagolysosomes proceeds through a sequential series of interactions with the endocytic apparatus. J Cell Biol. 1994b;124(5):677-688.

Desjardins M. Biogenesis of phagolysosomes: the 'kiss and run' hypothesis.Trends Cell Biol. 1995;5(5):183-186.

Desjardins M, Griffiths G. Phagocytosis: latex leads the way. Curr Opin Cell Biol. 2003;15(4):498-503.

Di A, Brown ME, Deriy LV, Li C, Szeto FL, Chen Y, Huang P, Tong J, Naren AP, Bindokas V, Palfrey HC, Nelson DJ. CFTR regulates phagosome acidification in macrophages and alters bactericidal activity. Nat Cell Biol. 2006;8(9):933-944.

Dridi L, Ahmed Ouameur A, Ouellette M. High affinity S-Adenosylmethionine plasma membrane transporter of Leishmania is a member of the folate biopterin transporter (FBT) family. J Biol Chem. 2010;285(26):19767-19775.

DiFranco M, Villarroel A, Ponte-Sucre A, Quinonez M, Drujan D, Dagger F. Incorporation of ion channels from the plasma membrane of Leishmania mexicana into planar bilayers. Acta Cient Venez. 1995;46(3):206-207.

Dong XP, Wang X, Xu H. TRP channels of intracellular membranes. J Neurochem. 2010;113(2):313-328.

Eder C, Klee R, Heinemann U. Pharmacological properties of Ca2+-activated K+ currents of ramified murine brain macrophages. Naunyn Schmiedebergs Arch Pharmacol. 1997;356(2):233-239.

Eilam Y, El-On J, Spira DT. Leishmania major excreted factor, calcium ions and the survival of amastigotes. Exp Parasitol. 1985;59(2):161-168.

Fajardo C, Villota Y, Forero ME, Camacho M. La infección por *Leishmania* promueve desactivación del macrófago y esto se ve reflejado en las propiedades eléctricas de la célula hospedera. Biomedica. 2007; 27(S2):146.

Felibertt P, Bermúdez R, Cervino V, Dawidowicz K, Dagger F, Proverbio T, Marín R, Benaim G. Ouabain-sensitive Na+,K(+)-ATPase in the plasma membrane of Leishmania mexicana. Mol Biochem Parasitol. 1995;74(2):179-87.

Figarella K, Uzcategui NL, Zhou Y, LeFurgey A, Ouellette M, Bhattacharjee H, Mukhopadhyay R. Biochemical characterization of Leishmania major aquaglyceroporin LmAQP1: possible role in volume regulation and osmotaxis. Mol Microbiol. 2007;65(4):1006-1017.

Fischer HG, Eder C. Voltage-gated K+ currents of mouse dendritic cells. FEBS Lett. 1995;373(2):127-30.

Forero ME, Marín M, Corrales A, Llano I, Moreno H, Camacho M. Leishmania amazonensis infection induces changes in the electrophysiological properties of macrophage-like cells. J Membr Biol. 1999;170(2):173-180.

Gagnon E, Duclos S, Rondeau C, Chevet E, Cameron PH, Steele-Mortimer O, Paiement J, Bergeron JJ, Desjardins M. Endoplasmic reticulum-mediated phagocytosis is a mechanism of entry into macrophages. Cell. 2002;110(1):119-131.

Gallin EK. Ion channels in leukocytes. Physiol Rev. 1991;71(3):775-811.

Gallin EK, Gallin JI. Interaction of chemotactic factors with human macrophages. Induction of transmembrane potential changes. J Cell Biol. 1977;75(1):277-89.

Gallin EK, Livengood DR. Demonstration of an electrogenic Na+-K+ pump in mouse spleen macrophages. Am J Physiol. 1983;245(3):C184-8.

Gallin EK, Sheehy PA. Differential expression of inward and outward potassium currents in the macrophage-like cell line J774.1. J Physiol. 1985;369:475-479.

Gamaley IA, Kirpichnikova KM, Klyubin IV. Superoxide release is involved in membrane potential changes in mouse peritoneal macrophages. Free Radic Biol Med. 1998;24(1):168-174.

Garzón C, Stuhmer W, Camacho M Corrientes aniónicas de Leishmania expresadas en ovocitos de Xenopus laevis luego de la inyección de mRNA. Biomedica. 2009;29(S):177-178.

Germain RN. Binding domain regulation of MHC class II molecule assembly, trafficking, fate, and function. Semin Immunol. 1995;7(6):361-372.

Gendelman HE, Ding S, Gong N, Liu J, Ramirez SH, Persidsky Y, Mosley RL, Wang T, Volsky DJ, Xiong H. Monocyte chemotactic protein-1 regulates voltage-gated K+ channels and macrophage transmigration. J Neuroimmune Pharmacol. 2009;4(1):47-59.

Ginsburg H, Stein WD. How many functional transport pathways does Plasmodium falciparum induce in the membrane of its host erythrocyte? Trends Parasitol. 2005;21(3):118-121.

Glaser TA, Mukkada AJ. Proline transport in Leishmania donovani amastigotes: dependence on pH gradients and membrane potential. Mol Biochem Parasitol. 1992;51(1):1-8.

Gomez MA, Li S, Tremblay ML, and Olivier M. NRAMP-1 Expression Modulates Protein-tyrosine Phosphatase Activity in Macrophages Impact on host cell signaling and functions. The J Biol Chem. 2007;282;(50):36190-36198

González U, Pinart M, Rengifo-Pardo M, Macaya A, Alvar J, Tweed JA. Interventions for American cutaneous and mucocutaneous leishmaniasis. Cochrane Database Syst Rev. 2009;(2):CD004834

Gordon S. Alternative activation of macrophages. Nat Rev Immunol. 2003;3:23-35.

Grabe M, Oster G. Regulation of organelle acidity. J Gen Physiol. 2001;117(4):329-344.

Grigore D, Meade JC. A COOH-terminal domain regulates the activity of Leishmania proton pumps LDH1A and LDH1B. Int J Parasitol. 2006;36(4):381-393.

Gourbal B, Sonuc N, Bhattacharjee H, Legare D, Sundar S, Ouellette M, Rosen BP, Mukhopadhyay R. Drug uptake and modulation of drug resistance in Leishmania by an aquaglyceroporin. J Biol Chem. 2004;279(30):31010-31017.

Hackam DJ, Rotstein OD, Zhang W, Gruenheid S, Gros P, and Grinstein S. Host Resistance to Intracellular Infection: Mutation of Natural Resistance-associated Macrophage

Protein 1(Nramp1) Impairs Phagosomal Acidification. J Exp Med. 1998;188(2):351-364.

Hamill OP, Marty A, Neher E, Sakmann B, Sigworth FJ. Improved patch-clamp techniques for high resolution current recording from cells and cell-free membrane patches. Pflügers Archv Eur J Physiol. 1981;391:85-100.

Hanley PJ, Musset B, Renigunta V, Limberg SH, Dalpke AH, Sus R, Heeg KM, Preisig-Müller R, Daut J. Extracellular ATP induces oscillations of intracellular Ca2+ and membrane potential and promotes transcription of IL-6 in macrophages. Proc Natl Acad Sci U S A. 2004;101(25):9479-9484.

Hara N, Ichinose M, Sawada M, Imai K, Maeno T. Activation of single Ca2(+)-dependent K+ channel by external ATP in mouse macrophages. FEBS Lett. 1990;267(2):281-284.

Haggie PM, Verkman AS. Cystic fibrosis transmembrane conductance regulator-independent phagosomal acidification in macrophages. J Biol Chem. 2007;282(43):31422-31428.

Harding CV. Phagocytic processing of antigens for presentation by MHC molecules. Trends Cell Biol. 1995;5(3):105-109.

Harvey WR, Wieczorekh. Animal plasma membrane energization by chemiosmotic H+ V-ATPases. J Exp Biol. 1997;200:203-216.

Hattori T, Kajikuri J, Katsuya H, Itoh T. Effects of H2O2 on membrane potential of smooth muscle cells in rabbit mesenteric resistance artery. Eur J Pharmacol. 2003;464(2-3):101-109.

Heinzen RA, Hackstadt T. The Chlamydia trachomatis parasitophorous vacuolar membrane is not passively permeable to low-molecular-weight compounds. Infect Immun. 1997;65(3):1088-1094.

Henriques C, Atella GC, Bonilha VL, de Souza W. Biochemical analysis of proteins and lipids found in parasitophorous vacuoles containing Leishmania amazonensis. Parasitol Res. 2003;89:123-133.

Holevinsky KO, Jow F, Nelson DJ. Elevation in intracellular calcium activates both chloride and proton currents in human macrophages. J Membr Biol. 1994;140(1):13-30.

Holevinsky KO, Nelson DJ. Simultaneous detection of free radical release and membrane current during phagocytosis. J Biol Chem. 1995;270(14):8328-8336.

Holevinsky KO, Nelson DJ. Membrane capacitance changes associated with particle uptake during phagocytosis in macrophages. Biophys J. 1998;75(5):2577-2586.

Hoyos M, Niño A, Camargo M, Díaz JC, León S, Camacho M. Separation of Leishmania-infected macrophages by step-SPLITT fractionation. J Chromatogr B Analyt Technol Biomed Life Sci. 2009;877(29):3712-3718.

Horta MF. Pore-forming proteins in pathogenic protozoan parasites. Trends Microbiol. 1997;5(9):363-366.

Huynh C, Sacks DL, Andrews NW. A Leishmania amazonensis ZIP family iron transporter is essential for parasite replication within macrophage phagolysosomes. J Exp Med. 2006;203(10):2363-2375.

Huynh C, Andrews NW. Iron acquisition within host cells and the pathogenicity of Leishmania. Cell Microbiol. 2008;10(2):293-300.

Ichinose M, Hara N, Sawada M, Maeno T. Activation of K+ current in macrophages by platelet activating factor. Biochem Biophys Res Commun. 1992;182(1):372-378.

Idone V, Tam C, Andrews NW. Two-way traffic on the road to plasma membrane repair. Trends Cell Biol. 2008;18(11):552-559.

Inbar E, Canepa GE, Carrillo C, Glaser F, Suter Grotemeyer M, Rentsch D, Zilberstein D, Pereira CA. Lysine transporters in human trypanosomatid pathogens. Amino Acids. 2010. [Epub ahead of print]

Informe epidemiológico nacional 2009, Instituto Nacional de Salud, Bogotá, Colombia

Jabado N, Jankowski A, Dougaparsad S, Picard V, Grinstein S, and Gros P. Natural Resistance to Intracellular Infections: Natural Resistance-associated Macrophage Protein 1(NRAMP1) Functions as a pH-dependent Manganese Transporter at the Phagosomal Membrane. J Exp Med. 2000;192(9): 1237-1247.

Jacques I, Andrews NW, Huynh C. Functional characterization of LIT1, the Leishmania amazonensis ferrous iron transporter. Mol Biochem Parasitol. 2010;170(1):28-36.

Jentsch TJ, Neagoe I, Scheel O. CLC chloride channels and transporters. Curr Opin Neurobiol. 2005 Jun;15(3):319-325.

Jentsch TJ. Chloride and the endosomal-lysosomal pathway: emerging roles of CLC chloride transporters. Physiol. 2007;578(Pt 3):633-640.

Jiang S, Anderson SA, Winget GD, Mukkada AJ. Plasma membrane K+/H(+)-ATPase from Leishmania donovani. J Cell Physiol. 1994;159(1):60-66.

Judge SI, Montcalm-Mazzilli E, Gallin EK. IKir regulation in murine macrophages: whole cell and perforated patch studies. Am J Physiol. 1994; 267(6 Pt 1):C1691-1698.

Lagos M LF, Moran O, Camacho M. Leishmania amazonensis: Anionic currents expressed in oocytes upon microinjection of mRNA from the parasite. Exp Parasitol. 2007;116(2):163-170.

Lamb FS, Moreland JG, Miller FJ Jr. Electrophysiology of reactive oxygen production in signaling endosomes. Antioxid Redox Signal. 2009;11(6):1335-1347.

Liew FY, Xu D, Chan WL. Immune effector mechanism in parasitic infections. Immunol Lett. 1999;65(1-2):101-104.

Liu JH, Bijlenga P, Fischer-Lougheed J, Occhiodoro T, Kaelin A, Bader CR, Bernheim L. Role of an inward rectifier K+ current and of hyperpolarization in human myoblast fusion. J Physiol. 1998;510(Pt2):467-476.

Liu C, Cotten JF, Schuyler JA, Fahlman CS, Au JD, Bickler PE, Yost CS. Protective effects of TASK-3 (KCNK9) and related 2P K channels during cellular stress. Brain Res. 2005;1031(2):164-73.

LeFurgey A, Ingram P, Blum JJ. Compartmental responses to acute osmotic stress in Leishmania major result in rapid loss of Na+ and Cl-. Comp Biochem Physiol A Mol Integr Physiol. 2001;128(2):385-394.

LeFurgey A, Gannon M, Blum J, Ingram P. Leishmania donovani amastigotes mobilize organic and inorganic osmolytes during regulatory volume decrease. J Eukaryot Microbiol. 2005;52(3):277-289.

Lingelbach K, Joiner KA. The parasitophorous vacuole membrane surrounding Plasmodium and Toxoplasma: an unusual compartment in infected cells. J Cell Sci. 1998;111(Pt11):1467-1475.

Link TM, Park U, Vonakis BM, Raben DM, Soloski MJ, Caterina MJ. TRPV2 has a pivotal role in macrophage particle binding and phagocytosis. Nat Immunol. 2010;11(3):232-239.

Lisi S, Sisto M, Acquafredda A, Spinelli R, Schiavone M, Mitolo V, Brandonisio O, Panaro M. Infection with Leishmania infantum Inhibits actinomycin D-induced apoptosis of human monocytic cell line U-937. J Eukaryot Microbiol. 2005;52(3):211-217.

Lozano Y, Gomez C, Posada ML, Camacho M. Canales de cloruro CLC de Leishmania. Biomedica. 2009;29(S):184.

Lozano Y, Posada ML, Camacho M Lozano Jiménez YY, Posada ML, Forero ME, Camacho M. El canal de cloruro putativo *LbrM01_V2.0210* de *Leishmania* genera corrientes de cloruro sensibles a pH y osmolaridad. Biomedica. 2011;31(S3):23-205:106.

Lu HG, Zhong L, Chang KP, Docampo R. Intracellular Ca2+ pool content and signaling and expression of a calcium pump are linked to virulence in Leishmania mexicana amazonesis amastigotes. J Biol Chem. 1997;272(14):9464-9473.

Mackenzie AB, Chirakkal H, North RA. Kv1.3 potassium channels in human alveolar macrophages. Am J Physiol Lung Cell Mol Physiol. 2003;285(4):L262-268.

Meade JC, Shaw J, Lemaster S, Gallagher G, Stringer JR. Structure and expression of a tandem gene pair in Leishmania donovani that encodes a protein structurally homologous to eucaryotic cation-transporting ATPases. Mol Cell Biol. 1987;7(11):3937-3946.

Maharjan M, Singh S, Chatterjee M, Madhubala R. Role of aquaglyceroporin (AQP1) gene and drug uptake in antimony-resistant clinical isolates of Leishmania donovani. Am J Trop Med Hyg. 2008;79(1):69-75.

Mandal D, Mukherjee T, Sarkar S, Majumdar S, Bhaduri A. The plasma-membrane Ca2+-ATPase of Leishmania donovani is an extrusion pump for Ca2+. Biochem J. 1997;322(Pt1):251-257.

Marchesini N, Docampo R- A plasma membrane P-type H(+)-ATPase regulates intracellular pH in Leishmania mexicana amazonensis. Mol Biochem Parasitol. 2002;119(2):225-236.

Martin AM, Liu T, Lynn BC, Sinai AP. The Toxoplasma gondii parasitophorous vacuole membrane: transactions across the border. J Eukaryot Microbiol. 2007;54(1):25-28.

Martin RE, Ginsburg H, Kirk K. Membrane transport proteins of the malaria parasite. Mol Microbiol. 2009;74(3):519-528.

McCann FV, Keller TM, Guyre PM. Ion channels in human macrophages compared with the U-937 cell line. J Membr Biol. 1987;96(1):57-64.

McCann FV, Cole JJ, Guyre PM, Russell JA. Action potentials in macrophages derived from human monocytes. Science. 1983; 219(4587):991-993.

McKinney LC, Gallin EK. Inwardly rectifying whole-cell and single-channel K currents in the murine macrophage cell line J774.1. J Membr Biol. 1988;103(1):41-53.

McKinney LC, Gallin EK. Effect of adherence, cell morphology, and lipopolysaccharide on potassium conductance and passive membrane properties of murine macrophage J774.1 cells. J Membr Biol. 1990;116(1):47-56.

McKinney LC, Gallin EK. G-protein activators induce a potassium conductance in murine macrophages. J Membr Biol. 1992;130(3):265-276.

McNally AK, Anderson JM. Multinucleated giant cell formation exhibits features of phagocytosis with participation of the endoplasmic reticulum. Exp Mol Pathol. 2005;79(2):126-135.

Meirelles MNL, De Souza W. Interaction of lysosomes with endocytic vacuoles in macrophages simultaneously infected with Trypanosoma cruzi and Toxoplasma gondii. J Submicrosc Cytol. 1983;(15):889-896

Miledi R, Parker I, Woodward RM. Membrane currents elicited by divalent cations in Xenopus oocytes. J Physiol. 1989;417:173-195.

Moreno SN, Docampo R. Calcium regulation in protozoan parasites. Curr Opin Microbiol. 2003;6(4):359-364

Mosser DM, Edwards JP. Exploring the full spectrum of macrophage activation. Nat Rev Immunol. 2008;8(12):958-969.

Mukherjee T, Mandal D, Bhaduri A. Leishmania plasma membrane Mg2+-ATPase is a H+/K+-antiporter involved in glucose symport. Studies with sealed ghosts and vesicles of opposite polarity. J Biol Chem. 2001;276(8):5563-5569.

Murray RZ, Wylie FG, Khromykh T, Hume DA, Stow JL. Syntaxin 6 and Vti1b form a novel SNARE complex, which is up-regulated in activated macrophages to facilitate exocytosis of tumor necrosis Factor-alpha.J Biol Chem. 2005a;280(11):10478-10483.

Murray RZ, Kay JG, Sangermani DG, Stow JL. A role for the phagosome in cytokine secretion. Science. 2005b;310(5753):1492-1495.

Nelson DJ, Jow B, Popovich KJ. Whole-cell currents in macrophages: II. Alveolar macrophages. J Membr Biol. 1990;117(1):45-55.

Noronha FS, Cruz JS, Beirão PS, Horta MF. Macrophage damage by Leishmania amazonensis cytolysin: evidence of pore formation on cell membrane. Infect Immun. 2000;68(8):4578-4584.

Novarino G, Weinert S, Rickheit G, Jentsch TJ. Endosomal chloride-proton exchange rather than chloride conductance is crucial for renal endocytosis. Science. 2010;328(5984):1398-1401.

Ogbunude PO, Dzimiri MM. Expression of a channel-like pathway for adenosine transport in Leishmania donovani promastigotes. Int J Parasitol. 1993;23(6):803-807.

Olivier M. Modulation of Host cell Intracellular Ca2+. Parasitol Today. 1996;12(4):145-150.

Ouellette M, Legare D, Papadopoulou B. Multidrug resistance and ABC transporters in parasitic protozoa. J Mol Microbiol Biotechnol. 2001;3(2):201-206.

Ouellette M, Drummelsmith J, El-Fadili A, Kündig C, Richard D, Roy G. Pterin transport and metabolism in Leishmania and related trypanosomatid parasites. Int J Parasitol. 2002;32(4):385-398.

Panyi G, Varga Z, Gáspár R. Ion channels and lymphocyte activation. Immunol Lett. 2004;92(1-2):55-66.

Park SA, Lee YC, Ma TZ, Park JA, Han MK, Lee HH, Kim HG, Kwak YG. hKv1.5 channels play a pivotal role in the functions of human alveolar macrophages. Biochem Biophys Res Commun. 2006(28);346(2):567-571.

Perez C, Hoyos M, Camacho M. Aislamiento de la vacuola parasitófora que contiene al parásito *Leishmania amazonensis* mediante la técnica SPLITT. Biomedica. 2007; 27(S2):144-145.

Perez C Estudio de corrientes iónicas de la membrana de la vacuola parasitófora que contiene al parásito *Leishmania amazonensis.* Masters thesis, Departamento de Química, Facultad de Ciencias, Universidad Nacional de Colombia; 2008.

Perez C, Stuhmer W, Camacho M Canales de cloruro y potasio en la membrana de la vacuola parasitófora de Leishmania amazonensis. Biomedica. 2009;29(S)193-194.

Philosoph H, Zilberstein D. Regulation of intracellular calcium in promastigotes of the human protozoan parasite Leishmania donovani. J Biol Chem. 1989;264(18):10420-10424.

Pitt A, Mayorga LS, Stahl PD, Schwartz AL. Alterations in the protein composition of maturing phagosomes. J Clin Invest. 1992;90(5):1978-1983.

Ponte-Sucre A, Campos Y, Fernandez M, Moll H, Mendoza-León A. Leishmania sp.: growth and survival are impaired by ion channel blockers. Exp Parasitol. 1998;88(1):11-19.

Prina E, Antoine JC, Wiederanders B, Kirschke H. Localization and activity of various lysosomal proteases in Leishmania amazonensis-infected macrophages. Infect Immun. 1990;58(6):1730-1737.

Quintana E, Torres Y, Alvarez C, Rojas A, Forero ME, Camacho M. Changes in macrophage membrane properties during early Leishmania amazonensis infection differ from those observed during established infection and are partially explained by phagocytosis. Exp Parasitol. 2010;124(3):258-264.

Rabinovitch M, Topper G, Cristello P, Rich A. Receptor-mediated entry of peroxidases into the parasitophorous vacuoles of macrophages infected with Leishmania Mexicana amazonensis. J Leukoc Biol. 1985;37(3):247-261.

Randriamampita C, Trautmann A. Ionic channels in murine macrophages. J Cell Biol. 1987;105(2):761-769.

Randriamampita C, Trautmann A. Biphasic increase in intracellular calcium induced by platelet-activating factor in macrophages. FEBS Lett. 1989;249(2):199-206.

Randriamampita C, Bismuth G, Trautmann A. Ca(2+)-induced Ca2+ release amplifies the Ca2+ response elicited by inositol trisphosphate in macrophages. Cell Regul. 1991;2(7):513-522.

Rodrigues CO, Scott DA, Docampo R. Presence of a vacuolar H+-pyrophosphatase in promastigotes of Leishmania donovani and its localization to a different compartment from the vacuolar H+-ATPase. Biochem J. 1999;340(Pt3):759-766.

Russell DG, Xu S, Chakraborty P. Intracellular trafficking and the parasitophorous vacuole of Leishmania mexicana-infected macrophages. J Cell Sci. 1992;103:1193-1210.

Russell DG. Mycobaterium and Leishmania: stowaways in the endosomal network. Trends Cell Biol. 1995;5:125-128

Schaible UE, Schlesinger PH, Steinberg TH, Mangel WF, Kobayashi T, Russell DG. Parasitophorous vacuoles of Leishmania mexicana acquire macromolecules from the host cell cytosol via two independent routes. J Cell Sci. 1999;112(Pt5):681-693.

Scheel O, Zdebik AA, Lourdel S, Jentsch TJ. Voltage-dependent electrogenic chloride/proton exchange by endosomal CLC proteins. Nature. 2005;436(7049):424-427.

Schwab JC, Beckers CJ, Joiner KA. The parasitophorous vacuole membrane surrounding intracellular Toxoplasma gondii functions as a molecular sieve. Proc Natl Acad Sci U S A. 1994;18;91(2):509-513.

Shaked-Mishan P, Suter-Grotemeyer M, Yoel-Almagor T, Holland N, Zilberstein D, Rentsch D. A novel high-affinity arginine transporter from the human parasitic protozoan Leishmania donovani. Mol Microbiol. 2006;60(1):30-38.

Singh A, Mandal D. A novel sucrose/H+ symport system and an intracellular sucrase in Leishmania donovani. Int J Parasitol. 2011;41(8):817-826.

Scott KD, Stafford JL, Galvez F, Belosevic M, Goss GG. Plasma membrane depolarization reduces nitric oxide (NO) production in P388D.1 macrophage-like cells during Leishmania major infection. Cell Immunol. 2003;222(1):58-68.

Sousa-Franco J, Araújo-Mendes E, Silva-Jardim I, L-Santos J, Faria DR, Dutra WO, Horta MF. Infection-induced respiratory burst in BALB/c macrophages kills Leishmania guyanensis amastigotes through apoptosis: possible involvement in resistance to cutaneous leishmaniasis. Microbes Infect. 2006;8(2):390-400.

Shepherd, VL, Stahl, PD, Bernd P, Rabinovitch, M. Receptor-mediated entry of beta-glucuronidase into the parasitophorous vacuoles of macrophages infected with Leishmania mexicana amazonensis. J Exp Med. 1983;157(5):1471-1482-1492.

Staines HM, Alkhalil A, Allen RJ, De Jonge HR, Derbyshire E, Egée S, Ginsburg H, Hill DA, Huber SM, Kirk K, Lang F, Lisk G, Oteng E, Pillai AD, Rayavara K, Rouhani S, Saliba KJ, Shen C, Solomon T, Thomas SL, Verloo P, Desai SA. Electrophysiological studies of malaria parasite-infected erythrocytes: current status. Int J Parasitol. 2007;37(5):475-482.

Steinberg TH, Newman AS, Swanson JA, Silverstein SC. ATP4- permeabilizes the plasma membrane of mouse macrophages to fluorescent dyes. J Biol Chem. 1987a;262(18):8884-8888.

Steinberg TH, Newman AS, Swanson JA, Silverstein SC. Macrophages possess probenecid-inhibitable organic anion transporters that remove fluorescent dyes from the cytoplasmic matrix. J Cell Biol. 1987b;105(6 Pt 1):2695-2702.

Steinberg TH, Mangel WF, Steinberg BE, Huynh KK, Brodovitch A, Jabs S, Stauber T, Jentsch TJ, Grinstein S. A cation counterflux supports lysosomal acidification. J Cell Biol. 2010;189(7):1171-1186.

Stiles JK, Kucerova Z, Sarfo B, Meade CA, Thompson W, Shah P, Xue L, Meade JC. Identification of surface-membrane P-type ATPases resembling fungal K(+)- and Na(+)-ATPases, in Trypanosoma brucei, Trypanosoma cruzi and Leishmania donovani. Ann Trop Med Parasitol. 2003;97(4):351-366.

Sturgill-Koszycki S, Schlesinger PH, Chakraborty P, Haddix PL, Collins HL, Fok AK, Allen RD, Gluck SL, Heuser J, Russell DG. Lack of acidification in Mycobacterium phagosomes produced by exclusion of the vesicular proton-ATPase. Science. 1994;263(5147):678-681.

Stühmer W: Electrophysiological recording from *Xenopus* oocytes in Rudy B, Iverson L (eds): Methods in Enzymology. Academic Press. 1992;207:318-339.

Tanabe K. Ion metabolism in malaria-infected erythrocytes. Blood Cells. 1990;16(2-3):437-49.

Tardieux I, Nathanson MH, Andrews NW. Role in host cell invasion of Trypanosoma cruzi-induced cytosolic-free Ca2+ transients. J Exp Med. 1994;179(3):1017-1022.

ter Kulle BH. Glucose and proline transport in kinetoplastids. Parasitol Today. 1993;9(6):206-210.

Tetaud E, Bringaud F, Chabas S, Barrett MP, Baltz T. Characterization of glucose transport and cloning of a hexose transporter gene in Trypanosoma cruzi. Proc Natl Acad Sci U S A. 1994;91(17):8278-8282.

Vargas R, Botero L, Lagos L, Camacho M. Bufo marinus oocytes as a model for ion channel protein expression and functional characterization with electrophysiological studies. Cel Physiol Biochem. 2004;14(4-6):197-202.

Veras PS, de Chastellier C, Rabinovitch M. Transfer of zymosan (yeast cell walls) to the parasitophorous vacuoles of macrophages infected with Leishmania amazonensis. J Exp Med. 1992;176(3):639-646.

Veras PS, de Chastellier C, Moreau MF, Villiers V, Thibon M, Mattei D, Rabinovitch M. Fusion between large phagocytic vesicles: targeting of yeast and other particulates to phagolysosomes that shelter the bacterium Coxiella burnetii or the protozoan Leishmania amazonensis in Chinese hamster ovary cells. J Cell Sci. 1994;107 107(Pt11):3065-3076.

Veras PS, Moulia C, Dauguet C, Tunis CT, Entry and survival of Leishmania amazonensis amastigotes within phagolysosome-like vacuoles that shelter Coxiella burnetii in Chinese hamster ovary cells. Thibon M, Rabinovitch M. Infect Immun. 1995;63(9):3502-3506.

Veras PS, Topilko A, Gouhier N, Moreau MF, Rabinovitch M, Pouchelet M. Fusion of Leishmania amazonensis parasitophorous vacuoles with phagosomes containing zymosan particles: cinemicrographic and ultrastructural observations. Braz J Med Biol Res. 1996;29(8):1009-1018.

Vercesi AE, Rodrigues CO, Catisti R, Docampo R. Presence of a Na(+)/H(+) exchanger in acidocalcisomes of Leishmania donovani and their alkalization by anti-leishmanial drugs. FEBS Lett. 2000;473(2):203-206.

Vicente R, Escalada A, Coma M, Fuster G, Sánchez-Tilló E, López-Iglesias C, Soler C, Solsona C, Celada A, Felipe A. Differential voltage-dependent K+ channel responses during proliferation and activation in macrophages. J Biol Chem. 2003;278(47):46307-46320.

Vicente R, Escalada A, Soler C, Grande M, Celada A, Tamkun MM, Solsona C, Felipe A. Pattern of Kv beta subunit expression in macrophages depends upon proliferation and the mode of activation. J Immunol. 2005;174(8):4736-4744.

Vicente R, Escalada A, Villalonga N, Texidó L, Roura-Ferrer M, Martín-Satué M, López-Iglesias C, Soler C, Solsona C, Tamkun MM, Felipe A. Association of Kv1.5 and Kv1.3 contributes to the major voltage-dependent K+ channel in macrophages. J Biol Chem. 2006;281(49):37675-37685.

Vicente R, Villalonga N, Calvo M, Escalada A, Solsona C, Soler C, Tamkun MM, Felipe A. Kv1.5 association modifies Kv1.3 traffic and membrane localization. J Biol Chem. 2008;283(13):8756-8764.

Vieira L, Lavan A, Dagger F, Cabantchik ZI. The role of anions in pH regulation of Leishmania major promastigotes. J Biol Chem. 1994;269(23):16254-16259.

Vieira LL, Cabantchik ZI. Amino acid uptake and intracellular accumulation in Leishmania major promastigotes are largely determined by an H(+)-pump generated membrane potential. Mol Biochem Parasitol. 1995;75(1):15-23.

Vieira LL, Lafuente E, Gamarro F, Cabantchik Z. An amino acid channel activated by hypotonically induced swelling of Leishmania major promastigotes. Biochem J. 1996;319:691-697.

Vieira LL, Lafuente E, Blum J, Cabantchik ZI. Modulation of the swelling-activated amino acid channel of Leishmania major promastigotes by protein kinases. Mol Biochem Parasitol. 1997;90(2):449-461.

Vignery A. Macrophage fusion: the making of osteoclasts and giant cells. J Exp Med. 2005;202(3):337-340.

Villalonga N, Escalada A, Vicente R, Sánchez-Tilló E, Celada A, Solsona C, Felipe A. Kv1.3/Kv1.5 heteromeric channels compromise pharmacological responses in macrophages. Biochem Biophys Res Commun. 2007;352(4):913-918.

Villalonga N, David M, Bielanska J, Vicente R, Comes N, Valenzuela C, Felipe A. Immunomodulation of voltage-dependent K+ channels in macrophages: molecular and biophysical consequences. J Gen Physiol. 2010;135(2):135-147.

Vinet AF, Fukuda M, Turco SJ, Descoteaux A. The Leishmania donovani lipophosphoglycan excludes the vesicular proton-ATPase from phagosomes by impairing the recruitment of synaptotagmin V. PLoS Pathog. 2009;5(10):e1000628.

Weinert S, Jabs S, Supanchart C, Schweizer M, Gimber N, Richter M, Rademann J, Stauber T, Kornak U, Jentsch TJ. Lysosomal pathology and osteopetrosis upon loss of H+-driven lysosomal Cl- accumulation. Science. 2010;328(5984):1401-1403.

Zilberstein D, Dwyer DM. Protonmotive force-driven active transport of D-glucose and L-proline in the protozoan parasite Leishmania donovani. Proc Natl Acad Sci U S A. 1985;82(6):1716-1720.

Zilberstein D. Transport of Nutrients and Ions across Membranes of Trypanosomatid Parasites. In Advances in Parasitology. 1993(32):261-291.

Zilberstein D, Shapira M. The role of pH and temperature in the development of Leishmania parasites. Annu Rev Microbiol. 1994;48:449-470.

Part 2

Advantages of Using Patch Clamp Technique

11

Cardiac Channelopathies: Disease at the Tip of a Patch Electrode

Brian P. Delisle
University of Kentucky
USA

1. Introduction

The ordered electrical excitation of the heart via the cardiac conduction system coordinates the efficient pumping of blood. Electrical impulses normally originate in the sinoatrial node and then propagate through the atria, atrioventricular node, and into the ventricles. The electrical activity of the heart can be measured non-invasively using an electrocardiogram (ECG). An ECG reflects the summed electrical activity of the individual contractile myocytes in the different regions of the heart over time. The first event is the "P wave" and corresponds to the depolarization of the atria, the next three waves "Q, R, and S" represent the progressive wave of depolarization through the ventricles, and the final wave of the cardiac cycle "T" corresponds to ventricular repolarization. The action potentials of contractile atrial and ventricular myoctes share several common characteristics: a stable resting or diastolic membrane potential (phase 4), a rapid depolarization (phase 0), an initial repolarization (phase 1), a plateau (phase 2), and a rapid repolarization to diastolic potential (phase 3). The different phases of the cardiac action potential reflect changes of the sarcolemma permeability to K^+, Na^+, and Ca^{2+}. Figure 1 shows the relationship between the ECG, a ventricular action potential, and several of the important ventricular K^+, Na^+, and Ca^{2+} currents (Nerbonne and Kass, 2005).

Arrhythmias are electrical disturbances that disrupt the normal initiation or propagation of the cardiac impulse. They cause abnormal impulse rates (bradycardia or tachycardia), block impulse propagation, or initiate the impulse to circle in a "reentry" loop. Atrial arrhythmias can result in the formation of blood clots and increase the risk of stroke, whereas ventricular arrhythmias can cause the inefficient pumping of blood, loss of consciousness, and sometimes death (Shah et al., 2006).

"Cardiac channelopathies" are caused by mutations in ion channel genes that encode the proteins that underlie the K^+, Na^+, or Ca^{2+} currents that shape the cardiac action potential (Figure 1). These mutations can cause syndromes that increase the risk for atrial and/or ventricular arrhythmias. Cardiac channelopathies typically cause a "gain- or loss-of-function" in one of these currents by altering channel synthesis, channel transport to and from the cell surface membrane (trafficking), channel gating, and/or single channel permeation (Figure 2) (Delisle et al., 2004). This chapter provides a description for several different cardiac syndromes that can be caused by channelopathies and the utilization of the patch-clamp technique to learn about functional phenotypes.

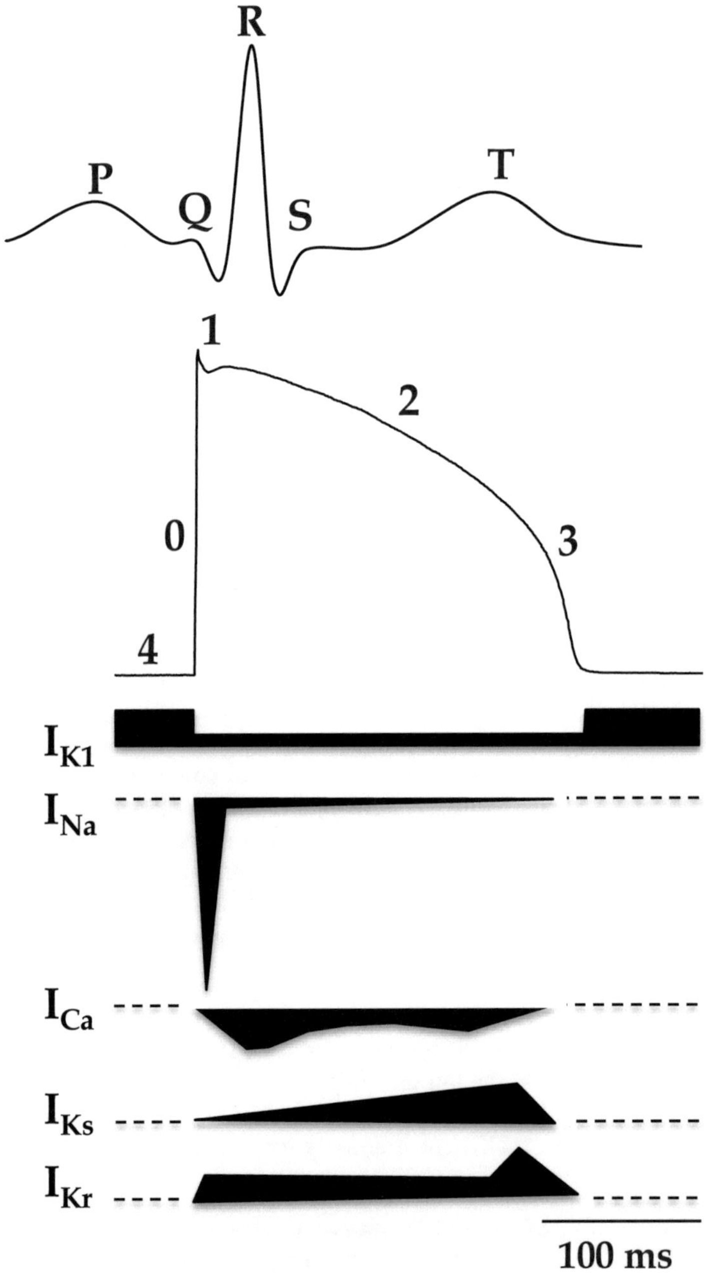

Fig. 1. The relationship between an ECG trace, a ventricular action potential, and important K^+, Na^+, and Ca^{2+} currents. The P, Q, R, S, T correspond to the different waves on the ECG, the 4, 0, 1, 2, 3 correspond to the different phases of the ventricular action potential. I_{K1} is the inwardly rectifying K^+ current, I_{Na} represents the Na^+ current, I_{Ca} represents the Ca^{2+} current, I_{Ks} is the slowly activating delayed rectifier K^+ current, and I_{Kr} is the rapidly activating delayed rectifier K^+ current.

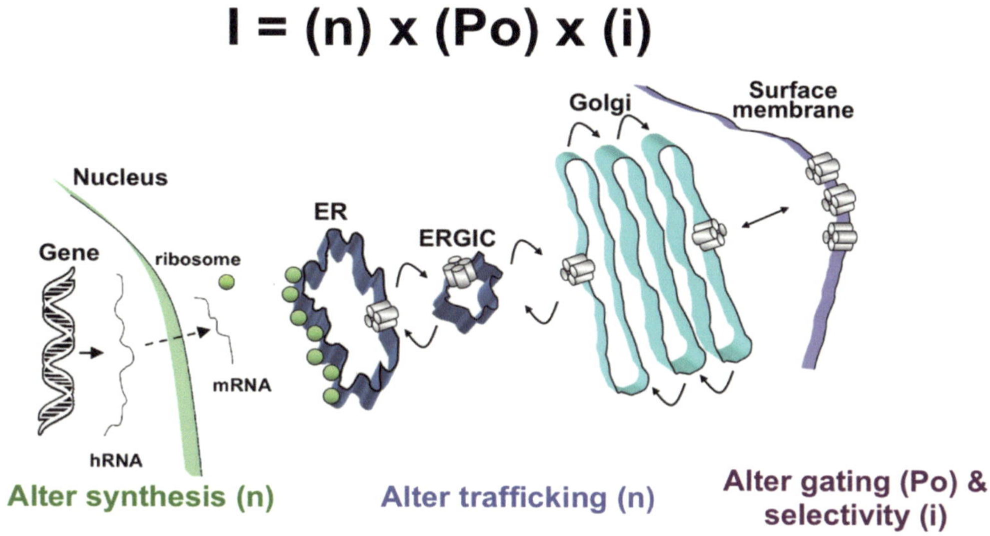

Fig. 2. A *cartoon* showing the synthesis, trafficking, and surface expression of ion channels. Macroscopic current (I) is a function of channel number at the cell surface (n), open probability (Po), and the amplitude of the single channel current (i). Cardiac channelopathy mutations can alter n by changing synthesis and trafficking, or they can alter channel function by changing gating and selectivity (hRNA = heterogeneous ribonucleic acid; mRNA = messenger ribonucleic acid; ER = Endoplasmic Reticulum; ERGIC = ER Golgi Intermediate Compartment).

2. Cardiac channelopathies

Hundreds of different mutations in over a dozen different genes that encode cardiac K^+, Na^+ and Ca^{2+} channels or their regulatory proteins are linked to several different congenital arrhythmia syndromes including Long QT syndrome, Short QT syndrome, Brugada Syndrome, and familial Atrial Fibrillation (Lehnart et al., 2007).

2.1 Long QT syndrome

Congenital Long QT syndrome (LQT) is typically characterized by an abnormally long corrected QT (QTc) interval on an ECG. It is one of the most common "monogenic" arrhythmia syndromes and occurs in ~1:2,500 healthy births (Lehnart et al., 2007; Crotti et al., 2008). LQT-linked mutations are postulated to cause a prolongation of the plateau phase of the ventricular action potential. This allows L-type Ca^{2+} channels to recover from inactivation and reopen, which leads to an early after depolarization. Early after depolarizations increase risk for polymorphic ventricular tachycardia (torsade de pointes), which can cause a loss of cardiac output, syncope, and sudden death. The majority of LQT-linked mutations follow a dominant inheritance pattern (Romano-Ward type of LQT), and it is linked to at least thirteen different genes (LQT1-LQT13; Table 1) (Delisle et al., 2004; Ruan et al., 2008); however, autosomal-recessive forms, Jervell and Lange-Nielsen syndrome (JLN) exist and are caused by homozygous or compound mutations in *KCNQ1* and *KCNE1*. Most LQT-linked mutations cause a prolongation of the ventricular action potential duration by either decreasing cardiac K^+ currents, or increasing the "late" Na^+ or Ca^{2+} current. LQT1-3 account for 70-75% of all LQT cases, and the other 9 forms contribute an additional 5% (Kapa et al., 2009). The types of LQT mutations include splice, nonsense, frameshift, deletions, and missense, the latter type represents ~70% of the case-linked mutations.

Disease	Current	Chromosome	Defective gene (protein)	Frequency	Key reference
LQT1	I_{Ks}, ↓ amplitude	11p15.5	*KCNQ1* (Kv7.1)	~30-35%	Wang, et al., 1996
LQT2	I_{Kr}, ↓ amplitude	7q35-q36	*KCNH2, hERG* (Kv11.1)	~30-35%	Curran et al., 1995
LQT3	I_{Na}, ↑ late amplitude	3p22.2	*SCN5a* (Nav1.5)	~7-10%	Wang et al., 1995
LQT4	cell Ca^{2+}	4q25-4q26	*ANKB*	Rare	Schott et al., 1995, Mohler et al., 2003
LQT5	I_{Ks}, ↓amplitude	21q22.1-22.2	*KCNE1* (MiNK1)	Rare	Splawski et al., 1997
LQT6	I_{Kr}, ↓ amplitude	21q22.11	*KCNE2* (MiRP1)	Rare	Abbott et al., 1999
LQT7	I_{K1}, ↓ amplitude	17q23-17q24.2	*KCNJ2* (Kir2.1)	Rare	Tristani-Firouzi et al., 2002; Andelfinger et al., 2002.
LQT8	I_{Ca}, ↑ amplitude	12p13.33	*CACNA1C* (Cav1.2)	Rare	Splawski et al., 2005
LQT9	I_{Na}, ↑ late current	3p25	*CAV3* (caveolin 3)	Rare	Vatta et al., 2006
LQT10	I_{Na}, ↑ late current	11q23	*SCN4B*	Rare	Medeiros-Domingo et al., 2007
LQT11	I_{Ks}, ↓ amplitude	7q21-q22	*AKAP9* (Yotiao)	Rare	Chen et al., 2007
LQT12	I_{Na}, ↑ late amplitude	20q11.2	*SNTA1* (syntrophin-α1)	Rare	Ueda et al., 2008
LQT13	I_{KATP}, ↓ amplitude	11q24.3	*KCNJ5* (Kir3.4)	Rare	Yang et al., 2010
JLN1	I_{Ks}, ↓ amplitude	11p15.5	*KCNQ1* (Kv7.1)	Rare	Neyroud et al., 1997
JLN2	I_{Ks}, ↓ amplitude	21q22.1-22.2	*KCNE1* (MiNK1)	Rare	Schultze-Bahr et al., 1997

Table 1. Long QT syndrome disease types. Autosomal dominant forms (LQT1-12) include Romano Ward Syndrome (LQT1-6), Anderson Tawill Syndrome (LQT7), and Timothy Syndrome (LQT8). The autosomal recessive form is Jervell Lange-Neilson syndrome (JLN1-2) and is associated with deafness.

2.2 Short QT syndrome

Short QT syndrome (SQT) is characterized by a short QTc interval and can cause syncope, paroxysmal atrial fibrillation, and ventricular fibrillation (Lehnart et al., 2007; Patel et al.,

Disease	Current	Chromosome	Defective gene (protein)	Key reference
SQT1	I_{Kr}, ↑ amplitude	7q35-q36	*KCNH2, hERG* (Kv11.1)	Brugada et al., 2004
SQT2	I_{Ks}, ↑ amplitude	11p15.5	*KCNQ1* (Kv7.1)	Bellocq et al., 2004 Hong et al., 2005
SQT3	I_{K1}, ↑ amplitude	17q23-17q24.2	*KCNJ2* (Kir2.1)	Priori et al., 2005
SQT4	I_{Ca}, ↓ amplitude	12p13.33	*CACNA1C* (Cav1.2)	Antzelevitch et al., 2007
SQT5	I_{Ca}, ↓ amplitude	10p12.33-p12.31	*CACNB2* (Cavβ2)	Antzelevitch et al., 2007

Table 2. Short QT syndrome disease types.

2010). Similar to LQT, SQT follows a dominant inheritance pattern, but it is extremely rare and only a small number of families with missense mutations have been identified. SQT mutations are predicted to shorten the ventricular action potential duration by increasing cardiac K^+ currents or decreasing Ca^{2+} current.

2.3 Brugada syndrome

Brugada syndrome (BrS) is characterized by ST segment elevation, conduction abnormalities (prolonged PR or QRS segments), and an increased risk of ventricular fibrillation (Lehnart et al., 2007). The prevalence of Brugada syndrome is slightly more common than LQT at ~1:2,000. The ECG changes are not always obvious but they can be usually seen after the administration of drugs that block the cardiac Na^+ current. The majority of BrS patients have missense, nonsense, deletion, or insertion mutations in *SCN5A* that cause a reduction in Na^+ current (Table 3). There appears to be much more genetic heterogeneity associates with Brugada syndrome, because ~2/3rds of the patients do not have a clear genotyped link. BrS-related cardiac events occur more often in adult men (this is likely because of gender differences in cardiac ion channel expression and hormone levels).

Disease	Current	Chromosome	Defective gene (protein)	Frequency	Key reference
BrS1	I_{Na}, ↓ amplitude	3p22.2	*ScN5a* (Nav1.5)	~10-30%	Chen, et al., 1998
BrS2	I_{Na}, ↓ amplitude	3p22.3	*GPDL1* (GPDL1)	Rare	London et al., 2007 Van Norstrand et al., 2007
BrS3	I_{Ca}, ↓ amplitude	12p13.33	*CACNA1C* (Cav1.2)	Rare	Antzelevitch et al., 2007
BrS4	I_{Ca}, ↓ amplitude	10p12.33-p12.31	*CACNB2* (Cavβ2)	Rare	Antzelevitch et al., 2007
BrS5	I_{Na}, ↓ amplitude	19q13.12	*SCN1B* (Navβ1)	Rare	Watanabe et al., 2008
BrS6	I_{TO}, ↑ amplitude	11q13.4	*KCNE3* (MiRP2)	Rare	Delpon et al., 2008
BrS7	I_{Na}, ↓ amplitude	11q24.1	*SCN3B* (Navβ3)	Rare	Hu et al., 2009
BrS8	I_F, ↑ amplitude	15q24.1	*HCN4* (HCN4)	Rare	Ueda et al., 2009
BrS9	I_{Ca}, ↓ amplitude	7q21.11	*CACNA2D1* (α2ζ)	Rare	Burashnikov et al., 2010

Table 3. Brugada syndrome disease types.

2.4 Familial Atrial Fibrillation

The most common arrhythmia is Atrial Fibrillation (AF), and it is usually secondary to structural changes in the heart (i.e. organ heart disease, valvular disease, hypertensive heart disease, cardiac hypertrophy, etc.). In some instances AF in the absence of structural heat disease (lone AF) occurs, and it is typically caused by genetic factors. Mutations in ion channel genes linked to familial forms of AF have been identified, but these mutations in these genes only account for a small number of cases (Lehnart et al., 2007).

Disease	Current	Chromosome	Defective gene (protein)	Key reference
ATFB3	I_{Ks}, ↑amplitude	11p15.5	*KCNQ1* (Kv7.1)	Chen et al., 2003
ATFB4	I_{Ks}, ↑amplitude	21q22.11	*KCNE2* (MiRP1)	Yang et al., 2004
ATFB7	I_{Kur}, ↓ amplitude	12p13.32	*KCNA5* (Kv1.5)	Olson et al., 2006 Yang et al., 2009
ATFB9	I_{K1}, ↑amplitude	17q24.3	*KCNJ2* (Kir2.1)	Xia et al., 2005
ATFB10	I_{Na}, ↑ or ↓amplitude	3p22.2	*SCN5a* (Nav1.5)	Darbar et al., 2008 Ellinor et al., 2008 Makiyama et al., 2008 Benito et al., 2008
ATFB11	I_{CNX40}, ↓ amplitude	1q21.2	*GJA5* (connexin 40)	Gollob et al., 2006
ATFB12	I_{KATP}, ↓ ADP current	12p12.1	*ABCC9* (Sur2A)	Olson et al., 2007

Table 4. Familial Atrial Fibrillation syndrome disease types linked to channelopathies.

2.5 Summary

Many different types of cardiac arrhythmias are linked to channelopathies. Thus far, mutations in over 20 different genes that encode cardiac ion channel α-subunits, auxiliary channel subunits, or channel modulatory proteins are linked to one or more arrhythmia syndromes. This genetic heterogeneity has identified ion channel and macromolecular signaling complexes that are important for normal cardiac excitability and arrhythmia susceptibility. Many of these mutations are predicted to increase or decrease the macroscopic K^+, Na^+ or Ca^{2+} currents to increase arrhythmia susceptibility. This occurs by a number of different mechanisms, including alterations in protein synthesis, trafficking to/from the membrane, channel gating, and/or permeation.

3. Patch clamp techniques to study disease mechanisms

This section focuses on using the patch clamp technique to study disease mechanisms linked to the genetic variants in *KCNQ1* or *KCNH2*, which encode the voltage-gated K^+ channel α-subunits that underlie I_{Ks} (Kv7.1) and I_{Kr} (Kv11.1), respectively. Each Kv α-subunit has six transmembrane (TM) segments that form a voltage-sensor (TM1-TM4) and a pore domain (TM5-TM6) (Figure 3).

Mutations in these genes are the most common cause of "monogenic" arrhythmia syndromes (Lehnart et al., 2007; Kapa et al., 2009). *KCNQ1* and *KCHN2* mutations can cause a loss- or gain-of-function (Wang et al., 1996; Neyroud et al., 1997; Chen et al., 2003). Loss of function mutations can be splice-site, nonsense, or frame-shift and result in haploinsufficiency; however, the vast majority of *KCNQ1* and *KCNH2* mutations linked to arrhythmia syndromes are missense. In syndromes that follow a dominant inheritance pattern (i.e. LQT, SQT, familial atrial fibrillation, etc) these mutant α-subunits may co-assemble with wild type to generate functional phenotypes that cause a more severe phenotype (Figure 4).

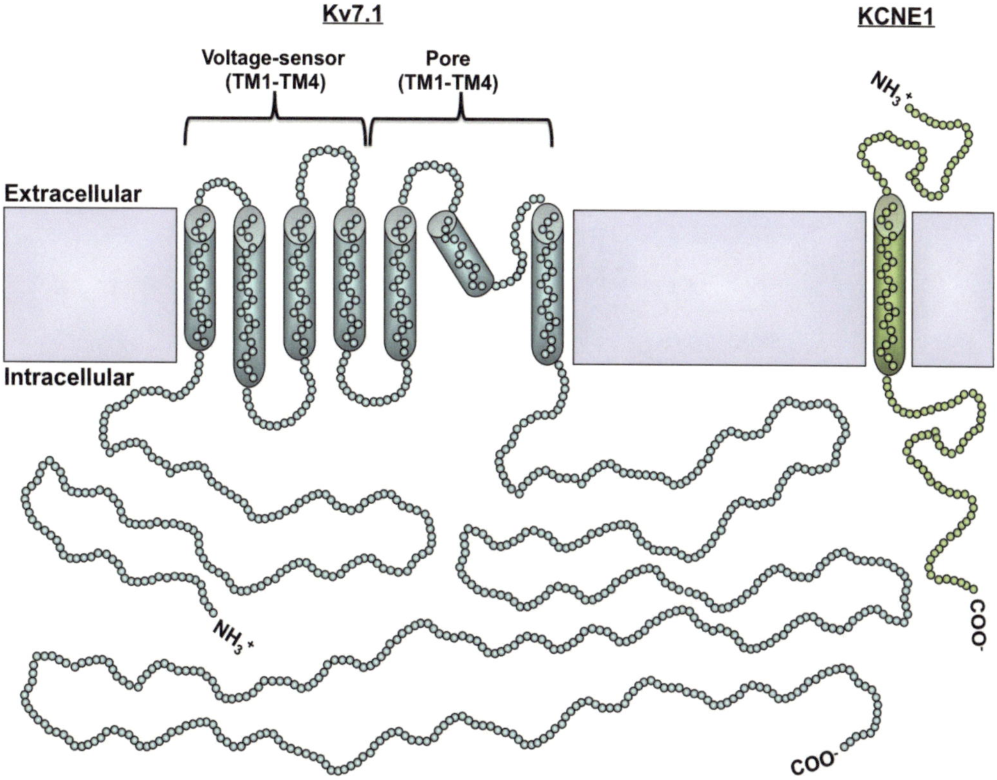

Fig. 3. A *cartoon* of a single Kv7.1 α-subunit and its putative auxiliary subunit KCNE1. The image shows a single Kv7.1 α-subunit and a single KCNE1 subunit. The membrane is represented by the purple rectangles, the individual amino acids are shown as small circles, and the secondary structures (α-helices) are shown as cylinders. The respective amino and carboxy termini are labeled NH_3^+ and COO^-, respectively. Heterolgous expression studies suggest that co-expression of Kv7.1 and KCNE1 are needed to generate macroscopic currents similar to native I_{Ks} in the heart (Barhanin et al., 1996; Sanguinetti et al., 1996).

3.1 Long QT type 1

Thus far, hundreds of different mutations in *KCNQ1* have been linked to LQT1. Kv7.1 tetramerizes to form the aqueous channel of the I_{Ks} in the heart (Kapa et al., 2009; http://www.fsm.it/cardmoc/). Mutations in *KCNQ1* account for ~30-35% of congenital LQT cases (Yuan et al., 2008; Kapa et al., 2009). Heterologous expression studies suggest that many LQT1 mutations throughout Kv7.1 cause a loss-of-function (Wollnik et al., 1997; Neyoud et al., 1998; Saarinen et al., 1998; Yamashita et al., 2001; Pippo et al., 2001; Deschênes et al., 2003; Yamaguchi et al., 2003; Thomas et al., 2005; Thomas et al., 2010).

LQT1 patients exhibit a wide-range of clinical phenotypes from asymptomatic to recurrent episodes of syncope and sudden cardiac arrest, and LQT1-related cardiac events are most often triggered by exercise (Priori et al., 2003; Napolitano et al., 2005; Schwartz et al., 2001; Priori et al., 2004; Ruan et al., 2008). Studies of multi-center LQT1 registries with hundreds of patients suggest that the risk for LQT1 related cardiac events is related to the type/location of the LQT1 mutation (Zereba et al., 2003; Moss et al., 2007; Goldenberg et al.,

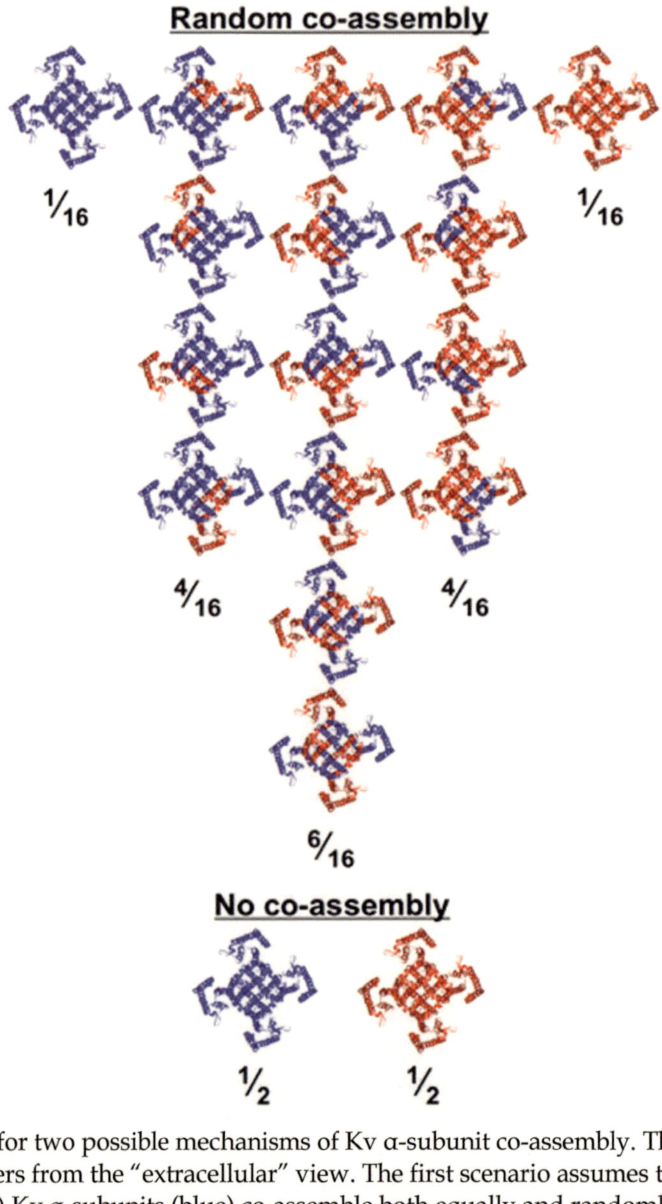

Fig. 4. A *cartoon* for two possible mechanisms of Kv α-subunit co-assembly. The images show Kv tetramers from the "extracellular" view. The first scenario assumes that WT (blue) and mutant (red) Kv α-subunits (blue) co-assemble both equally and randomly so only 1/16th of the total number of channels is all WT or mutant. The other 14/16ths of the channels are a combination of WT and mutant Kv α-subunits with varying stoichometries. If the WT and mutant Kv α-subunits do not co-assemble, then half of the channels are all WT or mutant (January et al., 2000). The structural model for the Kv7.1 α-subunits used to generate the channels shown was developed by Smith and colleagues (2007).

2008; Jons et al., 2009). About 80% of LQT1 mutations are missense, and genetic testing for LQT1 is commercially available, but the diagnostic and prognostic value of a genotype positive test is limited because it does not distinguish between relatively benign or pathogenic variants (Kapa et al., 2009). Current treatments for LQT1 include life-style modification, β-blockers, implantable cardioverter defibrillators (ICDs), and/or left cardiac sympathetic denervation (Ruan et al., 2008). A major challenge for clinicians is to determine which therapeutic strategies are most appropriate for an individual. While many considerations need to be made, *in vitro* patch clamp studies of individual LQT1 mutations will provide a wealth of information, including identifying dominant negative functional phenotypes that may predict a more severe LQT1 clinical phenotype (Moss et al., 2007).

The most common way to determine if LQT1 mutations cause dominant negative suppression is to co-express WT and mutant channels in the heterologous expression systems. Initially a common heterologous expression system was *Xenopus laevis* oocytes. Indeed, the first studies that measured the functionality of LQT1-linked mutation utilized this system with the two-electrode voltage-clamp (Wang et al., 1999; Shalaby et al., 1997). An advantage to the *Xenopus* system is that investigators can control the amount of WT and mutant cRNA injected into cell. Unfortunately *Xenopus* oocytes express a Kv7 isoform that can make experiments with exogenously added Kv7.1/KCNE1 difficult to interpret (Lesage et al., 1993). Additionally, *Xenopus* oocytes are cultured at room temperature and the trafficking of many different LQT-linked mutations are temperature sensitive, which can also complicate the interpretation of the data (Zhou et al., 1999; Ficker et al., 2000; Anderson et al., 2006).

The whole-cell patch clamp technique of transiently expressing mammalian is an effective way to determine whether LQT1 mutations cause dominant negative effects. Mammalian cells can be transfected using chemical or electroporation techniques with Kv7.1 cDNA in plasmid expression vectors that contain WT-Kv7.1 or mutant Kv7.1 constructs. A disadvantage to this technique is that the uptake of the plasmid DNA is uncontrolled. Scenarios that further complicate this is that the cDNA for KCNE1 auxiliary channel subunits is also needed for transfection in order to generate native-like I_{Ks} (Barhanin et al., 1996; Sanguinetti et al., 1996). For co-expression studies that utilize multiple plasmids, WT-Kv7.1 and mutant cDNA plasmids are typically transfected in a 50:50 ratio (to mimic heterozygousity) using the total equivalent cDNA plasmid concentrations as in parallel experiments with WT or mutant only transfections. To favor the selection of cells that express multiple cDNA plasmids, cells are typically co-transfected with a fluorescent protein cDNA expression plasmid (i.e. green fluorescent proteins or GFP). The assumption is that cells expressing the fluorescent protein are likely to express the other plasmids as well.

Bartos and colleagues recently determined the effect of co-expressing WT-Kv7.1 and the LQT1-linked missense mutations E160K-Kv7.1 in Human Embryonic Kidney 293 (HEK293) (Bartos et al., 2011). This mutation has been previously characterized (Silva & Rudy, 2009; Wu D et al., 2010; Silva & Rudy, 2010;). The appropriate nucleotide change to generate the E160K-Kv7.1 mutations was engineered in the WT Kv7.1 (WT-Kv7.1) cDNA subcloned in pcDNA3 mammalian expression vector using the QuickChange Site Directed Mutagenesis

Kit (Stratagene, La Jolla, CA). HEK293 cells were transfected using Superfect (Qiagen Valencia, CA). Cells transfected with WT-Kv7.1 (3 µg) plasmid DNA expressed small currents that did not resemble I_{Ks}, and so to recapitulate I_{Ks}-like currents, cells were also transfected with KCNE1 (3 µg) plasmid DNA (total concentration was 6 µg). The final ratio of KCNE1 to Kv7.1 cDNA ratio is ~2:1. For co-expression studies (to mimic heterozygousity), they transfected KCNE1 (3 µg), WT-Kv7.1 (1.5 µg), and E160K-Kv7.1 (1.5 µg) plasmid DNA. Additionally, They transfected a third set of cells with KCNE1 (3 µg) and E160K-Kv7.1 (3 µg). Cells were also transfected with enhanced GFP cDNA subcloned in pKR5 expression plasmid (0.3 µg). The cells were cultured in Minimum Essential Media (with 10% Fetal Bovine Serum) at 37 °C in 5% CO_2 and analyzed 24-30 hrs after transfection. Functional analyses were done using standard whole cell patch clamp technique on GFP positive cells. An Axopatch-200B patch clamp amplifier (Axon Instruments, Union City, CA) was used to measure macroscopic currents and cell capacitance. The uncompensated pipette resistances were 1-2 MΩ and series resistance was compensated between up to 95%. pCLAMP 10 software (Axon Instruments, Union City, CA) was used to acquire current signals and Origin (7.0, Microcal, Northhampton, MA) was used for performing data analyses. The external solution contained (in mM) 137 NaCl, 4 KCl, 1.8 CaCl2, 1 MgCl2, 10 glucose, and 10 HEPES (pH 7.4 with NaOH), and an internal pipette solution contained (in mM) 130 KCl, 1 MgCl2, 5 EDTA, 5 MgATP, 10 HEPES (pH 7.2 with KOH). Figure 5 shows representative families of whole-cell currents measured from cells expressing KCNE1 with WT-Kv7.1, WT- and E160K-Kv7.1, E160K-Kv7.1. The currents were measured by applying "step" pulses from -80 mV to 70 mV in 10 mV increments for 5 s, followed by a "tail" pulse to -50 mV for 5 s. The intersweep holding potential and duration was -80 mV and 10 s, respectively. Cells expressing WT-Kv7.1 exhibited large macroscopic currents that resembled native I_{Ks}. Cells expressing WT- and E160K-Kv7.1 showed currents with similar characteristics as I_{Ks}, but the amplitudes were smaller than cells expressing WT-Kv7.1. Similar to that previously reported, cells expressing E160K-Kv7.1 did not generate hardly any current at all (Silva et al., 2009; Wu et al., 2010; Silva et al., 2010; Bartos et al., 2010). The mean peak tail current

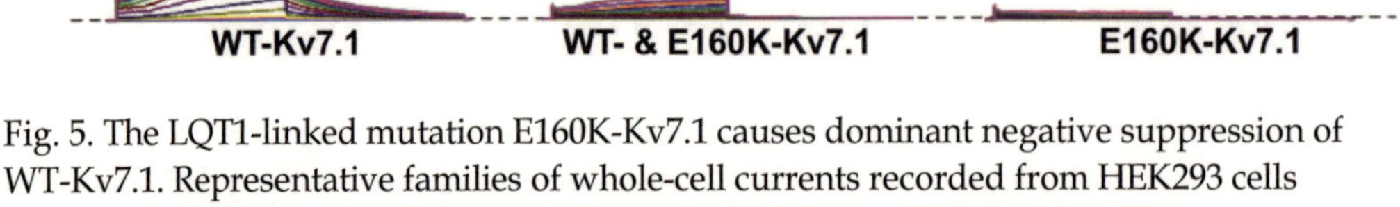

Fig. 5. The LQT1-linked mutation E160K-Kv7.1 causes dominant negative suppression of WT-Kv7.1. Representative families of whole-cell currents recorded from HEK293 cells expressing KCNE1 and WT-Kv7.1 , WT- & E160K-Kv7.1, or E160K-Kv7.1. The corresponding voltage protocol used to measure the currents is also shown. Several mechanistic and computational studies suggest that E160K-Kv7.1 causes a loss of function by alter gating (Silva et al., 2009; Wu D et al., 2010; Silva et al., 2010; Bartos et al, 2011).

data for cells expressing WT-Kv7.1 or WT- and E160K-Kv7.1 were described using the Boltzmann function:

$$I = (I_{MIN}-I_{MAX})/(1+e(V-V_{½})/k) + I_{MAX} \quad (1)$$

where I_{MIN} is the minimally activated current; I_{MAX} is the maximally activated current; $V_{½}$ is the mid-point potential for half maximal activation; and k is the slope factor. Compared to cells expressing WT-Kv7.1, cells expressing WT- and E160K-Kv7.1 did not alter I_{MIN}, $V_{½}$, or k, but it did reduce I_{MAX} by more than half that of cells expressing WT-Kv7.1. These data suggest that E160K-Kv7.1 causes dominant negative suppression of WT-Kv7.1. Additionally, this mutation is associated with a relatively severe LQT1-related clinical phenotype (Silvia et al., 2009).

3.2 Long QT type 2

Similar to LQT1, LQT2 is linked to hundred of different *KCNH2* mutations (Kapa et al., 2009; http://www.fsm.it/cardmoc/); mutations in *KCNH2* account for ~30-35% of congenital LQT cases (Yuan et al., 2008; Kapa et al., 2009); and heterologous expression studies suggest that many LQT2 mutations throughout the Kv11.1 α-subunit cause a loss-of-function (Sanguinetti et al., 1996; Zhou et al., 1998; Zhou et al., 1999; Chen et al., 1999; Furutani et al., 1999; Nakajima et al., 1999; Lees-Miller et al., 2000; Huang et al., 2001; Paulussen et al., 2002; Hayashi et al., 2002; Yang et al., 2002; Johnson et al., 2003; Hayashi et al., 2004; Rossenbacker

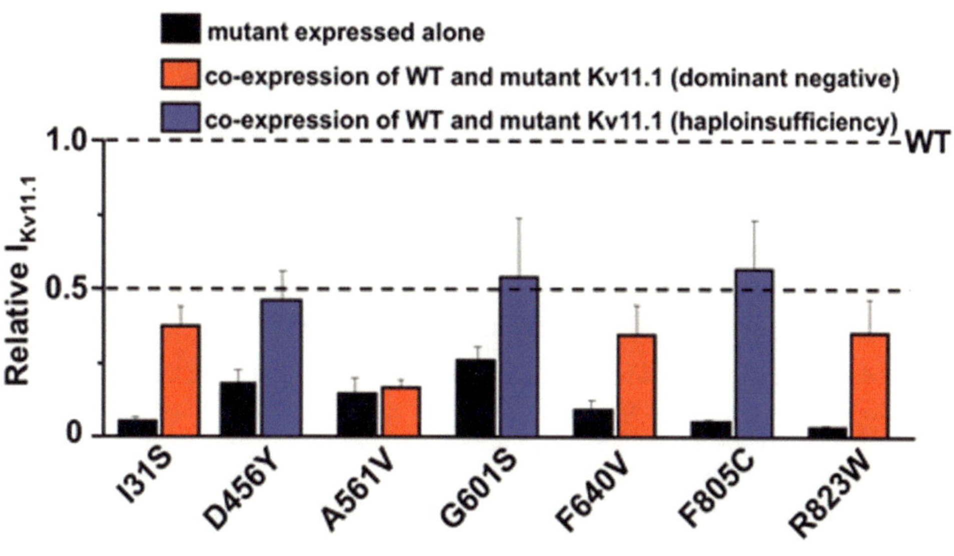

Fig. 6. Trafficking-deficient LQT2-linked mutations generate a spectrum of functional phenotypes. Maximally activated currents were measured from cells expressing WT-Kv11.1, mutant Kv11.1, or WT and mutant Kv11.1 (Modified from Anderson et al., 2006). The relative amplitudes compared to WT-Kv11.1 are shown. Mutations expressed alone all reduced maximally activated current by > 0.5 (black). Co-expression of WT and Kv11.1 mutations suggested that are caused dominant negative effects and reduced maximally activated current by >0.5 (red), whereas others resulted in haploinsufficiency and reduced maximally activated current by ~0.5 (blue).

et al., 2005; Anderson et al., 2006; Gong et al., 2007). Studies suggest that approximately 90% of LQT2 missense mutations cause a loss-of-function preventing the trafficking of the mutant Kv11.1 to the cell surface membrane (Anderson et al., 2006). These "trafficking-deficient" mutations can cause dominant negative effects by inhibiting the trafficking of WT-Kv11.1 (Ficker et al., 2000). Using similar techniques as described above co-expressing WT-Kv7.1 and LQT1 mutations, Anderson and colleagues (2006) co-expressed WT-Kv11.1 with different trafficking-deficient LQT2 mutations in HEK293 cells. Cells were depolarized to 50 mV for 5 seconds, and the maximally activated peak tail $I_{Kv11.1}$ was measured at a test pulse to −120 mV for 3 seconds. Figure 6 shows the relative mean peak amplitude of tail current measured from cells expressing only the trafficking-deficient LQT2 mutations or co-expressing WT-Kv11.1. The data suggest some trafficking-deficient LQT2 mutations can cause dominant negative effects whereas others cause haploinsufficiency.

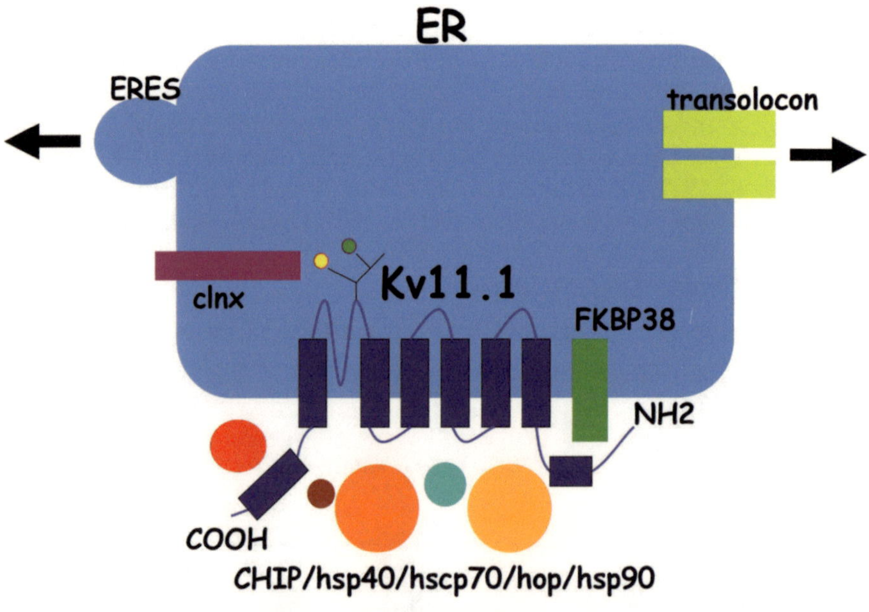

Fig. 7. A *cartoon* showing the different chaperones that regulate the trafficking of Kv11.1 in the secretory pathway. In the ER, several different chaperones and co-chaperones, including calnexin (clns), CHIP, heat shock proteins 40, 70 ,90 (hsp40, hscp70, hsp90), hop, and FKBP38 associate with Kv11.1. Misfolded Kv11.1 is prevented from trafficking into the secretory pathway in Endoplasmic Reticulum Exit Sites (ERES) and is degraded in the transolocan/proteasomal pathway.

Trafficking-deficient LQT2 channels are retained in the Endoplasmic Reticulum (ER). Several studies suggest that the mutations cause "misofolding" of Kv11.1 and cellular quality control mechanisms prevent its export into the secretory pathway (Ficker 2003; Delisle et al., 2005, Gong et al, 2005; Gong et al., 2006; Walker et al., 2007; Walker et al., 2010; Li et al., 2010). Indeed, many different molecular chaperone proteins that regulate the trafficking of WT-Kv11.1 and LQT2 mutations have been identified (Figure 7). Misfolded Kv11.1 undergoes degradation through the proteasome pathway. To understand how

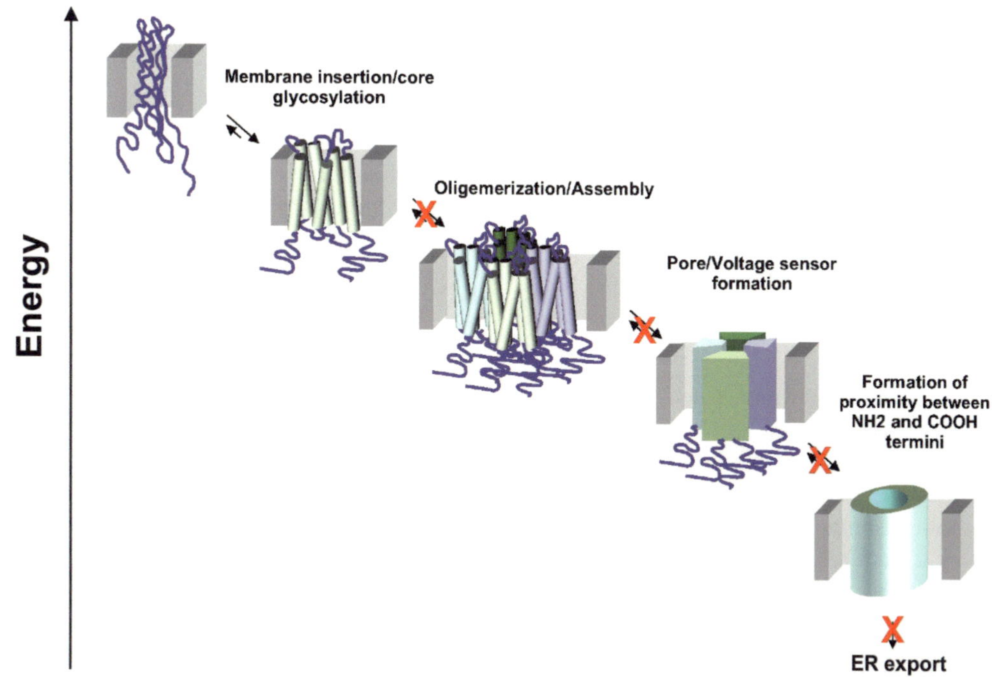

Fig. 8. A *cartoon* showing how some trafficking-deficient LQT2 mutations may disrupt distinct steps in Kv11.1 biogenesis. Trafficking-deficient LQT2 mutations may cause haploinsufficiency by preventing the oligomerization or co-assembly of WT- and mutant Kv11.1 α-subunits. Alternatively, mutant Kv11.1 α-subunits may cause dominant negative effects by co-assembling with WT-Kv11.1 and disrupt pore/voltage sensor formation or formation of proximity between the amino (NH2) or carboxy (COOH) termini. The cartoon shows the different stages of biogenesis, from the formation of secondary structures (cylinders) to the a folded channel (large cylinder with a circle).

trafficking-deficient LQT2 mutations may cause dominant negative or haploinsufficiency, the steps that underlie the assembly and folding of Kv channel proteins during biogenesis need to be understood. Schulteis and colleagues (1998) proposed the following sequence of events during Shaker K^+ channel biogenesis in the ER: 1) membrane insertion and core-glycosylation, 2) assembly of Shaker α and β subunit channel proteins, 3) formation of the pore and voltage sensor, and 4) formation of proximity between adjacent amino and carboxy termini. Assuming a similar sequence of events for Kv11.1 (Figure 8), then trafficking-deficient LQT2 mutations may be capable of disrupting different steps in Kv11.1 channel biogenesis to cause either dominant negative or haploinsufficiency. LQT2 mutations that do not co-assemble do not cause dominant negative suppression of WT-Kv11.1 channel proteins (Ficker et al., 2002) and this would suggest that some trafficking-deficient LQT2 mutations disrupt Kv11.1 biogenesis before the co-assembly step(s). Dominant negative trafficking-deficient LQT2 disrupt Kv11.1 biogenesis steps after co-assembly. The latter group may disrupt Kv11.1 biogenesis steps such as the folding of the voltage sensor/pore, or the formation of proximity between adjacent amino and carboxy termini.

3.3 Summary

The most common cardiac channelopathies are caused by mutations in *KCNQ1* (Kv7.1) and *KCNH2* (Kv11.1) that follow a dominant inheritance pattern and cause LQT1 and LQT2, respectively. The vast majority of these mutations are missense, and they can co-assemble with WT α-subunits to cause effects that range from less than or more than (dominant negative) haploinsufficiency. *In vitro* techniques that utilize the whole cell patch clamp can identify dominant negative mutations that suppress channel function and/or trafficking. We conclude that the patch clamp technique is useful for identifying particularly pathogenic LQT1 and LQT2 mutations.

4. Conclusions

Cardiac channelopathies have informed us about the channels and proteins that are important for atrial and ventricular excitability. Correlations between *in vitro* and clinical data suggest that patch clamp technique can provide prognostic information about individual mutations. We expect advancements in genotyping and high throughput molecular, tissue culture, and patch clamp techniques may enable clinicians to assess the risk of individual patient mutations identified through genotyping using *in vitro* patch clamp assays.

5. Acknowledgements

To Jennifer L. Smith, Daniel C. Bartos, Don E. Burgess, and Allison Reloj for their work in preparation of this chapter, and the National Heart Lung and Blood Institute grant R01 HL087039.

6. References

Abbott. G.W., Sesti, F., Splawski, I., Buck, M.E., Lehmann, M.H., Timothy, K.W., Keating, M.T., & Goldstein, S.A. MiRP1 forms IKr potassium channels with HERG and is associated with cardiac arrhythmia. *Cell.* 1999;97:175-187.

Andelfinger, G., Tapper, A.R., Welch, R.C., Vanoye, C.G., George, A.L. Jr, & Benson, D.W.KCNJ2 mutation results in Andersen syndrome with sex-specific cardiac and skeletal muscle phenotypes. *Am J Hum Genet.* 2002;71:663-668.

Anderson, C.L., Delisle, B.P., Anson, B.D., Kilby, J.A., Will, M.L., Tester, D.J., Gong, Q., Zhou, Z., Ackerman, M.J., & January, C.T. Most LQT2 mutations reduce Kv11.1 (hERG) current by a class 2 (trafficking-deficient) mechanism. *Circulation.* 2006;113:365-373.

Antzelevitch, C., Pollevick, G. D., Cordeiro, J. M., Casis, O., Sanguinetti, M. C., Aizawa, Y., Guerchicoff, A., Pfeiffer, R., Oliva, A., Wollnik, B., Gelber, P., & Bonaros, E. P., Jr., and 11 others. Loss-of-function mutations in the cardiac calcium channel underlie a new clinical entity characterized by ST-segment elevation, short QT intervals, and sudden cardiac death. *Circulation.* 2007;115: 442-449.

Barhanin, J., Lesage, F., Guillemare, E., Fink, M., Lazdunski, M., & Romey, G. K(V)LQT1 and lsK (minK) proteins associate to form the I(Ks) cardiac potassium current. *Nature.*1996;384:78-80.

Bartos, D.C., Duchatelet, S., Burgess, D.E., Klug, D., Denjoy, I., Peat, R., Lupoglazoff, J.M., Fressart, V., Berthet, M., Ackerman, M.J., January, C.T., Guicheney, P., & Delisle, B.P. R231C mutation in KCNQ1 causes long QT syndrome type 1 and familial atrial fibrillation. *Heart Rhythm.* 2011;8:48-55.

Bellocq, C., van Ginneken, A.C., Bezzina, C.R., Alders, M., Escande, D., Mannens, M.M., Baró, I., & Wilde, A.A. Mutation in the KCNQ1 gene leading to the short QT-interval syndrome. *Circulation.* 2004;109:2394-2397.

Benito, B., Brugada, R., Perich, R.M., Lizotte, E., Cinca, J., Mont, L., Berruezo, A, Tolosana, J.M., Freixa, X., Brugada, P., & Brugada, J. A mutation in the sodium channel is responsible for the association of long QT syndrome and familial atrial fibrillation. *Heart Rhythm.* 2008;5:1434-1440.

Brugada, R., Hong, K., Dumaine, R., Cordeiro, J., Gaita, F., Borggrefe, M., Menendez, T.M., Brugada, J., Pollevick, G.D., Wolpert, C., Burashnikov, E., Matsuo, K., Wu, Y.S., Guerchicoff, A., Bianchi, F., Giustetto, C., Schimpf, R., Brugada, P., & Antzelevitch, C. Sudden death associated with short-QT syndrome linked to mutations in HERG. *Circulation.* 2004;109:30-35.

Burashnikov, E., Pfeiffer, R., Barajas-Martinez, H., Delpón, E., Hu, D., Desai, M., Borggrefe, M., Häissaguerre, M., Kanter, R., Pollevick, G.D., Guerchicoff, A., Laiño, R., Marieb, M., Nademanee, K., Nam, G.B., Robles, R., Schimpf, R., Stapleton, D.D., Viskin, S., Winters, S., Wolpert, C., Zimmern, S., Veltmann, C., & Antzelevitch, C. Mutations in the cardiac L-type calcium channel associated with inherited J-wave syndromes and sudden cardiac death. *Heart Rhythm.* 2010;7:1872-1882.

Chen, J., Zou A., Splawski, I., Keating, M.T., & Sanguinetti, M.C. Long QT syndrome-associated mutations in the Per-Arnt-Sim (PAS) domain of HERG potassium channels accelerate channel deactivation. *J Biol Chem.* 1999;274:10113-10118.

Chen, L., Marquardt, M.L., Tester, D.J., Sampson, K.J., Ackerman, M.J., & Kass, R.S. Mutation of an A-kinase-anchoring protein causes long-QT syndrome. *Proc Natl Acad Sci U S A.* 2007;104:20990-20995.

Chen, Q., Kirsch, G. E., Zhang, D., Brugada, R., Brugada, J., Brugada, P., Potenza, D., Moya, A., Borggrefe, M., Breithardt, G., Ortiz-Lopez, R., Wang, Z., Antzelevitch, C., O'Brien, R. E., Schulze-Bahr, E., Keating, M. T., Towbin, J. A., & Wang, Q. Genetic basis and molecular mechanism for idiopathic ventricular fibrillation. *Nature.* 1998;392: 293-295.

Chen, Y.-H., Xu, S.-J., Bendahhou, S., Wang, X.-L., Wang, Y., Xu, W.-Y., Jin, H.-W., Sun, H., Su, X.-Y., Zhuang, Q.-N., Yang, Y.-Q., Li, Y.-B., Liu, Y., Xu, H.-J., Li, X.-F., Ma, N., Mou, C.-P., Chen, Z., Barhanin, J., & Huang, W. KCNQ1 gain-of-function mutation in familial atrial fibrillation. *Science.* 2003; 299: 251-254.

Crotti, L., Celano, G., Dagradi, F., & Schwartz, P.J. Congenital long QT syndrome. *Orphanet J Rare Dis.* 2008;3:18.

Curran, M.E., Splawski, I., Timothy, K.W., Vincent, G.M., Green, E.D., & Keating, M.T. A molecular basis for cardiac arrhythmia: HERG mutations cause long QT syndrome. *Cell.* 1995;80:795-803.

Darbar, D., Kannankeril, P. J., Donahue, B. S., Kucera, G., Stubblefield, T., Haines, J. L., George, A. L., Jr., & Roden, D. M. Cardiac sodium channel (SCN5A) variants associated with atrial fibrillation. *Circulation.* 2008;117: 1927-1935.

Deschênes, D., Acharfi, S., Pouliot, V., Hegele, R., Krahn, A., Daleau, P., & Chahine, M. Biophysical characteristics of a new mutation on the KCNQ1 potassium channel (L251P) causing long QT syndrome. *Can J Physiol Pharmacol.* 2003;81:129-134.

Delisle, B.P., Anson, B.D., Rajamani, S., & January, C.T. Biology of cardiac arrhythmias: ion channel protein trafficking. *Circ Res.* 2004;94:1418-1428.

Delisle, B.P., Slind, J.K., Kilby, J.A., Anderson, C.L., Anson, B.D., Balijepalli, R.C., Tester, D.J., Ackerman, M.J., Kamp, T.J., & January, C.T..Intragenic suppression of trafficking-defective KCNH2 channels associated with long QT syndrome. *Mol Pharmacol.* 2005;68:233-240.

Delpon, E., Cordeiro, J. M., Nunez, L., Thomsen, P. E. B., Guerchicoff, A., Pollevick, G. D., Wu, Y., Kanters, J. K., Larsen, C. T., Burashnikov, E., Christiansen, M., & Antzelevitch, C. Functional effects of KCNE3 mutation and its role in the development of Brugada syndrome. *Circ. Arrhythmia Electrophysiol.* 2008;1: 209-218.

Ellinor, P. T., Nam, E. G., Shea, M. A., Milan, D. J., Ruskin, J. N., & MacRae, C. A. Cardiac sodium channel mutation in atrial fibrillation. Heart Rhythm 2008;5: 99-105.

Ficker, E., Dennis, A.T., Obejero-Paz, C.A., Castaldo, P., Taglialatela, M., & Brown, A.M. Retention in the endoplasmic reticulum as a mechanism of dominant-negative current suppression in human long QT syndrome. *J Mol Cell Cardiol.* 2000;32:2327-2337.

Ficker, E., Dennis, A.T., Wang, L., & Brown, A.M. Role of the cytosolic chaperones Hsp70 and Hsp90 in maturation of the cardiac potassium channel HERG. *Circ Res.* 2003;92:e87-e100.

Ficker, E., Thomas, D., Viswanathan, P.C., Dennis, A.T., Priori, S.G., Napolitano, C., Memmi, M., Wible, B.A., Kaufman, E.S., Iyengar, S., Schwartz, P.J., Rudy, Y., & Brown, A.M. Novel characteristics of a misprocessed mutant HERG channel linked to hereditary long QT syndrome. *Am J Physiol Heart Circ Physiol.* 2000;279:H1748-H1756.

Furutani. M., Trudeau, M.C., Hagiwara, N., Seki, A., Gong, Q., Zhou, Z., Imamura, S., Nagashima, H., Kasanuki, H., Takao, A., Momma, K., January, C.T., Robertson, G.A., & Matsuoka, R. Novel mechanism associated with an inherited cardiac arrhythmia: defective protein trafficking by the mutant HERG (G601S) potassium channel. *Circulation.* 1999;99:2290-2294.

Goldenberg, I. & Moss, A.J. Long QT syndrome. *J Am Coll Cardiol.* 2008;51:2291-3000.

Gollob, M. H., Jones, D. L., Krahn, A. D., Danis, L., Gong, X.-Q., Shao, Q., Liu, X., Veinot, J. P., Tang, A. S. L., Stewart, A. F. R., Tesson, F., Klein, G. J., Yee, R., Skanes, A. C., Guiraudon, G. M., Ebihara, L., & Bai, D. Somatic mutations in the connexin 40 gene (GJA5) in atrial fibrillation. *New Eng. J. Med.* 2006; 354: 2677-2688.

Gong, Q., Jones, M.A., & Zhou, Z. Mechanisms of pharmacological rescue of trafficking-defective hERG mutant channels in human long QT syndrome. *J Biol Chem.* 2006;281(7):4069-4074.

Gong Q, Keeney DR, Molinari M, Zhou Z. Degradation of trafficking-defective long QT syndrome type II mutant channels by the ubiquitin-proteasome pathway. *J Biol Chem.* 2005;280:19419-19425.

Gong, Q., Zhang, L., Vincent, G.M., Horne, B.D., & Zhou, Z. Nonsense mutations in hERG cause a decrease in mutant mRNA transcripts by nonsense-mediated mRNA decay in human long-QT syndrome. *Circulation.* 2007;116:17-24.

Hayashi, K., Shimizu, M., Ino, H., Yamaguchi, M., Mabuchi, H., Hoshi, N., & Higashida, H. Characterization of a novel missense mutation E637K in the pore-S6 loop of HERG in a patient with long QT syndrome. *Cardiovasc Res.* 2002;54:67-76.

Hayashi, K., Shimizu, M., Ino, H., Yamaguchi, M., Terai, H., Hoshi, N., Higashida, H., Terashima, N., Uno, Y., Kanaya, H., & Mabuchi, H. Probucol aggravates long QT syndrome associated with a novel missense mutation M124T in the N-terminus of HERG. *Clin Sci (Lond).* 2004;107:175-182.

Hong, K., Piper, D.R., Diaz-Valdecantos, A., Brugada, J., Oliva, A., Burashnikov, E., Santos-de-Soto, J., Grueso-Montero, J., Diaz-Enfante, E., Brugada, P., Sachse, F., & Sanguinetti, M.C., Brugada R. De novo KCNQ1 mutation responsible for atrial fibrillation and short QT syndrome in utero. *Cardiovasc Res.* 2005;68(3):433-440.

Hu, D., Barajas-Martinez, H., Burashnikov, E., Springer, M., Wu, Y., Varro, A., Pfeiffer, R., Koopmann, T. T., Cordeiro, J. M., Guerchicoff, A., Pollevick, G. D., & Antzelevitch, C. A mutation in the beta-3 subunit of the cardiac sodium channel associated with Brugada ECG phenotype. *Circ. Cardiovasc. Genet.* 2009;2: 270-278.

Huang, F.D., Chen, J., Lin, M., Keating, M.T., & Sanguinetti, M.C. Long-QT syndrome-associated missense mutations in the pore helix of the HERG potassium channel. *Circulation.* 2001;104:1071-1075.

January, C.T., Gong, Q., & Zhou, Z. Long QT syndrome: cellular basis and arrhythmia mechanism in LQT2. *J Cardiovasc Electrophysiol.* 2000;11:1413-1418

Johnson, W.H. Jr, Yang, P., Yang, T., Lau, Y.R., Mostella, B.A., Wolff, D.J., Roden, D.M., & Benson, D.W. Clinical, genetic, and biophysical characterization of a homozygous HERG mutation causing severe neonatal long QT syndrome. *Pediatr Res.* 2003;53:744-748.

Jons, C., Moss, A.J., Lopes, C.M., McNitt, S., Zareba, W., Goldenberg, I., Qi, M., Wilde, A.A., Shimizu, W., Kanters, J.K., Towbin, J.A., Ackerman, M.J., & Robinson, J.L. Mutations in conserved amino acids in the KCNQ1 channel and risk of cardiac events in type-1 long-QT syndrome. *J Cardiovasc Electrophysiol.* 2009;20:859-865.

Kapa, S., Tester, D.J., Salisbury, B.A., Harris-Kerr, C., Pungliya, M.S., Alders, M., Wilde, A.A., & Ackerman, M.J. Genetic testing for long-QT syndrome: distinguishing pathogenic mutations from benign variants. *Circulation.* 2009;120:1752-1760.

Lees-Miller, J.P., Duan, Y., Teng, G.Q., Thorstad, K., & Duff, H.J.. Novel gain-of-function mechanism in K(+) channel-related long-QT syndrome: altered gating and selectivity in the HERG1 N629D mutant. *Circ Res.* 2000 ;86:507-513.

Lehnart, S.E., Ackerman, M.J., Benson, D.W., Jr, Brugada, R., Clancy, C.E., Donahue, J.K., George, A.L. Jr, Grant, A.O., Groft, S.C., January, C.T., Lathrop, D.A., Lederer, W.J., Makielski, J.C., Mohler, P.J., Moss, A., Nerbonne, J.M., Olson, T.M., Przywara, D.A., Towbin, J.A., Wang, L.H., & Marks, A.R. Inherited arrhythmias: a National Heart, Lung, and Blood Institute and Office of Rare Diseases workshop consensus report about the diagnosis, phenotyping, molecular mechanisms, and therapeutic approaches for primary cardiomyopathies of gene mutations affecting ion channel function. *Circulation.* 2007;116:2325-2345.

Lesage, F., Attali, B., Lakey, J., Honoré, E., Romey, G., Faurobert, E., Lazdunski, M., & Barhanin, J. Are Xenopus oocytes unique in displaying functional IsK channel heterologous expression? *Receptors Channels.* 1993;1(2):143-152.

Li, P., Ninomiya H, Kurata, Y., Kato, M., Miake, J., Yamamoto, Y., Igawa, O,. Nakai, A., Higaki, K., Toyoda, F., Wu, J., Horie, M,. Matsuura, H., Yoshida, A., Shirayoshi, Y., Hiraoka, M., & Hisatome, I. Reciprocal control of hERG stability by Hsp70 and Hsc70 with implication for restoration of LQT2 mutant stability. *Circ Res.* 2011;108:458-468.

London, B., Michalec, M., Mehdi, H., Zhu, X., Kerchner, L., Sanyal, S., Viswanathan, P. C., Pfahnl, A. E., Shang, L. L., Madhusudanan, M., Baty, C. J., Lagana, S., Aleong, R., Gutmann, R., Ackerman, M. J., McNamara, D. M., Weiss, R., & Dudley, S. C., Jr. Mutation in glycerol-3-phosphate dehydrogenase 1-like gene (GPD1-L) decreases cardiac Na+ current and causes inherited arrhythmias. *Circulation* 2007;116: 2260-2268.

Makiyama, T., Akao, M., Shizuta, S., Doi, T., Nishiyama, K., Oka, Y., Ohno, S., Nishio, Y., Tsuji, K., Itoh, H., Kimura, T., Kita, T., & Horie, M. A novel SCN5A gain-of-function mutation M1875T associated with familial atrial fibrillation. *J Am Coll Cardiol.* 2008;52:1326-1334.

Medeiros-Domingo, A., Kaku, T., Tester, D.J., Iturralde-Torres, P., Itty, A., Ye, B., Valdivia, C., Ueda, K., Canizales-Quinteros, S., Tusié-Luna, M.T., Makielski, J.C., & Ackerman, M.J. SCN4B-encoded sodium channel beta4 subunit in congenital long-QT syndrome. *Circulation.* 2007;116:134-142.

Mohler, P.J., Schott, J.J., Gramolini, A.O., Dilly, K.W., Guatimosim, S., duBell, W.H., Song, L.S., Haurogné, K., Kyndt, F., Ali, M.E., Rogers, T.B., Lederer, W.J., Escande, D., Le Marec, H., & Bennett, V. Ankyrin-B mutation causes type 4 long-QT cardiac arrhythmia and sudden cardiac death. *Nature.* 2003;421:634-639.

Moss, A.J., Shimizu, W., Wilde, A.A., Towbin, J.A., Zareba, W., Robinson, J.L., Qi, M., Vincent, G.M., Ackerman, M.J., Kaufman, E.S., Hofman, N., Seth, R., Kamakura, S., Miyamoto, Y., Goldenberg, I., Andrews, M.L., & McNitt, S. Clinical aspects of type-1 long-QT syndrome by location, coding type, and biophysical function of mutations involving the KCNQ1 gene. *Circulation.* 2007;115:2481-2489.

Nakajima, T., Furukawa, T., Hirano, Y., Tanaka, T., Sakurada, H., Takahashi, T., Nagai, R., Itoh, T., Katayama, Y., Nakamura, Y., & Hiraoka, M. Voltage-shift of the current activation in HERG S4 mutation (R534C) in LQT2. *Cardiovasc Res.* 1999;44:283-293.

Napolitano, C., Priori, S.G., Schwartz, P.J., Bloise, R., Ronchetti, E., Nastoli, J., Bottelli, G., Cerrone, M., & Leonardi, S. Genetic testing in the long QT syndrome: development and validation of an efficient approach to genotyping in clinical practice. *JAMA.* 2005;294:2975-2980.

Nerbonne, J.M. & Kass, R.S. Molecular physiology of cardiac repolarization. *Physiol Rev.* 2005;85:1205-1253.

Neyroud, N., Denjoy, I., Donger, C., Gary, F., Villain, E., Leenhardt, A., Benali, K., Schwartz, K., Coumel, P., & Guicheney, P. Heterozygous mutation in the pore of potassium channel gene KvLQT1 causes an apparently normal phenotype in long QT syndrome. *Eur J Hum Genet.* 1998;6:129-133.

Neyroud, N., Tesson, F., Denjoy, I., Leibovici, M., Donger, C., Barhanin, J., Fauré, S., Gary, F., Coumel, P., Petit, C., Schwartz, K., & Guicheney, P. A novel mutation in the potassium channel gene KVLQT1 causes the Jervell and Lange-Nielsen cardioauditory syndrome. *Nat Genet.* 1997;15:186-189.

Olson, T. M., Alekseev, A. E., Liu, X. K., Park, S., Zingman, L. V., Bienengraeber, M., Sattiraju, S., Ballew, J. D., Jahangir, A., & Terzic, A. Kv1.5 channelopathy due to KCNA5 loss-of-function mutation causes human atrial fibrillation. *Hum. Molec. Genet.* 2006;15: 2185-2191.

Olson, T. M., Alekseev, A. E., Moreau, C., Liu, X. K., Zingman, L. V., Miki, T., Seino, S., Asirvatham, S. J., Jahangir, A., & Terzic, A. K(ATP) channel mutation confers risk for vein of Marshall adrenergic atrial fibrillation. *Nat. Clin. Pract. Cardiovasc. Med.* 2007;4: 110-116.

Paulussen, A., Raes, A., Matthijs, G., Snyders, D.J., Cohen, N., & Aerssens, J. A novel mutation (T65P) in the PAS domain of the human potassium channel HERG results in the long QT syndrome by trafficking deficiency. *J Biol Chem.* 2002;277:48610-48616. Epub 2002 Sep 26.

Piippo, K., Swan, H., Pasternack, M., Chapman, H., Paavonen, K., Viitasalo, M., Toivonen, L., & Kontula, K. A founder mutation of the potassium channel KCNQ1 in long QT syndrome: implications for estimation of disease prevalence and molecular diagnostics. *J Am Coll Cardiol.* 2001;37:562-568.

Priori, S.G., Napolitano, C., Schwartz, P.J., Grillo, M., Bloise, R., Ronchetti, E., Moncalvo, C., Tulipani, C., Veia, A., Bottelli, G., & Nastoli, J. Association of long QT syndrome loci and cardiac events among patients treated with beta-blockers. *JAMA.* 2004;292:1341-1344.

Priori, S.G., Pandit, S.V., Rivolta, I., Berenfeld, O., Ronchetti, E., Dhamoon, A., Napolitano, C., Anumonwo, J., di Barletta, M.R., Gudapakkam, S., Bosi, G., Stramba-Badiale, M., & Jalife, J. A novel form of short QT syndrome (SQT3) is caused by a mutation in the KCNJ2 gene. *Circ Res.* 2005;96:800-807.

Priori, S.G., Schwartz, P.J., Napolitano, C., Bloise, R., Ronchetti, E., Grillo, M., Vicentini, A., Spazzolini, C., Nastoli, J., Bottelli, G., Folli, R., & Cappelletti, D. Risk stratification in the long-QT syndrome. *N Engl J Med.* 2003;348:1866-1874.

Rossenbacker, T., Mubagwa, K., Jongbloed, R.J., Vereecke, J., Devriendt, K., Gewillig, M., Carmeliet, E., Collen, D., Heidbüchel, H., & Carmeliet, P. Novel mutation in the Per-Arnt-Sim domain of KCNH2 causes a malignant form of long-QT syndrome. *Circulation.* 2005;111:961-968.

Ruan, Y., Liu, N., Napolitano, C., & Priori, S.G. Therapeutic strategies for long-QT syndrome: does the molecular substrate matter? *Circ Arrhythm Electrophysiol.* 2008;1:290-297.

Saarinen, K., Swan, H., Kainulainen, K., Toivonen, L., Viitasalo, M., & Kontula, K. Molecular genetics of the long QT syndrome: two novel mutations of the KVLQT1 gene and phenotypic expression of the mutant gene in a large kindred. *Hum Mutat.* 1998;11:158-165.

Sanguinetti, M.C., Curran, M.E., Spector, P.S., & Keating, M.T. Spectrum of HERG K+-channel dysfunction in an inherited cardiac arrhythmia *Proc Natl Acad Sci U S A.* 1996;93:2208-2212.

Sanguinetti, M.C., Curran, M.E., Zou, A., Shen, J., Spector, P.S., Atkinson, D.L., & Keating, M.T. Coassembly of K(V)LQT1 and minK (IsK) proteins to form cardiac I(Ks) potassium channel. *Nature.*1996;384:80-83.

Schott, J.J., Charpentier, F., Peltier, S., Foley, P., Drouin, E., Bouhour, J.B., Donnelly, P., Vergnaud, G., Bachner, L., Moisan, J.P., Le Marec, H., & Pascal, O. Mapping of a gene for long QT syndrome to chromosome 4q25-27. *Am J Hum Genet.* 1995;57:1114-1122.

Schulteis, C.T., Nagaya, N., & Papazian, D.M. Subunit folding and assembly steps are interspersed during Shaker potassium channel biogenesis. *J Biol Chem.* 1998;273:26210-26217.

Schulze-Bahr, E., Haverkamp, W., Wedekind, H., Rubie, C., Hördt, M., Borggrefe, M., Assmann, G., Breithardt, G., & Funke, H. Autosomal recessive long-QT syndrome (Jervell Lange-Nielsen syndrome) is genetically heterogeneous. *Hum Genet.* 1997;100:573-576.

Schwartz, P.J., Priori, S.G., Spazzolini, C., Moss, A.J., Vincent, G.M., Napolitano, C., Denjoy, I., Guicheney, P., Breithardt, G., Keating, M.T., Towbin, J.A., Beggs, A.H., Brink, P., Wilde, A.A., Toivonen, L., Zareba, W., Robinson, J.L., Timothy, K.W., Corfield, V., Wattanasirichaigoon, D., Corbett, C., Haverkamp, W., Schulze-Bahr, E., Lehmann, M.H., Schwartz, K., Coumel, P., & Bloise, R. Genotype-phenotype correlation in the long-QT syndrome: gene-specific triggers for life-threatening arrhythmias. *Circulation.* 2001;103:89-95.

Shah, M., Akar, F.G., & Tomaselli, G.F. Molecular basis of arrhythmias. *Circulation.* 2005;112:2517-2529.

Shalaby, F.Y., Levesque, P.C., Yang, W.P., Little, W.A., Conder, M.L., Jenkins-West, T., & Blanar, M.A. Dominant-negative KvLQT1 mutations underlie the LQT1 form of long QT syndrome. *Circulation.* 1997;96:1733-1736.

Silva, J.R., Pan, H., Wu, D., Nekouzadeh, A., Decker, K.F., Cui, J., Baker, N.A., Sept, D., & Rudy, Y. A multiscale model linking ion-channel molecular dynamics and electrostatics to the cardiac action potential. *Proc Natl Acad Sci U S A.* 2009;106:11102-11106.

Silva, J.R. & Rudy, Y. Multi-scale electrophysiology modeling: from atom to organ. *J Gen Physiol.* 2010;135:575-581

Smith, J.L., McBride, C.M., Nataraj, P.S., Bartos, D.C., January, C.T., & Delisle, B.P. Trafficking-deficient hERG K^+ channels linked to long QT syndrome are regulated by a microtubule-dependent quality control compartment in the ER. *Am J Physiol Cell Physiol.* 2011;301:C75-C85.

Smith, J.A., Vanoye, C.G., George, A.L. Jr, Meiler, J., & Sanders, C.R.. Structural models for the KCNQ1 voltage-gated potassium channel. *Biochemistry.* 2007;46:14141-14152.

Splawski, I., Timothy, K.W., Decher, N., Kumar, P., Sachse, F.B., Beggs, A.H., Sanguinetti, M.C., & Keating, M.T. Severe arrhythmia disorder caused by cardiac L-type calcium channel mutations. *Proc Natl Acad Sci U S A.* 2005;102:8089-8096.

Splawski, I., Tristani-Firouzi, M., Lehmann, M.H., Sanguinetti, M.C., & Keating, M..T. Mutations in the hminK gene cause long QT syndrome and suppress IKs function. *Nat Genet.* 1997;17:338-340.

Thomas, D., Khalil, M., Alter, M., Schweizer, P.A., Karle, C.A., Wimmer, A.B., Licka, M., Katus, H.A., Koenen, M., Ulmer, H.E., & Zehelein, J. Biophysical characterization of KCNQ1 P320 mutations linked to long QT syndrome 1. *J Mol Cell Cardiol.* 2010;48:230-237.

Thomas, D., Wimmer, A.B., Karle, C.A., Licka, M., Alter, M., Khalil, M., Ulmer, H.E., Kathöfer, S., Kiehn, J., Katus, H.A., Schoels, W., Koenen, M., & Zehelein, J.

Dominant-negative I(Ks) suppression by KCNQ1-deltaF339 potassium channels linked to Romano-Ward syndrome. *Cardiovasc Res.* 2005;67:487-497.

Tristani-Firouzi M, Jensen JL, Donaldson MR, Sansone V, Meola G, Hahn A, Bendahhou S, Kwiecinski H, Fidzianska A, Plaster N, Fu YH, Ptacek LJ, Tawil R.Functional and clinical characterization of KCNJ2 mutations associated with LQT7 (Andersen syndrome). *J Clin Invest.* 2002;110:381-388.

Ueda, K., Hirano, Y., Higashiuesato, Y., Aizawa, Y., Hayashi, T., Inagaki, N., Tana, T., Ohya, Y., Takishita, S., Muratani, H., Hiraoka, & M., Kimura, A. Role of HCN4 channel in preventing ventricular arrhythmia. *J. Hum. Genet.* 2009;54: 115-121, 2009.

Ueda, K., Valdivia C, Medeiros-Domingo, A., Tester, D.J., Vatta, M., Farrugia, G., Ackerman, M.J., & Makielski, J.C. Syntrophin mutation associated with long QT syndrome through activation of the nNOS-SCN5A macromolecular complex. *Proc Natl Acad Sci U S A.* 2008;105:9355-9360.

Van Norstrand, D. W., Valdivia, C. R., Tester, D. J., Ueda, K., London, B., Makielski, J. C., & Ackerman, M. J. Molecular and functional characterization of novel glycerol-3-phosphate dehydrogenase 1-like gene (GPD1-L) mutations in sudden infant death syndrome. *Circulation.* 2007;116: 2253-2259, 2007.

Vatta, M., Ackerman, M.J., Ye, B., Makielski, J.C., Ughanze, E.E., Taylor, E.W., Tester, D.J., Balijepalli, R.C, Foell, J.D., Li, Z., Kamp, T.J., & Towbin, J.A. Mutant caveolin-3 induces persistent late sodium current and is associated with long-QT syndrome. *Circulation.* 2006;114:2104-2112.

Walker, V.E., Atanasiu, R., Lam, H., & Shrier, A. Co-chaperone FKBP38 promotes HERG trafficking. *J Biol Chem.* 2007 A;282:23509-23016.

Walker, V.E., Wong, M.J., Atanasiu, R., Hantouche, C., Young, J.C., & Shrier, A. Hsp40 chaperones promote degradation of the HERG potassium channel. *J Biol Chem.* 2010;285:3319-3329.

Wang, Q., Curran, M.E., Splawski, I., Burn, T.C., Millholland, J.M., VanRaay, T.J., Shen, J., Timothy, K.W., Vincent, G.M., de Jager, T., Schwartz, P.J., Toubin, J.A., Moss, A.J., Atkinson, D.L., Landes, G.M., Connors, T.D., & Keating, M.T. Positional cloning of a novel potassium channel gene: KVLQT1 mutations cause cardiac arrhythmias. *Nat Genet.* 1996;12:17-23.

Wang, Q., Shen, J., Splawski, I., Atkinson, D., Li, Z., Robinson, J.L., Moss, A.J., Towbin, J.A., & Keating, M.T. SCN5A mutations associated with an inherited cardiac arrhythmia, long QT syndrome. *Cell.* 1995;80:805-811.

Wang, Z., Tristani-Firouzi, M., Xu, Q., Lin, M., Keating, M.T., & Sanguinetti, M.C. Functional effects of mutations in KvLQT1 that cause long QT syndrome. J Cardiovasc Electrophysiol. 1999 Jun;10(6):817-26.

Watanabe, H., Koopmann, T. T., Le Scouarnec, S., Yang, T., Ingram, C. R., Schott, J.-J., Demolombe, S., Probst, V., Anselme, F., Escande, D., Wiesfeld, A. C. P., Pfeufer, A., Kaab, S., Wichmann, H.-E., Hasdemir, C., Aizawa, Y., Wilde, A. A. M., Roden, D. M., & Bezzina, C. R. Sodium channel beta-1 subunit mutations associated with Brugada syndrome and cardiac conduction disease in humans. *J. Clin. Invest.* 118: 2260-2268, 2008.

Wollnik, B., Schroeder, B.C., Kubisch, C., Esperer, H.D., Wieacker, P., & Jentsch, T.J. Pathophysiological mechanisms of dominant and recessive KVLQT1 K+ channel mutations found in inherited cardiac arrhythmias. Hum Mol Genet. 1997;6:1943-

1949.Xia, M., Jin, Q., Bendahhou, S., He, Y., Larroque, M.-M., Chen, Y., Zhou, Q., Yang, Y., Liu, Y., Liu, B., Zhu, Q., Zhou, Y., et al. A Kir2.1 gain-of-function mutation underlies familial atrial fibrillation. *Biochem. Biophys. Res. Commun.* 2005;332: 1012-1019.

Wu, D., Delaloye, K., Zaydman, M.A., Nekouzadeh, A., Rudy, Y., & Cui, J. State-dependent electrostatic interactions of S4 arginines with E1 in S2 during Kv7.1 activation. *J Gen Physiol.* 2010;135:595-606.

Yamaguchi, M., Shimizu, M., Ino, H., Terai, H., Hayashi, K., Mabuchi, H., Hoshi, N., & Higashida, H. Clinical and electrophysiological characterization of a novel mutation (F193L) in the KCNQ1 gene associated with long QT syndrome. *Clin Sci (Lond).* 2003;104:377-382.

Yamashita, F., Horie, M., Kubota, T., Yoshida, H., Yumoto, Y., Kobori, A., Ninomiya, T., Kono, Y., Haruna, T., Tsuji, K., Washizuka, T., Takano, M., Otani, H., Sasayama, S., & Aizawa, Y. Characterization and subcellular localization of KCNQ1 with a heterozygous mutation in the C terminus. *J Mol Cell Cardiol.* 2001;33:197-207.

Yang, P., Kanki, H., Drolet, B., Yang, T., Wei, J., Viswanathan, P.C., Hohnloser, S.H., Shimizu, W., Schwartz, P.J,. Stanton, M., Murray, K.T., Norris, K., George, A.L., & Jr, Roden, D,M,. Allelic variants in long-QT disease genes in patients with drug-associated torsades de pointe. *Circulation.* 2002;105:1943-2948s.

Yang, Y., Li, J., Lin, X., Yang, Y., Hong, K., Wang, L., Liu, J., Li, L., Yan, D., Liang, D., Xiao, J., Jin, H., Wu, J., Zhang, Y., & Chen, Y.-H. Novel KCNA5 loss-of-function mutations responsible for atrial fibrillation. *J. Hum. Genet.* 2009;54: 277-283.

Yang, Y., Xia, M., Jin, Q., Bendahhou, S., Shi, J., Chen, Y., Liang, B., Lin, J., Liu, Y., Liu, B., Zhou, Q., Zhang, D., & 11 others. Identification of a KCNE2 gain-of-function mutation in patients with familial atrial fibrillation. *Am. J. Hum. Genet.* 2004;75: 899-905.

Yang, Y., Yang, Y., Liang, B., Liu, J., Li, J., Grunnet, M., Olesen, S.P., Rasmussen, H.B., Ellinor, P.T., Gao, L., Lin, X., Li, L., Wang, L., Xiao, J., Liu, Y., Liu, Y., Zhang, S., Liang, D., Peng, L., Jespersen, T., & Chen, Y.H. Identification of a Kir3.4 mutation in congenital long QT syndrome. *Am J Hum Genet.* 2010;86:872-880.

Zareba, W., Moss, A.J,. Sheu, G., Kaufman, E.S., Priori, S., Vincent, G.M., Towbin, J.A., Benhorin, J., Schwartz, P.J., Napolitano, C., Hall, W.J., Keating, M.T., Qi, M., Robinson, J.L., & Andrews, M.L. Location of mutation in the KCNQ1 and phenotypic presentation of long QT syndrome. *J Cardiovasc Electrophysiol.* 2003;14:1149-1153.

Zhou, Z., Gong, Q., Epstein, M.L., & January, C.T. HERG channel dysfunction in human long QT syndrome. Intracellular transport and functional defects. *J Biol Chem.* 1998;273:21061-21066.

Zhou, Z., Gong, Q., & January, C.T. Correction of defective protein trafficking of a mutant HERG potassium channel in human long QT syndrome. Pharmacological and temperature effects. *J Biol Chem.* 1999;274:31123-31126.

http://www.fsm.it/cardmoc/

12

Gating Charge Movement in Native Cells: Another Application of the Patch Clamp Technique

Oscar Vivas, Isabel Arenas and David E. García
Department of Physiology, School of Medicine, Universidad Nacional Autónoma de México, UNAM. México, D.F., México

1. Introduction

Since the invention of the patch clamp technique by Neher and Sakmann (Hamill et al., 1981) macroscopic ionic currents can be recorded in neurons of a small diameter. Cortical neurons from mammals (Edwards et al., 1989) and even synaptic ends (Stuart and Sakmann, 1994) have been successfully patched in order to study many physiological phenomena which are sensitive to voltage changes. In addition, the activity of just one protein has been observed (Hamill and Sakmann, 1981; Sakmann et al., 1980). Unitary currents from a single voltage dependent channel can be recorded in a small patch of the plasma membrane. Both macroscopic and unitary currents allow understand the processes occurring once channels are opened.

However, there are various closed states before a channel opens and therefore, processes taking place during these closed states that to date are partially understood by either macroscopic or unitary currents. That is why gating charge movement occurring during transitional states of the channels should be studied. A case of a process taking place during closed states is the regulation of voltage dependent channels by G-proteins, which has begun to be studied by recording gating charge movement.

Voltage dependent calcium, potassium and even sodium channels are modulated by neurotransmitters (Hille, 1994; Ikeda and Dunlap, 1999). They exert its effect by activating G-protein coupled receptors (GPCRs). Neurotransmitter-induced activation of GPCRs can in turn activate several signaling pathways having pleiotropic effects on neurons (Hille, 1994; Hille et al., 1995) The molecule responsible for ion channel regulation has been identified to be βγ G-protein subunits ($G_{\beta\gamma}$) (Herlitze et. al., 1996; Ikeda, 1996). Binding of $G_{\beta\gamma}$ to the channel induces characteristic changes. First, the current amplitude is reduced (Dunlap and Fischbach, 1981). Second, the voltage dependence of the channel is modified (Bean, 1989). Third, the activation kinetics of the current is slowed (Marchetti et. al., 1986). And fourth, $G_{\beta\gamma}$ unbinds from the channels by a strong depolarizing prepulse (Elmslie et. al., 1990; Grassi and Lux, 1989). However, it is still controversial whether $G_{\beta\gamma}$ dimer diminishes the number of available channels by either changing the unitary conductance or modifying the structure of the channel through conformational changes, therefore altering its function.

The aim of this chapter is to show that recording of the gating charge movement is a powerful tool to address ion channel regulation during transitional states. Firstly, we will overview briefly the fundamentals of the charge movement to point out the methodology required to successfully investigate gating charge movement. Then, we will discuss why cell lines or native cells can be chosen for recording gating charge movement. Following, the procedure to separate charge movement into components has to be carried out in the case of native cells. Finally, an application of studying charge movement to understand ion channel regulation should be addressed.

2. Fundamentals of charge movement recording

Many physiological functions are regulated by changes in membrane voltage through regulation of different proteins. Several proteins involved in membrane transport mechanisms (including ion channels, transporters and pumps) and G-protein coupled receptors are voltage dependent (Ben-Chaim et al., 2006; Hodgkin and Huxley, 1952). The molecular events that link changes in membrane voltage and the structural changes of these proteins, and thus to the regulation of physiological functions, seem to began with the displacement or reorientation of some amino acids capable of sensing the membrane voltage. This voltage sensor is the machinery required to transduce membrane potential to responses such as action potential conduction, neurotransmitter release, muscle contraction, secretion from endocrine cells and agonist binding sensitivity (Almers, 1978). Therefore, facilitating or hampering the displacement of the voltage sensor could be related to the regulation of these physiological functions.

During the study of the voltage dependence of action potential generation in nerve, Hodgkin and Huxley (1952) predicted, for the first time, the existence of molecules with charge or dipole movement capable of distribute or orientate in response to changes in the electric field. Today it is well accepted that gating charge movement correlates well with the opening and closing of sodium and potassium ion channels (Almers and Armstrong, 1980; Armstrong, 1975; Armstrong and Bezanilla, 1974; Bezanilla et al., 1982), which are responsible for the permeability changes occurring during an action potential. Positive charge amino acids, which constitute the voltage sensor of ion channels, are located in the fourth segment (S4) of each domain of the protein and S4 mutations alter gating charge movement and voltage dependence of the channels (Perozo et al., 1994; Stühmer et al., 1989). As a consequence of the movement of the voltage sensor a current is produced. This current is the so-called gating charge movement (Figure 1). Since the movement of these charged molecules is the first step in the transduction of voltage changes, gating charge movement precedes the opening and follows the closure of ion channels.

After Hodgkin and Huxley, there was an intense search of the current related to the opening and closing of the channels to corroborate the presence of polar molecules in the bilayer of cell membranes either in nerve or in muscle. This was firstly accomplished before the patch clamp technique invention in muscle (Schneider and Chandler, 1973) and in the squid giant axon (Armstrong and Bezanilla, 1973) by using an axial wire to clamp the voltage along the fiber. Gating charge movement is observed solely under favorable conditions since it is masked almost completely by capacitive and ionic currents, which are at least one order of magnitude bigger. For gating charge movement recording cells have to be bathed constantly with appropriated solutions designed to eliminate ionic components such as sodium, potassium and calcium currents.

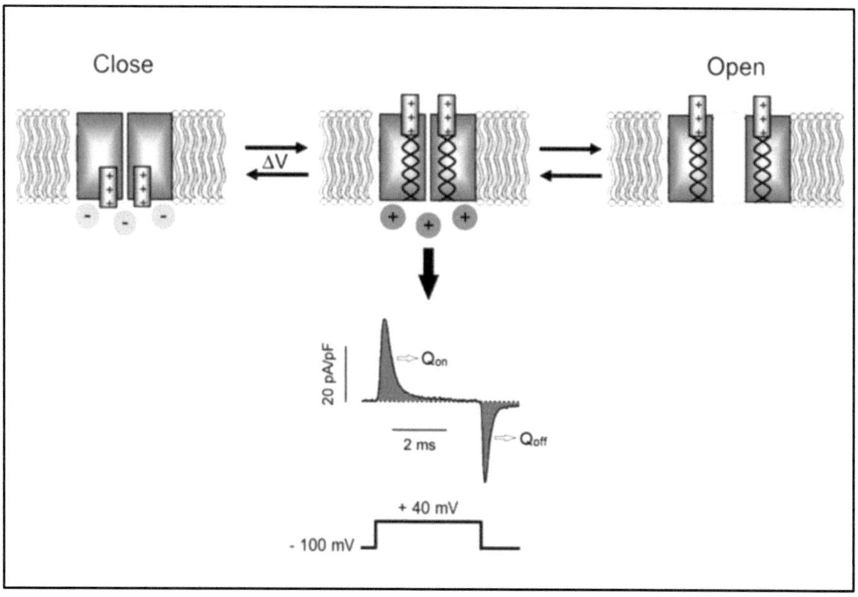

Fig. 1. Model that summarizes the origin of gating charge movement. According to the classical gating model (Horn, 2004; Starace and Bezanilla, 2004) a narrow portion of the S4 is in the gating pore where the membrane electrical field (dashed area surrounding voltage sensor) exerts its effect. Membrane depolarization (ΔV) exerts an electrostatic force on S4's positive charges that are within or near the membrane electrical field and thus can sense changes in the membrane voltage (the so-called gating charges). This causes S4 movements, resulting in a transfer of gating charges (coiled-coils into the electrical field) through the gating pore (gating charge transfer).

Ionic currents can be abolished by replacing permeant ions on both sides of the membrane by species that do not pass through the channels present in the membrane. Some typical replacements are tetraethylammonium (TEA), N-methylglucamine (NMG), some organic cations such as aspartate or glutamate and cesium. In addition to substitute permeant ions, it is common to use blockers such as tetrodotoxin (TTX), nifedipine and cadmium. One point to consider is the effect of some ions or blockers on the movement of the voltage sensor themselves since they would lead to a misinterpretation of the results. For example, we have designed an internal solution to study gating charge movement in neurons from superior cervical ganglion after several trials and experimental errors. It has been designed with cesium as the non permeant ion in the internal solution. Cesium based solution improves the sealing and extends the duration of the experiment by blocking ionic currents without altering the kinetics or voltage dependence of gating charge movement. Additionally TTX, cadmium and lanthanum are added to the external solution to block residual ionic currents (Hernandez-Ochoa et al., 2007).

When a depolarizing voltage pulse (test pulse) is given to a cell bathed with the solution stated above, the recording is comprised of a capacitive current, a small leak current and the gating charge movement. However the signal of the gating charge movement cannot be observed because it is under the portion of the slow linear capacitive component. This component does not change with time and voltage as it can be measured from the resting membrane potential. Taking advantage of the linear properties of the capacitive and leak currents, gating charge movement can be unmasked by means of a subtraction protocol applied before or after the test pulse. Figure 2A shows a protocol intended to subtract the linear components from the whole signal with 5 pulses of the fifth amplitude of the test pulse from a holding potential that does not elicit an ionic current (P/-5). Figure 2B shows that voltage steps of opposite polarities, which are applied at the range of subtraction pulses, are totally cancelled by subtraction confirming the linear properties at these potentials. By contrast, gating charge movement behaves nonlinearly with voltage. There are a finite number of charges sensing voltage in the membrane, so that gating charge movement strongly depends on voltage and saturates (Figure 2C). Therefore the subtraction of a signal obtained at the range of voltage where only capacitive and leak currents are elicited let us obtain the recording of gating charge movement, whatever subtraction protocol is used (Figure 2D).

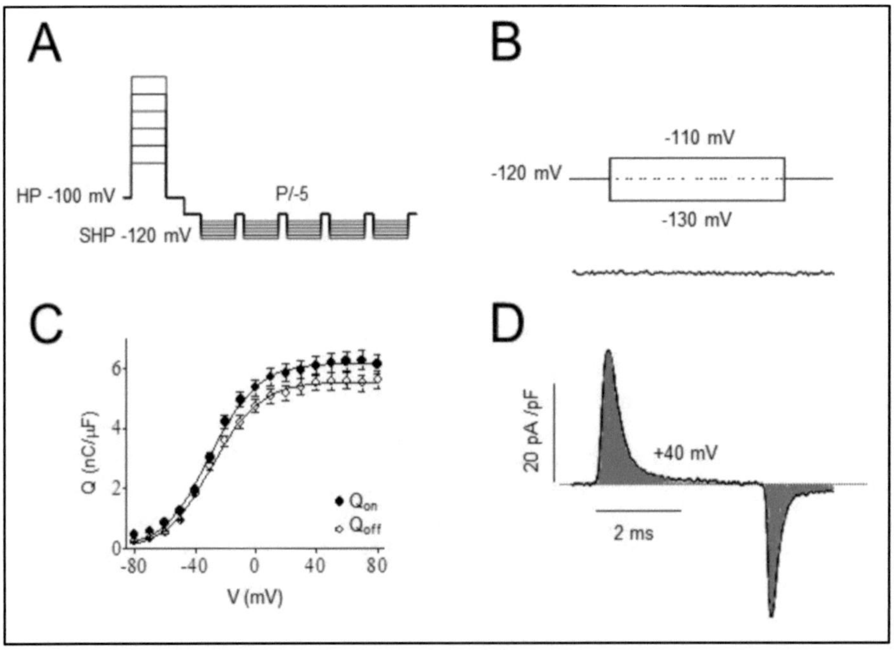

Fig. 2. General procedure to measure charge movement. A: a typical protocol for linear-components subtraction. B: absence of intramembrane charge movement at more negative potentials. C: Q-V relationship illustrating voltage dependence and saturation. D: typical recording of asymmetric current after linear-components subtraction. Shaded area under the curve represents charge movement. (Modified from Hernandez-Ochoa et al., 2007).

3. Why native cells or cell lines?

As stated before gating charge movement is at least one order of magnitude smaller than capacitive and ionic currents. In addition, the magnitude of gating charge movement depends on the number of voltage sensors moving in the cell membrane (Armstrong and Bezanilla, 1997; Noceti et al., 1996). Thus, the ideal system to study gating charge movement is a cell with a small capacitance but with a high density of channels. As an example of the latter, it has been successfully studied gating charge movement of only one type of channel in an expression system Xenopus oocytes (Bezanilla et al., 1991; Conti and Stühmer, 1989; Neely et al., 1994) or HEK cells (Starace et al., 1997).

Heterologous channel expression has led to significant advances in the molecular and functional understanding of gating charge movement. Channel expression in Xenopus oocytes has allowed to establish that gating charge movement is the consequence of protein conformational changes (Villalba-Galea et al., 2009) where each structural transition involves a quantum of charge (Conti and Stühmer, 1989), that the voltage sensor of the channels is the 4 transmembrane segment, S4 (Logothetis et al., 1992; Perozo et al., 1994), and that every transition from the closed to open state has a unique constant rate (Schoppa and Sigworth, 1998).

On the other hand, charge movement recording in native cells has the advantage of studying related phenomena in unaltered conditions. However, the study of charge movement in native cells is complicated by the presence of different types of channels, making difficult to know the contribution of each type of channel in the total charge movement recording (Chameau et al., 1995). In the case of the classical preparation of squid axons, the study of gating charge movement did not represent a problem due to the presence of only two population of channels at very high densities, sodium and potassium, and they could be readily separated by kinetic and temperature changes (Bezanilla, 1985; Gilly and Armstrong, 1980). In contrast, in mammalian ventricular cardiac cells were found that gating charge movement involving sodium and calcium channel populations, could not be separated (Bean and Rios, 1989).

4. Separation of gating charge movement components in native cells

In a previous report we show how to isolate and identify channels that are contributing to the total gating charge movement from a sympathetic neuron by two different approaches. The first approach analyzes gating charge movement kinetics. The hypothesis behind is that if sodium (Na_V), calcium, (Ca_V) and potassium (A-type, K_V) channels with different constant times are contributing to the total gating charge movement, every component should be observed separated in time. This approach is essentially a kinetic separation of components of the gating charge movement. It has been used classically to establish the components of gating charge movement in skeletal muscle (Adrian and Peres, 1979; Francini et al., 2001; Huang, 1982; Huang, 1988).

An additional analysis compares the gating charge movement immobilization with voltage-gated channel inactivation. Immobilization of gating charge movement can be associated with fast voltage-dependent inactivation of ion currents, such as I_{Na} or $I_{K(A)}$ (Armstrong and Bezanilla, 1977; Bezanilla, 2000). Time- or voltage-dependent immobilization leads to differences in the charges moved between depolarization (Q_{on}) or repolarization (Q_{off}). To test this possibility the voltage dependence of charge immobilization is compared to the voltage dependence of the steady-state inactivation for both ionic currents (Figure 3). Moreover, the parameters of a Boltzmann distribution fitted to gating charge movement immobilization and ionic current inactivation can help to determine which channel is mainly involved in the total gating charge movement. Following this approach we found that there is a component of gating charge movement that immobilizes with the same parameters of the I_{Na} inactivation. Half-voltage was -67.2 mV and slope factor was 10.6 mV for charge immobilization; half-voltage was -67.4 mV and slope factor was 6.7 mV for I_{Na} inactivation. This component is the two-thirds of the total charge movement. The other component, which does not immobilize probably arise from $Ca_V2.2$ calcium channels in SCG neurons. Bean and Rios (1989) also described two components of the total gating charge movement in rat and rabbit cardiac cells. With the same approach, they found that the 40% of the gating charge movement arises from sodium channels and hypothesized that the other 60% most probably arises from calcium channels.

The second approach uses non-stationary fluctuation analysis of currents. This analysis can be used to estimate the number of channels in the cell body of the neuron (Alvarez et al., 2002; Sigworth, 1980). Then, the maximal charge transferred during the gating charge movement (Q_{max}) is compared to the charge transferred by one type of channel. To this end, ensembles of currents are generated by a serial of 80 to 200 identical and consecutive voltage pulses. The temporal course of the variance (σ^2) and mean current of the ensemble is calculated. Since changes in σ^2 arise not only from stochastic gating of channels but also from thermal background noise, the temporal course of σ^2 at holding potential is subtracted. After correction of the σ^2, the relationship of mean current and σ^2 can be fitted to the following equation:

$$\sigma^2 = i(I) - (I)^2/N$$

Where *I* is the mean current, *N* is the number of channels and *i* is the unitary current. This theory is based on three general assumptions about voltage-gated channels: 1) Channels are composed of a homogeneous population 2) Channels population has a fixed number of independent gating channels and 3) Channels are assumed to exist in either a conducting or non conducting state.

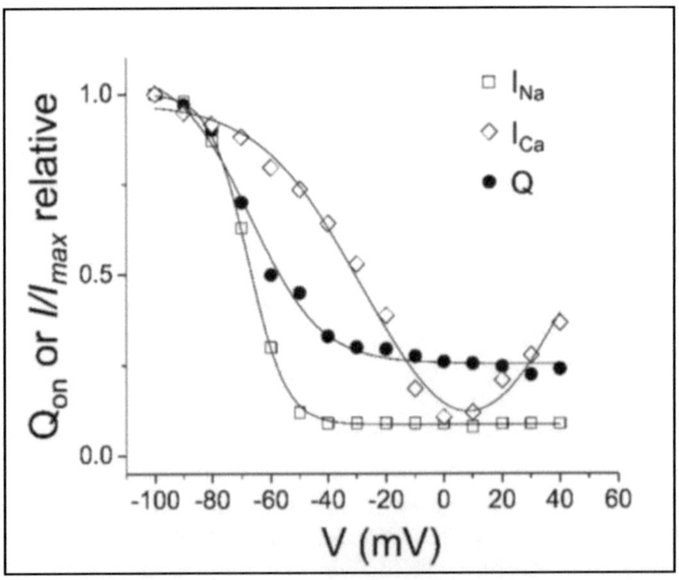

Fig. 3. Relationship between charge movement and two ionic-channel currents in a native cell (Modified from Hernandez-Ochoa et al., 2007).

When the probability in the conducting state is maximal, the theory predicts a parabolic relation between σ^2 and the mean current that can be used to estimate the number of channels and the unitary current amplitude. We found that 3.8 x 10^4 calcium channels are in the cell body of one-day cultured SCG neuron of 40 pF, which it is equal to 9.5 channels/μm^2. Considering 12 elementary charges (e_0) equivalent gating charge per channel (Noceti et al., 1996) and that Q_{max} is equal to 6.2 nC/cm^2 or, translated into elementary charges, 390 e_0 it would be 32 channels/μm^2 in rat sympathetic neurons. Thus, gating charge movement arising from calcium channels comprises approximately one-third of total gating charge movement. This value correlates well with our first approach in which two-third of total gating charge movement arises from Na_V channels while one-third from Ca_V channels.

5. Ion channel regulation during closed states studied with gating charge movement recording

Ion channels are regulated by a wide range of neurotransmitters through G-protein coupled receptors (Hille, 1994). Effector molecules of this neurotransmitter mediated regulation are $\beta\gamma$ subunits of heterotrimeric G-proteins (Herlitze et al., 1996; Ikeda, 1996). It has been proposed that G-protein regulation is carried out at the closed states transiting the channel toward its opening (Bean, 1989; Boland and Bean, 1993). Whole cell recording has been the most frequently used technique to study the effect of G-proteins on channels after they have opened, however channels are already regulated before ionic currents come into view. Thus, the effect of G-proteins during transitional states of the channels is still unclear. Since the transition between closed states involves the movement of the voltage sensor, regulation of ion channels at these states has been better understood through gating charge movement recording. The hypothesis that ion channels are regulated during closed states is strongly supported by the finding that the activation of G-proteins reduces gating charge movement (Hernandez-Ochoa et al., 2007). GTPγS dialysis, a non-hydrolysable activator of G-proteins, has three main effects on the voltage dependence of the total charge in SCG neurons: 1) a 34% decrease in Q_{max}, 2) a 10 mV shift of half-voltage toward positive voltage and 3) a 63% increase in the slope factor. These observations suggest that non-selective G-protein activation modifies both the voltage dependence and the number of available channels present in the cell body of SCG neurons. In addition, the activation of a specific signaling pathway by application of neurotransmitters such as noradrenaline or angiotensin II also reduces the gating charge movement.

Changes in gating charge movement recording under modulated conditions suggest that the voltage sensor of the channels is altered in such conditions. However, how voltage sensor activity can be modified by G-proteins? It has been proposed different mechanisms for modulating the properties of voltage sensors. All these studies have used gating charge movement recording to understand ion channel modulation.

Immobilization of gating charge movement is related with inactivation of channels, an intrinsic process of regulation of some voltage dependent ion channels (Armstrong and Bezanilla, 1977; Armstrong et al., 1973; Bezanilla and Armstrong, 1977; Bezanilla et al., 1991; Nonner, 1980). However, inactivation seems to be not the only process by which gating charge movement is immobilized. Recently it has been shown that modification of the lipid milieu induces immobilization of gating charge movement on potassium channels. The enzymatic removal of phosphate head groups of sphingomyelin diminishes gating charge movement by 90% without changes in the voltage dependence of the charge (Xu et al., 2008). In addition, toxins from several groups of animals alter ion channels (Sheets and Hanck, 1999; Sheets et al., 1999) by constraining conformational changes between transitional states (Catterall et al., 2007; Sokolov et al., 2008). This mechanism is called trapping. In molecular terms, it can be interpreted as withholding of the voltage sensor in resting or active position and thus the channel is modulated. Toxins can inhibit ion channel activation, deactivation or inactivation since they entrap the channel in the open, closed or inactivated conformation.

Another process that modulates ion channels, mainly potassium channels, is phosphorylation (Kaczmarek, 1988; Levitan, 1985). Throughout studying the effect of phosphorylation on gating charge movement of potassium channels it has been discovered that gating charge movement is reduced and that voltage dependence of the charge shifted toward positive potentials (Augustine and Bezanilla, 1990). The modulation of potassium channels is mediated by electrostatic interactions between voltage sensor and phosphate groups (Perozo and Bezanilla, 1990). In this case it can be interpreted as a screening of the potential profile.

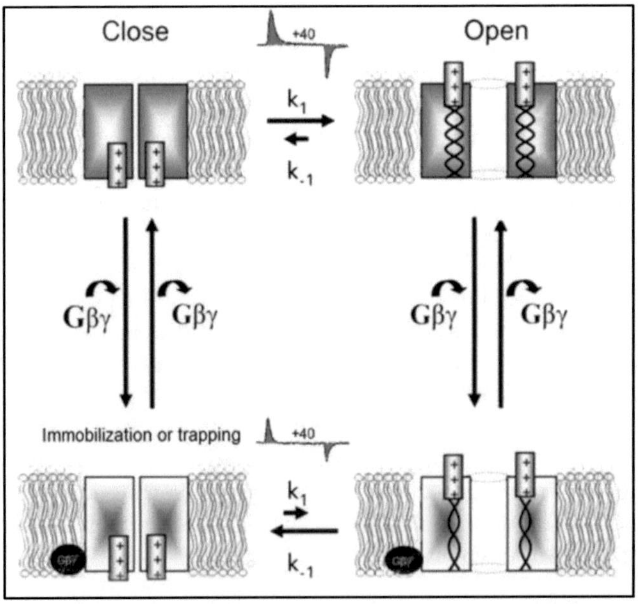

Fig. 4. Allosteric model to explain modulation of gating charge movement by G-proteins. This model postulates that the change in the number of gating charges (coiled-coils into the electrical field) due to G-protein βγ subunits binding to the channel is by means of a redistribution of the local electric field that surrounds the voltage sensor (shown as a change in the color pattern) as a result of the allosteric modulation of the gating charge movement. (Modified from Rebolledo-Antunez et al., 2009).

The other molecules affecting gating charge movement are accessory subunits of the ion channels and ruthenium complexes. While accessory subunits of calcium channels improve the coupling between excitation and opening (Lacinova and Klugbauer, 2004), ruthenium complexes uncouples voltage sensor movement with the opening transition (Jara-Oseguera et al., 2011). All these possible changes: immobilization, trapping, screening, coupling or uncoupling, are examples of how ion channels can be modulated by means of changes in properties of the gating charge movement. Thus gating charge movement recording is a powerful tool to understand how ion channels are modulated.

Concerning G-protein modulation of ion channels is still unclear which mechanism could explain the inhibition of gating charge movement by neurotransmitters in SCG neurons. Inhibition of gating charge movement is commonly accompanied by a shift and an increase of the slope factor of the voltage dependence of the charge. In addition we have observed that during ion channel modulation, the effective charge is also reduced (Rebolledo-Antunez et al., 2009). Reductions in the effective charge, which is transferred during activation of the channel with a potential change (Figure 4), support the idea that G-proteins exert an allosteric modulation (Herlitze et al., 2001; Monod et al., 1965) of the gating charge movement, condition in which voltage sensors respond differentially to changes in the membrane potential.

6. Conclusions

This chapter offers an updated overview on the origin of the gating charge movement and points out the powerful application of this tool in understanding mechanistic processes underlying ion channel regulation. Furthermore, it illustrates a remarkable example using a model of ion channel regulation by G-proteins.

7. Acknowledgment

Supported by UNAM-DGAPA-PAPIIT IN200710 grant, and a grant from The Alexander von Humboldt Stiftung, Germany to DEG.

8. References

Adrian, RH., & Peres, A. (1979). Charge movement and membrane capacity in frog muscle. *J Physiol.* 289:83-97.

Almers, W. (1978). Gating currents and charge movements in excitable membranes. *Rev Physiol Biochem Pharmacol.* 82:96-190.

Almers, W., & Armstrong, CM. (1980). Survival of K+ permeability and gating currents in squid axons perfused with K+-free media. *J Gen Physiol.* 75:61-78.

Alvarez, O., Gonzalez, C., and Latorre, R. (2002). Counting channels: a tutorial guide on ion channel fluctuation analysis. *Adv Physiol Educ.* 26:327-341.

Armstrong, CM. (1975). Currents associated with the ionic gating structures in nerve membrane. *Ann N Y Acad Sci.* 30:264:265-77.

Armstrong, CM., & Bezanilla, F. (1973). Currents related to movement of the gating particles of the sodium channels. *Nature.* 242:459-61.

Armstrong, CM., & Bezanilla, F. (1974). Charge movement associated with the opening and closing of the activation gates of the Na channels. *J Gen Physiol.* 63:533-52.

Armstrong, CM., & Bezanilla, F. (1977). Inactivation of the sodium channel. II. Gating current experiments. *J Gen Physiol.* 70:567-590.

Armstrong, CM., Bezanilla, F., & Rojas, E. (1973). Destruction of sodium conductance inactivation in squid axons perfused with pronase. *J Gen Physiol.* 62:375-391.

Augustine, CK., & Bezanilla, F. (1990). Phosphorylation modulates potassium conductance and gating current of perfused giant axons of squid. *J Gen Physiol*. 95:245-271.

Bean, BP. (1989). Neurotransmitter inhibition of neuronal calcium currents by changes in channel voltage dependence. *Nature*. 340:153-156.

Bean, BP., & Rios, E. (1989). Nonlinear charge movement in mammalian cardiac ventricular cells. Components from Na and Ca channel gating. *J Gen Physiol*. 94:65-93

Ben-Chaim, Y., Chanda, B., Dascal, N., Bezanilla, F., Parnas, I., & Parnas, H. (2006). Movement of 'gating charge' is coupled to ligand binding in a G-protein-coupled receptor. *Nature*. 444:106-9.

Bezanilla, F., White, MM., & Taylor, RE. (1982). Gating currents associated with potassium channel activation. *Nature*. 296:657-9.

Bezanilla, F. (1985). Gating of sodium and potassium channels. *J Membr Biol*. 88:97-111.

Bezanilla, F. (2000). The voltage sensor in voltage-dependent ion channels. *Physiol Rev*. 80:555-592.

Bezanilla, F., & Armstrong, CM. (1977). Inactivation of the sodium channel. I. Sodium current experiments. *J Gen Physiol*. 70:549-566.

Bezanilla, F., Perozo, E., Papazian, DM., & Stefani, E. (1991). Molecular basis of gating charge immobilization in Shaker potassium channels. *Science*. 254:679-683.

Boland, LM., & Bean, BP. (1993). Modulation of N-type calcium channels in bullfrog sympathetic neurons by luteinizing hormone-releasing hormone: kinetics and voltage dependence. *J Neurosci*. 13:516-533.

Catterall, WA., Cestele, S., Yarov-Yarovoy, V., Yu, FH., Konoki, K., & Scheuer, T. (2007). Voltage-gated ion channels and gating modifier toxins. *Toxicon*. 49:124-141.

Chameau, P., Bournaud, R., & Shimahara, T. (1995). Asymmetric intramembrane charge movement in mouse hippocampal pyramidal cells. *Neurosci Lett*. 201:159-162.

Conti, F., & Stühmer, W. (1989). Quantal charge redistributions accompanying the structural transitions of sodium channels. *Eur Biophys J*. 17:53-9.

Dunlap, K., & Fischbach, GD. (1981). Neurotransmitters decrease the calcium conductance activated by depolarization of embryonic chick sensory neurones. *J Physiol*. 317:519-35.

Edwards, FA., Konnerth, A., Sakmann, B., &Takahashi, T. (1989). A thin slice preparation for patch clamp recordings from neurones of the mammalian central nervous system. *Pflugers Arch*. 414:600-612.

Elmslie, KS., Zhou, W., & Jones, SW. (1990). LHRH and GTP-gamma-S modify calcium current activation in bullfrog sympathetic neurons. *Neuron*. 5:75-80.

Francini, F., Bencini, C., Piperio, C., & Squecco, R. (2001). Separation of charge movement components in mammalian skeletal muscle fibres. *J Physiol*. 537:45-56.

Gilly, WF., & Armstrong, CM. (1980). Gating current and potassium channels in the giant axon of the squid. *Biophys J*. 29:485-492.

Grassi, F., & Lux, HD. (1989). Voltage-dependent GABA-induced modulation of calcium currents in chick sensory neurons. *Neurosci Lett*. 105:113-9.

Hamill, OP., Marty, A., Neher, E., Sakmann, B., & Sigworth, FJ. (1981). Improved patch-clamp techniques for high-resolution current recording from cells and cell-free membrane patches. *Pflugers Arch*. 391:85-100.

Hamill, OP., & Sakmann, B. (1981). Multiple conductance states of single acetylcholine receptor channels in embryonic muscle cells. *Nature*. 294:462-464.

Herlitze, S., Garcia, DE., Mackie, K., Hille, B., Scheuer, T., & Catterall, WA. (1996). Modulation of Ca2+ channels by G-protein beta gamma subunits. *Nature*. 380:258-262.

Herlitze, S., Zhong, H., Scheuer, T., & Catterall, WA. (2001). Allosteric modulation of Ca2+ channels by G proteins, voltage-dependent facilitation, protein kinase C, and Ca(v)beta subunits. *Proc Natl Acad Sci U S A*. 98:4699-704.

Hernandez-Ochoa, EO., Garcia-Ferreiro, RE., & Garcia, DE. (2007). G protein activation inhibits gating charge movement in rat sympathetic neurons. *Am J Physiol Cell Physiol*. 292:C2226-2238.

Hille, B. (1994). Modulation of ion-channel function by G-protein-coupled receptors. *Trends Neurosci*. 17:531-536.

Hille, B., Beech, DJ., Bernheim, L., Mathie, A., Shapiro, MS., & Wollmuth, LP. (1995). Multiple G-protein-coupled pathways inhibit N-type Ca channels of neurons. *Life Sci*. 56:989-92.

Hodgkin, AL., & Huxley, AF. (1952). A quantitative description of membrane current and its application to conduction and excitation in nerve. *J. Physiol*. 117:500-544.

Horn, R. (2004). How S4 segments move charge. Let me count the ways. *J Gen Physiol*. 123:1-4.

Huang, CL. (1982). Pharmacological separation of charge movement components in frog skeletal muscle. *J Physiol*. 324:375-387.

Huang, CL. (1988). Intramembrane charge movements in skeletal muscle. *Physiol Rev*. 68:1197-1147.

Ikeda, SR. (1996). Voltage-dependent modulation of N-type calcium channels by G-protein beta gamma subunits. *Nature*. 380:255-258.

Ikeda, SR., & Dunlap, K. (1999). Voltage-dependent modulation of N-type calcium channels: role of G protein subunits. *Adv Second Messenger Phosphoprotein Res*. 33:131-51.

Jara-Oseguera, A., Ishida, IG., Rangel-Yescas, GE., Espinosa-Jalapa, N., Perez-Guzman, JA., Elias-Vinas, D., Le Lagadec, R., Rosenbaum, T., & Islas, LD. (2011). Uncoupling charge movement from channel opening in voltage-gated potassium channels by ruthenium complexes. *J Biol Chem*. 286:16414-16425.

Kaczmarek, LK. (1988). The regulation of neuronal calcium and potassium channels by protein phosphorylation. *Adv Second Messenger Phosphoprotein Res*. 22:113-138.

Lacinova, L., & Klugbauer, N. (2004). Modulation of gating currents of the Ca(v)3.1 calcium channel by alpha 2 delta 2 and gamma 5 subunits. *Arch Biochem Biophys*. 425:207-213.

Levitan, IB. (1985). Phosphorylation of ion channels. *J Membr Biol*. 87:177-190.

Logothetis, DE., Movahedi, S., Satler, C., Lindpaintner, K., & Nadal-Ginard, B. (1992). Incremental reductions of positive charge within the S4 region of a voltage-gated K+ channel result in corresponding decreases in gating charge. *Neuron*. 8:531-40.

Marchetti, C., Carbone, E., & Lux, HD. (1986). Effects of dopamine and noradrenaline on Ca channels of cultured sensory and sympathetic neurons of chick. *Pflugers Arch*. 406:104-11.

Monod, J., Wyman, J., Changeux, JP. (1965). On the nature of allosteric transitions: a plausible model. *J Mol Biol*. 12:88-118.

Neely, A., Olcese, R., Wei, X., Birnbaumer, L., & Stefani, E. (1994). Ca(2+)-dependent inactivation of a cloned cardiac Ca2+ channel alpha 1 subunit (alpha 1C) expressed in Xenopus oocytes. *Biophys J*. 66:1895-903.

Noceti, F., Baldelli, P., Wei, X., Qin, N., Toro, L., Birnbaumer, L., & Stefani, E. (1996). Effective gating charges per channel in voltage-dependent K+ and Ca2+ channels. *J Gen Physiol*. 108:143-155.

Nonner, W. (1980). Relations between the inactivation of sodium channels and the immobilization of gating charge in frog myelinated nerve. *J Physiol*. 299:573-603.

Perozo, E., & Bezanilla, F. (1990). Phosphorylation affects voltage gating of the delayed rectifier K+ channel by electrostatic interactions. *Neuron*. 5:685-690.

Perozo, E., Santacruz-Toloza, L., Stefani. E., Bezanilla, F., & Papazian, DM. (1994). S4 mutations alter gating currents of Shaker K channels. *Biophys J*. 66:345-54.

Rebolledo-Antunez, S., Farias, JM., Arenas, I., & Garcia, DE. (2009). Gating charges per channel of Ca(V)2.2 channels are modified by G protein activation in rat sympathetic neurons. *Arch Biochem Biophys*. 486:51-57.

Sakmann, B., Patlak, J., & Neher, E. (1980). Single acetylcholine-activated channels show burst-kinetics in presence of desensitizing concentrations of agonist. *Nature*. 286:71-73.

Schneider, MF., & Chandler, WK. (1973). Voltage dependent charge movement of skeletal muscle: a possible step in excitation-contraction coupling. *Nature*. 242:244-6.

Schoppa, NE., & Sigworth, FJ. (1998). Activation of shaker potassium channels. I. Characterization of voltage-dependent transitions. *J Gen Physiol*. 111:271-94.

Sheets, MF., & Hanck, DA. (1999). Gating of skeletal and cardiac muscle sodium channels in mammalian cells. *J Physiol*. 514 (Pt 2):425-436.

Sheets, MF., Kyle, JW., Kallen, RG., & Hanck, DA. (1999). The Na channel voltage sensor associated with inactivation is localized to the external charged residues of domain IV, S4. *Biophys J*. 77:747-757.

Sigworth, FJ. (1980). The variance of sodium current fluctuations at the node of Ranvier. *J Physiol*. 307:97-129.

Sokolov, S., Kraus, RL., Scheuer, T., & Catterall, WA. (2008). Inhibition of sodium channel gating by trapping the domain II voltage sensor with protoxin II. *Mol Pharmacol*. 73:1020-1028.

Starace, DM., & Bezanilla, F. (2004). A proton pore in a potassium channel voltage sensor reveals a focused electric field. *Nature*. 427:548-53.

Starace, DM., Stefani, E., & Bezanilla, F. (1997). Voltage-dependent proton transport by the voltage sensor of the Shaker K+ channel. *Neuron*. 19:1319-27.

Stuart, GJ., & Sakmann, B. (1994). Active propagation of somatic action potentials into neocortical pyramidal cell dendrites. *Nature*. 367:69-72.

Stühmer, W., Conti, F., Suzuki, H., Wang, XD., Noda, M., Yahagi, N., Kubo, H., & Numa, S. (1989).

Structural parts involved in activation and inactivation of the sodium channel. *Nature*. 339:597-603.

Villalba-Galea, CA., Sandtner, W., Dimitrov, D., Mutoh, H., Knöpfel, T., & Bezanilla, F. (2009).Charge movement of a voltage-sensitive fluorescent protein. *Biophys J*. 96:L19-21.

Xu, Y., Ramu, Y., & Lu, Z. (2008). Removal of phospho-head groups of membrane lipids immobilizes voltage sensors of K+ channels. *Nature*. 451:826-829.

13

Use of Patch Clamp Electrophysiology to Identify Off-Target Effects of Clinically Used Drugs

Lioubov I. Brueggemann and Kenneth L. Byron
Loyola University Chicago, Dept. of Molecular Pharmacology & Therapeutics
USA

1. Introduction

Most drugs have effects attributed to actions at sites other than those that are intended. In many cases these off-target effects have adverse consequences, though in some instances these effects may be neutral or even beneficial. Many off-target effects involve either direct or indirect actions on ion channels. Hence, electrophysiological approaches can be employed to screen drugs for effects on ion channels and thereby predict their off-target actions. The pharmaceutical industry routinely uses cellular expression systems and cloned channels to quickly screen thousands of compounds to eliminate those that have well known adverse ion channel effects, such as inhibition of the Kv11.1 potassium channel encoded by the human *Ether-à-go-go* related gene (hERG). However, these methods are not well-suited to predicting many other off-target effects mediated by actions on ion channels natively expressed in specific tissues. We have employed a more directed electrophysiological approach to evaluate a small number of compounds (e.g. drugs with known or predicted adverse effects) to identify ion channel targets that might explain their actions. This chapter will describe this approach in some detail and illustrate its use with some specific examples.

2. Approach

Our general approach to evaluating ion channel effects of a specific drug on a particular cell type involves the following steps:

2.1 Establish an adequate single cell physiological model to evaluate the drug of interest

While expression systems such as the human embryonic kidney (HEK) cell line or Chinese hamster ovary (CHO) cell line over-expressing individual ion channels are useful tools for initial drug screening they may not adequately reflect the functional roles of ion channels in their native tissues. Immortalized or primary cell culture models that retain expression of the same ion channels that are natively expressed in the tissue under investigation should be considered. Some examples include neonatal cardiomyocytes for studying cardiac ion channel function (Markandeya *et al.*, 2011), rat superior cervical ganglion (SCG) neurons for natively expressed neuronal ion channels (Kim *et al.*, 2011; Zaika *et al.*, 2011), and embryonic rat aortic (A7r5 and A10) cell lines for investigating vascular smooth muscle electrophysiology (Roullet *et al.*, 1997; Brueggemann *et al.*, 2005; Brueggemann *et al.*, 2007). The advantages of using cultured cells compared with freshly dissociated cells from the native tissue include their accessibility, ease of maintenance, high experimental reproducibility, and susceptibility to molecular interventions. However, there are also disadvantages in the use of cultured cells. In particular, the expression pattern of ion channels, receptors, and signaling proteins may differ between cultured cells and native tissues due to differences in proliferative phenotype, absence of surrounding tissues in cell culture and developmental stage from which the cells were derived. Hence, the results obtained using cultured cells should be interpreted with caution and, whenever possible, supplemented by studies performed on freshly dispersed cells and/or by functional assays using intact tissues or live animals.

2.2 Select the patch-clamp mode (e.g. ruptured or perforated patch) for electrophysiological recording

Selection of the patch-clamp mode of recording is generally based on the known properties of the ion channels to be studied. Important considerations include their regulation by phosphatidylinositol 4,5-bisphosphate (PIP_2) and soluble second messengers. The open state of many types of ion channels is known to be stabilized by membrane PIP_2 (Hilgemann & Ball, 1996; Loussouarn *et al.*, 2003; Zhang *et al.*, 2003; Bian & McDonald, 2007; Rodriguez *et al.*, 2010; Suh *et al.*, 2010). With conventional (ruptured) patch-clamp recording, the levels of PIP_2 decrease over time, which can cause irreversible rundown of the currents. Inclusion of Mg-ATP in the internal solution may slow rundown of the PIP_2-dependent currents in excised or ruptured patch recordings (Ribalet *et al.*, 2000), but only the use of the perforated patch configuration enables extended recording of stable whole cell currents for tens of minutes. Regulation of channels via the actions of soluble second messengers may be altered in the ruptured patch configuration as cytosolic solutes may be lost by dialysis into the relatively large volume of the pipette solution. Use of the perforated patch configuration prevents dialysis of signaling molecules and loss of PIP_2 from the membrane. But the ruptured patch-clamp configuration is technically less demanding and so is often preferable if signaling mechanisms are not a concern or if the channels of interest are less dependent on PIP_2 for their activity. Ruptured patch techniques are commonly used for recording currents from voltage-gated sodium channels, Ca_v3 (T type) calcium channels, potassium channels of Kv1, Kv2, Kv3 and K2P families, cystic fibrosis transmembrane receptor (CFTR)-type chloride channels, TRPC family of non-selective cationic channels and ORAI1 store-operated channels. The ruptured patch mode also enables faster data collection, saving the investigator the extra 15-30 min required for patch perforation in each experiment.

The choice of pore-forming agent used for patch perforation is often a matter of personal preference. Pores formed by amphotericin B and nystatin in the membrane under the patch are selectively permeable to monovalent ions (such as K^+, Na^+, Cs^+, Cl^-) preserving cytosolic Ca^{2+} and Mg^{2+} concentrations and all soluble cytosolic signaling molecules (Horn & Marty, 1988; Rae *et al.*, 1991). Use of gramicidin for patch perforation also preserves intracellular Cl^- concentration as gramicidin pores are impermeable to Cl^- (Ebihara *et al.*, 1995). It is possible to record stable currents for several hours in perforated-patch mode from a single cell when appropriate pipette and bath solution compositions are used with continuous bath perfusion.

When using the perforated-patch recording technique, attention should be paid to the value of access resistance achieved. The value of the access resistance (or series resistance as it is also known) will likely exceed pipette resistance by 2- to 5-fold and when current amplitudes are in the nanoampere range will introduce significant error into the true membrane voltage-clamped value. The amount of voltage error can be estimated by multiplying series resistance by current amplitude. If the error exceeds a few millivolts, series resistance compensation should be used.

2.3 Determine the appropriate composition of the internal (pipette) and external (bath) solutions

After choosing a physiological cell model that mimics as closely as possible cells in the intact tissue, it is logical to use intracellular and extracellular solutions with compositions similar to body fluids, at least for initial drug testing. Recipes for different extracellular physiological saline solutions (PSS) such as Krebs-Henseleit solution, Hank's balanced salt solution (HBSS), artificial cerebral spinal fluid (CSF) are readily available in the relevant scientific literature. The pH of the external solution is typically 7.3-7.4.

The composition of the internal (pipette) solution should also closely match known cytosolic ionic composition: high in K^+ (usually in the range of 135-140 mM), low in Na^+ (normally from 0-5 mM). The concentration of Cl^- may vary depending on cell type. For many cultured cells, such as A7r5 cells, stable recordings are most easily obtained using relatively low $[Cl^-]_{in}$, in the range of 30-45 mM, in combination with large impermeable anions such as gluconate or aspartate to balance K^+. For other cell types, including freshly dispersed smooth muscle cells, pipette solutions with 135-140 mM KCl are preferable. If the internal solution will be used in perforated patch-clamp mode, inclusion of Mg^{2+} and buffering of cytosolic Ca^{2+} is not required, as amphotericin and nystatin pores are impermeable to Ca^{2+} and Mg^{2+} (Horn & Marty, 1988; Rae *et al.*, 1991).

For ruptured patch recording, free Mg^{2+} concentration should be set within the range of 1-2 mM and free Ca^{2+} concentration should be approximately 100 nM. To accomplish this, Ca^{2+} buffers such as EGTA, EDTA, or BAPTA should be included in the pipette solution. Free Ca^{2+} concentration will depend on the concentration of the buffers (usually 0.1-10 mM), their binding constants for Ca^{2+} and Mg^{2+} and the amounts of added Ca^{2+} and Mg^{2+}. MAXCHELATOR is a series of programs freely available online (http://maxchelator.stanford.edu/) that can be used for determining the free Ca^{2+} concentration in the presence of Ca^{2+} buffers. Mg- or Na-ATP (1-5 mM) should also be included in the internal solution for ruptured patch recording. The range of pH for internal solutions may vary from 7.2-7.4, usually buffered with HEPES (1-10 mM).

Attention should be paid to the osmolality of both internal and external solutions. Osmolality of body fluids is approximately 275-290 mOsM; the osmolality of the internal and external solutions should be measured with an osmometer, adjusted to the physiological range, and balanced (within 1 or 2 mOsM) between internal and external solutions.

Use of approximately physiological external and internal solutions for patch-clamp experiments enables recording of a mix of ionic conductances for initial evaluation of drug effects on the cell type under investigation.

2.4 Use appropriate voltage clamp protocols (e.g. voltage steps or ramps) to record drug effects on total currents

Design of the voltage protocol should be based on biophysical properties of the ion channels expressed in the cells under investigation. It is very useful for initial drug screening to select a holding voltage close to the resting membrane voltage measured or reported for that particular cell type. A voltage protocol designed to apply a family of long test voltage steps (1-5 s) in both negative and positive directions from the resting membrane voltage allows the investigator to record a mix of both rapidly and slowly activating/inactivating currents through voltage-dependent and voltage-independent ion channels.

The time between voltage steps should be sufficient for channel deactivation.

The stability of the measured currents should be established before testing the effects of drugs on the currents. For example, applying the same series of voltage steps should generate approximately equal currents on successive trials in the same cell. Voltage-ramp protocols can also be used to record instantaneous (voltage-independent) or rapidly activating currents; the ramp can be applied at regular intervals to monitor the stability of the currents over time. For slowly activating currents, use of a single voltage step applied at regular intervals is generally more appropriate for time course measurements.

When stable recordings of total ionic conductances are achieved, it is possible to test the effects of a drug, usually applied at varying concentrations. Drugs may affect the amplitudes of the conductances and the kinetics of their responses to the applied voltage protocols as well as their voltage-dependence of activation. The drug effects are generally time-dependent and vary with drug dose in a reproducible manner. Careful evaluation of the drug effects on total membrane currents provides important clues to the types of ionic conductances that may be affected. It is then desirable to record the drug-targeted ionic conductances in isolation.

2.5 Adjust recording conditions to isolate drug-sensitive currents

To record specific currents among the mix of total cellular ionic conductances, a tailored voltage protocol should be used in combination with internal and external solutions and pharmacological approaches that are rationally chosen to enhance or maintain the current of interest while minimizing other conductances. The voltage protocol should reflect the specific biophysical properties of the channels under investigation. If the data are available, consider the voltage dependence of activation and time constants of activation, inactivation and deactivation of the currents. For example, store-operated currents are known to be inwardly-rectifying and highly Ca^{2+}-selective, with fast Ca^{2+}-dependent inactivation at negative voltages (Parekh & Putney, 2005). To isolate the highly Ca^{2+}-selective store-operated currents from other conductances, consider using an external solution containing 10-20 mM Ca^{2+} and replacing all monovalent ions (K^+, Na^+ and Cl^-) with impermeant ions such as N-methyl D-glucamine and aspartate. A voltage protocol comprised of a 0 mV holding voltage with 100 ms ramps from +100 to -100 mV can be applied every 5-20 s to record the time course of current activation in response to store depletion (often induced by dialyzing cells with EGTA- or BAPTA-containing pipette solution in ruptured patch mode or by application of thapsigargin or cyclopiazonic acid, which block the ability of the cells to sequester Ca^{2+} in the endoplasmic/sarcoplasmic reticulum (Brueggemann *et al.*, 2006)).

In general, the isolation of broad classes of ionic conductances (i.e. Ca^{2+} currents, K^+ currents, nonselective cation currents, or Cl^- currents) may be achieved by using bath and pipette solutions containing ions that cannot permeate or that block the movement of ions through other classes of ion channels. For example, to record Ca^{2+} conductance in isolation, Cs^+ can be used to replace K^+ because it blocks most if not all K^+ channels and thereby minimizes contributions of outward K^+ currents to the recording. Similarly, replacing Cl^- with aspartate, gluconate or sulfonate will minimize contributions of Cl^- conductances.

It is much more difficult to isolate ion currents within the same class. This may require the use of pharmacological agents that are selective for a particular class of channels. For example, if it is desired to isolated T-type Ca^{2+} current from L-type Ca^{2+} current, specific L-type Ca^{2+} channel blockers like verapamil can be used. In this case, an alternative (or adjunct) approach is the use of the ruptured patch mode, which leads to rundown of L-type Ca^{2+} current over time; other Ca^{2+} conductances (e.g. T-type Ca^{2+} currents) that have less tendency to run-down in the ruptured patch configuration, may then be recorded in isolation. The use of pharmacological ion channel blockers to eliminate unwanted conductances should be employed with caution unless the specificity of the drugs has been thoroughly established.

Probably most difficult is the isolation of specific K^+ currents because many different potassium channels are normally expressed in each cell. Several highly specific toxins are available for certain subfamilies of potassium channels (hongotoxin and margatoxin for Kv1.1, Kv1.2, Kv1.3 (Koschak *et al.*, 1998), hanatoxin for Kv2 (Swartz & MacKinnon, 1995), K-dendrotoxin for homo- and heteromeric channels containing Kv1.1 (Robertson *et al.*, 1996)). These can be used to eliminate a subtype of K^+ current or determine the contribution of that subtype to the larger mix of K^+ currents. It is important to consider that different members within a subfamily of K^+ channels can combine to form functional heteromeric channels, which may vary in their sensitivities to toxins depending on the subunit composition (Tytgat *et al.*, 1995; Plane *et al.*, 2005) In some cases, a combination of pharmacological approaches and voltage protocols that take advantage of the unique biophysical properties of the K^+ channels expressed in the cell type under investigation can effectively isolate a specific subtype of K^+ conductance (see example below).

Recording a specific current 'in isolation' from other currents is never fully achieved, but conditions may be established that provide a reasonable signal to noise ratio to evaluate contributions of a subset of ion channels. Specific pharmacological ion channel blockers or activators may be useful to confirm that the currents measured are largely attributable to a particular type of channel, but molecular knockdown approaches are often the best way to determine what fraction of the currents measured are mediated by a specific channel subtype. When conditions have been optimized for recording isolated currents, the effects of the drug on those currents can be tested.

2.6 Evaluate the actions of the drug of interest at its physiologically or clinically relevant concentrations

An appropriate dose-response range of the drug of interest should be based on consideration of physiological or clinically achieved plasma concentrations and doses used in vitro from previously published studies. Dose-dependent effects can be evaluated both under physiological ionic conditions as well as under recording conditions that isolate specific currents. Stable recording of currents in the absence of drug should be established by applying voltage steps or ramps at regular intervals and measuring similar current amplitudes for several minutes. Increasing concentrations of the drug are then applied, starting at a dose that has little or no effect and increasing in 10-fold or smaller increments to at least the maximum clinical or physiological drug concentration. Be aware that repetitive drug administration or incrementally increasing doses may induce tachyphylaxis. Applying a single dose acutely to a naïve cell may provide the best assessment of the effect of that dose of the drug.

To evaluate whether the presence of the drug changes the biophysical properties of the channel, such as its gating kinetics or voltage-dependence of activation, specific voltage protocols may be applied when steady-state effects of a particular dose of the drug have been achieved. For example, a tail current voltage protocol can be used to evaluate the effects of a drug on voltage-dependence of channel activation. This protocol should be applied at the end of the control recording (before drug application); two successive voltage protocols that yield similar currents establish the stability of the control recording. The same successive voltage protocols should then be repeated when measurement of the time course of current amplitude indicates that the current amplitude has reached a new plateau in the presence of the drug.

To determine the reversibility of drug effects, it is important to measure the currents during drug application and during washout of the drug. It may require tens of minutes to achieve a stable reversal of a drug effect and in some cases the effects will not be reversed within a practical time frame. Reversibility, when it is achieved, provides convincing evidence that the effect measured was specifically due to the drug and not simply due to time-dependent changes such as run-up or run-down of currents. It is also important to include vehicle and time controls to assure that effects are due to the presence of the drug rather than the time of recording or the solvent in which the drug is dissolved.

Reproducible effects of a drug on the amplitude or biophysical characteristics of a particular current in the cultured cell model may provide important clues to the drug's effect on a particular tissue. However, whenever possible, results based on cultured cells should be confirmed using freshly isolated cells from the tissue from which the cultured cells were derived. The electrophysiological characteristics of the drug-sensitive currents may suggest one or more specific ion channel subtypes as the drug targets. Molecular biological approaches may then be used to confirm the identity of the drug-sensitive ion channel.

2.7 Apply molecular biological approaches such as knock-down and overexpression as necessary to confirm an ion channel drug target

As was noted above, cultured cells are often suitable for molecular biological interventions. Knock-down of expression of specific ion channels may be achieved by treatment with short hairpin RNA (shRNA) or small interfering RNA (siRNA). Alternatively, expression of dominant-negative ion channel subunits may specifically abrogate the function of particular ion channels. These molecular constructs can be introduced into the cultured cells using transfection techniques with appropriate plasmids or by infecting the cells with viral vectors engineered to express the constructs. Inactive constructs (e.g. scrambled shRNA) should be used as a control. Biochemical techniques such as RT-PCR or Western blotting and/or immunohistochemistry are required to confirm the effectiveness of knock-down.

Knock-down of expression or function of a specific ion channel can reveal how much that channel type contributes to the currents measured and whether a drug effect can be attributed to specific actions on that channel type. In electrophysiological recordings, knock-down of a specific channel type should eliminate the contribution of those channels to the currents measured. In other functional assays, loss of the drug effect when the channel is knocked down would provide evidence that the functional effects of the drug can be specifically attributed to its actions on that channel type. Alternatively, if the effects persist even after knockdown of a particular channel then that particular channel is unlikely to be the primary drug target.

Another way to implicate an ion channel as a drug target is to over-express the ion channel and test the effects of the drug on the currents. This is best done in the same cellular environment known to express that type of ion channel endogenously because the cellular environment dictates many properties of ion channels, including regulation by signaling pathways that may at times mediate or modulate drug effects. Overexpression typically results in much larger currents that can be unambiguously attributed to the overexpressed channels. If these channels are direct or indirect targets for the drug, then drug application should have effects on the currents similar to the effects observed on native currents.

Potential pitfalls of these molecular biological strategies include changes in expression or function of other molecules that may compensate for the increased or decreased channel expression or otherwise alter the measured currents. It is important to keep in mind that the effects of drugs on ion currents may not be via a direct interaction with the channel itself, but instead mediated by other mechanisms, including activation of cellular signaling molecules whose expression may or may not be altered when channel expression levels change.

2.8 Establish a multicellular functional system or animal model for final proof of principle

To determine the physiological significance of drug effects on particular types of ion channels, the drug can be tested on in vitro (ex vivo) and/or in vivo functional models. Examples of in vitro functional models include the isolated Langendorff heart preparation (Skrzypiec-Spring *et al.*, 2007), muscle strips of various origins, aortic or bronchial rings, brain slices, lung slices, pressurized artery preparations, etc. These more complex experimental systems more closely mimic physiological conditions, but also introduce additional factors that may complicate interpretation of drug effects. It is important to consider whether changes in tissue function in the presence of a drug can be attributed primarily to the drug's effects on ion channels in a particular cell type. There may be multiple effects on multiple cell types within the tissue.

In vivo drug testing adds a further level of complexity, but it is the ultimate test of how a drug will affect whole animal physiology. Many different animal models have been developed and are described in the literature. To determine whether an effect of a drug in vivo can be attributed to its actions on a specific ion channel, it may be possible to compare its effects with the effects of another drug that is known to have the same or opposite effects on that ion channel.

3. Example: Cyclooxygenase-2 inhibitor effects on vascular smooth muscle ion channels

The following example illustrates how we have employed the approaches described above in an attempt to elucidate the mechanisms underlying differential adverse cardiovascular risk profiles among clinically used drugs of the same class. Selective cyclooxygenase-2 (COX-2) inhibitors, such as celecoxib (Celebrex®), rofecoxib (Vioxx®), and diclofenac, are non-steroidal anti-inflammatory drugs (NSAIDs) commonly used for the treatment of both acute and chronic pain. About five years after celecoxib and rofecoxib were approved for use in the United States, rofecoxib (Vioxx®) was voluntarily withdrawn from the market because of adverse cardiovascular side effects (Dajani & Islam, 2008). The ensuing investigation of the cardiovascular side effects of this drug class revealed differential risk profiles, with celecoxib being relatively safe, compared with rofecoxib and diclofenac (Cho *et al.*, 2003; Hermann *et al.*, 2003; Aw *et al.*, 2005; Hinz *et al.*, 2006; Dajani & Islam, 2008). Early reports suggested that these differences might relate to pro-hypertensive effects of COX-2 inhibition (Cho *et al.*, 2003; Hermann *et al.*, 2003; Aw *et al.*, 2005; Hinz *et al.*, 2006) that were offset by vasodilatory effects of celecoxib (Widlansky *et al.*, 2003; Klein *et al.*, 2007). However, the mechanisms underlying the vasodilatory effects of celecoxib remained elusive.

We employed the following strategies to investigate whether celecoxib might exert its vasodilatory actions via effects on ion channels in vascular smooth muscle cells (VSMCs). Additional details of these studies were published previously (Brueggemann *et al.*, 2009).

3.1 Vascular smooth muscle cell model

The embryonic rat aortic A7r5 cell line was chosen as the cell model to compare the effects of celecoxib on vascular smooth muscle ion channels with those of rofecoxib and diclofenac. The immortalized A7r5 cell line retains a differentiated smooth muscle phenotype including expression of L-type and T-type Ca^{2+} channels (Qar *et al.*, 1988; Brueggemann *et al.*, 2005), several types of potassium channels (Kv7.5 (Brueggemann *et al.*, 2007), K(Ca)3.1, K(Ca)1.1 (Si *et al.*, 2006) and Kv1.2 (Byron & Lucchesi, 2002)), several members of TRPC family of non-selective cation channels (TRPC6, TRPC4, TRPC1, TRPC7 (Soboloff *et al.*, 2005; Brueggemann *et al.*, 2006; Maruyama *et al.*, 2006)) as well as several Gq-coupled receptors (e.g. V1a vasopressin receptors (Thibonnier *et al.*, 1991), 5-HT2 serotonin receptors (Weintraub *et al.*, 1994), and ET_A endothelin-1 receptors (Bucher & Taeger, 2002). A7r5 cells proliferate in cell culture until they form a confluent monolayer of the cells that are electrically coupled by gap junctions formed by connexins (Cxs) 40 and 43 (Beyer *et al.*, 1992).

3.2 Patch clamp mode

To investigate the vasodilator actions of celecoxib, it was important to consider two types of ion channels that are perhaps the most important in determining the contractile state of vascular smooth muscle cells: Kv7 channels that determine the resting membrane voltage

(Mackie & Byron, 2008), and L-type voltage-gated Ca^{2+} channels, activation of which induces Ca^{2+} influx, smooth muscle contraction, and vasoconstriction (Jackson, 2000). Both of these types of channels are known to be regulated by PIP_2 (Suh & Hille, 2008; Suh *et al.*, 2010). We therefore chose to use the perforated patch-clamp configuration to record currents in voltage-clamp mode (200 μg/ml amphotericin B in internal solution was used for membrane patch perforation).

3.3 Internal and external solutions

For the initial test of celecoxib, rofecoxib and diclofenac, total currents were recorded in A7r5 cells under approximately physiological ionic conditions. The standard bath solution contained (in mM): 5 KCl, 130 NaCl, 10 HEPES, 2 $CaCl_2$, 1.2 $MgCl_2$, 5 D-glucose, pH 7.3. Standard internal (pipette) solution contained (in mM): 110 K gluconate, 30 KCl, 5 HEPES, 1 K_2EGTA, 2 Na_2ATP, pH 7.2. Osmolality was adjusted to 268-271 mOsm/l with D-glucose.

3.4 Voltage clamp protocols

We used a 5s voltage step protocol from -74 mV holding potential to test voltages ranging from -94- +36 mV. After each test pulse the voltage was returned to -74 mV for 10 s to allow full deactivation before the next voltage step was applied. This protocol enabled us to simultaneously record the current-voltage (I-V) relationship for L-type Ca^{2+} channels (based on peak inward currents recorded at the beginning of the voltage steps; see inset on Fig. 1A) and for Kv7 channels (based on steady-state outward K^+ currents recorded at the end of the voltage steps). The evaluation of L-type currents at the beginning of the voltage steps was only possible because of the absence of rapidly-activating K^+ currents in the voltage range used. The long (5s) voltage steps enabled relative isolation of Kv7 currents at the end of the voltage steps because Kv7 channels do not inactivate, whereas most other K^+ channels do inactivate when stepped to a constant activating voltage for 5 s. Representative current traces and I-V relationships are shown on Fig. 1A and 1B.

The I-V voltage protocol requires approximately 4 min to complete all the 5 s voltage steps with a 10 s interval between each step. We repeated this three times to determine that the currents were stable (the I-V curves were approximately superimposable). When the currents were stable we initiated a voltage protocol designed to record the time course of drug application. The time course voltage protocol combined 100 ms voltage ramps (from a -74 mV holding potential to +36 mV) to record the rapidly-activating Ca^{2+} current (as the peak inward current) followed by 5 s voltage steps to -20 mV to record slowly-activating and non-inactivating Kv7 current (measured as the average steady-state current recorded at the end of the voltage step; Figure 1C). The time course protocol was applied every 15 s. Ca^{2+} and K^+ currents were recorded for at least 5 min before application of celecoxib (10 μM). Celecoxib was then applied until a stable drug effect was achieved (approximately 15 min). Then the I-V voltage-step protocol was applied again (twice in succession) to record I-V relationships of the Ca^{2+} and K^+ channels in the presence of the drug. The time course protocol was then re-initiated to monitor the effects of washout of celecoxib. These experiments revealed that celecoxib induced a reversible enhancement of Kv7 current and inhibition of L-type Ca^{2+} current—both of these effects could potentially contribute to the vasodilatory actions of celecoxib.

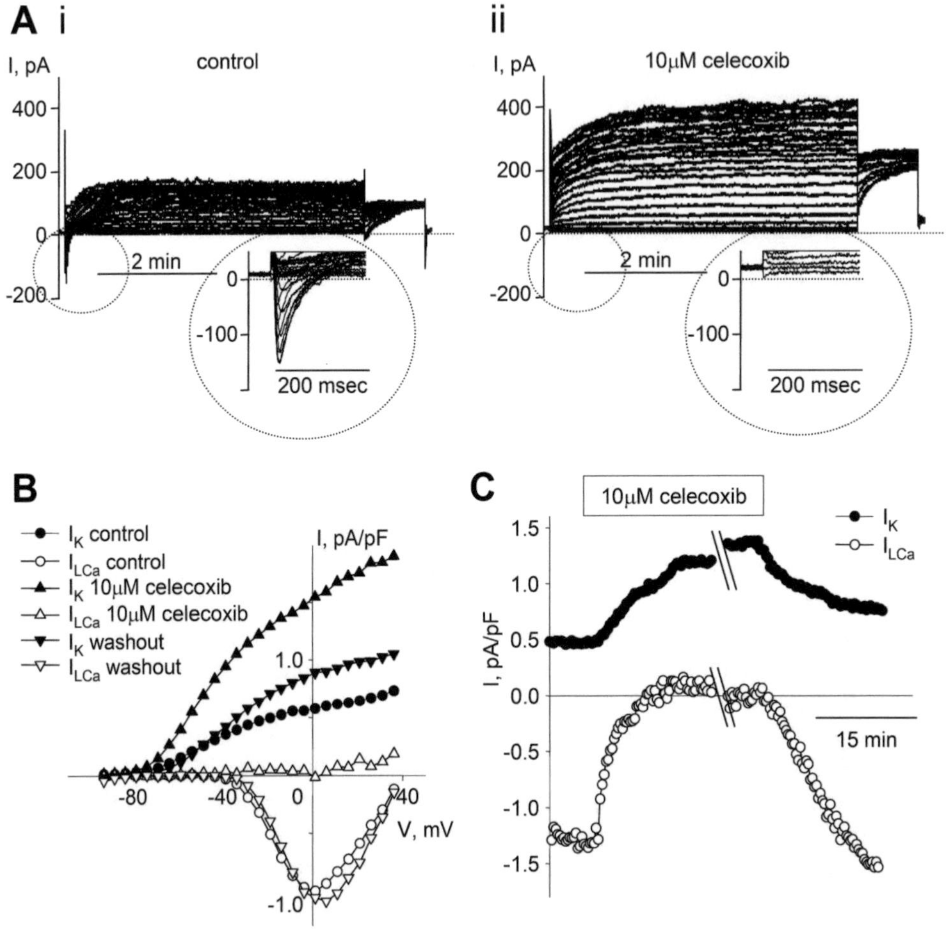

Fig. 1. Enhancement of K^+ current and inhibition of Ca^{2+} current by celecoxib in A7r5 cells.

A, representative traces of whole-cell K^+ and Ca^{2+} currents measured in a single A7r5 cell; i, control; ii, in the presence of 10 μM celecoxib. Inward Ca^{2+} currents, activated at the beginning of the voltage steps, are shown in insets on an expanded scale for clarity. B, I-V curves, corresponding to traces in A, for steady-state K^+ current (filled symbols) and peak inward Ca^{2+} current (open symbols) in control (circles), in the presence of 10 μM celecoxib (triangles), and after washout of celecoxib (inverted triangles). C, corresponding time course of inhibition of the peak inward Ca^{2+} current and activation of K^+ current. Reproduced with permission from Brueggemann *et al.*, (2009).

Similar experiments were conducted to evaluate the effects of other NSAIDs (rofecoxib and diclofenac), 2,5-dimethylcelecoxib (a celecoxib analog lacking COX-2 inhibitory activity), as well as verapamil (a known inhibitor of L-type Ca^{2+} channels) and flupirtine (a known activator of Kv7.2- Kv7.5 channels) (Brueggemann *et al.*, 2009). The effects of these drugs were compared with celecoxib-induced effects on L-type Ca^{2+} currents and Kv7 currents. From these studies it was apparent that neither rofecoxib nor diclofenac mimicked celecoxib in its actions on either L-type Ca^{2+} currents or Kv7 currents; on the other hand, 2,5-dimethylcelecoxib was indistinguishable from celecoxib in its effects (Brueggemann *et al.*, 2009).

3.5 Isolation of L-type Ca^{2+} currents and Kv7 currents

To evaluate in more detail the actions of celecoxib on L-type Ca^{2+} currents and Kv7 currents, each type of current was recorded in isolation. To record Ca^{2+} currents in isolation, a Cs^{+}-containing internal solution was used (for A7r5 cells, the internal solution contained (in mM): 110 Cs aspartate, 30 CsCl, 5 HEPES, 1 Cs-EGTA, pH 7.2). Isolated Ca^{2+} currents were recorded with a 300 ms voltage step protocol from -90 mV holding potential. To isolate Kv7 currents, 100 μM $GdCl_3$, sufficient to block L- and T-type Ca^{2+} channels and non-selective cation channels, was added to the external solution. Isolated Kv7 currents were recorded with the same 5s voltage-step protocol used to record a mix of currents (see above).

3.6 Effects of celecoxib at therapeutic concentrations

Celecoxib dose-response curves for L-type Ca^{2+} currents and Kv7 currents were obtained by measuring the currents during successive applications of increasing concentrations of celecoxib (ranging from 0.1 μM to 30 μM), each time waiting until the drug effect had stabilized before application of the next dose. The celecoxib concentrations selected for dose-response determinations were based on a ± 10-fold range of mean therapeutic concentrations typically achieved in the plasma of patients treated with celecoxib (1-3 μM) (Hinz *et al.*, 2006). Our estimated IC_{50} value for suppression of L-type Ca^{2+} currents was 8.3 ± 1.3 μM (Brueggemann *et al.*, 2009).

To extend the findings to a more physiological model system, the effects of celecoxib were also examined using freshly dispersed mesenteric artery myocytes. Celecoxib inhibited Ca^{2+} currents and enhanced Kv7 currents recorded in isolation in mesenteric artery myocytes, just as had been observed in A7r5 cells (Brueggemann *et al.*, 2009).

3.7 Molecular biological approaches to evaluate Kv7.5 as a target of celecoxib

Kv7 currents measured in A7r5 cells had previously been attributed to Kv7.5 (KCNQ5) channel activity based on expression studies and on elimination of the currents by shRNA treatment targeting the Kv7.5 (KCNQ5) mRNA transcripts (Brueggemann *et al.*, 2007; Mani *et al.*, 2009). To determine whether Kv7.5 was a specific target for celecoxib, we measured the effects of celecoxib on overexpressed human Kv7.5 channels, using the A7r5 cells as an expression system. Celecoxib robustly enhanced the overexpressed Kv7.5 currents (Brueggemann *et al.*, 2009).

3.8 Functional assays to evaluate how ion channel targeting by celecoxib affects cell and tissue physiology

As noted above, Kv7 channel activity is believed to stabilize negative resting membrane voltages in arterial myocytes and thereby opposes the activation of L-type voltage-gated Ca^{2+} channels. The latter mediate Ca^{2+} influx, smooth muscle contraction, and vasoconstriction. Drugs that enhance Kv7 channel activity or that directly inhibit L-type Ca^{2+} activity would therefore be expected to reduce cytosolic Ca^{2+} concentration, relax the arterial myocytes, and dilate arteries. To test the hypothesis that the effects of celecoxib on arterial smooth muscle ion channels contributes to its vasodilatory actions, three different functional assays were used:

a. Arginine-vasopressin (AVP) is a vasoconstrictor hormone that has been shown to induce Ca^{2+} oscillations in confluent monolayers of A7r5 cells. We therefore loaded A7r5 cells with the fluorescent Ca^{2+} indicator fura-2 and examined the effects of celecoxib (10 μM) in comparison with rofecoxib (10 μM) on AVP-induced Ca^{2+} oscillations. In support of our hypothesis, celecoxib opposed the actions of the vasoconstrictor hormone, essentially abolishing AVP-stimulated Ca^{2+} oscillations. Rofecoxib, in contrast, had no effect (Figure 2A). Inhibition of AVP-stimulated Ca^{2+} oscillations was also observed using known L-type Ca^{2+} channel blockers or activators of Kv7 channels (not shown, but see (Byron, 1996; Brueggemann *et al.*, 2007)).

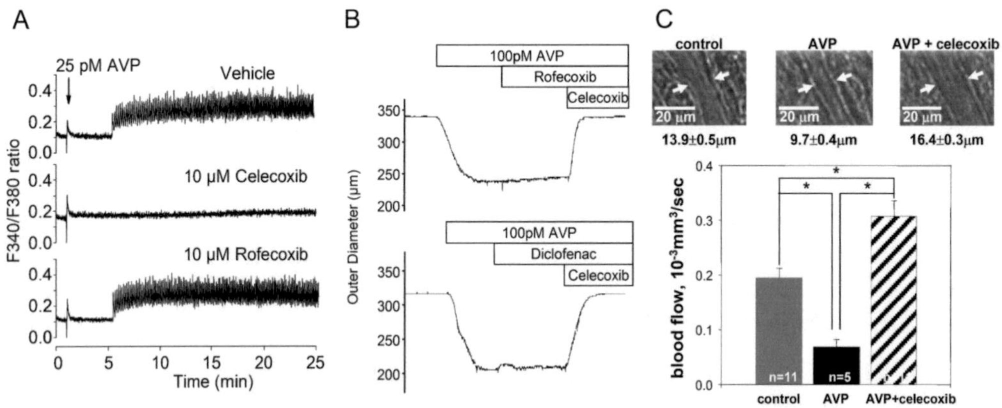

Fig. 2. Functional assays to evaluate celecoxib ion channel actions.

A, Celecoxib, but not rofecoxib, abolishes AVP-induced Ca^{2+} oscillations in A7r5 cells. Confluent monolayers of fura-2-loaded A7r5 cells were treated with 25 pM AVP (arrow). Representative traces show the absence of AVP-induced Ca^{2+} oscillations with simultaneous addition of celecoxib (10 μM, middle) but not with addition of vehicle (top) or rofecoxib (10 μM, bottom). B, representative traces from rat mesenteric artery pressure myography illustrating the inability of 20 μM rofecoxib (top) and 20 μM diclofenac (bottom) to dilate arteries preconstricted with 100 pM AVP. Celecoxib (20 μM) fully dilated the same arteries when added after either rofecoxib or diclofenac. C, measurement of arteriolar blood flow in vivo using intravital microscopy reveals that AVP (100 pM) significantly constricts arterioles (top panels) and reduces blood flow (bar graph), but this effect is more than fully reversed by the addition of 10 μM celecoxib. Panels A and B reproduced with permission from Brueggemann *et al.*, (2009).

b. The constriction and dilation of arteries can be measured in vitro using pressure myography. Small segments of artery are cannulated at either end, pressurized to their normal physiological pressure, and maintained at physiological temperatures and ionic balance; arterial diameter is monitored continuously by digital image analysis while drugs are applied to the bath. We used these methods to test the ability of celecoxib to dilate pressurized mesenteric artery segments that were pre-constricted with AVP (100 pM). In support of our hypothesis, celecoxib induced concentration-dependent, endothelium-independent dilation of pre-constricted mesenteric arteries. Similar effects were obtained using known L-type Ca^{2+} channel blockers or activators of Kv7 channels (not shown, but see (Henderson & Byron, 2007; Mackie *et al.*, 2008)). The maximum dilatory effect of celecoxib was achieved at a concentration of 20 μM; neither rofecoxib nor diclofenac induced significant artery dilation at the same concentration (Figure 2B).

c. Finally, it is important to evaluate the effects of the drug in an in vivo model. We therefore examined the ability of celecoxib to increase blood flow in mesenteric arterioles of live anesthetized rats using intravital microscopy and intravenous perfusion of fluorescent microspheres. AVP (100 pM) superfused over the exposed portion of the mesenteric vasculature induced significant arteriolar constriction and reduced blood flow (determined from the velocity of fluorescent microspheres moving through the arterioles). Application of celecoxib (10 μM) in addition to AVP more than fully restored both arteriolar diameter and blood flow (Figure 2C).

The combined functional assays provided strong evidence supporting the hypothesis that celecoxib, but not other NSAIDs of the same class, exerts vasodilatory effects via combined activation of Kv7 potassium channels and inhibition of L-type voltage-gated Ca^{2+} channels in arterial smooth muscle cells. These results may explain the differential risk of adverse cardiovascular events in patients taking these different NSAIDs.

4. Conclusion

Carefully designed and executed electrophysiological experiments can provide important insights into the mechanisms of drug actions, including their off-target effects on specific tissues.

5. References

Aw TJ, Haas SJ, Liew D & Krum H. (2005). Meta-analysis of cyclooxygenase-2 inhibitors and their effects on blood pressure, *Arch Intern Med* 165.(5): 490-496.

Beyer EC, Reed KE, Westphale EM, Kanter HL & Larson DM. (1992). Molecular cloning and expression of rat connexin40, a gap junction protein expressed in vascular smooth muscle, *J Membr Biol* 127.(1): 69-76.

Bian J-S & McDonald T. (2007). Phosphatidylinositol 4,5-bisphosphate interactions with the HERG K^+ channel, *Pflügers Archives European Journal of Physiology* 455.(1): 105-113.

Brueggemann LI, Mackie AR, Mani BK, Cribbs LL & Byron KL. (2009). Differential effects of selective cyclooxygenase-2 inhibitors on vascular smooth muscle ion channels may account for differences in cardiovascular risk profiles, *Mol Pharmacol* 76.(5): 1053-1061.

Brueggemann LI, Markun DR, Henderson KK, Cribbs LL & Byron KL. (2006). Pharmacological and electrophysiological characterization of store-operated currents and capacitative Ca^{2+} entry in vascular smooth muscle cells, *J Pharmacol Exp Ther* 317.(2): 488-499.

Brueggemann LI, Martin BL, Barakat J, Byron KL & Cribbs LL. (2005). Low voltage-activated calcium channels in vascular smooth muscle: T-type channels and AVP-stimulated calcium spiking, *Am J Physiol Heart Circ Physiol* 288.(2): H923-H935.

Brueggemann LI, Moran CJ, Barakat JA, Yeh JZ, Cribbs LL & Byron KL. (2007). Vasopressin stimulates action potential firing by protein kinase C-dependent inhibition of KCNQ5 in A7r5 rat aortic smooth muscle cells, *Am J Physiol Heart Circ Physiol* 292.(3): H1352-H1363.

Bucher M & Taeger K. (2002). Endothelin-receptor gene-expression in rat endotoxemia, *Intensive Care Medicine* 28.(5): 642-647.

Byron KL. (1996). Vasopressin stimulates Ca^{2+} spiking activity in A7r5 vascular smooth muscle cells via activation of phospholipase A_2, *Circulation Research* 78.(5): 813-820.

Byron KL & Lucchesi PA. (2002). Signal transduction of physiological concentrations of vasopressin in A7r5 vascular smooth muscle cells. A role for PYK2 and tyrosine phosphorylation of K^+ channels in the stimulation of Ca^{2+} spiking, *J Biol Chem* 277.(9): 7298-7307.

Cho J, Cooke CE & Proveaux W. (2003). A retrospective review of the effect of COX-2 inhibitors on blood pressure change, *Am J Ther* 10.(5): 311-317.

Dajani EZ & Islam K. (2008). Cardiovascular and gastrointestinal toxicity of selective cyclo-oxygenase-2 inhibitors in man, *J Physiol Pharmacol* 59.(Suppl 2): 117-133.

Ebihara S, Shirato K, Harata N & Akaike N. (1995). Gramicidin-perforated patch recording: GABA response in mammalian neurones with intact intracellular chloride, *The Journal of Physiology* 484.(Pt 1): 77-86.

Henderson KK & Byron KL. (2007). Vasopressin-induced vasoconstriction: two concentration-dependent signaling pathways, *Journal of Applied Physiology* 102.(4): 1402-1409.

Hermann M, Camici G, Fratton A, Hurlimann D, Tanner FC, Hellermann JP, Fiedler M, Thiery J, Neidhart M, Gay RE, Gay S, Luscher TF & Ruschitzka F. (2003). Differential Effects of Selective Cyclooxygenase-2 Inhibitors on Endothelial Function in Salt-Induced Hypertension, *Circulation* 108.(19): 2308-2311.

Hilgemann DW & Ball R. (1996). Regulation of cardiac Na^+,Ca^{2+} exchange and K_{ATP} potassium channels by PIP_2, *Science* 273.(5277): 956-959.

Hinz B, Dormann H & Brune K. (2006). More pronounced inhibition of cyclooxygenase 2, increase in blood pressure, and reduction of heart rate by treatment with diclofenac compared with celecoxib and rofecoxib, *Arthritis and Rheumatism* 54.(1): 282-291.

Horn R & Marty A. (1988). Muscarinic activation of ionic currents measured by a new whole-cell recording method, *J Gen Physiol* 92.(2): 145-159.

Jackson WF. (2000). Ion channels and vascular tone, *Hypertension* 35.(1 Pt 2): 173-178.

Kim Y-H, Nam T-S, Ahn D-S & Chung S. (2011). Modulation of N-type Ca^{2+} currents by moxonidine via imidazoline I_1 receptor activation in rat superior cervical ganglion neurons, *Biochemical and Biophysical Research Communications* 409.(4): 645-650.

Klein T, Eltze M, Grebe T, Hatzelmann A & Komhoff M. (2007). Celecoxib dilates guinea-pig coronaries and rat aortic rings and amplifies NO/cGMP signaling by PDE5 inhibition, *Cardiovascular Research* 75.(2): 390-397.

Koschak A, Bugianesi RM, Mitterdorfer Jr, Kaczorowski GJ, Garcia ML & Knaus H-Gn. (1998). Subunit Composition of Brain Voltage-gated Potassium Channels Determined by Hongotoxin-1, a Novel Peptide Derived from Centruroides limbatus Venom, *Journal of Biological Chemistry* 273.(5): 2639-2644.

Loussouarn G, Park KH, Bellocq C, Baro I, Charpentier F & Escande D. (2003). Phosphatidylinositol-4,5-bisphosphate, PIP_2, controls KCNQ1/KCNE1 voltage-gated potassium channels: a functional homology between voltage-gated and inward rectifier K^+ channels, *EMBO J* 22.(20): 5412-5421.

Mackie AR, Brueggemann LI, Henderson KK, Shiels AJ, Cribbs LL, Scrogin KE & Byron KL. (2008). Vascular KCNQ potassium channels as novel targets for the control of mesenteric artery constriction by vasopressin, based on studies in single cells, pressurized arteries, and in vivo measurements of mesenteric vascular resistance, *J Pharmacol Exp Ther* 325.(2): 475-483.

Mackie AR & Byron KL. (2008). Cardiovascular KCNQ (Kv7) Potassium Channels: Physiological Regulators and New Targets for Therapeutic Intervention, *Mol Pharmacology* 741171-1179.

Mani BK, Brueggemann LI, Cribbs LL & Byron KL. (2009). Opposite regulation of KCNQ5 and TRPC6 channels contributes to vasopressin-stimulated calcium spiking responses in A7r5 vascular smooth muscle cells., *Cell Calcium* 45.(4): 400-411.

Markandeya YS, Fahey JM, Pluteanu F, Cribbs LL & Balijepalli RC. (2011). Caveolin-3 regulates protein kinase A modulation of the Ca(V)3.2 ($alpha_{1H}$) T-type Ca^{2+} channels, *J Biol Chem* 286.(4): 2433-2444.

Maruyama Y, Nakanishi Y, Walsh EJ, Wilson DP, Welsh DG & Cole WC. (2006). Heteromultimeric TRPC6-TRPC7 channels contribute to arginine vasopressin-induced cation current of A7r5 vascular smooth muscle cells, *Circ Res* 98.(12): 1520-1527.

Parekh AB & Putney JW. (2005). Store-Operated Calcium Channels, *Physiological Reviews* 85.(2): 757-810.

Plane F, Johnson R, Kerr P, Wiehler W, Thorneloe K, Ishii K, Chen T & Cole W. (2005). Heteromultimeric Kv1 channels contribute to myogenic control of arterial diameter, *Circ Res* 96.(2): 216-224.

Qar J, Barhanin J, Romey G, Henning R, Lerch U, Oekonomopulos R, Urbach H & Lazdunski M. (1988). A novel high affinity class of Ca^{2+} channel blockers, *Molecular Pharmacology* 33.(4): 363-369.

Rae J, Cooper K, Gates P & Watsky M. (1991). Low access resistance perforated patch recordings using amphotericin B, *Journal of Neuroscience Methods* 37.(1): 15-26.

Ribalet B, John SA & Weiss JN. (2000). Regulation of Cloned ATP–sensitive K Channels by Phosphorylation, MgADP, and Phosphatidylinositol Bisphosphate (PIP_2) A Study of Channel Rundown and Reactivation, *The Journal of General Physiology* 116.(3): 391-410.

Robertson B, Owen D, Stow J, Butler C & Newland C. (1996). Novel effects of dendrotoxin homologues on subtypes of mammalian Kv1 potassium channels expressed in Xenopus oocytes, *FEBS Letters* 383.(1-2): 26-30.

Rodriguez N, Amarouch MY, Montnach J, Piron J, Labro AJ, Charpentier F, Mérot J, Baró I & Loussouarn G. (2010). Phosphatidylinositol-4,5-Bisphosphate (PIP_2) Stabilizes the Open Pore Conformation of the Kv11.1 (hERG) Channel, *Biophysical Journal* 99.(4): 1110-1118.

Roullet J-B, Luft UC, Xue H, Chapman J, Bychkov R, Roullet CM, Luft FC, Haller H & McCarron DA. (1997). Farnesol Inhibits L-type Ca^{2+} Channels in Vascular Smooth Muscle Cells, *Journal of Biological Chemistry* 272.(51): 32240-32246.

Si H, Grgic I, Heyken WT, Maier T, Hoyer J, Reusch HP & Kohler R. (2006). Mitogenic modulation of Ca^{2+} -activated K^+ channels in proliferating A7r5 vascular smooth muscle cells, *Br J Pharmacol* 148.(7): 909-917.

Skrzypiec-Spring M, Grotthus B, Szelag A & Schulz R. (2007). Isolated heart perfusion according to Langendorff--Still viable in the new millennium, *Journal of Pharmacological and Toxicological Methods* 55.(2): 113-126.

Soboloff J, Spassova M, Xu W, He LP, Cuesta N & Gill DL. (2005). Role of endogenous TRPC6 channels in Ca^{2+} signal generation in A7r5 smooth muscle cells, *Journal of Biological Chemistry* 280.(48): 39786-39794.

Suh BC & Hille B. (2008). PIP2 is a necessary cofactor for ion channel function: how and why?, *Annu Rev Biophys* 37175-195.

Suh BC, Leal K & Hille B. (2010). Modulation of high-voltage activated Ca^{2+} channels by membrane phosphatidylinositol 4,5-bisphosphate, *Neuron* 67.(2): 224-238.

Swartz KJ & MacKinnon R. (1995). An inhibitor of the Kv2.1 potassium channel isolated from the venom of a Chilean tarantula, *Neuron* 15.(4): 941-949.

Thibonnier M, Bayer AL, Simonson MS & Kester M. (1991). Multiple signaling pathways of V1-vascular vasopressin receptors of A7r5 cells, *Endocrinology* 129.(6): 2845-2856.

Tytgat J, Debont T, Carmeliet E & Daenens P. (1995). The α-Dendrotoxin Footprint on a Mammalian Potassium Channel, *Journal of Biological Chemistry* 270.(42): 24776-24781.

Weintraub WH, Cleveland-Wolfe P & Fewtrell C. (1994). Paracrine Ca^{2+} signaling in vitro: serotonin-mediated cell-cell communication in mast cell/smooth muscle cocultures, *J Cell Physiol* 160.(2): 389-399.

Widlansky ME, Price DT, Gokce N, Eberhardt RT, Duffy SJ, Holbrook M, Maxwell C, Palmisano J, Keaney JF, Jr., Morrow JD & Vita JA. (2003). Short- and Long-Term COX-2 Inhibition Reverses Endothelial Dysfunction in Patients With Hypertension, *Hypertension* 42.(3): 310-315.

Zaika O, Zhang J & Shapiro MS. (2011). Combined phosphoinositide and Ca^{2+} signals mediating receptor specificity toward neuronal Ca^{2+} channels, *Journal of Biological Chemistry* 286.(1): 830–841.

Zhang H, Craciun LC, Mirshahi T, Rohacs T, Lopes CM, Jin T & Logothetis DE. (2003). PIP(2) activates KCNQ channels, and its hydrolysis underlies receptor-mediated inhibition of M currents, *Neuron* 37.(6): 963-975.

14

Role of Non-Steroidal Anti-Inflammatory Drugs (NSAIDs) in Modulating Vascular Smooth Muscle Cells by Activating Large-Conductance Potassium Ion Channels

Donald K. Martin[1,3], Christopher G. Schyvens[1], Kenneth R. Wyse[1], Jane A. Bursill[1], Robert A. Owe-Young[2], Peter S. Macdonald[1] and Terence J. Campbell[1]

[1]*Department of Medicine, University of NSW, & Victor Chang Cardiac Research Institute, St. Vincent's Hospital, Sydney, N.S.W.*

[2]*Centre for Immunology, St. Vincent's Hospital, Sydney, N.S.W.*

[3]*Fondation « Nanosciences aux Limites de la Nanoélectronique » Université Joseph Fourier, TIMC-GMCAO, Pavillon Taillefer, Faculté de Médecine, La Tronche*

[1,2]*Australia*

[3]*France*

1. Introduction

In this chapter we propose to discuss the role of K^+ ion channels in stimulating vasodilatation by altering the membrane potential of vascular smooth muscle cells. We present evidence that the K^+ channels are modulated by a direct action of non-steroidal anti-inflammatory drugs (NSAIDs) to activate the K^+ ion channels.

The primary cellular action of non-steroidal anti-inflammatory drugs (NSAIDs) is thought to be through inhibition of pathways involving cyclo-oxygenase (COX). COX catalyses the conversion of arachidonic acid to prostaglandin endoperoxides (Vane, 1971) which are the precursors of both prostacyclin and thromboxane A_2 (Moncada et al., 1976). Such an action of NSAIDs may be expected to be vasoconstrictive and lead to increased blood pressure, which is a possibility that has been suggested by meta-analyses of clinical studies (Johnson et al., 1994). However, early reports indicated that chronic administration of indomethacin or other NSAIDs had varying effects on blood pressure (Lopez-Ovejero et al., 1978; Ylitalo et el., 1978). For example, whilst indomethacin and naproxen are associated with increases in blood pressure, NSAIDs such as sulindac, aspirin, piroxicam or ibuprofen have negligible effects (Pope et al., 1993). Moreover, in a direct study of the effects of NSAIDs in patients with mild essential hypertension, it was found that ibuprofen increased systolic blood pressure but neither aspirin nor sulindac had any significant effect on systolic or diastolic blood pressure (Minuz et al., 1990). In this chapter we report our investigations of the hypothesis that the variable effect on blood pressure was due to NSAIDs inducing vasodilatation.

Vasodilatation can be mediated by contributions from any one of several independent cellular mechanisms which include release of COX metabolites and nitric oxide (NO), release of an endothelium-derived hyperpolarising factor (EDHF), or activation of ATP-sensitive potassium channels (Feletou & Vanhoutte, 2000; Pinheiro & Malik, 1993). In addition, the particular mechanism underlying the eventual vasodilatation can be related to its initiating chemical mediator. For example, endothelium-dependent vasorelaxation does not appear to be mediated by COX products and is critically dependent on NO (Pinheiro & Malik, 1993), although it may also involve cell hyperpolarisation via the opening of ATP-sensitive potassium channels (Sakuma et al., 1993). Some authors have suggested that the particular type of potassium channel that is activated to produce cell hyperpolarisation may not be confined to the classical ATP-sensitive channel (Seigel et al., 1992). For example, the NO-independent coronary vasodilator effect of bradykinin was found to utilise a Ca^{2+}-activated potassium channel (Fulton et al., 1994).

It has been reported that NSAIDs of the fenamate family, which include mefenamic acid, niflumic acid and flufenamic acid, activated Ca^{2+}-activated potassium channels (Farrugia et al., 1993; Ottolia & Toro, 1994). We examined the possibility, using patch-clamp electrophysiology, that other NSAIDs may also activate Ca^{2+}-activated potassium channels in aortic smooth muscle cells since not all NSAIDs have been shown to cause significant increases in blood pressure. Furthermore, we used the enantiomers of flurbiprofen to separate the COX-mediated effects from those related to potassium channel activation in organ bath experiments where we recorded constrictor responses of the aorta to phenylephrine. It is important to note that R-flurbiprofen has negligible effects on COX pathways compared to S-flurbiprofen, which does inhibit COX pathways (Peskar et al., 1991). Also, R-flurbiprofen does not convert to S-flurbiprofen in biological systems, unlike an enantiomer such as R-ibuprofen. We report that low concentrations of several NSAIDs were found to activate a Ca^{2+}-sensitive and ATP-activated K^+ channel (K_{AC}) in vascular smooth muscle cells, leading to cell hyperpolarisation and vasodilatation. Our results indicate that several NSAIDs may cause vasodilatation which would explain the clinical reports that some NSAIDs have negligible effects, or even reductions, in blood pressure.

2. Experimental procedures and methods

To investigate the effect of NSAIDs, we correlated the results from patch-clamp electrophysiology and physiological organ-bath investigations utilizing rings of vascular tissue. The results from experiments using those techniques were used to test our hypothesis that the variable effect on blood pressure which is reported in the clinical literature is due to the spectrum of potency of NSAIDs to activate K^+ channels (thereby inducing vasodilatation) in synergy with the classical NSAID effect on intracellular pathways that involve cyclo-oxygenase (which usually results in a vasoconstriction).

For the experiments we used the following drugs: acetylcholine, phenylephrine, ATP, ADP, AMP, aspirin, indomethacin, flufenamic acid, niflumic acid, mefenamic acid, pinacidil, TEA, and collagenase were purchased from Sigma. Porcine pancreatic elastase was from Calbiochem-Novabiochem (Sydney). Glibenclamide was from RBI (Natlick MA, U.S.A.). R- and S- isomers of flurbiprofen were a kind donation from the Boots Company (UK) by Dr Ken Williams, St Vincent's Hospital Sydney. Other chemicals used for the intracellular, extracellular, and Krebs bicarbonate solutions were of AR grade.

2.1 Animal studies

This protocol was approved by the Garvan Institute of Medical Research/St Vincent's Hospital Animal Ethics Committee. The study complied with the guidelines published by the Australian National Health and Medical Research Council for the care and conduct of experiments using animals in research, and conformed to the Guide for the Care and Use of Laboratory Animals published by the US National Institutes of Health (NIH Publication No. 85-23, revised 1996).

2.2 Preparation of RASM myocytes for patch-clamp electrophysiology

New Zealand white rabbits of either sex, weighing 1 to 2 kg, were anaesthetised with pentobarbital sodium via an ear vein (Nembutal, Boehringer Ingelheim, 45 mg/kg). Heparin (500 Units) was infused at the same time. A short section of the thoracic aorta was dissected and the vessel was washed several times in Hanks Balanced Salt Solution (GIBCO, Life Technologies, Melbourne) and incubated in 1000 U/ml collagenase (Type II, Sigma #C6885) for 30 minutes at 37°C to remove endothelial cells. Strips of media were carefully peeled off using jewellers forceps, diced and incubated in a solution containing 1000 U/ml collagenase (Type II) and 60 U/ml porcine pancreatic elastase (Calbiochem-Novabiochem, Sydney) for 2-3 hours at 37°C with periodic trituration. Either the tissue explants or dispersed cells were seeded in Dulbecco's Modified Eagles Medium (DMEM) supplemented with 100 U/ml penicillin (GIBCO), 100 μg/ml streptomycin (GIBCO), 4 mM fresh L-glutamine (GIBCO) and 10% foetal bovine serum (PA Biologicals, Sydney). The cells ($2{\times}10^5$/well) at passage 2-3 were plated onto glass coverslips at least 3 days before an experiment and serum-deprived for 24 hours in 4% Monomed (Commonwealth Serum Laboratories, Melbourne) in DMEM. Cultured cells expressed smooth muscle actin and were negative for Factor VIII:RAg, an endothelial cell marker.

2.3 Recording and analysis of K^+ ion channels using patch-clamp electrophysiology

Standard patch-clamp techniques that we have used previously on cardiovascular cells (Martin et al., 1994) were used to record single ion channel activity, at 37°C, in the inside-out, cell-attached and whole-cell configurations from the rabbit aortic smooth muscle (RASM) cells. The channel currents were amplified and filtered at 1kHz (–3dB point) using an Axopatch 1D amplifier (Axon Instruments, Union City, CA, U.S.A.) and sampled on-line by a microcomputer (IBM 486 compatible) using commercial software and associated A/D hardware (pClamp 6.0/Digidata 1200, Axon Instruments and Scientific Solutions Inc., Foster City, CA, U.S.A.). The single-channel open probability was calculated from the areas of Gaussian curves fitted to amplitude histograms compiled from 2 minute channel recordings. We calculated changes in the channel activity following the addition of NSAIDs by dividing the open probability in the presence of the drug by that recorded before the drug was applied. All data are presented as mean±SE with the number of observations in parenthesis (n).

Ion currents are referred to the trans-patch potential (V_m). For cell-attached patches, this was determined from the pipette potential (V_p), cell resting potential (E_m) and the liquid junction potential (E_L) between the bath and pipette solutions. The liquid junction potential (E_L) was calculated using commercial software (Barry, 1994), and was typically around 4mV. Thus

$$V_m = (E_m - V_p) + E_L \quad (1)$$

For inside-out patches, the V_m was only determined by V_p and E_L, with E_m=0 in equation (1), thus

$$V_m = V_p - E_L \quad (2)$$

For recording from inside-out membrane patches an extracellular solution was used in the pipette and for superfusion of the cells during seal formation, which contained (mM): NaCl (130), KCl (4.8), $MgCl_2$ (1.2), NaH_2PO_4 (1.2), N-[2-Hydroxyethyl]piperazine-N'-[2-ethanesulphonic acid] (HEPES) (10), glucose (12.5), $CaCl_2$ (1.0), and Bovine Albumin (0.5mg/ml, fraction V, Sigma, #A7888) (pH=7.4, with NaOH). After a gigaseal was formed this extracellular superfusing solution was replaced with an intracellular solution that contained (mM): KCl (140), $MgCl_2$ (1.2), ethylene glycol-bis (b-amino ethyl ether) tetraacetic acid (EGTA) (5), and HEPES (10) (pH=7.2, with KOH) and an inside-out membrane patch was excised from the cell. The various NSAIDs were added to this intracellular solution.

2.4 Recording and analysis of the contractile properties of arterial rings

Eight 2 mm wide rings were cut from sections of the thoracic aorta of the rabbits. In some experiments the endothelium was removed by gently rotating a wooden swab stick in the lumen of the vessel. Removal of the endothelium was confirmed by the absence of acetylcholine (1μM) induced relaxation in rings that had been pre-constricted with phenylephrine (PE) (1μM). The 8 rings were suspended in individual 10 ml water-jacketed (37°C) organ baths filled with freshly prepared Krebs-Bicarbonate solution, which had the following composition (mM): NaCl (118), KCl (4.7), KH_2PO_4 (1.2), $MgSO_4.7H_2O$ (1.18), glucose (5.0), $NaHCO_3$ (25.0), $CaCl_2.2H_2O$ (2.54). The aortic rings were pre-loaded with a basal tension of 2.00±0.05g. Isometric tension in each ring was measured using a Grass FT03 force transducer (Quincy, MA, USA), the output of which was multiplexed and sampled on-line by a microcomputer (Macintosh IIsi) using commercial software and associated A/D hardware (MACLAB, Analog Digital Instruments, Sydney, Australia).

The design of the organ bath experiments were as follows: *(i)* challenge with a submaximal contractile dose of KCl (40mM), which was used to normalise all subsequent responses to PE in the presence of the NSAID; *(ii)* determine a PE dose-response curve from each vascular ring in order to produce a control response. After suitable washout and re-equilibration periods of 30 minutes, the NSAIDs were added to the organ baths 15 minutes prior to subsequent challenge with PE; *(iii)* time mediated changes in responsiveness of the vascular rings were analysed by repeated challenges/trials with PE during a complete experiment; *(iv)* PE and NSAIDs were added to the organ baths in cumulative concentrations; *(v)* concentrations noted in all figures reflect the final concentration in the organ baths.

Contraction of vascular rings is presented either as tension (g), or a contraction relative to the contraction elicited by KCl (40mM) which was expressed as a percentage (%). Differences between dose-response curves were assessed using 2-way analysis of variance (ANOVA). Further comparisons of individual data-points were tested using unpaired Student's t-tests with the appropriate Bonferroni correction for multiple comparisons (SPSS v10.0, Chicago, Ill).

Separate stock solutions of phenylephrine (10mM), acetylcholine perchlorate (10mM), and KCl (4M) were prepared with deionized water (Milli-Q, Millipore Corporation, Bedford MA, USA). Subsequent dilutions for each drug were made with freshly prepared Krebs bicarbonate solution. R- and S-flurbiprofen were prepared as a 100mM stock with 0.1M Na_2CO_3 solution, and subsequently diluted with freshly prepared Krebs bicarbonate solution. As a control, no change in contraction was recorded after exposing rabbit aortic rings to Na_2CO_3 (0.1M) for 15 minutes.

3. Results from experiments

In summary, the following results of the experiments describe a large-conductance K^+ channel in smooth muscle cells that is activated by intracellular ATP and Ca^{2+}. Furthermore, the K^+ channel is activated by some NSAIDs and pinacidil. We have designated this channel K_{AC}, since it does not have the normal characteristics of the classical ATP-sensitive K^+ channel, and the K_{AC} channels have additional features compared to the previously reported maxi-K channels (Kuriyama et al., 1998). The RASM smooth muscle cells were hyperpolarised by aspirin, an NSAID that potently activated the K_{AC} channel. Both R- and S-flurbiprofen antagonised constrictor responses of the rabbit aorta to PE, suggesting that relaxation occurred via a mechanism other than inhibition of cyclo-oxygenase pathways. Our results allow us to conclude that NSAIDs are potent openers of a Ca^{2+}-activated phosphorylation-dependent potassium channel in vascular smooth muscle cells leading to cell hyperpolarisation and vessel dilatation. The activation of potassium channels is thought to be significant in controlling excitability of vascular smooth muscle cells and regulation of myogenic tone (Brayden & Nelson, 1992), an idea that has been corroborated in coronary arteries (Scornik et al., 1993) and in rabbit aorta (Gelband & McCullough, 1993). The K_{AC} channel that we describe in this paper thus provides a novel target to control excitability of vascular smooth muscle cells and regulate myogenic tone.

3.1 Characteristics of K^+ ion channels in RASM cells

The predominant channel recorded from inside-out membrane patches, with a NaCl-rich extracellular pipette and a KCl-rich intracellular bath solution, had a single-channel conductance of 128±6 pS (n=31) as shown in Figure 1.

Under these conditions the reversal potential of the current-voltage relation was –60 mV (95% confidence interval of -69 mV to -52 mV) indicating that it was permeable to potassium ions (E_K = –85 mV). This was confirmed in experiments with 7 other patches where half of the KCl in the superfusing intracellular (bath) solution was replaced with K-gluconate. As would be expected for a mainly K^+-permeable channel, neither the reversal potential (–47 mV with KCl and –49 mV with K-gluconate) nor the single-channel conductance (140±9 pS with KCl and 146±13 pS with K-gluconate) was altered following this ion substitution. With symmetrical KCl solutions bathing the inside-out patches, the single-channel conductance was 259±18 pS (n=6) and the reversal potential shifted toward zero (+4.3 mV).

The pooled distribution of channel openings from 11 inside-out patches with the pipette filled with the extracellular solution and the bath filled with the intracellular solution and no applied voltage (trans-patch potential of 0mV) was fitted by a single exponential with a time constant of 2.7±0.0 ms (2,567 events). The average single-channel open probability from these inside-out patches was 0.0125±0.0053 (n=11). The single-channel open probability was increased by 64±21 % (n=4) when the concentration of Ca^{2+} was increased from <0.001 μM to 1 μM at the cytosolic face of the inside-out patches. The open probability of the channel was also increased following the addition of adenosine nucleotides (all 5 mM), with the sequence of potency adenosine 5′-triphosphate (ATP) > adenosine 5′-diphosphate (ADP) > adenosine 5′-monophosphate (AMP) (Figure 2).

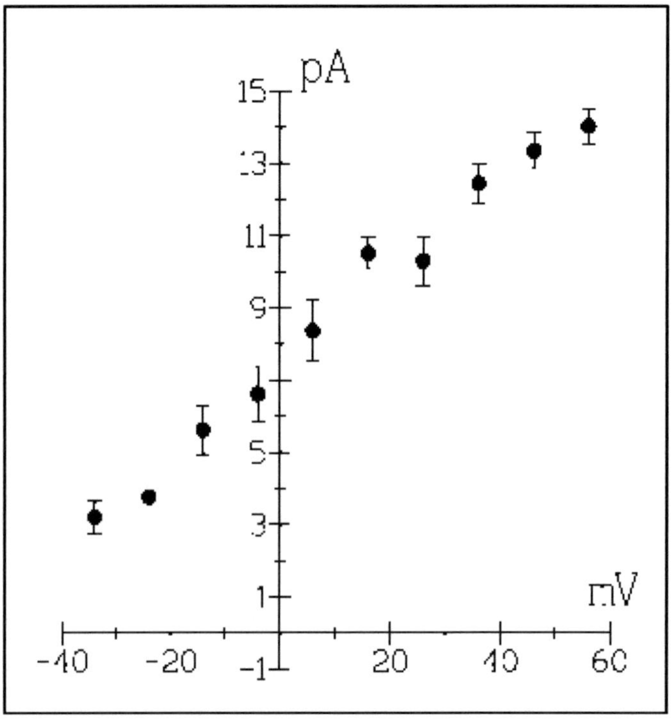

Fig. 1. Current-voltage relation of the predominant K_{AC} channel in rabbit aorta vascular smooth muscle cells recorded from inside-out membrane patches. The pipette solution (extracellular) contained (mM): NaCl (130), KCl (4.8), $MgCl_2$ (1.2), NaH_2PO_4 (1.2), HEPES (10), glucose (12.5), $CaCl_2$ (1.0), and Bovine Albumin (0.5mg/ml) (pH=7.4, with NaOH). The superfusing solution (intracellular) contained (mM): KCl (140), $MgCl_2$ (1.2), EGTA (5), and HEPES (10) (pH=7.2, with KOH). The single-channel conductance with these solutions is 128±6 pS (n=31).

The activation of the 259 pS K^+ channel by ATP and Ca^{2+} led us to use the abbreviation K_{AC} for this channel. The open probability of the K_{AC} channel was unaffected when the ATP-sensitive K^+ channel blocker glibenclamide, in the range from 3 μM to 60 μM, was added to the cytosolic face of inside-out membrane patches. Pinacidil, which is known to activate the classical ATP-sensitive K^+ channel, also increased the open probability of the K_{AC} channel (Figure 3).

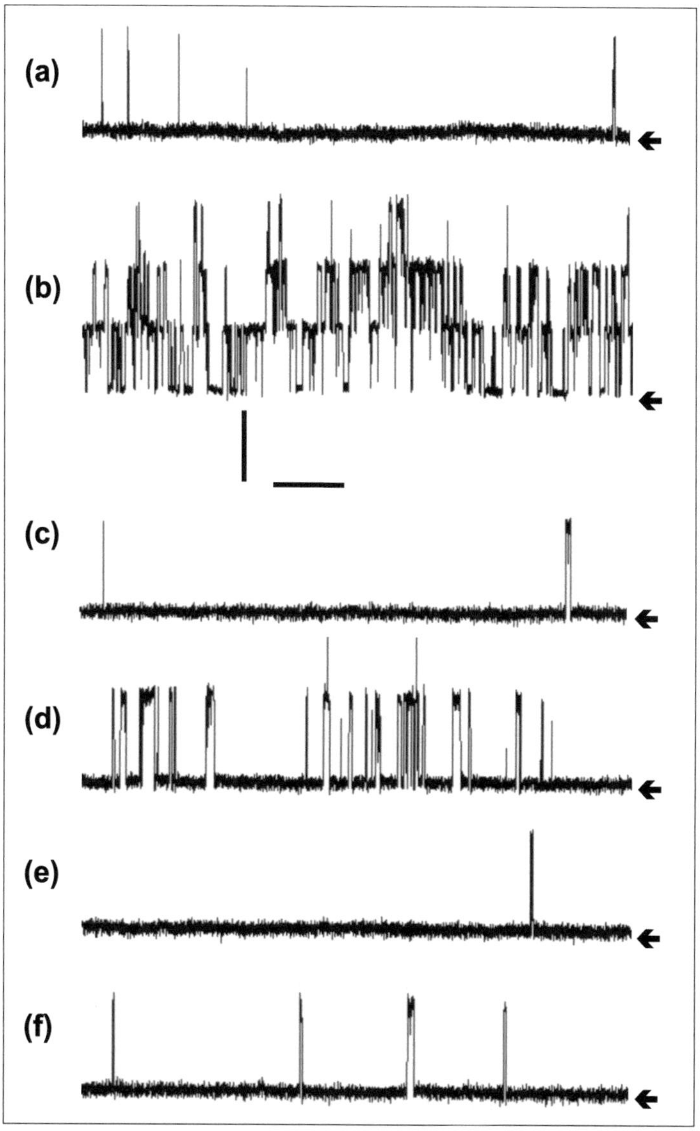

Fig. 2. Activation of the K_{AC} channel in RASM by adenosine nucleotides (all 5mM) applied to the cytoplasmic face of inside-out membrane patches. (a) superfusing solution alone, (b) ATP added to superfusing solution, (c) washout with superfusing solution, (d) ADP added to superfusing solution, (e) washout with superfusing solution, (f) AMP added to superfusing solution. The pipette solution (extracellular) contained (mM): NaCl (130), KCl (4.8), $MgCl_2$ (1.2), NaH_2PO_4 (1.2), HEPES (10), glucose (12.5), $CaCl_2$ (1.0), and Bovine Albumin (0.5mg/ml) (pH=7.4, with NaOH).The superfusing solution (intracellular) contained (mM): KCl (140), $MgCl_2$ (1.2), EGTA (5), and HEPES (10) (pH=7.2, with KOH). The activation of the channel was in the sequence ATP > ADP > AMP. The trans-patch potential was 0 mV and channel openings are upward, with the zero-current (baseline) level indicated with the solid arrow. Scale bars are shown between traces (b) and (c). Vertical bar is 15pA. Horizontal bar is 400 ms.

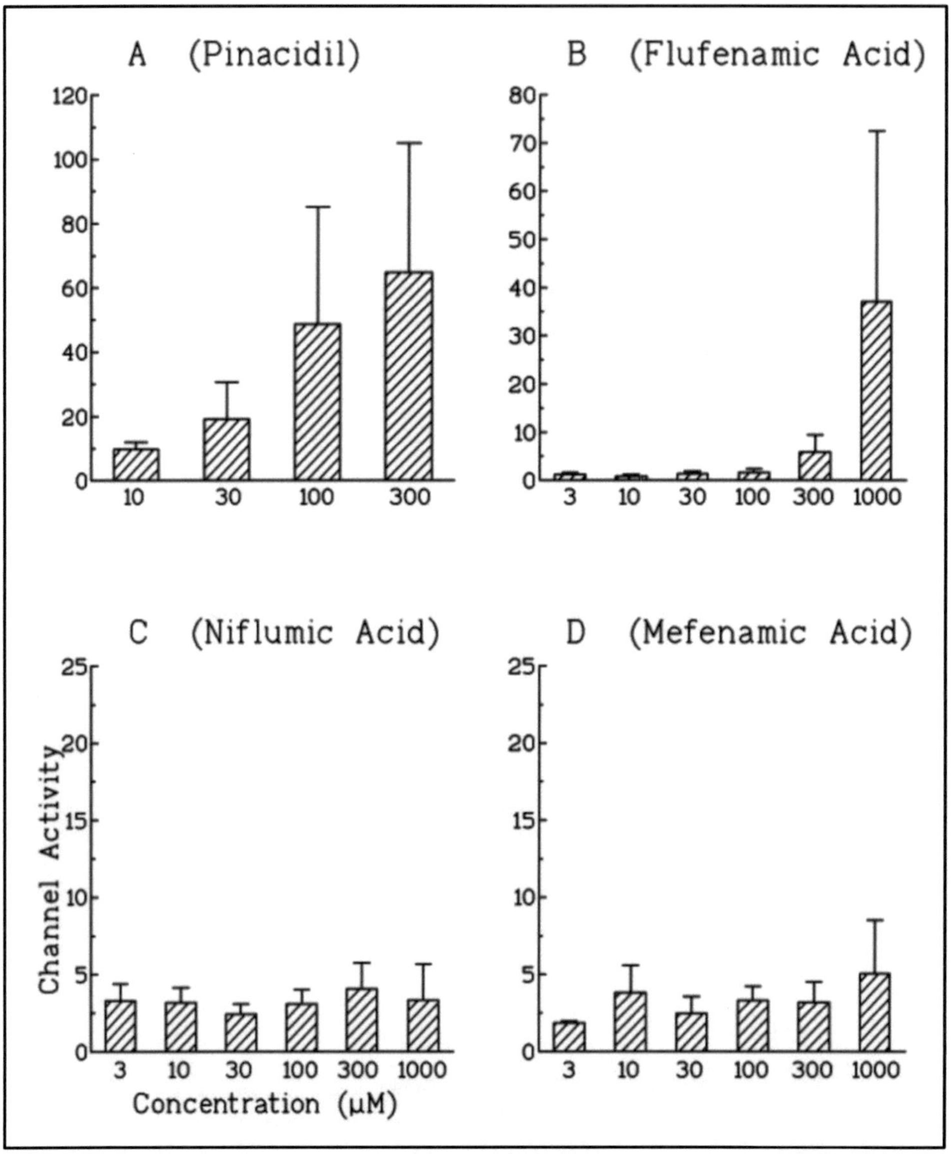

Fig. 3. Effects of (A) pinacidil (n=5), (B) flufenamic acid (n=6), (C) niflumic acid (n=4) and (D) mefenamic acid (n=4) on the open probability of the KAC channel in RASM. In all panels the y-axis represents KAC channel activity, which was calculated by dividing the open probability in the presence of the drug by that recorded before the drug was applied. In all panels the x-axis represents concentration of the drug in μM.

3.2 Effects of NSAIDs on K_{AC} ion channels

There was a dose-dependent increase in the open probability when either aspirin, R-flurbiprofen, S-flurbiprofen, indomethacin or flufenamic acid was added to the cytosolic face of inside-out membrane patches (Figures 3, 4).

Aspirin was the most potent of these NSAIDs, with the order of potency being aspirin > R-flurbiprofen = S-flurbiprofen > indomethacin > flufenamic acid. Those NSAIDs were more potent activators of the KAC channel than pinacidil (Figure 5). The NSAIDs niflumic acid and mefenamic acid were tested, but had no effects on the open probability of the K_{AC} channel.

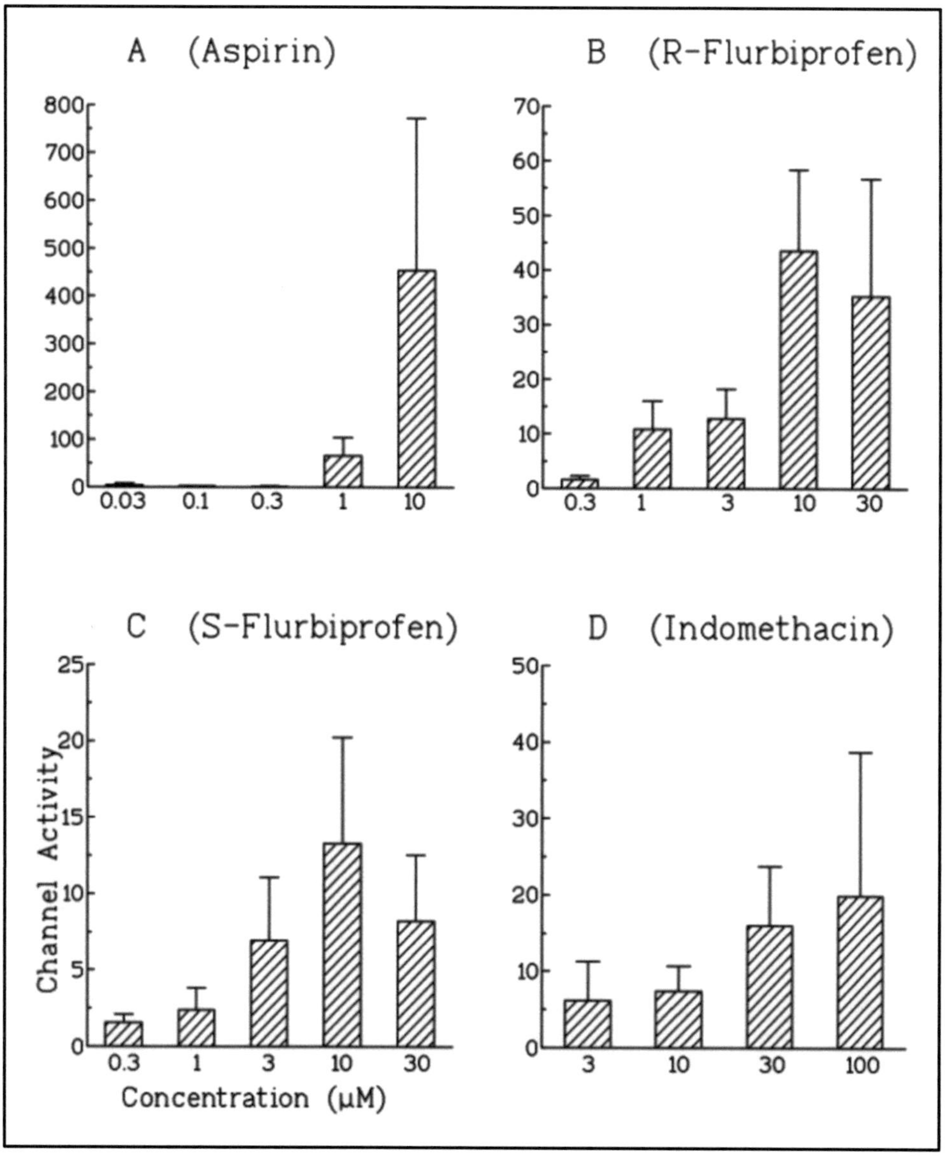

Fig. 4. Effects of (A) aspirin (n=5), (B) R-flurbiprofen (n=6), (C) S-flurbiprofen (n=4) and (D) indomethacin (n=4) on the open probability of the KAC channel in RASM. In all panels the y-axis represents KAC channel activity, which was calculated by dividing the open probability in the presence of the NSAID by that recorded before the NSAID was applied. In all panels the x-axis represents concentration of the drug in μM.

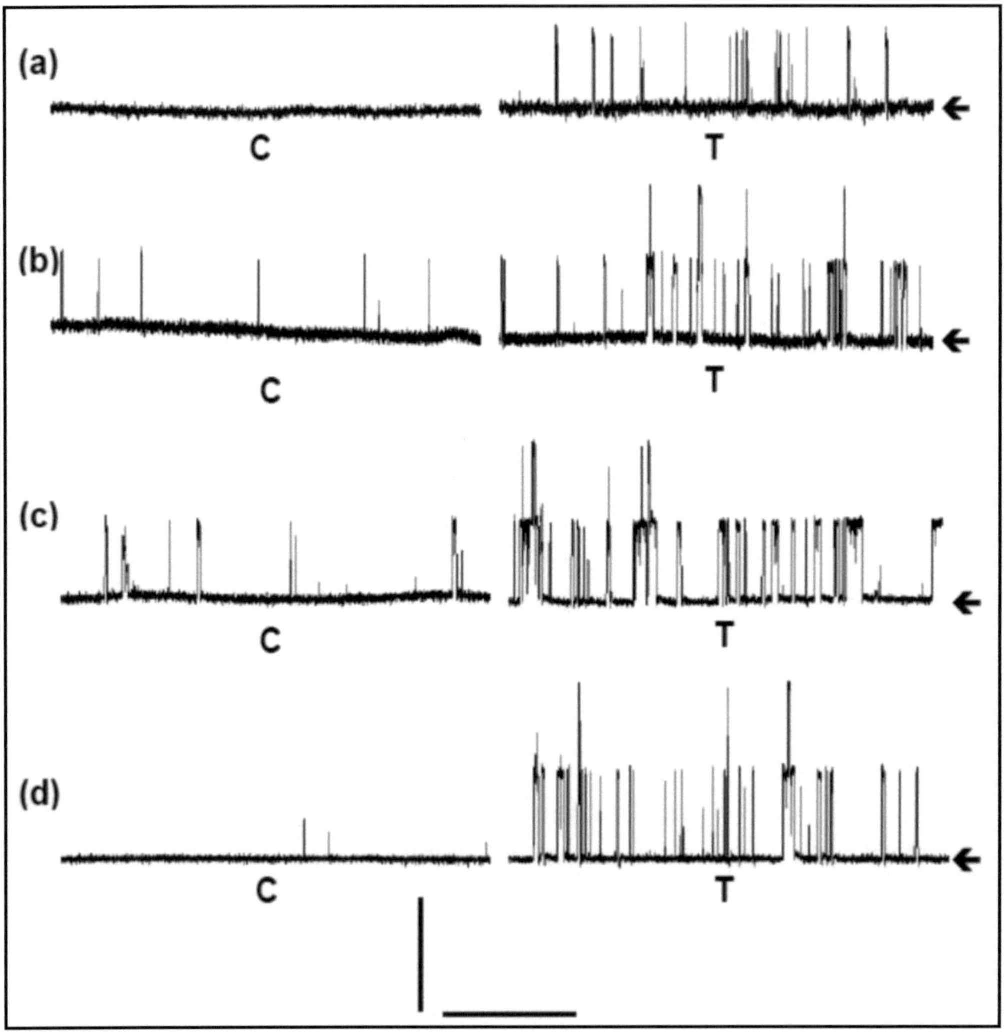

Fig. 5. Activation of the K_{AC} channel in RASM by the NSAIDs (a) aspirin, (b) R-flurbiprofen, (c) S-flurbiprofen, and by (d) pinacidil applied to the cytosolic face of inside-out membrane patches. The trans-patch potential was 0 mV and channel openings are upward, with the zero-current (baseline) level indicated with the solid arrow. In all panels trace C represents the control condition of no drug added to the superfusing solution and trace T represents the recording when the NSAID was added to the superfusing solution. For recording from inside-out membrane patches an extracellular solution was used in the pipette which contained (mM): NaCl (130), KCl (4.8), $MgCl_2$ (1.2), NaH_2PO_4 (1.2), N-[2-Hydroxyethyl]piperazine-N′-[2-ethanesulphonic acid] (HEPES) (10), glucose (12.5), $CaCl_2$ (1.0), and Bovine Albumin (0.5mg/ml, fraction V, Sigma, #A7888) (pH=7.4, with NaOH). The superfusing solution contained (mM): KCl (140), $MgCl_2$ (1.2), ethylene glycol-bis (b-amino ethyl ether) tetraacetic acid (EGTA) (5), and HEPES (10) (pH=7.2, with KOH) and an inside-out membrane patch was excised from the cell. Vertical bar is 15pA. Horizontal bar is 400 ms.

3.3 Effects of NSAIDs on membrane potential

The activation of the KAC channel by the NSAIDs may be expected to hyperpolarise the smooth muscle cells. To test this hypothesis, we recorded the membrane potential while current-clamping (whole-cell mode) the smooth muscle cells used after adding aspirin 1 μM extracellularly. This experiment was repeated on 12 cells that satisfied the criteria of remaining in a stable whole-cell configuration for at least 5 minutes with no depolarising shifts in membrane potential, and the membrane potential was at least –10 mV at the end of this stabilisation period.

The average resting membrane potential for the 12 cells was –19.4±1.9 mV. After 5 minute exposure to aspirin 1 μM in the superfusing solution, the membrane potential was hyperpolarised in the majority (10/12) of these cells. The average hyperpolarisation shift in membrane potential induced by aspirin was by –5.8±1.4 mV (n=12), which was significant when tested with a Wilcoxon Rank-Sum test (p=0.001).

3.4 Effects of NSAIDs on contractile properties of rabbit aorta

The patch-clamp electrophysiology results indicate that the activation of the K_{AC} channel leads to hyperpolarisation of the rabbit aorta smooth muscle cells, which may also lead to relaxation of the intact blood vessel. We tested this hypothesis by measuring the effect of both the R- and S- enantiomers of flurbiprofen (1 mM) on the contraction of rings of rabbit aorta in response to cumulative concentrations of PE. We utilised enantiomers of flurbiprofen in these experiments so that any effects due to K_{AC} channel activation could be distinguished from effects on cyclo-oxygenase pathways. Both enantiomers of flurbiprofen activated the K_{AC} channels. It is important to note that R-flurbiprofen has negligible effects on cyclo-oxygenase pathways compared to S-flurbiprofen, which does inhibit these pathways (Peskar et al., 1991). Also, R-flurbiprofen does not convert to S-flurbiprofen in biological systems, unlike an enantiomer such as R-ibuprofen.

Both R- and S- enantiomers of flurbiprofen shifted the PE dose-response curve to the right and reduced the maximum contraction induced by PE (Figure 6). The EC_{50} values were calculated from non-linear regression of the data in figure 6 using a Gompertz 3-parameter sigmoidal equation (SigmaPlot 2000, Chicago, Ill) and are shown in table 1. The effect of either R- or S- flurbiprofen was to increase the EC_{50} value for PE-induced contraction by approximately 2½ times. The effect of flurbiprofen (R- and S- combined) was to increase significantly the EC_{50} for the PE-induced contraction to 527.0 ± 5.5 nM from the control value of 214.4 ± 9.2 nM ($p<0.05$, t = 23.976, Student's t-test).

Ring treatment	R-flurbiprofen		S-flurbiprofen	
	control	treatment	control	treatment
+ endothelium	207.0	517.6 (2.50)	241.5	518.0 (2.14)
- endothelium	200.0	533.0 (2.67)	209.1	539.5 (2.58)

Table 1. The effect of R- or S-flurbiprofen on the phenylephrine EC_{50} derived from rabbit thoracic aortic rings, with (+) or without endothelium (-).Values shown in brackets indicate magnitude of increase from control to treatment EC_{50}. The calculated EC_{50} (nM) is derived from the data shown in figure 6, using non-linear regression of a Gompertz 3-parameter sigmoidal equation (SigmaPlot 6.0).

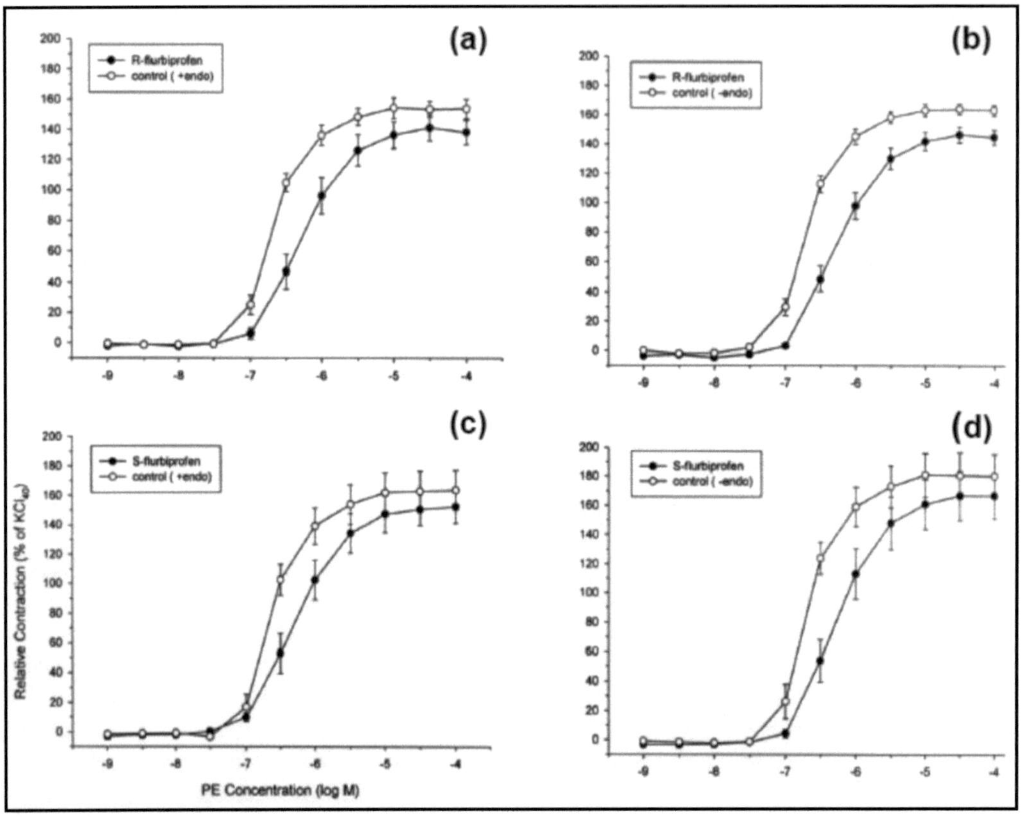

Fig. 6. Effect of enantiomers of flurbiprofen on pre-constricted rings dissected from rabbit aorta. (a) R-flurbiprofen added to rings with intact endothelium, (b) R-flurbiprofen added to rings with no endothelium, (c) S-flurbiprofen added to rings with intact endothelium, and (d) S-flurbiprofen added to rings with no endothelium. Both R- and S-flurbiprofen antagonised the constrictor responses of the aorta to PE. Note that, in comparison to S-flurbiprofen, R-flurbiprofen has a very weak effect on cyclo-oxygenase pathways. The data-points represent mean±SE. In all panels the x-axis represents the PE concentration (log M) and the y-axis represents the relative contraction (%) of the aortic rings relative to the contraction to KCl (40mM).

4. Discussion and conclusion

We describe a large conductance Ca^{2+}-activated K^+ channel (K_{AC}) in RASM that has the unusual property of being activated by ATP applied intracellularly. Investigations of a similar channel in smooth muscle cells of the rat pulmonary artery suggested that phosphorylation is important for its activation (Robertson et al., 1992). A subsequent report from the Kozlowski research group described this type of channel being present in isolated smooth muscle cells from the aorta, mesenteric and basilar arteries of the rat (Hartley & Kozlowsky, 1996). Such large ATP- and Ca^{2+}-activated K^+ channels may represent a link between cellular metabolism and hypoxia (Pinheiro & Malik, 1993), since the glycolysis inhibitor, 2-deoxy-D-glucose, has been shown to inhibit K^+ currents in rat pulmonary arterial smooth muscle cells (Hartley & Kozlowski, 1996).

The novel results that we present are that K_{AC} is activated by several NSAIDs, including R-flurbiprofen, S-flurbiprofen, indomethacin and aspirin. In contrast to the previously published effect of fenamates to activate large conductance Ca^{2+}-activated K^+ channels (Farrugia et al., 1993; Ottolia & Toro, 1994) or to block classical ATP-sensitive K^+ channels (Li et al., 2007), the K_{AC} channels that we report are not sensitive to mefenamic acid, niflumic acid nor flufenamic acid. This insensitivity to fenamates and a sensitivity to intracellular ATP distinguish the K_{AC} channels from the large-conductance Ca^{2+}-activated channels described previously (Farrugia et al;? 1993; Ottolia & Toro, 1994). Activating the K_{AC} channels with aspirin also hyperpolarised the RASM. Furthermore, we report that pinacidil also activated the K_{AC} channels. Pinacidil has been previously thought to only activate classical K_{ATP} channels.

There are some reports of large conductance Ca^{2+}-activated K^+ ("maxi-K") channels as being a target for the fenamate class of NSAIDs. Ottilia & Toro (1994) reported that niflumic acid, flufenamic acid and mefenamic acid rapidly and reversibly activated a large conductance Ca^{2+}-activated K^+ channels. Greenwood & Large (1995) confirmed that those fenamates activated a large conductance Ca^{2+}-activated K^+ channel in rabbit portal vein, and those authors reported that the Ca^{2+}-activated K^+ current was inhibited by TEA but not by glibenclamide. Also, the cardiac delayed rectifier K^+ channel HERG is activated by the fenamates flufenamic and niflumic acids (Malykhina et al., 2002) and flufenamic acid activated the maxi-K channel in the trabecular meshwork of both human and bovine origin (Stumpff et al., 2001). We found that neither niflumic acid, mefenamic acid nor flufenamic acid (but only in very high concentration) could activate the K_{AC} channel. The K_{AC} channels differ from the maxi-K class of channels in at least the lack of a fenamate binding site.

Further insights into the nature of the NSAID binding site on the K_{AC} channels can be obtained from the range of drugs used to activate these channels. K_{AC} was activated by representative NSAIDs from the salicylate (aspirin), propionic acid (flurbiprofen) and indole (indomethacin) families of NSAIDs. The potency of the drugs in activating the K_{AC} channels was aspirin > R-flurbiprofen = S-flurbiprofen > indomethacin > pinacidil > flufenamic acid (very weak effect) > niflumic acid = mefenamic acid (both no effect). The drugs with the most potent activating effect on K_{AC} were generally the more acidic. This implies a different extracellular conformation of the K_{AC} channels compared to the maxi-K channels, such that the nature of the residues available at the NSAID binding site may be different for these two types of large-conductance K^+ channels.

In our organ bath experiments R-flurbiprofen, an NSAID that activates the K_{AC} channels but does not inhibit prostaglandin synthesis, antagonised PE-induced contraction of rings of aorta. In an earlier report, Pallapies et al (1994) found that both R- and S-flurbiprofen relaxed rat aorta that had been pre-contracted with PE. Furthermore, McGrath et al (1990) indicated that flurbiprofen inhibited acetylcholine-induced contractions of rabbit saphenous vein in which the endothelium was intact. In direct measurements of pressure in the perfused vascular bed of isolated rabbit lungs, aspirin was found to inhibit the PE-induced increase in arterial pressure (Delaunois et al., 1994). Our results demonstrate the effect of flurbiprofen was not influenced by the presence of the endothelium, which suggested that flurbiprofen was acting directly on the smooth muscle cells of the aorta rings. The similarity in responses from R- and S-flurbiprofen suggested that the relaxation was due to activation of the K_{AC} channel, rather than through effects on cyclo-oxygenase pathways. Our results provide evidence that K_{AC} channels provide a mechanism underlying the vasodilatory effect of NSAIDs.

The primary cellular action of non-steroidal anti-inflammatory drugs (NSAIDs) is thought to be inhibition of cyclo-oxygenase pathways, the enzyme that catalyses the conversion of arachidonic acid to prostaglandin endoperoxides (Vane, 1971) which are the precursors of both prostacyclin and thromboxane A_2 (Moncada et al., 1976). Such an action may be expected to be vasoconstrictive and lead to increased blood pressure, a possibility that has been suggested by meta-analyses of clinical studies (Johnson et al., 1994). However, early reports indicated that chronic administration of indomethacin or other NSAIDs had varying effects on blood pressure (Lopez-Ovejero et al., 1978; Ylitalo et al., 1978). On the contrary, whilst indomethacin and naproxen are associated with increases in blood pressure, NSAIDs such as sulindac, aspirin, piroxicam or ibuprofen have negligible effects (Pope et al., 1993). Moreover, in a direct study of the effects of NSAIDs in patients with mild essential hypertension, it was found that ibuprofen increased systolic blood pressure but neither aspirin nor sulindac had any significant effect on systolic or diastolic blood pressure (Minuz et al., 1990). Those clinical effects of different NSAIDs on blood pressure correlate well with our results on the potency of NSAIDs to activate K_{AC}, which induces hyperpolarisation of the smooth muscle cells. Those clinical reports support our hypothesis that the variable effect on blood pressure was due to some NSAIDs, especially non-fenamates, inducing vasodilatation through the mechanism of activating K_{AC}.

The activating effect of NSAIDs on ion channels may be an important therapeutic effect in other parts of the body. Liu et al (2005) reported that diclofenac was able to activate transient outward potassium, I(A), channels in neurons. More recently, it has been reported that diclofenac (1 mg/mL) may exert a "local anaesthetic-like" action by reducing the excitability of muscle nociceptors without involving the opening of K_{ATP} channels (Cairns et al., 2008). The mechanism for such an effect could be hyperpolarisation of the muscle nociceptors by diclofenac activating channels such as those we describe. However, in very high concentrations salicylate (1 mM) applied to rat pyramidal neurons in the auditory cortex was reported to increase the firing rate of neurons and enhance neuronal excitability, with the mechanism apparently to inhibit ion currents including the voltage-gated sodium current, the delayed rectifier potassium current and the L-type voltage-gated calcium current (Liu et al., 2007).

In summary, we describe a large-conductance K^+ channel in smooth muscle cells that is activated by intracellular ATP and Ca^{2+}, and which is activated by some NSAIDs and pinacidil. We have designated this channel K_{AC}, as it does not have the characteristics of the classical ATP-sensitive K^+ channel, and the K_{AC} channels have additional features compared to the previously reported maxi-K channels (Kuriyama et al., 1998). The smooth muscle cells were hyperpolarised by aspirin, an NSAID that potently activated the K_{AC} channel. Both R- and S-flurbiprofen antagonised constrictor responses of the rabbit aorta to PE, suggesting that relaxation occurred via a mechanism other than inhibition of cyclo-oxygenase pathways. We conclude that NSAIDs are potent openers of a Ca^{2+}-activated phosphorylation-dependent potassium channel in vascular smooth muscle cells leading to cell hyperpolarisation and vessel dilatation. The activation of potassium channels is thought to be significant in controlling excitability of vascular smooth muscle cells and regulation of myogenic tone (Brayden & Helson, 1993), an idea that has been corroborated in coronary arteries (Scornik et al., 1993) and in rabbit aorta (Gelband & McCullough, 1993). The K_{AC} channel that we describe in this paper thus provides a novel target to control excitability of vascular smooth muscle cells and regulate myogenic tone.

5. Acknowledgment

This work was supported by the National Heart Foundation of Australia, the Clive & Vera Ramaciotti Foundation of Australia, the National Health & Medical Research Council of Australia, St Vincent's Clinic Foundation and St Vincent's Hospital. Dr Ken Williams, Department of Clinical Pharmacology at St Vincent's Hospital Sydney kindly donated the R- and S- isomers of flurbiprofen. Prof. D.K. Martin is supported by a Chaire d'Excellence from the Fondation Nanosciences (RTRA), France.

6. References

Barry, PH. (1994). JPCalc, a Software Package for Calculating Liquid Junction Potential Corrections in Patch-clamp, Intracellular, Epithelial and Bilayer Measurements and for Correcting Junction Potential Measurements. *Journal of Neuroscience Methods,* Vol.51, No.1, (January 1994), pp. 107-116, ISSN 0165-0270.

Brayden, JE. & Nelson, MT. (1992). Regulation of Arterial Tone by Activation of Calcium-dependent Potassium Channels. *Science,* Vol.256, No.5056, (24 April 1992), pp. 532-535, ISSN 0036-8075.

Cairns, BE.; Mann, MK., Mok, E., Dong, XD. & Svensson, P. (2008). Diclofenac Exerts Local Anesthetic-like Actions on Rat Masseter Muscle Afferent Fibers. *Brain Research,* Vol.1194, (15 February 2008), pp. 56-64, ISSN 0006-8993.

Delaunois, A.; Gustin, P., Dessy-Doize, C. & Ansay, M. (1994). Modulatory Effect of Neuropeptide Y on Acetylcholine-induced Oedema and Vasoconstriction in Isolated Perfused Lungs of Rabbit. *British Journal of Pharmacology,* Vol.113, No.3, (November 1994), pp. 973-981, ISSN 0007-1188.

Farrugia, G.; Rae, JL., Sarr, MG. & Szurszewski, JH. (1993). Potassium Current in Circular Smooth Muscle of Human Jejunum Activated by Fenamates. *American Journal of Physiology – Gastrointestinal and Liver Physiology,* Vol.265, No.5(Pt 1), (November 1993), pp. G873-G879, ISSN 0193-1857.

Feletou, M. & Vanhoutte, PM. (2000). Endothelium-dependent Hyperpolarization of Vascular Smooth Muscle Cells. *Acta Pharmacologica Sinica,* Vol.21, No.1, (January 2000), pp. 1-18, ISSN 1671-4083.

Fulton, D.; McGiff, JC. & Quilley, J. (1994). Role of K^+ Channels in the Vasodilator Response to Bradykinin in the Rat Heart. *British Journal of Pharmacology,* Vol.113, No.3, (November 1994), pp. 954-958, ISSN 0007-1188.

Gelband, GH. & McCullough, JR. (1993). Modulation of Rabbit Aortic Ca^{2+}-activated K^+ Channels by Pinacidil, Cromakalim, and Glibenclamide. *American Journal of Physiology – Cell Physiology,* Vol.264, No.5(Pt 1), (May 1993), pp. C1119-C1127, ISSN 0363-6143.

Greenwood, IA. & Large, WA. (1995). Comparison of the Effects of Fenamates on Ca-activated Chloride and Potassium Currents in Rabbit Portal Vein Smooth Muscle

Cells. *British Journal of Pharmacology,* Vol.116, No.7, (December 1995), pp. 2939-2948, ISSN 0007-1188.

Hartley, SA. & Kozlowski, RZ. (1996). ATP Increases Ca^{2+}-activated K^+ Channel Activity in Isolated Rat Arterial Smooth Muscle Cells. *Biochimica et Biophysica Acta - Biomembranes,* Vol.1283, No.2, (September 1996), pp. 192-198, ISSN 0005-2736.

Johnson, AG.; Nguyen, TV. & Day, RO. (1994). Do Nonsteroidal Anti-inflammatory Drugs Affect Blood Pressure? A Meta-analysis. *Annals of Internal Medicine,* Vol.121, No.4, (August 1994), pp. 289-300, ISSN 0003-4819.

Kuriyama, H.; Kitamura, K., Itoh, T. & Inoue, R. (1998). Physiological Features of Visceral Smooth Muscle Cells, with Special Reference to Receptors and Ion Channels. *Physiology Reviews,* Vol.78, pp. 811-920.

Li, J.; Zhang, N., Yei, B., Ju, W., Orser, B., Fox, JEM., Wheeler, MB., Wang, Q. & Liu, W-Y. (2007). Non-steroidal Anti-inflammatory Drugs Increase Insulin Release from Beta Cells by Inhibiting ATP-sensitive Potassium Channels. *British Journal of Pharmacology,* Vol.151, No.4, (June 2007), pp. 483-493, ISSN 0007-1188.

Liu, LY.; Fei, XW., Li, ZM., Zhang, ZH. & Mei, YA. (2005). Diclofenac, a Nonsteroidal Anti-inflammatory Drug, Activates the Transient Outward K^+ Current in Rat Cerebellar Granule Cells. *Neuropharmacology,* Vol.48, No.6, (May 2005), pp. 918-926, ISSN 0028-3908.

Liu, Y.; Zhang, H., Li, X., Wang, Y., Lu, H., Qi, X., Ma, C. & Liu, J. (2007). Inhibition of Voltage-gated Channel Currents in Rat Auditory Cortex Neurons by Salicylate. *Neuropharmacology,* Vol.53, No.7, (December 2007), pp. 870-880, ISSN 0028-3908.

Lopez-Ovejero, JH.; Weber, MA., Drayer, JI., Sealey, JE. & Laragh JH. (1979). Effects of Indomethacin alone and during Diuretic or Beta-adrenoreceptor-blockade Therapy on Blood Pressure and in the Renin System in Essential Hypertension. *Clinical Science and Molecular Medicine - Supplement,* Vol.4, No.1, (December 1979), pp. 203s-205s, ISSN 0144-4107.

Malykhina, AP.; Shoeb, F. & Akbaralia, HI. (2002). Fenamate-induced Enhancement of Heterologously Expressed HERG Currents in Xenopus Oocytes. *European Journal of Pharmacology,* Vol.452, No.3, (October 2002), pp. 269-277, ISSN 0014-2999.

Martin, DK.; Nakaya, Y., Wyse, KR. & Campbell, TJ. (1994). Effects of Disopyramide and Flecainide on the Kinetics of Inward Rectifier Potassium Channels in Rabbit Heart Muscle. *British Journal of Pharmacology,* Vol.111, No.3, (March 1994), pp. 873-879, ISSN 0007-1188.

McGrath, JC.; Monaghan, S., Templeton, AG. & Wilson, VG. (1990). Effects of Basal and Acetylcholine-induced Release of Endothelium-derived Relaxing Factor on Contraction to Alpha-adrenoceptor Agonists in a Rabbit Artery and Corresponding Veins. *British Journal of Pharmacology,* Vol.99, No.1, (January 1990), pp. 77-86, ISSN 0007-1188.

Minuz, P.; Barrow, SE., Cockcroft, JR. & Ritter, JM. (1990). Effects of Non-steroidal Anti-inflammatory Drugs on Prostacyclin and Thromboxane Biosynthesis in Patients with Mild Essential Hypertension. *British Journal of Clinical Pharmacology,* Vol.30, No.4, (October 1990), pp. 519-526, ISSN 0306-5251.

Moncada, S.; Gryglewski, R., Bunting, S. & Vane, JR. (1976). An Enzyme Isolated from Arteries Transforms Prostaglandin Endoperoxides to an Unstable Substance that

Inhibits Platelet Aggregation. *Nature*, Vol.263, No.5579, (21 October 1976), pp. 663-665, ISSN 0028-0836.

Ottolia, M. & Toro, L. (1994). Potentiation of Large Conductance K_{Ca} channels by Niflumic, Flufenamic, and Mefenamic Acids. *Biophysical Journal*, Vol.67, No.6, (December 1994), pp. 2272-2279, ISSN 0006-3495.

Pallapies, D.; Peslar, BA., Brune, K. & Zeihofer, HU. (1994). Modulation of Nitric Oxide Effects by Flurbiprofen Enantiomers and Nefopam and its Relation to Antinociception. *European Journal of Pharmacology*, Vol.271, No.2-3, (December 1994), pp. 335-340, ISSN 0014-2999.

Peskar, BM.; Kluge, S., Peskar, BA., Menzel-Soglowek, S. & Brune, K. (1991). Effects of Pure Enantiomers of Flurbiprofen in Comparison to Racemic Flurbiprofen on Eicosanoid Release from Various Rat Organs ex vivo. *Prostaglandins*, Vol.42, No.6, (December 1991), pp. 515-531, ISSN 1098-8823.

Pinheiro, JM. & Malik, AB. (1993). Mechanisms of Endothelin-1-induced Pulmonary Vasodilatation in Neonatal Pigs. *Journal of Physiology- London*, Vol.469, (September 1993), pp. 739-752, ISSN 0022-3751.

Pope, JE.; Anderson, JJ. & Felson, DT. (1993). A Meta-analysis of the Effects of Nonsteroidal Anti-inflammatory Drugs on Blood Pressure. *Archives of Internal Medicine*, Vol.153, No.4, (22 February 1993), pp. 477-484, ISSN 0003-4819.

Robertson, BE.; Corry, PR., Nye, PCG. & Kozlowski, RZ. (1992). Ca^{2+} and Mg-ATP Activated Potassium Channels from Rat Pulmonary Artery. *Pflügers Archiv – European Journal of Physiolog*, Vol.421, No.2-3, (June 1992), pp. 94-96, ISSN 0031-6768.

Sakuma, I.; Asajima, H., Fukao, M., Tohse, N., Tamura, M. & Kitabatake, A. (1993). Possible Contribution of Potassium Channels to the Endothelin-induced Dilatation of Rat Coronary Vascular Beds. *Journal of Cardiovascular Pharmacology*, Vol.22, (Supplement 8), pp. S232-S234, ISSN 0160-2446.

Scornik, FS., Codina, J., Birnbaumer. L. & Toro, L. (1993). Modulation of Coronary Smooth Muscle K_{Ca} Channels by G_s alpha Independent of Phosphorylation by Protein Kinase A. *American Journal of Physiolology – Heart and Circulatory Physiology*, Vol.265, No.4 (Pt 2), (October 1993), pp. H1460-1465, ISSN 0363-6135.

Seigel, G.; Emden, J., Wenzel, K., Mironneau, J. & Stock, G. (1992). Potassium Channel Activation in Vascular Smooth Muscle, In: *Advances in Experimental Medicine and Biology*, G.B. Frank, C.P. Bianchi, . Keurs (Eds.), Vol.311, pp. 53-72, Springer, ISBN 978-0-306-44194-3, New York.oxberger

Stumpff, F.; Boxberger, M., Thieme, H., Strauss, O. & Wiederholt, M. (2001). Flufenamic Acid Enhances Current Through Maxi-K Channels in the Trabecular Meshwork of the Eye. *Current Eye Research*, Vol.22, No.6, (June 2001), pp. 427-437, ISSN 0271-3683.

Vane, JR. (1971). Inhibition of Prostaglandin Synthesis as a Mechanism for Action of Aspirin-like Drugs. *Nature*, Vol.231, No.25, (23 June 1971), pp. 232-235, ISSN 0028-0836.

Ylitalo, P.; Pitkajarvi, T., Metsa-Ketela, T. & Vapaatalo, H. (1978). The Effect of Inhibition of Prostaglandin Synthesis on Plasma Renin Activity and Blood Pressure in Essential Hypertension. *Prostaglandins and Medicine*, Vol.1, No.6 'December 1978), pp. 479-488, ISSN 0952-3278.

Yuan, X-J.; Tod, ML., Rubin, LJ. & Blaustein, MP. (1994). Deoxyglucose and Reduced Glutathione Mimic Effects of Hypoxia on K^+ and Ca^{2+} Conductances in Pulmonary Artery Cells. *American Journal of Physiology – Lung Cellular and Molecular Physiology*, Vol.267, No.1 (Pt 1), (July 1994), pp. L52-L63, ISSN 1040-0605.

15

Perforated Patch Clamp in Non-Neuronal Cells, the Model of Mammalian Sperm Cells

Jorge Parodi[1] and Ataúlfo Martínez-Torres[2]
[1]*Escuela de Medicina Veterinaria, Facultad de Recursos Naturales, Núcleo de Producción Alimentaria, Universidad Católica de Temuco, Temuco*
[2]*Departamento de Neurobiología Celular y Molecular, Laboratorio de Neurobiología Molecular y Celular, Instituto de Neurobiología, Campus UNAM-Juriquilla, Querétaro,*
[1]*Chile*
[2]*México*

1. Introduction

This chapter deals with the steps required to obtain perforated patch-clamp recordings from mammalian sperm cells. (in spite of the fact that these cells) In spite of the fact that these cells are not electrically excitable, they possess a number of conductances due to their expression of ion channels and even neurotransmitter-gated ion-channels, many of which remain to be explored functionally and structurally. Detailed methods for obtaining cells suitable for electrophysiological recordings and protocols to perform patch-clamp recordings are outlined in the text.

2. Perforated patch recording

Classic whole-cell recordings (Hamill et al., 1981) could dilute or wash out crucial elements of intracellular signaling cascades and even completely replace the intracellular milieu. In this technique, occasionally it is necessary to apply negative pressure to break down the plasma membrane and allow the continuity between the recording electrode and the cytoplasm. With the advent of perforated patch recordings in neurons (Ebihara et al., 1995), a compromise between good electrical signals and preservation of the intracellular milieu became available, and importantly for this review, the method is also suitable for non-neuronal cells.

The principle of perforated patch clamping relies on the action of drugs, such as classic antifungals (Akaike and Harata, 1994; Akaike, 1996). These molecules form holes in the plasma membrane, which are permeable to ions but do not allow the traffic of larger molecules. One of the most common molecules used in this protocol is nystatin, and others such as gramicidin and amphotericin B are also widely used. One of the key elements to obtain successful patch recordings is the proper application of the perforating drug; during this critical period of time is necessary to control the right time of drug application must be carefully controlled, because if the drug is added before the electrode forms a tight gigaseal with the plasma membrane, the membrane could be irreversible damaged. Depending on the cell type, some time should be taken to make the seal: a significant time interval (around 35 minutes) should be allowed for seal formation to ensure that the perforating drug is exposed only to the area of the plasma membrane that forms the seal with the recording electrode (Ueno et al., 1992; Lippiat, 2008). The most practical method to induce the perforated patch is by filling the pipette tip with intracellular recording solution containing the perforating drug. The drug will gradually diffuse to the membrane, and development of the whole-cell current response can be monitored.

3. Procedure

3.1 The sperm cell patch-clamp

The main tool used to investigate the characteristics and distribution of ion channels in the plasma membrane is the "patch-clamp" technique first described by Neher and Sakmann in 1976 and modified by Hamill in 1981. However, at that point patch-clamp recordings were of limited use in electrophysiological studies in other complex cells like cells like the mammalian sperm due to its shape and small size (Linares-Hernandez et al., 1998; Gorelik et al., 2002).

To deal with this technical problem several alternatives have been devised. Some researchers determined that high resistance seals with hyposmotic solutions are ideal, especially for the sea urchin sperm; however, the cell-attached configuration and high resistance seals last only for a few minutes, making it difficult to record the currents generated by ion channels (Sanchez et al., 2001). The second alternative is the electrophysiological study of ion channels in the membranes of spermatogonial cells in late stages of development (Munoz-Garay et al., 2001). Technically, the procedure is easier in these cells than in mature sperm because the spermatogonial cells are much larger (Arnoult et al., 1996; Santi et al., 1996; Darszon et al., 1999). The plasma membrane of spermatogonial cells is functionally and structurally similar to the membrane of mature sperm, but there are many differences in types and cell distribution of ion channels (Serrano et al., 1999). The third alternative is the reconstitution of ion channels in artificial lipid bilayers; however, this technique removes the ion-channels from their natural lipidic environment and therefore, their biophysical properties are not precisely the same as in the cell plasma membrane (Lievano et al., 1990). The probability of obtaining a high resistance seal of the sperm head by microscopy using perforated patch clamp is about 45%; this is 15 times greater than the probability reported in studies using conventional "patch-clamp". In our personal experience, a tight seal is successfully obtained in 1 out of 10 trials (Guerrero et al., 1987; Navarrete et al., 2010). The time interval of the cell-attached recordings averages 35 min, which suffices to obtain many valuable biophysical data (Marconi et al., 2008; Navarrete et al., 2010).

3.1.1 Sperm selection

The sperm selection method chosen is "swim up" (WHO, 1999). Briefly, the sample is collected by manual manipulation and incubated for 40 min at 37 °C to liquefy the semen. Then 1 ml of semen is suspended in a Falcon tube with 4 ml of medium (DMEM), spun down at 1200 rpm (200 *g*) for 10 min, and the supernatant removed. The pellet is dissolved in 4 ml of medium and centrifuged for 5 min at 1200 rpm (200 *g*). The supernatant is removed and the pellet is incubated with 1 ml of medium for 1 h at 37 ° C tilted at 45°. Finally, the upper portion of the suspension that is rich in motile sperms is collected (Navarrete et al., 2010).

3.2 "Perforated patch-clamp" in mammalian sperm

An aliquot (200 µl) of the upper portion of sperm suspension is incubated with 1800 µl of hyposmotic medium (see section Preparation of solutions) for 20 min, then centrifuged at 1200 rpm (200 *g*) for 5 min. The supernatant is discarded, and the pellet is resuspended in 3 ml of bath solution (150 mM KCl) and centrifuged at 1200 rpm (200 *g*) for 5 min. After removing the supernatant, the pellet is resuspended in 2 ml of bathing solution (see section Preparation of solutions). Subsequently 1 ml of suspension is placed in the "patch clamp" recording chamber previously prepared with Pegotina® (peptides with adherent properties, US patent 20110062047) or another suitable adhesive, such as poly-lysine, laminin etc. The sperm suspension is placed in the chamber and left undisturbed for 10 min to secure the interaction and permit attachment of the cells, with the bioadhesive.

Micropipettes are made of borosilicate capillaries (Sutter Instrument Co., CA, USA is one of several options in a vertical or horizontal puller). The level of stretching and heat is adjusted to obtain a tip between 0.5 µm and 1.5 µm of diameter, and then the tip is heat polished.

A silver cylinder of 2 mm diameter and 3 cm long is used as reference electrode (Ag / AgCl); previously bathed it is pretreated in a bath of sodium hypochlorite and placed in a polyethylene tube whose tip is shaped to form a fine point. Inside the tube is a bed of agar-KCl (2% agar in 150 mM KCl) to minimize the junction potential of the solution-electrode interface. One electrode is connected to a current-voltage converter "probe" and to the amplifier. A second electrode is a thin silver chloride wire inserted into the adapter micropipette "holder" and connected to the "probe". Through the "holder", positive pressure is applied to the pipette to prevent adhesion of particles and, in addition, negative pressure is essential to obtain high-resistance seals.

3.3 "Cell attached" configuration

The "patch clamp" is basically a piece of electric isolate to record unitary currents that flow through one or more ion channels present in a fragment of membrane set to a given potential by the researcher. To achieve a "patch clamp" it is necessary that the seal between the tip of the micropipette and the membrane has a resistance on the order of GΩ. Initially, the micropipette inserted into the "holder" will be immersed in the solution, and the resistance is determined through a current, monitored in a data acquisition system, in response to a rectangular pulse voltage of 5 mV, 110 ms, and 5 Hz generated by the same amplifier. Depending on the diameter of the pipette, the resistance should be between 6 and 12 MΩ when measured in symmetrical solution concentrations of 150 mM KCl. The micropipette is then positioned with a micromanipulator in an inverted optical microscope and pressed against the surface of the sperm membrane; this maneuver leads to an increase of resistance.

By applying negative pressure in the pipette, the resistance and formation of a seal will be reached: the resistance will increase and a seal will be formed, then capacitive current spikes will appear at the beginning and end of the voltage pulse. These peaks are minimized by the capacitive compensation circuit before beginning the experiment. The configuration with the pipette attached to the surface of the plasma membrane is called the "cell attached" (Figure 1) and allows the recording of unitary currents through the electrically isolated "patch" at the tip of the pipette. This configuration is also suitable to evaluate the current response to changes in the perfusion medium.

3.4 Perforating the plasma membrane.

The antifungals must be dissolved in their appropriate solvent (DMSO, ethanol) and then diluted in the intracellular recording solution. The pipette tip is filled with intracellular solution using different approaches, either a syringe with a long, fine needle, or by capillarity, or even with a syringe whose tip has been melted. Typical concentrations for nystatin or amphotericin B are 5–20 mgr/ml, from a 1000-fold stock (Mistry and Hablitz, 1990; Akaike and Harata, 1994; Rhee et al., 1994). After the formation of the gigaseal the patch is left intact, and the current response to test pulses is monitored. If all is well, the cell capacitance should be evident after 10 min, with decreasing access resistance (faster capacitive transient). After 20-30 min the capacitive transient should be stable, and the experiment can begin (Figure 1).

3.5 Final step, perforated patch-clamp

When the cell-attached configuration is reached, some time is needed to see changes in the capacitive peak, and to observe the openings of the membrane (Figure 1). When a change is detected in the capacitive peak, as evidenced by modifications in the decay slope (figure 1, arrows), it indicates that the seal has been opened. The velocity of the event depends on the concentration of the perforating molecules, and some time is required to fully calibrate the system. No more steps are needed after the cell is attached (Parodi et al., 2010; Sepulveda et al., 2010).

Yet another critical factor to achieve perforated patch recordings is the time that the molecules in the intracellular solution take to form a strong gigaseal. High concentrations of the molecule can be toxic and induce alterations in the recordings, whereas low concentrations may take longer to form the gigaseal (and affect the viability of the cells), reaching times of declining cell viability. To assure that the gigaseal is properly developing, the passive properties of the membrane should be continually monitored.

3.6 Quality control

Below, we present a list of criteria to get proper whole-cell recordings:

1. The seal resistance must be higher than 1 G at the beginning of the experiment.
2. The series resistance must be lower than 20 MΩ and stay that way throughout the recording. A high series resistance is undesirable because the voltage clamp of the cell membrane is adversely affected.
3. The time constant (τ) of the capacitive transient in the current response should be proportional to the series resistance, so a slow capacitive transient is a bad sign. A common problem is gradual resealing of the patch after breakthrough, apparently by spreading of the capacitive transients.

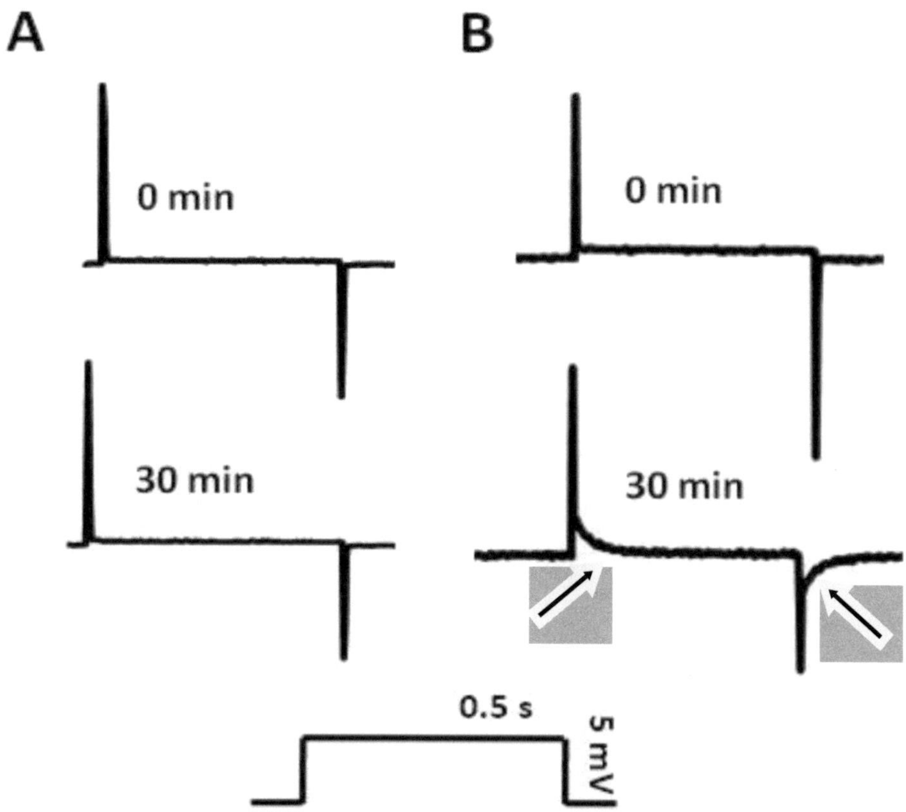

Fig. 1. Establishment of perforated patch mature bovine sperm cells. A. A membrane seal without perforating molecules. The lower recording shows the cell before applying negative pressure in the pipette. B. shows the time course of seal formation with perforating agent. The gigaseal formed 30 min after inclusion of the drug (lower trace). The arrow indicated, the changes in the slope, when the seal are open.

4. The membrane potential must be more negative than -50 mV if a high-potassium intracellular solution is used. This value varies depending on the cell type; thus, laboratory conditions for these which just the potential parameters should be controlled.
5. Cell capacitance and resistance must be stable.

3.7 General conclusions and applications

The perforated patch-clamp technique has allowed the study of functional properties in several cell models. In particular, the physiology of sperm has been widely studied by Dr. Darszon´s group in Mexico. Using this experimental approach along with other techniques of cell biology, this group described the presence and function of CatSper channels in mammalian sperm cells and showed their importance in basic functional processes such as sperm "capacitation" (Darszon et al., 1999)

Dr. Romero's group in Chile continues to explore other ion currents present in sperm cells and has described outward potassium currents of the Kv type (Marconi et al., 2008). Using this approach, the same group recently showed that this current was sensitive to peptides isolated from spider venom (Parodi et al., 2010) and that this modulation generates functional changes that alter the acrosome reaction of the sperm (Navarrete et al., 2010) and lead to changes in the relationship between sperm and oviductal cells (Navarrete et al., 2011, in press Andrology). Recently, this information was reviewed and suggestions raised that indicate that CatSper channels modulate several other cellular functions in sperm cells

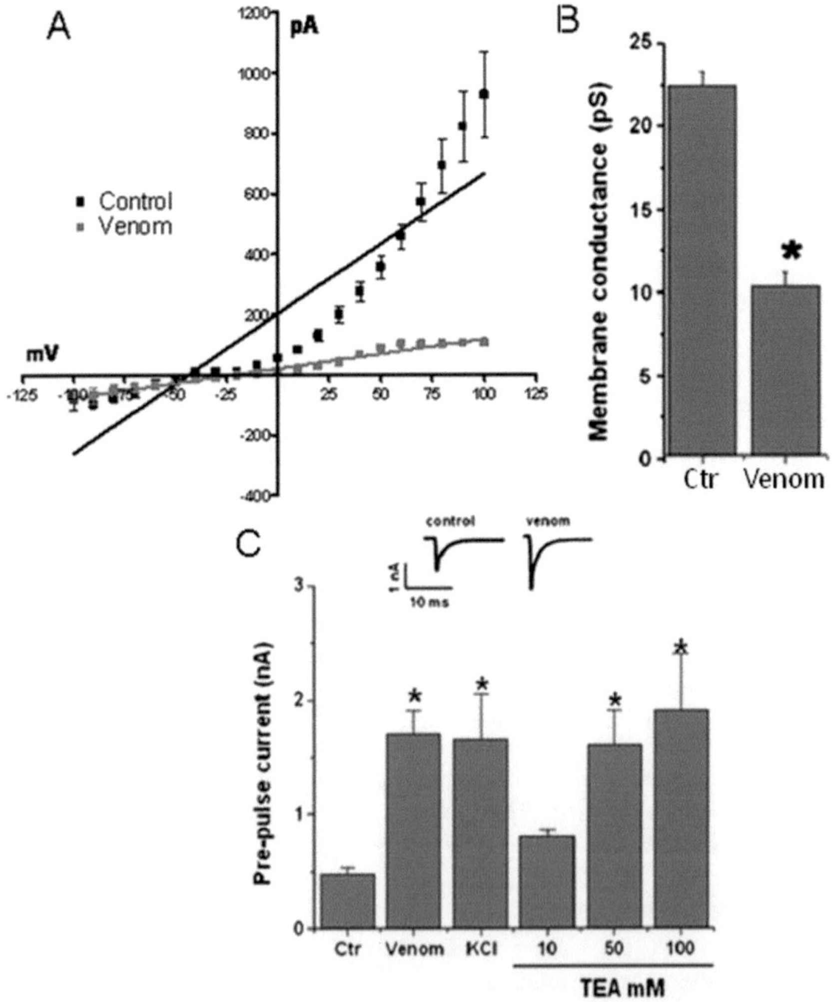

Fig. 2. Effects of Chilean *L. mactans* venom extract on electrophysiological properties of mammalian spermatozoa. A. Current-voltage relations in the absence or presence of 7.5 μg/mL venom. B. Membrane conductance of sperm exposed to the venom. C. The effect of different conditions on pre-pulse currents. The sperm cells were held in the whole-cell configuration, perforated patch-clamp. The bars are means ± SD of 6 different experiments, * indicates significant difference, $p < 0.05$. Modified, from Parodi et al., 2010.

(Navarrete et al., 2011). A sample recording obtained by the perforated patch-clamp technique in sperm is observed in figure 2, in which the typical Kv current is evidenced.

A new approach to study the impact of toxic agents was devised by Dr. Aguayo´s group in Chile, who modified the intracellular recording solutions to introduce aggregates of β-amyloid, a peptide highly accumulated in the brains of Alzheimer's disease patients. Following seal formation, as we have described above, and without the use of a traditional penetrating agent, this group suggested that β-amyloid induces pores in the plasma membrane, thus allowing perforated patch-clamp recordings in several cell types (Sepulveda et al., 2010) and US Patent 02908-2007.

4. Preparation of solutions

To prepare solution with BAPTA tetracesium salt* in internal solution with CsCl:

1. Pour filtered solution into 50-ml beaker.
2. Add 0.502 g BAPTA for 50 ml of 10 mM solution (1.004 for 100 ml).
3. Adjust pH to 7.4 with CsOH.
4. Measure osmolarity.
5. Adjust osmolarity to 290-310 with sucrose (≈ 0.5 g for 50 ml).
6. Pour into 50-ml flask and complete to 50 ml with ultra pure water.
7. Store as 5-ml aliquots.

* A cell-impermeant chelator, highly selective for Ca^{2+} over Mg^{2+}

To prepare solution with EGTA in internal solution with KCl:

1. Pour filtered solution into 50-ml beaker.
2. Add 0.3804 g EGTA (for 10 mM).
3. Adjust pH to 7.4 with KOH.
4. Measure osmolarity and adjust to 290-310 with sucrose.
5. Pour into 50-ml flask and complete to 50 ml with ultra pure water.
6. Aliquot, cover with aluminum foil, label, and put in freezer.

PATCH CLAMP

Internal Cesium Solution (without Na_2ATP, 2 mM)

	mM	FW	g/50 ml	g/100 ml
CsCl	120	168.36	**1.01**	2.02
HEPES	10	238.30	**0.12**	0.24
$MgCl_2 6H_2O$	4	203.31	**0.04**	0.08

Internal Potassium Solution (without Na^2ATP, 2 mM)

	mM	FW	g/50 ml	g/100 ml
KCl	120	74.56	**0.45**	0.89
Hepes	10	238.30	**0.12**	0.24
$MgCl_2 6H_2O$	4	203.31	**0.04**	0.08

External bathing solution

	mM	FW	g/LT	g/500 ml
NaCl	140	58.5	190	4095
KCl	5	74.6	373	186.5
$CaCl_2$	1	112.2	112.2	56.1
$MgCl_2$	1	95.5	95.5	47.75
Hepes	10	238.3	2383	1191.5
Glucose	10	180	1800	900

External solution, hyposmotic

	mM	FW	g/LT	g/500 ml
KCl	35	74.6	2611	1305.5
Hepes	10	238.3	2383	1191.1

5. Acknowledgments

This work was partially supported by grants from CONACYT 101851 and UNAM-PAPIIT 204806 (to AM-T and RM). J.P. is a postdoctoral fellow from CTIC-UNAM. JP has a travel grant from MECESUP - PUC/0708 of "Pontificia Universidad Catolica de Chile".

We are in debt to Dr. Dorothy Pless for editing the manuscript.

6. References

Akaike N (1996) Gramicidin perforated patch recording and intracellular chloride activity in excitable cells. Prog Biophys Mol Biol 65:251-264.

Akaike N, Harata N (1994) Nystatin perforated patch recording and its applications to analyses of intracellular mechanisms. Jpn J Physiol 44:433-473.

Arnoult C, Grunwald D, Villaz M (1996) Novel postfertilization inward Ca2+ current in ascidian eggs ensuring a calcium entry throughout meiosis. Dev Biol 174:322-334.

Darszon A, Labarca P, Nishigaki T, Espinosa F (1999) Ion channels in sperm physiology. Physiol Rev 79:481-510.

Ebihara S, Shirato K, Harata N, Akaike N (1995) Gramicidin-perforated patch recording: GABA response in mammalian neurones with intact intracellular chloride. J Physiol 484 (Pt 1):77-86.

Gorelik J, Gu Y, Spohr HA, Shevchuk AI, Lab MJ, Harding SE, Edwards CR, Whitaker M, Moss GW, Benton DC, Sanchez D, Darszon A, Vodyanoy I, Klenerman D, Korchev YE (2002) Ion channels in small cells and subcellular structures can be studied with a smart patch-clamp system. Biophys J 83:3296-3303.

Guerrero A, Sanchez JA, Darszon A (1987) Single-channel activity in sea urchin sperm revealed by the patch-clamp technique. FEBS Lett 220:295-298.

Hamill OP, Marty A, Neher E, Sakmann B, Sigworth FJ (1981) Improved patch-clamp techniques for high-resolution current recording from cells and cell-free membrane patches. Pflugers Arch 391:85-100.

Lievano A, Vega-SaenzdeMiera EC, Darszon A (1990) Ca2+ channels from the sea urchin sperm plasma membrane. J Gen Physiol 95:273-296.

Linares-Hernandez L, Guzman-Grenfell AM, Hicks-Gomez JJ, Gonzalez-Martinez MT (1998) Voltage-dependent calcium influx in human sperm assessed by simultaneous optical detection of intracellular calcium and membrane potential. Biochim Biophys Acta 1372:1-12.

Lippiat JD (2008) Whole-cell recording using the perforated patch clamp technique. Methods Mol Biol 491:141-149.

Marconi M, Sanchez R, Ulrich H, Romero F (2008) Potassium current in mature bovine spermatozoa. Syst Biol Reprod Med 54:231-239.

Mistry DK, Hablitz JJ (1990) Nystatin-perforated patch recordings disclose NMDA-induced outward currents in cultured neocortical neurons. Brain Res 535:318-322.

Munoz-Garay C, De la Vega-Beltran JL, Delgado R, Labarca P, Felix R, Darszon A (2001) Inwardly rectifying K(+) channels in spermatogenic cells: functional expression and implication in sperm capacitation. Dev Biol 234:261-274.

Navarrete P, Ormeño D, Miranda A, Sánchez R, Romero R, Parodi J. (2011) Molecular characterization, electrophysiological and contraceptive effect of Chilean Latrodectus venom. International Journal of Morphology 29(3): 733-741.

Navarrete P, Martinez-Torres A, Gutierrez RS, Mejia FR, Parodi J (2010) Venom of the Chilean Latrodectus mactans alters bovine spermatozoa calcium and function by blocking the TEA-sensitive K(+) current. Syst Biol Reprod Med 56:303-310.

Neher E, Sakmann B (1976) Single-channel currents recorded from membrane of denervated frog muscle fibres. Nature 260:799-802.

Parodi J, Navarrete P, Marconi M, Gutierrez RS, Martinez-Torres A, Mejias FR (2010) Tetraethylammonium-sensitive K(+) current in the bovine spermatozoa and its blocking by the venom of the Chilean Latrodectus mactans. Syst Biol Reprod Med 56:37-43.

Rhee JS, Ebihara S, Akaike N (1994) Gramicidin perforated patch-clamp technique reveals glycine-gated outward chloride current in dissociated nucleus solitarii neurons of the rat. J Neurophysiol 72:1103-1108.

Sanchez D, Labarca P, Darszon A (2001) Sea urchin sperm cation-selective channels directly modulated by cAMP. FEBS Lett 503:111-115.

Santi CM, Darszon A, Hernandez-Cruz A (1996) A dihydropyridine-sensitive T-type Ca2+ current is the main Ca2+ current carrier in mouse primary spermatocytes. Am J Physiol 271:C1583-1593.

Sepulveda FJ, Parodi J, Peoples RW, Opazo C, Aguayo LG (2010) Synaptotoxicity of Alzheimer beta amyloid can be explained by its membrane perforating property. PLoS One 5:e11820.

Serrano CJ, Trevino CL, Felix R, Darszon A (1999) Voltage-dependent Ca(2+) channel subunit expression and immunolocalization in mouse spermatogenic cells and sperm. FEBS Lett 462:171-176.

Ueno S, Ishibashi H, Akaike N (1992) Perforated-patch method reveals extracellular ATP-induced K+ conductance in dissociated rat nucleus solitarii neurons. Brain Res 597:176-179.

16

Drug Screening and Drug Safety Evaluation by Patch Clamp Technique

Yan Long and Zhiyuan Li
Guangzhou Institutes of Biomedicine and Health, Chinese Academy of Sciences
China

1. Introduction

Ion channels are transmembrane proteins with great importance for the human physiology and become highly attractive molecular drug targets. Ever since the development of patch clamp technique in the late 1970s, it remains the gold standard approach for functional analysis, providing both real time and mechanistic information. Currently, that nearly 400 potential ion channels are found, indicating large-scale ion channel screening of wanted and unwanted drug effects is required (Venter et al., 2001). Unfortunately, it is limited by the lack of adequate screening approach, because available methods put a tradeoff between high-throughput and high-information content. The advent of automated patch clamp platforms revolutionize ion channel screening and enable investigations from a more functional perspective to a much more higher throughput. Compared with conventional patch clamp technique, the current status of automated patch clamp platforms also has some limitations. This chapter will provide the basic concepts of drug screening and drug safety evaluation by patch clamp technique and it moving from conventional to automated trend for high-throughput screening of ion channels.

2. Ion channels as drug targets

Ion channels are membrane proteins which form pores allowing ions to pass through impermeable cell membranes mediating and regulating crucial electrical functions throughout the body. Some channels are sensitive to membrane potential alteration and such channels are termed voltage-gated channels. Others have ligand binding sites to which a ligand can bind and change membrane permeability. They are so-called ligand-gated channels. It is estimated that there are more than 400 ion channel genes have been identified in the human genome, including about 170 potassium channels, 38 calcium channels, 29 sodium channels, 58 chloride channels, and 15 glutamate receptors. The remaining are genes encoding other channels such as inositol triphosphate (IP_3) receptors, transient receptor potential (TRP) channels and others (Venter et al., 2001; Fermini, 2008).

In a range of disorders including pain, epilepsy, depression, stroke, bipolar disorder, COPD, arrhythmia, hypertension, autoimmune disorders and diabetes, over 55 different inherited ion channel diseases, known as 'channelopathies' have now been identified, suggesting ion channels are important therapeutic targets (Table 1) (Ashcroft, 2006). Recently, the recognized importance of these proteins in physiological and pathological conditions has led to an extremely active search for ion channel targets in the worldwide drug discovery market. Approximately 13.4% of known drugs have their primary therapeutic action at ion channels, making them the second largest target class after G-protein coupled receptors (GPCRs) (Overington et al., 2006). These include the nicotinic, $GABA_A$ (γ-aminobutyric acid A) and NMDA (*N*-methyl-d-aspartate) ligand-gated channels, the voltage-gated Na^+, K^+ and Ca^{2+} channels, and the ryanodine, inositol 1,4,5-trisphosphate (IP_3) and transient receptor potential Ca^{2+} channel families (Imming et al., 2006). Worldwide sales of ion channel drugs are estimated to be in excess of $12 billion. For instance, ion channel modulators generated

Drug	Commercial Name	Target channel	Disease Indication
Amlodipine	Norvasc	L-Cav	Angina, hypertension, arrhythmia (Hogg et al., 2006)
Felodipine	Plendil	L-Cav	Hypertension (Hogg et al., 2006)
Nifedipine	Adalat	L-Cav	Angina, hypertension (Hogg et al., 2006)
Verapamil	Calan/ Verelan	L-Cav	Angina, hypertension, arrhythmia (Hogg et al., 2006)
Diltiazem	Cardizem/ Tiazac	L-Cav	Angina, hypertension, arrhythmia (Hogg et al., 2006)
Nimodipine	Nimotop	L-Cav	Angina, hypertension, arrhythmia (Hogg et al., 2006)
Nifedipine	Procardia	L-Cav	Angina, hypertension, arrhythmia (Hogg et al., 2006)
Carbamazepine	Tegretol	Nav	Epilepsy (Chouinard, 2006)
Topiramate	Topamax	Nav	Epilepsy (Chouinard, 2006)
Lamotrigine	Lamictal	Nav	Epilepsy (Chouinard, 2006)
Flecainide	Tambocor	Nav1.5	Arrhythmia (Hogg et al., 2006)
Lidocaine	Lidocaine	Nav	Local anesthesia (Hogg et al., 2006)
Gabapentin	Neurontin	Cavα2δ	Pain (Hamer et al., 2002; Mathew et al., 2001)
Glimepiride	Amaryl	KATP	Type-II diabetes (Proks & Lippiat, 2006)
Glibenclimide	Glimepiride	KATP	Diabetes (Proks & Lippiat, 2006)
Nateglimide	Starlix	KATP	Type-II diabetes (Proks & Lippiat, 2006)
Zaleplon	Sonata	$GABA_A$	Insomnia (Kozlowski, 1999)
Benzodiazepine	Diazepam	GABA	Depression (Kozlowski, 1999)
Ziconotide	Prialt	Cav2.2	Severe chronic pain (Hogg et al., 2006)
Lubiprostone	Amitiza	CLC-2	Chronic idiopathic constipation (Kozlowski, 1999)
Pregabalin	Lyrica	GABA	Neuropathic pain (Kozlowski, 1999)
Sotalol	Betapace	hERG	Arrhythmia (Hogg et al., 2006)
Amiodarone	Cordorone	hERG	Arrhythmia (Hogg et al., 2006)
Phenytion	Dilantin	Nav	Epilepsy (Chouinard, 2006)
Flupirtine	Flupirtine	KCNQ2/3	Epilepsy (Chouinard, 2006)

Table 1. Current Ion Channel Drugs

over $1.2 billion sales in 2002. Amlodipine besylate, an antihypertensive of Pfizer inc., generated around $4 billion in revenue in 2003. Ambien and Xanax, chloride channel modulators, generated nearly $2 billion together in annual sales (Li et al., 2005).

The current ion channel drugs are concluded in table 1, the table indicates that the pharmaceutical industry has successfully developed considerable drugs which targeted at selected calcium, potassium, sodium and chloride channels. Given the number of ion channel genes identified in the human genome project, it appears that there remains a significant pool of unexploited channels for tractable pharmacological target.

3. The impact of hERG for drug toxicity evaluation

In addition to being drug targets, as transmembrane proteins, ion channels are easily accessible for small molecules treatment and, in some cases, are also the direct cause of

Name	Indicator (Channel)	Potency (IC50, μM)	Tissue (Species-Type)	Cell (Type)
Astemizole (Salata et al., 1995)	I_{kr} hERG	0.0015 0.069	GP-VM	XO
Terfenadine Astemizole (Salata et al., 1995)	I_{kr} hERG hERG	0.05 0.431 0.35	Ibid.	Ibid. Ibid.
MK-499 (Spector et al., 1996)	hERG WT-hERG	0.123 0.034		Ibid. Ibid.
BRL-37872 (Thomas et al., 2001)	hERG hERG	0.241 0.0198		Ibid. HEK293
Clofilium (Suessbrich et al., 1997b)	hERG	0.25		XO
LY97241 (Suessbrich et al., 1997b)	hERG	0.019		Ibid.
Haloperidol (Suessbrich et al., 1997a)	hERG	0.024		Ibid.
E4031 (Sanguinetti & Jurkiewicz, 1990)	I_{kr}	0.497	GP-VM	
Quinidine (Woosley, 1996)	I_{kr}	0.2	F-CM	
Desmethylastemizole (Vorperian et al., 1996; Zhou et al., 1999)	I_{kr} hERG	0.02 0.001	Rab-VM	HEK293
Cisapride (Walker et al., 1999)	hERG	0.0236		CHO
Pimozide (Kang et al., 2000)	hERG	0.018		Ibid.
Bepridil (Chouabe et al., 1998)	hERG	0.55		COS
Droperidol (Drolet et al., 1999)	I_{kr} hERG	0.0276 0.032	GP-VM	HEK293
Sertindole (Rampe et al., 1998)	hERG	0.014		L-cells
Tedisamil (Zolotoy et al., 2003)	TdP *	3		
Dofetilide (Yang et al., 1995)	I_{kr}	0.012		AT-1
Ibutilide (Yang et al., 1995)	I_{kr}	0.02		AT-1
Verapamil (Chouabe et al., 2000; Waldegger et al., 1999; Zhang et al., 1999)	hERG hERG hERG	0.14 0.83 3.8		HEK293 XO

Table 2. Potency of Strong HERG and I_{kr} Blockers

unwanted side-effects. In particular, the discovery of hERG channel and its underling role in prolongation of QT interval brings considerable attention on the drug safety liability issue and leads to the withdrawal or restricted use of several hERG-related drugs since 1997, including antihistamine Seldane, gastric prokinetic drug cisapride, antipsychotic sertindole, and even the antibiotic grepafloxacin (De Bruin et al., 2005; Vitola et al., 1998). Physiologically, hERG channel is closely related to the phase 3 repolarization in the action potential of human ventricular myocytes. The blockade of hERG leads to malfunction in cardiac excitability, clinically defines as either cardiac arrhythmia (torsade de pointes) or long QT syndrome, a cardiac repolarization disorder. In accordance with the requirement of FDA and the European Medicines Agency, before clinical trials, drug testing of known ion channels, especially hERG, are needed in order to rule out possible side effects (Morganroth, 2004).

Compounds that have been validated with strong potency of hERG block are concluded in Table 2

4. Conventional patch clamp technique

Sakmann and Neher's patch clamp electrophysiology technique is a highly accurate and sensitive recording method, which revolutionized ion channel research and remains the gold standard approach for functional analysis (Neher & Sakmann, 1992). Patch clamp technique can be operated in either voltage clamp or current clamp mode. Voltage clamp allows an experimenter to 'clamp' the cell potential at a chosen value. This makes it possible to measure voltage specific activity of ionic channels. The voltage-gated channels are primarily studied in this mode. Current clamp records the membrane potential changes in response to it. This technique is used to study how a cell responds when electrical current enters a cell. This type of recording are performed to study action potential firing by the neuron.

Currently, there are a number of different patch clamp configurations and systems exist. The 'whole-cell' configuration is the most common method for majority cultured cells and tissue preparations. Conventionally, the technique uses a glass microelectrode, sometimes called 'patch pipette', containing an electrolyte resembling the fluid normally found within the cell (tip resistance 1-5 MΩ), which is positioned on the cell membrane with the aid of a microscope and micromanipulators, followed by gentle suction through the pipette to start seal formation. Once a high resistance (>1GΩ) seal between the rim of recording pipette and cell membrane is formed, further suction is applied to rupture the 'clamped' patch, allowing electrical access to the whole cell. The electrolyte freely dialyses from pipette to cell interior within minutes and the ionic gradients achieves equilibrium between the microelectrode and the cell. Patch clamp amplifier with low noise such as EPC form HEKA Electronik and Axopatch from Axon Instruments is utilized to monitor membrane potential with reference to a bath ground electrode, inject current to clamp the membrane potential to any voltage that we desire and detect current flow. Patch clamp technique can be operated in voltage and current clamp mode, which is, voltage steps or ramps, or current steps are applied to activate channels. Whole cell configuration monitors response of all ion channels within a cell membrane, currents as small as 10 pA, representing signals from just a few hundred open channels, can be recorded with sub-millisecond temporal resolution (Penner, 1995).

Perforated patch clamp technique is a variant of whole cell recording method. It differs in which electrical access to the cell interior is obtained via membrane permeabilizing reagents, typically nystatin or amphotericin, or with more potent gramicidin and saponin β–Escin. After these antibiotics added to the internal recording solution, it takes 10-30 minutes to perforate cell. Compared with whole cell approach, it has some advantages, including produces minimal disruption to the intracellular milieu, offers reduced dialysis and generally allows for highly stable recordings, but the access resistance is higher. The main disadvantages are decreased current resolution and increased noise (Penner, 1995).

Cell-attached, inside-out and outside-out patch clamp are called single channel recording technique. They sample channels from a small area (<10μm²) of the membrane rather than from the entire cell and study the behavior of ion channels on the section of membrane attached to the pipette. The cell-attached configuration is obtained by simply lowering the pipette onto the cell surface and measures currents arising from the channels which located within the diameter of the tip of recording pipette. A major disadvantage of this configuration is the entirely non-invasive of plasmalemma, making the application of pharmacological modulators to either side of the membrane becomes difficult, for this reason, it is largely restricted to biophysical analysis. By gently withdrawing the electrode from the cell after obtaining gigaohm seal, excising the patch inside the pipette tip, it is possible to form the inside-out recording configuration. With the internal aspect of the cell membrane now is exposed to the bath solution, compound acts on the inner facing channel can be readily achieved. This configuration is useful for studying the effects of alternative intracellular environment on single ion channel function. Outside-out patches are generated by first performing the whole cell recording and then slowly withdrawn the pipette from the cell membrane, a bleb of membrane separates from cell, creating a patch where the external aspect of cell membrane facing the bath solution. This is particularly attractive when studying the effects on single ion channel function of compounds applied to the extracellular solution (Fig 1) (Penner, 1995).

Typically, a conventional patch clamp setup comprises of components as follow (Fig 2) (Kornreich, 2007):

Micropipette puller: used to pull electrode from borosilicate glass capillary tubes. This process usually requires several pulls and may be either horizontal or vertical depending on protocol with user preference. Once pipettes have been pulled, they can be further fire-polished by Pipette Beveler.

Air suspension table: uses pressurized cylinders to 'float' the table so as to dampen the microscopic movements and vibrations that may interrupt the recording, and adds to achieve high resistance pipette membrane seals.

Faraday cage: shields the sensitive patch clamp preamplifier from stray electromagnetic fields that may introduce noise into electrophysiological recordings. The microscope and headstage are also most commonly placed inside the Faraday cage. Ideally composed of a copper mesh, the Faraday cage is most effective when connected to a ground source.

Inverted microscope: used to visualize the acutely dissociated cells or cultured cell lines being studied. This is important in order to obtain both good visualization of the cells and unhinder the flexible access of pipettes from the top.

Micromanipulator: used to manipulate the electrode (mounted on the headstage) to contact the acutely dissociated cells or cultured cell lines being studied. Micromanipulators are commercially available in mechanical, hydraulic, motorized and piezoelectric drives.

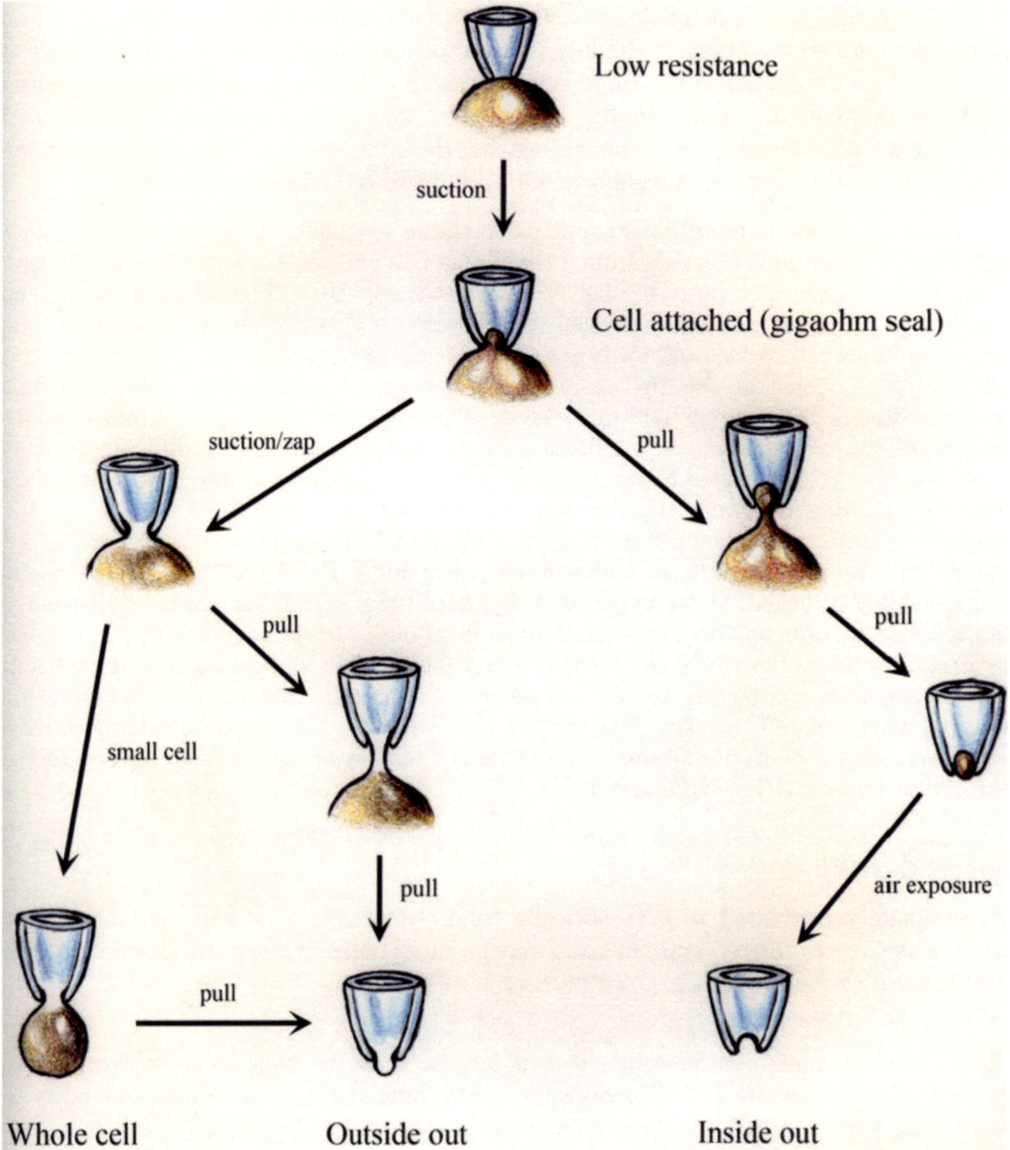

Fig. 1. Formation of different types of patch (Kornreich, 2007)

Amplifier: used to record current or voltage being measured and/or to control membrane voltage (voltage clamp) or current (current clamp) depending on experimental protocol. Additional functions also include filtering of input signal, capacitance compensation, and square wave generation etc. Using suitable software, the patch clamp recordings can be analyzed on a personal computer. For example, Pclamp (Axon), Pulse (Heka) and Jhon Dampaster's WCP (UK) are some of the widely used software.

Digital-analog converter: converts the analog signal recorded by the electrode to digital data, transforms digital output of computer to analog signal at headstage for control of membrane voltage/current, delivers stimulus protocols generated by computer to clamp ion channels.

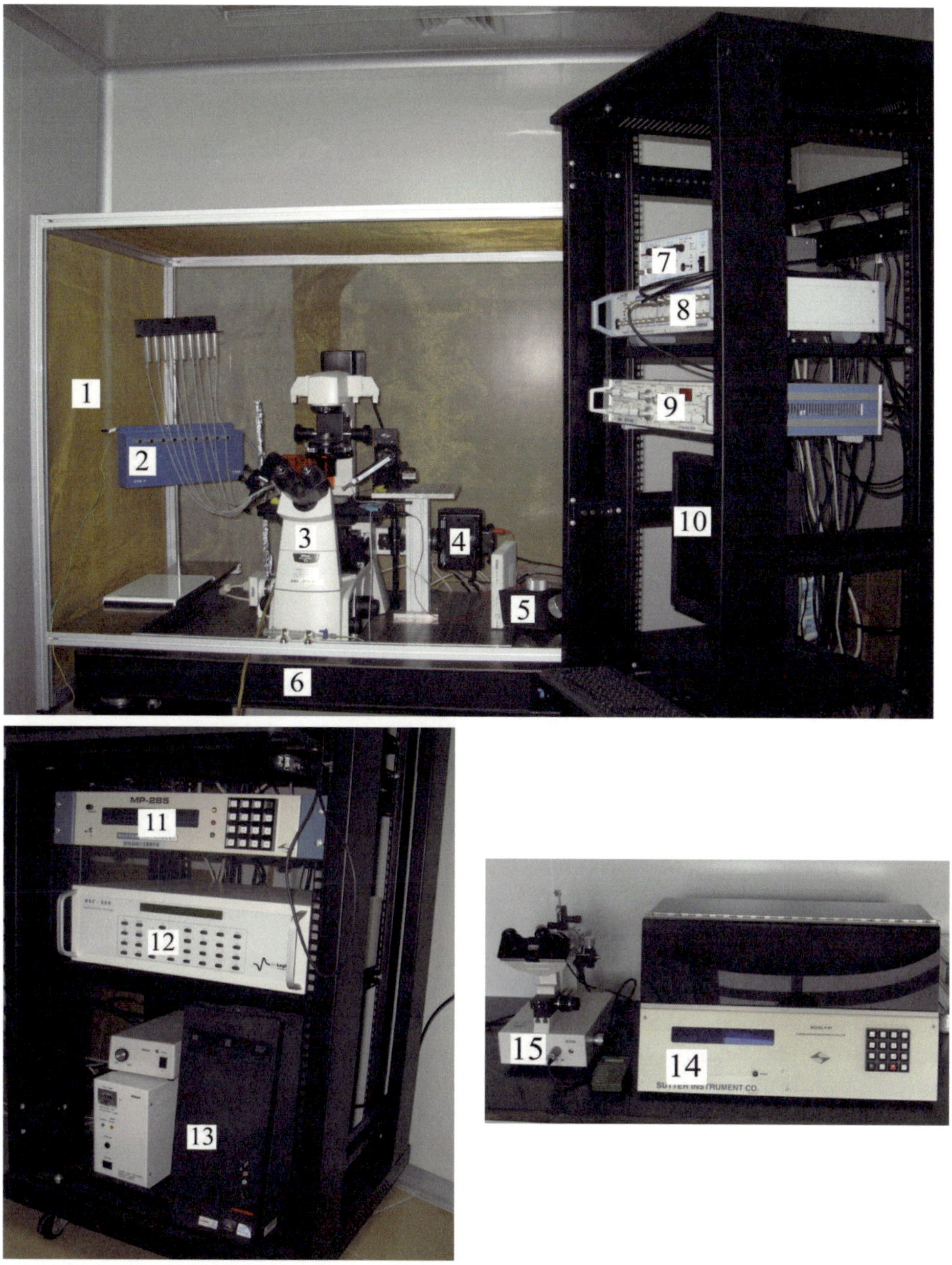

Fig. 2. A conventional patch clamp setup comprising of 1) Faraday cage; 2) Solution perfusion system (EVH-9); 3) Inverted microscope; 4) CCD camera; 5) Motorized micromanipulator controller; 6) Air suspension table; 7) Temperature controller; 8) Digidata; 9) Amplifier (Axopatch 200B); 10) Computer display; 11) Micromanipulator; 12) Rapid solution changer command box (RSC-200); 13) Computer monitor; 14) Micropipette puller; 15) Micropipette Beveler.

Bath chamber: a small chamber with volume of 500 µl-1 ml used to mount on microscope, during experiments, the cells can be placed in this chamber. A thin glass slide is used to stick the bottom of chamber.

Headstage: it is usually mounted on a micromanipulator, which connects the recording pipette and ground wire and contacts to the amplifier for data acquisition, control of membrane voltage/current, and stimulation release.

Computer: uses suitable software packages to acquire and analyze data. These packages are used to control membrane voltage/current, to process the input/output signals from the amplifier (filtering, capacitance compensation, etc.), to perform stimulus protocols, and to analyze acquired data.

5. Automated patch clamp technique

As the state-of-the-art technology for studying ion channels, the patch clamp technique is limited only in the later stages of drug discovery and development, including hits-to-leads, safety testing and basic research, owing to the ultra-low-throughput and time-consuming process. Because of FDA's new recommendations, at the preclinical stage, every drug candidate needs to be evaluated for their potential of hERG channel blockade. Therefore, a high-throughput drug screening approach is required as to accelerate the primary screening of ion channel modulators at the preliminary phase. Using a conventional patch clamp a sophisticated operator can study typically less than 10 cells per day, which is inefficient for large-scale screening for LQTS. Thus, there is a compelling need of high-throughput predictors of ion channel modulation to be developed.

Automated patch clamp setup suitable to high-throughput screening has been marketed. The complicated process of seal formation and current recording has been partially or completely automated. This method utilizes planar chips with small holes, seal forming substrates can either be glass, silicon nitride or plastic. The cells approach to the chips based on computerized feedback controlled application of suction. According to the preprogrammed protocols, the automated patch clamp robots accurately execute operations, in turns of filling the chip with electrolyte solutions, adding cell suspensions, sealing cells onto the holes, breaking through to the whole-cell configuration and performing different voltage-protocols in the absence and presence of drugs while simultaneously recording the ionic currents. The benefit of automated technique is obvious, for instance, the manual handling has been removed from the whole process, machine operation is very straightforward, making it suitable to the nonelectrophysiologists and increasing data reproducibility. Besides, the technology is scalable to array formats, thus parallel recordings of many individual cells can be achieved.

6. Limitations of automated patch clamping

Although there are various advantages over the conventional technique, automated technique itself has some limitations. Firstly, it is suitable mainly for stable and highly expressing cell lines. Transiently transfected cell lines are unsuitable for screening. Because cells are captured randomly from cell suspension, which means that the ion channel expression must be homogenous among cells and most cells must be healthy. Primary cultured cells such as cardiac myocytes and neurons have different cell shapes, and do not fit the criteria of the platform's random loading system. In addition, the cultured primary cells often contain different cell types and that the suspensions are not 'clean' enough. However, there are some publications reported that the Ionworks and Patchliner have been used for primary cells (Estes et al., 2008). Other issues include some systems can not be used to study of ligand-gated ion channels, temperature control is unavailable, and when screening tens of thousands of compounds, the cost of the tips and chips is considerable high. Currently, most automated patch clamp systems are really limited in their applications and are not particularly useful for academic researches. The main practical application of automated patch clamp is secondary drug screening and safety profiling of hERG ion channel block.

7. Available automated patch clamp systems

There are several automated platforms commercially available, the first one appear into market was Ionworks HT and later Ionworks Quattro (Molecular Devices Corporation), succeeded by the Patch Xpress (Molecular Devices Corporation.) and Nanion's Port-a-Patch. Other patch providers include Sophion's QPatch-16 and -48, Flyion's Flyscreen, Nanion's Patchliner platform, Cytocentrics's Cytopatch and Cellectricon's Dynaflow ProII, providing 16 to even 384 parallel recordings on a chip. Some of which including the PatchXpress, Port-a-Patch, Patchliner, Flyscreen and the QPatch are the true giga-seal patch clamp recording platforms (Table 3).

The above systems differ in some aspects, for instance, they adopt different solutions according to different materials used for the wells such as glass, silica or polymer, the number of wells on a chip can be various, commonly from 8 to 384, some systems utilize the technology of laminar solution flow in order to exchange different microfluids containing targeted compounds. The quality of the laminar flow is important because when evaluating the effects of different candidates or different concentrations of a candidate on ion channels, the fluids have to be completely exchanged without mixing (Neubert, 2004). Currently, there are two kinds of approaches with cells loading, one is the Flyscreen's 'Flip-the-tip' technology, in this technique the cells are loaded in the specially designed pipette named 'Flip-tips' or 'Chip-tips', and the cells are allowed to reach the tip. In contrast to the conventional method, external solution is filled in the pipette. The other is 'Planar array', which refers to the use of multi-well configurations in either a plate-based or a chip-based format to record multiple cells in parallel. In this technique, cells are loaded abruptly in the well and applying suction below the hole of the plates. Schematic representations of system assemblies (including bath, perfusion, cell trapping, and electrodes) are provided for each in Fig. 3.

8. Ionworks

Flyscreen: patch clamp robot Flyscreen 8500 from Flyion is a fully automated patch clamp system which is now commercially available in academic researches since 2003 (Lepple-Wienhues et al., 2003). It uses flip-the-tip technology with glass capillaries rather than the planar-array based approach, in this system, cells are patch clamped on the inside rim of the pipette tip, different from conventional patch clamp. The so-called 'flip-tips' or 'chip-tips' are used to perform recordings. When cells travel to the pipette tips by gravity, only slight suction can lead to the formation of a gigaseal. The Flyscreen is also available with temperature control and allows recording at physiological temperature. This instrument is designed to be scalable of asynchronously handling 1-6 recording positions and claims to have a throughput of 100-500 data points per day.

IonWorks HT and IonWorks Quattro: the IonWorks HT system from Molecular Devices was the first marketed automated screening system with planar array-based approach in 2003 (Schroeder et al., 2003). It uses a planar, multi-well plate named PatchPlate™ to replace the patch pipette (Fig. 3a). According to Molecular devices' reports, this disposable 'electrode' contains 384 wells, however, the recording head has only 48 amplifiers, therefore the IonWorks HT is capable of sampling cells sequentially, imposing discontinuity in the recording of currents. In 2006, the upgraded IonWorks Quattro system was launched (Finkel et al., 2006). This technology belongs to the population patch clamp, differs in the IonWorks HT system's recording mode of one cell in one hole in each well, it has 64 cells in 64 holes (in an 8×8 array) for each well (Fig. 3b), hence measured ion current from each well would be averaged and eliminates the cell-to-cell differences during recording.

A major drawback of either the HT version or the Quattro version is that they do not support giga-seal recordings, the seal resistance is on the order of 100MΩ. This low seal resistance induces low data quality, owing to lower signal-to-noise ratio. Furthermore, an antibiotic amphotericin B, rather than suction is used to permeabilize the cell membrane and results in a drastically higher access resistance compared to conventional patch clamping, which in turn leads to poor voltage control over the membrane, possibly altering current amplitude and creates errors in quantification the pharmacology of the ion channels and drugs under investigation. Besides, another drawback of this system is the discontinuous recording during compound administration, which makes it not an ideal platform which suited to recording fast transient currents or ligand-gated currents. However, the Ionworks reports that more than 95% success rate has proved to be obtained, and it measures about 3000 data points per day, this throughput has known to be the highest and is unmatched by other currently marketed devices, hence making it suitable for primary screening, secondary screening and safety assessment.

PatchXpress 7000A: the second developed planar-array based automated screening system was the PatchXpress from Molecular Devices, it was also the first automated device on the market supporting true giga-seal recording and can analyze both ligand-gated and voltage-gated ion channels (Tao et al., 2004; Xu et al., 2003). Their in-house developed software has versatile programming features, which is useful in testing operation and data analysis, and allows for continuous recordings during multiple drug concentrations' applications. This system uses Aviva Bioscience's SealChips, a 16-well disposable that has been reported to patch more than 12 cells (one cell per well) simultaneously for more than 15 min (Fig. 3d). Because of its high giga-seal rates, the longevity of seals comparable to glass pipettes, its versatile software allows the user to program any experimental protocol and its versatility of suitable to all kinds of ion channels. The PatchXpress has been considered as the golden standard in the field for automated electrophysiology. However, its throughput is stated to be 250-300 data points per day, compared with the throughputs of other available systems, is too low for true high-throughput screening.

Port-a-Patch and Patchliner: in 2003, Nanion introduced its first automated patch clamp system, the Port-a-Patch which is the third system marketed to automate voltage clamping. It is currently the world's smallest patch clamp setup with low maintenance setup. This system uses borosilicate glass chips which named NPC-chips as recording substrates. Port-a-Patch is a one-cell-at-a-time system. In 2006, Nanion attempted a higher-throughput device, the Patchliner, which supports true giga-seal recordings from up to eight cells simultaneously. It also expands the microfluidic exchange capability of both the internal and the external solutions (Farre et al., 2007). Three NPC-16 chips can be mounted on the chip loading area which means that up to 48 experiments can be preprogrammed and executed, where patch clamp measurements are possible in the whole-cell, cell-attached and perforated-patch recording configurations, and temperature controller in these two platforms supports experiments at physiological temperatures (Fig 3e). The Port-a-Patch has an approximate date throughput of 50 data points per day, whereas the Patchliner is capable of 500 data points per day. The throughput for this device was insufficient to compete successfully on the market and has remained unpopular. Recently, Nanion developed the SyncroPatch 96, which uses 96-well chips and has 96 simultaneous recording heads, with a throughput comparable to that of the IonWorks Quattro.

CytoPatch: the Cytocentrics's CytoPatch was the fourth system launched, this platform also uses planar-based array approach, but its designer has a slight difference from previous systems (Stett et al., 2003). It adopts Cytopatch™ Chips and Cytopatch™ Automat as electrode tip, which containing two opening, one for applying negative pressure to suction and to position the cells while simultaneously applying positive pressure to the "electrode tip" to keep it clean, the other for patch clamping and recording (Fig. 3f). These chips can be manufactured in large numbers at low cost and provide the formation of gigaohm seal. Although this device was released in 2003, there was no validation data available till now.

QPatch: the Sophion's QPatch was the fifth automated giga-seal system released (Fig. 3c) in 2004, which can be applied to both voltage- and ligand-gated channels (Mathes, 2006). This workstation can be manipulated and data can be analyzed with in-house developed software in computers. The first marketed system is QPatch 16 which uses QPlates as recording chips. This system provides 16 continuous recordings with 16 simultaneous headstages. To get a further higher throughput, Sophion recently launched the QPatch-HT, increased the number of headstages to 48 providing parallel recordings of 48 cells at a time. The stated throughput of the 16-amplifier system is 250-1200 data points and a significant enhancement of 750-3500 data points per day for the HT-system. Because of the high-quality recordings and the increased throughput, QPatch now has received quite good market acceptance.

Between 2007 and 2009, Fluxion Biosciences employed the so-called lateral patch clamp principle and introduced the IonFlux, a 16× precursor and the upcoming IonFlux HT (Fig. 3g)(Pearson, 2009). IonFlux uses PDMS-based microfluidic chips, according to the company, producing 8,000-10,000 data per day with a 96× consumable. In collaborate with AstraZeneca and The Automation Partnership (TAP), Cellectricon employs their laminar-flow solution control to develop a recently announced Dynaflow®HT (Fig. 3h) (Dabrowski, 2009). The system uses 96-well microfluidic chips made from silicone which enable to rapid solution exchange. The system achieved over 7,500 data points per day at first, but with the introduction of a new chip in 2009, the Dynaflow HT system generates 18,000 data points per day. Consistent with the IonFlux HT workstaion and many other high-throughput planar devices, Dynaflow HT has been validated with multiple cell lines including both voltage- and ligand-gated ion channels.

Method	Configuration	Format	Advantages	Shortcomings
QPatch (Sophion Bioscience)				
Planar QPlate with embedded recording and ground contacts	WC; gigaohm seals	16 or 48 wells in parallel; 1 cell/hole/well	• Laminar solution flow • Ligand addition during recordings • Cumulative/multiple compound additions • Continuous voltage clamp and current vs. time experiment flow • Validated for human ERG • Electrophysiological responses similar to those of conventional techniques	• Drug EC_{50} influenced by adherence to plate surfaces • No intracellular perfusion • No simultaneous visualization of channels
PatchXpress (Molecular Devices)				
Planar SealChip with electrodes separate from chip	WC; gigaohm seals	16 wells in parallel; 1 cell/hole/well	• Continuous voltage clamp vs. time plots • Cumulative/multiple compound additions • Ligand addition during recording • Simultaneous visualization of all 16 channels • Validated for human ERG • Electrophysiological responses similar to those of conventional techniques	• Drug EC_{50} influenced by adherence to plate surfaces • No intracellular perfusion
IonWorks HT and Quattro (Molecular Devices)				
Planar 384-well PatchPlate	Perforated WC; ~100 MΩ (HT) or 30–50	384 wells; 48 channels sequentially; 1	• Highest-throughput unit • Allows for collection	• Drug EC_{50} influenced by adherence to

Method	Configuration	Format	Advantages	Shortcomings
	MΩ (Quattro) seals	cell/hole/well (HT) or 64 cells/holes/well (Quattro)	of 384 data points: ~60% success rate for IonWorks HT and >95% for IonWorks Quattro	plate surfaces • No high-resistance membrane seal • No simultaneous ligand addition and recording • No intracellular perfusion • No voltage clamp between reads
Patchliner (Nanion Technologies)				
Planar NPC-16 chip	WC; gigaohm seals	2, 4, or 8 wells in parallel; 1 cell/hole/well (Port-a-Patch)	• Borosilicate chip surface • Primary cell use possible (single-well version) • Internal and external perfusion • Continuous voltage vs. time plots • Ligand addition during recordings • Cumulative/multiple compound additions	• System performance information limited
SyncroPatch96 (Nanion Technologies)				
Planar chip	WC; gigaohm seals	96-well plates; 16 channels sequentially; 1 cell/hole/well	• Borosilicate chip surface • Ligand addition during recordings • Internal and external perfusion • Continuous voltage vs. time plots • Cumulative/multiple compound additions	• Requires a proprietary add-in to solutions to achieve higher seal rate • Though meant for 96x, does not contain 96 amplifiers

Method	Configuration	Format	Advantages	Shortcomings
CytoPatch (Cytocentrics)				
Planar, electrode tip shape surrounded by aperture in borosilicate glass surface	WC	1 cell/hole/well	• Constant laminar flow • Positive pressure on electrode is independent of suction in surrounding aperture (mimics manual patch clamp)	• Not commercially available; used only for in-house screening
IonFlux HT (Fluxion Biosciences)				
Microfluidic well plate	WC; megaohm seals	96- or 384-well; 16 or 64 (HT) amplifier arrays; 20 cell ensemble recordings	• Laminar solution flow • Ligand addition during recordings • Cumulative/multiple compound additions • Continuous voltage vs. time plots • Tested with voltage- and ligand-gated channels • Separate cell and ligand wells	• Drug EC_{50} influenced by adherence to polymer plate surfaces? • No high-resistance membrane seal
Dynaflow HT (Cellectricon / The Automation Partnership / Astra Zeneca)				
Microfluidic well plate	WC; megaohm seals	96 channels/plate	• Ligand addition during recordings • Cumulative/multiple compound additions • Silica microfluidic chip minimizes compound adherence and consequent EC_{50} shifting • Validated for human ERG and $GABA_A$	• No high-resistance membrane seal

Table 3. Automated planar-array based platforms (Carmelle et al., 2011; Dunlop et al., 2008).

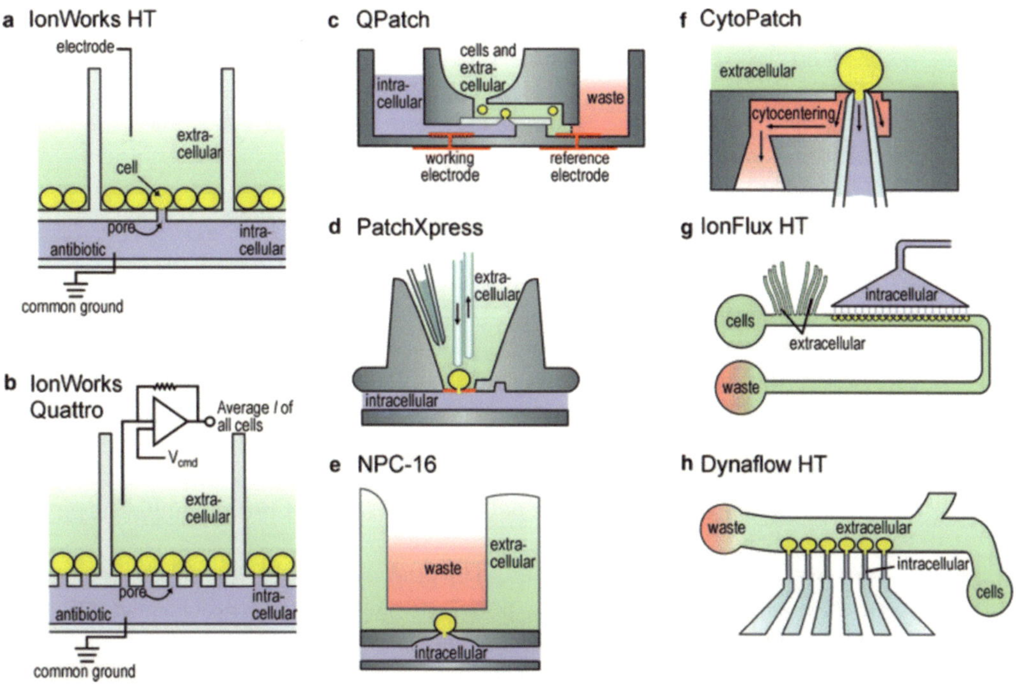

Fig. 3. Chamber configurations for automated and planar patch clamp technologies. (a) PatchPlate used in IonWorks HT. (b) PatchPlate population chip used in IonWorks Quattro. (c) QPlate used in QPatch. (d)*Seal*Chip used in PatchXpress. (e) NPC-16 chip used in Patchliner NPC-16. (f) CytoPatch chip used in CytoPatch. (g) Microfluidic bath used in the IonFlux HT system. (h) Microfluidic bath used in the Dynaflow HT syste (Carmelle, 2011)

9. Future directions for new drug discovery

The development of new drugs targeted to ion channels is a fast-growing market. Along with the costs of automated screening drop over time and the modification of patch clamp devices gradually live up to expectations, the manufacturers of patch clamp equipment have already acquired considerable number of orders during the past few years from pharmaceutical companies and numerous screening service providers. Recently, patch clamp combined with fluorescent techniques are developed, this technology has achieved in a small number of channels, allowing to simultaneously recording of fluorescence and currents, hence providing correlation between changes in channel structure and alterations in channel function (Zheng, 2006). These progresses open a new visual field in the upcoming era of ion channels studying as well as drug discovery.

10. References

Ashcroft, F.M. (2006). From molecule to malady. Nature 440, 440-447

Carmelle, V., Remillard & Jason, X.J. Yuan. (2011) Conventional patch clamp techniques and high-throughput patch clamp recordings on a chip for measuring ion channel activity. *Textbook of pulmonary vascular disease*, part 2, 495-510, DOI: 10.1007/978-0-387-87429-6_34

Chouabe, C., Drici, M.D., Romey, G. & Barhanin, J. (2000). Effects of calcium channel blockers on cloned cardiac K^+ channels IKr and IKs. Therapie 55, 195-202

Chouabe, C., Drici, M.D., Romey, G., Barhanin, J. & Lazdunski, M. (1998). HERG and KvLQT1/IsK, the cardiac K^+ channels involved in long QT syndromes, are targets for calcium channel blockers. Mol Pharmacol 54, 695-703

Chouinard, G. (2006). The search for new off-label indications for antidepressant, antianxiety, antipsychotic and anticonvulsant drugs. J Psychiatry Neurosci 31, 168-176

Dabrowski, M. (2009) Global ion channel initiative. *Ion Channel Retreat* June 29-July 1 See also the Cellectricon web site

De Bruin, M.L., Pettersson, M., Meyboom, R.H., Hoes, A.W. & Leufkens, H.G. (2005). Anti-HERG activity and the risk of drug-induced arrhythmias and sudden death. Eur Heart J 26, 590-597

Drolet, B., Zhang, S., Deschenes, D., Rail, J., Nadeau, S., Zhou, Z., January, C.T. & Turgeon, J. (1999). Droperidol lengthens cardiac repolarization due to block of the rapid component of the delayed rectifier potassium current. J Cardiovasc Electrophysiol 10, 1597-1604

Dunlop, J., Bowlby, M., Peri, R., Vasilyev, D. & Arias, R. (2008). High-throughput electrophysiology: an emerging paradigm for ion-channel screening and physiology. Nat Rev Drug Discov 7, 358-368

Estes, D.J., Memarsadeghi, S., Lundy, S.K., Marti, F., Mikol, D.D., Fox, D.A. & Mayer, M. (2008). High-throughput profiling of ion channel activity in primary human lymphocytes. Anal Chem 80, 3728-3735

Farre, C., Stoelzle, S., Haarmann, C., George, M., Bruggemann, A. & Fertig, N. (2007). Automated ion channel screening: patch clamping made easy. Expert Opin Ther Targets 11, 557-565

Finkel, A., Wittel, A., Yang, N., Handran, S., Hughes, J. & Costantin, J. (2006). Population patch clamp improves data consistency and success rates in the measurement of ionic currents. J Biomol Screen 11, 488-496

Hamer, A.M., Haxby, D.G., McFarland, B.H. & Ketchum, K. (2002). Gabapentin use in a managed medicaid population. J Manag Care Pharm 8, 266-271

Hogg, D.S., Boden, P., Lawton, G. & Kozlowski, R. (2006) Ion channel drug targets-unlocking the potential. *Drug discov world,* Vol.7, No.3, (August 2006), pp. 83-93, ISSN 1469-4344

Imming, P., Sinning, C. & Meyer, A. (2006). Drugs, their targets and the nature and number of drug targets. Nat Rev Drug Discov 5, 821-834

Kang, J., Wang, L., Cai, F. & Rampe, D. (2000). High affinity blockade of the HERG cardiac K(+) channel by the neuroleptic pimozide. Eur J Pharmacol 392, 137-140

Kornreich, B.G. (2007). The patch clamp technique: principles and technical considerations. J Vet Cardiol 9, 25-37

Kozlowski, R.Z. (1999) Chloride channels: Potential therapeutic targets, In: *Chloride Channels,* R.Z. Kozlowski, (Ed.), 177-186, ISIS Medical Media Ltd, Oxford, UK

Lepple-Wienhues, A., Ferlinz, K., Seeger, A. & Schafer, A. (2003). Flip the tip: an automated, high quality, cost-effective patch clamp screen. Receptors Channels 9, 13-17

Li, S., Gosling, M., Poll, C.T., Westwick, J. & Cox, B. (2005). Therapeutic scope of modulation of non-voltage-gated cation channels. Drug Discov Today 10, 129-137

Mathes, C. (2006). QPatch: the past, present and future of automated patch clamp. Expert Opin Ther Targets 10, 319-327

Mathew, N.T., Rapoport, A., Saper, J., Magnus, L., Klapper, J., Ramadan, N., Stacey, B. & Tepper, S. (2001). Efficacy of gabapentin in migraine prophylaxis. Headache 41, 119-128.

Morganroth, J. (2004). A definitive or thorough phase 1 QT ECG trial as a requirement for drug safety assessment. J Electrocardiol 37, 25-29

Neher, E. & Sakmann, B. (1992). The patch clamp technique. Sci Am 266, 44-51

Neubert, H.J. (2004). Patch clamping moves to chips. Anal Chem 76, 327A-330A

Overington, J.P., Al-Lazikani, B. & Hopkins, A.L. (2006). How many drug targets are there? Nat Rev Drug Discov 5, 993-996

Pearson, S. (2009) Investigating and focusing on ion channels as drug targets. Genetic Eng Biotech News Vol.29, No.11

Penner, R. (1995) A practical guide to patch clamping, in: *Single-Channel Recording*, B. Sakmann & E. Neher, eds., (2nd ed). Plenum Press, NY, pp.3-30

Proks, P. & Lippiat, J.D. (2006). Membrane ion channels and diabetes. Curr Pharm Des 12, 485-501

Rampe, D., Murawsky, M.K., Grau, J. & Lewis, E.W. (1998). The antipsychotic agent sertindole is a high affinity antagonist of the human cardiac potassium channel HERG. J Pharmacol Exp Ther 286, 788-793

Salata, J.J., Jurkiewicz, N.K., Wallace, A.A., Stupienski, R.F., 3rd, Guinosso, P.J., Jr. & Lynch, J.J., Jr. (1995). Cardiac electrophysiological actions of the histamine H1-receptor antagonists astemizole and terfenadine compared with chlorpheniramine and pyrilamine. Circ Res 76, 110-119

Sanguinetti, M.C. & Jurkiewicz, N.K. (1990). Two components of cardiac delayed rectifier K^+ current. Differential sensitivity to block by class III antiarrhythmic agents. J Gen Physiol 96, 195-215

Schroeder, K., Neagle, B., Trezise, D.J. & Worley, J. (2003). Ionworks HT: a new high-throughput electrophysiology measurement platform. J Biomol Screen 8, 50-64

Spector, P.S., Curran, M.E., Keating, M.T. & Sanguinetti, M.C. (1996). Class III antiarrhythmic drugs block HERG, a human cardiac delayed rectifier K^+ channel. Open-channel block by methanesulfonanilides. Circ Res 78, 499-503

Stett, A., Burkhardt, C., Weber, U., van Stiphout, P. & Knott, T. (2003). CYTOCENTERING: a novel technique enabling automated cell-by-cell patch clamping with the CYTOPATCH chip. Receptors Channels 9, 59-66

Suessbrich, H., Schonherr, R., Heinemann, S.H., Attali, B., Lang, F. & Busch, A.E. (1997a). The inhibitory effect of the antipsychotic drug haloperidol on HERG potassium channels expressed in Xenopus oocytes. Br J Pharmacol 120, 968-974

Suessbrich, H., Schonherr, R., Heinemann, S.H., Lang, F. & Busch, A.E. (1997b). Specific block of cloned Herg channels by clofilium and its tertiary analog LY97241. FEBS Lett 414, 435-438

Tao, H., Santa Ana, D., Guia, A., Huang, M., Ligutti, J., Walker, G., Sithiphong, K., Chan, F., Guoliang, T., Zozulya, Z., et al. (2004). Automated tight seal electrophysiology for assessing the potential hERG liability of pharmaceutical compounds. Assay Drug Dev Technol 2, 497-506

Thomas, D., Wendt-Nordahl, G., Rockl, K., Ficker, E., Brown, A.M. & Kiehn, J. (2001). High-affinity blockade of human ether-a-go-go-related gene human cardiac potassium channels by the novel antiarrhythmic drug BRL-32872. J Pharmacol Exp Ther 297, 753-761

Venter, J.C., Adams, M.D., Myers, E.W., Li, P.W., Mural, R.J., Sutton, G.G., Smith, H.O., Yandell, M., Evans, C.A., Holt, R.A., et al. (2001). The sequence of the human genome. Science 291, 1304-1351

Vitola, J., Vukanovic, J. & Roden, D.M. (1998). Cisapride-induced torsades de pointes. J Cardiovasc Electrophysiol 9, 1109-1113

Vorperian, V.R., Zhou, Z., Mohammad, S., Hoon, T.J., Studenik, C. & January, C.T. (1996). Torsade de pointes with an antihistamine metabolite: potassium channel blockade with desmethylastemizole. J Am Coll Cardiol 28, 1556-1561

Waldegger, S., Niemeyer, G., Morike, K., Wagner, C.A., Suessbrich, H., Busch, A.E., Lang, F. & Eichelbaum, M. (1999). Effect of verapamil enantiomers and metabolites on cardiac K^+ channels expressed in Xenopus oocytes. Cell Physiol Biochem 9, 81-89

Walker, B.D., Singleton, C.B., Bursill, J.A., Wyse, K.R., Valenzuela, S.M., Qiu, M.R., Breit, S.N. & Campbell, T.J. (1999). Inhibition of the human ether-a-go-go-related gene (HERG) potassium channel by cisapride: affinity for open and inactivated states. Br J Pharmacol 128, 444-450

Woosley, R.L. (1996). Cardiac actions of antihistamines. Annu Rev Pharmacol Toxicol 36, 233-252

Xu, J., Guia, A., Rothwarf, D., Huang, M., Sithiphong, K., Ouang, J., Tao, G., Wang, X. & Wu, L. (2003). A benchmark study with sealchip planar patch-clamp technology. Assay Drug Dev Technol 1, 675-684

Yang, T., Snyders, D.J. & Roden, D.M. (1995). Ibutilide, a methanesulfonanilide antiarrhythmic, is a potent blocker of the rapidly activating delayed rectifier K^+ current (IKr) in AT-1 cells. Concentration-, time-, voltage-, and use-dependent effects. Circulation 91, 1799-1806

Zhang, S., Zhou, Z., Gong, Q., Makielski, J.C. & January, C.T. (1999). Mechanism of block and identification of the verapamil binding domain to HERG potassium channels. Circ Res 84, 989-998

Zheng, J. (2006). Patch fluorometry: shedding new light on ion channels. Physiology (Bethesda) 21, 6-12

Zhou, Z., Vorperian, V.R., Gong, Q., Zhang, S. & January, C.T. (1999). Block of HERG potassium channels by the antihistamine astemizole and its metabolites desmethylastemizole and norastemizole. J Cardiovasc Electrophysiol 10, 836-843

Zolotoy, A.B., Plouvier, B.P., Beatch, G.B., Hayes, E.S., Wall, R.A. & Walker, M.J. (2003). Physicochemical determinants for drug induced blockade of HERG potassium channels: effect of charge and charge shielding. Curr Med Chem Cardiovasc Hematol Agents 1, 225-241

17

Enhanced Patch-Clamp Technique to Study Antimicrobial Peptides and Viroporins, Inserted in a Cell Plasma Membrane with Fully Inactivated Endogenous Conductances

Marco Aquila[1,2,*], Mascia Benedusi[1,*], Alberto Milani[1] and Giorgio Rispoli[1]
[1]*Dipartimento di Biologia ed Evoluzione, Sezione di Fisiologia e Biofisica, Università di Ferrara, Ferrara*
[2]*Institute for Maternal and Child Health – IRCCS "Burlo Garofolo"-Trieste*
Italy

1. Introduction

1.1 Chapter outline

Many short peptides (Table 1) selectively permeabilize the bacteria plasma membrane (Fig. 1), leading to their lyses and death: they are therefore a source of antibacterial molecules, and inspiration for novel and more selective drugs. Another class of short (<100 residues) membrane proteins called viroporins, because they are coded by viral genes (Table 1), permeabilizes the membrane of susceptible cells during infection of by most animal viruses (Carrasco, 1995; Fig. 1). The permeabilization leads to host cell lyses and the release of the virus mass, replicated at host cell expense, to propagate the infection. Detailed knowledge of the permeabilization properties of these proteins would allow to design, for instance, selective blockers of these pores, that would contrast the spread of the viral infection.

In this chapter, the patch-clamp technique is employed to study the mechanism of membrane permeabilization induced by the pore-forming peptides, under strict physiological conditions. This goal is achieved by recording the ion current through the channels formed by these peptides, once inserted in a cell plasma membrane. To avoid contamination by the cell membrane currents, all the endogenous current sources must be blocked. It has been found that the photoreceptor rod outer segment mechanically isolated from the retina of low vertebrates (OS; Vedovato & Rispoli, 2007a; Fig. 2) was the most suitable cell to carry on the above studies, because it was possible to fully block all its endogenous currents without using any drug (such as TTX, TEA, dihydropyridines, etc.), that could obstruct the peptide pores or interfere with the pore formation (Fig. 4). The peptides were applied to (and removed from) the extracellular OS side in ~50 ms with a computer-controlled microperfusion system, in which every perfusion parameter (as the rate of solution flow, the temporal sequence of solution changes or the number of automatic,

* These authors contributed equally to this work.

self-washing cycles) was controlled by a user-friendly interface. This system allowed rapid application and removal of ions, drugs and peptides on the cells with a controlled timing, so that the ion channel characteristics (as its selectivity, blockade and gating) and the dynamics of pore formation could be precisely assessed. On the basis of the electrophysiological recordings obtained with representative peptides and with selected analogs, as alamethicin F50/5 (Crisma et al., 2007; Vedovato & Rispoli, 2007a; Vedovato et al., 2007; Fig. 5-8), the cecoprine-mellitin hybrid peptide (CM15; Milani et al., 2009; Fig. 5-7), and a 20-amino acids long fragment of the viroporin poliovirus 2B (Madan et al., 2007; Fig. 9), it will be shown that the membrane pore formation occurs according to the barrel and stave, toroidal, and carpet model, respectively (Fig. 1), that are the most widely-accepted mechanisms of membrane permeabilization.

When recording large currents (produced for instance by high concentrations of peptides and/or highly permeable peptides), it is necessary to minimize series resistance, to reduce time constant of charging the cell membrane capacitance and error in membrane potential control. A second problem arises from the asymmetry of the plasma membrane: it is possible that the permeabilization properties of a particular peptide could be different depending upon the side of the membrane to which it is applied. For example, it is conceivable that viroporins are optimized to insert in the intracellular face of the plasma membrane, because they are synthesized in host cell cytosol. These two problems could be circumvented by widening the patch pipette shank, through the calibrated combination of heat and air pressure. These pipettes dramatically reduce series resistance, and allow at the same time to insert pulled quartz or plastic tubes very close to the pipette tip (Benedusi et al., 2011; Goodman et al., 2000; Johnson et al., 2008), making it possible the delivery of large molecules to the cytosol with a controlled timing (Fig. 3). Finally, it is presented here a simple procedure to consistently attain seals with conventional or pressure polished pipettes, made from just one glass type, on a wide variety of cell types, isolated from different amphibian, reptilian, fish, and mammalian tissues, and on artificial membranes made with many different lipid mixtures.

1.2 Antimicrobial peptides

The increasing number of common and emerging cases of infection due to antibiotic-resistant bacteria, occurring for instance in hospitals and in high frequented spaces, requires new and reliable sources of innovative antimicrobials. A promising source of these compounds are the pore-forming peptides, that permeabilize the bacterial membrane up to a point to induce the lyses and death of the pathogen. These peptides are an evolutionarily conserved component of the innate immune response found among all classes of life, and they are one of the oldest form of defense against pathogens. This chapter is focused on two of them: the F50/5 and the CM15 (Table 1). The former peptide is the major component of the neutral fraction of the peptaibol antibiotic alamethicin. The peptaibols (Toniolo et al., 2001) are members of a group of naturally occurring short peptides produced by fungi of the genus *Trichoderma,* characterized by an N-terminal acyl group, a C-terminal 1,2-aminoalcohol and a high content of the non-proteinogenic C^{α}-tetrasubstituted α-amino acid Aib (α-aminoisobutyric acid). Key amino acids of the F50/5 peptide sequence were also modified to understand their implication on the biophysical characteristics of the pore

(Vedovato et al., 2007). The CM15 is a small 15-residue synthetic hybrid peptide, first described by Andreu et al. (1992), composed of the first seven residues of the silk moth cecropin A and residues 2-9 from the bee venom melittin.

1.3 Viroporins

Many RNA viruses encode some 60-120 amino acids integral membrane proteins, called viroporins, which are localized primarily within the endoplasmic reticulum and plasma membrane of host cells. About a dozen proteins that qualify as viroporins have been discovered so far, including poliovirus 2B, alphavirus 6K, HIV-1 Vpu, hepatitis C virus (HCV) p7 protein, influenza virus M2, and the coronavirus E (CoVE) protein from the severe acute respiratory syndrome (SARS) virus (Gonzalez & Carrasco, 2003). Viroporins are not essential for virus replication, but they boost virus growth by taking part in several viral functions, and by perturbing several mechanisms of host cell, as glycoprotein and vesicle turnover, and plasma membrane permeability. The latter occurs because viroporin oligomerization gives rise to hydrophilic pores at the membranes of host cell. These pores are thought to play a crucial role for viral infectivity, possibly by causing the lyses and death of host cell (letting therefore the replicated virus particles to escape the cell), with a mechanism analogous to the bacterial lyses produced by the antimicrobial peptides. This chapter is focused on elucidating the mechanism of membrane permeation produced by the poliovirus 2B (Table 1).

1.4 Peptides as bioactive molecules and archetypal ion channel

Besides the development of anti-bacterial and anti-viral drugs, the pore-forming peptides, being the most ancestral form of membrane pores, provide also a simple model system to understand the structure-function relationships of channels and the molecular basis of peptide/protein oligomerization in lipid membranes. Moreover, these peptides are a powerful molecular model on which to build custom molecules with wide-ranging biotechnological application, as ion channel modulators (Hille, 2001), anti-tumorigenic agents (Hoskin & Ramamoorthy, 2008; Papo & Shai, 2005), biosensors for many different analytes (which could deliver a configurable binding site for substrates encoded in a readily measured electrical or optical signal; Aili & Stevens, 2010), and provide a pharmacological approach to cure channelopaties (reviewed in Wilde, 2008). In the latter case, a synthetic channel could be inserted in cells expressing an aberrant ion channel (Wallace et al., 2000) to restore the physiological ion flow. Bacterial and viral peptides may finally provide insights about the evolution of channel selectivity and gating and, more generally, of ion pumps and exchangers. Recent data from many laboratories indicate indeed that secondary transporters are basically pores with an highly specialized gate system, rather than proteins with a transport mechanism completely different from the ion channels one (reviewed in Gadsby, 2009).

1.5 Mechanism of membrane permeabilization

Regardless their origin and purpose, all the membrane permeabilizing peptides adhere first parallel to the lipid bilayer (Fig. 1*a* and *b*), they then orient perpendicular to the membrane, and finally they bind together and/or reorganize the lipid bilayer to form transmembrane pores. Because of their amino acid composition, amphipathicity, and cationic charge, three distinct mechanisms have been proposed to explain membrane permeabilization: "barrel and stave", "toroidal", and "carpet" (Fig. 1; reviewed in Brogden, 2005).

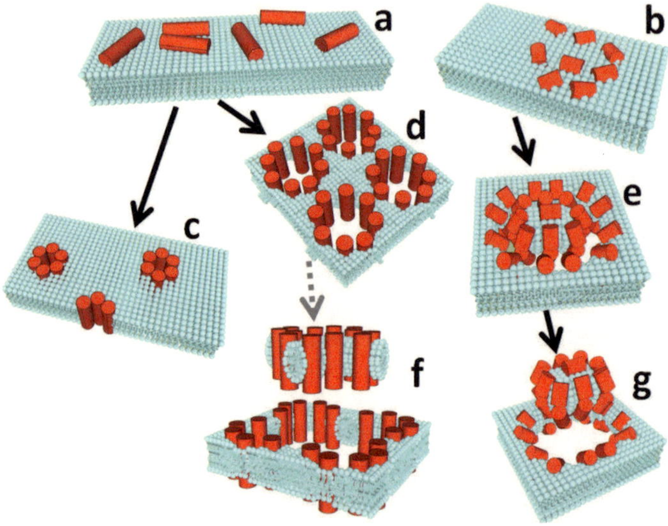

Fig. 1. Mechanisms of permeabilization induced by pore-forming peptides. After adhering on the external face of the membrane (*a* and *b*), the peptide could insert in the membrane according to a barrel and stave (*c*), toroidal (*d* and *e*), or carpet mechanism (*f* and *g*).

The first mechanism requires that the peptide monomers, weekly bound to the membrane surface (Fig. 1*a*), orient perpendicularly to the latter and bind together (as staves) around a central pore (the barrel; Fig. 1*c*). The peptide hydrophobic segments face the bilayer core, while their hydrophilic segments face the pore lumen. In the toroidal pore, the polar segments of the peptides interact with the polar head groups of the lipids more strongly than in the case of the barrel and stave pore, so that the lipids are forced to tilt from the lamellar normal up to connect the two leaflets of the membrane, forming a continuous bend from one side to the other of the membrane, in the fashion of a toroidal hole (peptides spanning the entire membrane: Fig. 1*d*; peptide shorter than the membrane thickness: Fig. 1*e*). Differently from the barrel and stave mechanism, the peptides are always associated with the lipid head groups, even when they are perpendicularly inserted in the lipid bilayer. In the carpet mechanism, the strong electrostatic interactions between peptides and the phospholipid head groups result in the peptide carpeting of the membrane (Fig. 1*b*). The peptides then assemble with the lipids to form transiently toroidal pores (Fig. 1*e*), allowing more and more peptides to cross the bilayer and eventually cover the other membrane leaflet. Finally, the peptides may isolate a micelle that detaches from the membrane to head to the water phase (Fig. 1*g*), leading to the bilayer disintegration in a detergent-like manner. It is also possible that the contour of a micelle may be delimited by several toroidal pores that happen to group together (Fig. 1*d*), and this micelle may eventually separate from the membrane and solubilize (Fig. 1*f*). Moreover, it is conceivable that the same peptide may permeabilize the membrane with one mechanism or with another (Brogden, 2005), depending upon the surrounding conditions, as the lipid environment, the temperature, the peptide concentration, the presence of some other endogenous or exogenous molecules (Noshiro et al., 2010), etc.

2. Methods: Studying peptide-induced permeabilization under strict physiological conditions

2.1 Rationale

Most of the research on the pore-forming peptides has the long-term goal to understand at the molecular level how the peptide-lipid interactions produce structural changes in both, and optimal peptide orientation, to eventually produce a raise in membrane conductance (Chen et al., 2001; Saint et al., 2002). With this aim, a widely used technique consists in recording the ion current flowing through the pore formed by these peptides, inserted in an artificial lipid bilayer. To have an acceptable signal-to-noise ratio in these experiments, it is necessary to apply non-physiological voltages and ionic gradients, that however could affect the peptides themselves, or even the membrane in which they were inserted (Bockmann et al., 2003). Moreover, to our knowledge, no studies aimed to measure the aggregation time course of peptides to form a functioning pore in the plasma membrane of a living cell, and the kinetics of their disaggregation.

In this chapter, the biophysical characteristics and the pore formation dynamics of selected antimicrobial peptides and viroporins, and some of their analogues, were studied under strict physiological conditions. This was accomplished by inserting these (or other) peptide monomers in the plasma membrane of a living cell, and recording the current flowing through the peptide channels at a desired voltage by using the whole-cell, voltage-clamp technique. To make sure that the entire current was flowing through these channels, it was looked for a particular cell system where all the endogenous conductances could be blocked without using any drug, that could affect the peptide pore as well. The outer segment of the photoreceptor rod (OS) mechanically isolated from low vertebrate retinae (Rispoli et al., 1993) is such a cell system, because the only channel type of this cell fragment is fully closed by bright light. The OS of *Rana esculenta* have been found particularly suitable to carry on this study, because of their large size (Fig. 2*a*, *Inset*) and for the commercial availability and low cost of this edible frog species.

To have a precise control of pore formation (and disaggregation), a custom-made, computer controlled microperfusion system was employed to rapidly apply (and remove) the peptides under study onto a cell (Vedovato & Rispoli, 2007a). Peptide formation and disaggregation dynamics were therefore gathered, respectively, from the time course of the development and fall of the exogenous current (at a given potential, V_h), following peptide application and removal. This system is also capable to apply and remove ions, drugs and peptides on any isolated cell or tissue fragment in ~50 ms with an accurately controlled timing. This allowed to study in detail the biophysical characteristics of the channels, that are of key importance to assess the peptide performances and its potential biotherapeutic activity.

The pipette tapered shank and the small tip opening give high access resistances (R_a) and constitute the dominant barrier to molecular diffusion between pipette and cell cytosol in the whole-cell recording configuration (Pusch & Neher, 1988). This shortcoming:

1. gives errors in membrane potential control, due to the voltage drop across the R_a in the presence of large membrane currents;
2. impedes the precise measure of current onset and offset kinetics, due to the often too large time constant of charging the cell membrane capacitance through the R_a;

3. causes intracellular ion accumulation or depletion, and it slows down the rate of exogenous molecules incorporation via the patch pipette.

It is however possible to enlarge the pipette shank (Goodman et al., 2000; Johnson et al., 2008), through the calibrated combination of heat and air pressure, with a custom made inexpensive set-up (Benedusi et al., 2011). Besides improving the electrical recordings, the enlarged tip geometry of the pressure polished pipettes could accommodate pulled quartz or plastic perfusion tubes close to the pipette tip, allowing the fast and controlled cytosolic incorporation of exogenous molecules.

Another problem of the patch-clamp recording is to gain consistently tight seals: although this technique has been widely used by more than 40 years, little is still known about the nature of the molecular interactions underlying the generation of the seal between membrane and glass (Suchyna et al., 2009). This problem has been addressed here, and it has been found a simple method that allows to consistently attain seals on natural or artificial membranes, by using conventional or pressure polished pipettes made from just one glass type (Benedusi et al., 2011).

2.2 OS preparation and view

OS were mechanically isolated from the retina of *Rana esculenta* in the dark, using infrared illumination and an infrared viewer. Methods are described in detail elsewhere (Vedovato & Rispoli, 2007a). The retina was "peeled" from an eyecup piece and was gently triturated in Ringer (~5 ml), using a fire-polished Pasteur pipette to obtain the OS; a Ringer drop containing the OS was then transferred to the recording chamber placed on the microscope (TE 300, Nikon, Tokyo, Japan) stage. The OS were illuminated with an ultrabright infrared LED (900 nm) and focused on a fast digital camera (C6790-81, Hamamatsu Photonics, Tokyo, Japan) coupled to the microscope. The OS were then viewed in the frame grabber window generated by the AquaCosmos software package (version 2.5.3.0; Hamamatsu Photonics), which controlled all the camera parameters (as gain, frame rate, binning, etc.) as well via a PCI board (PCDIG, Dalsa, Waterloo, ON, Canada).

2.3 Solutions and electrophysiological recordings

OS were recorded using the whole-cell configuration of the patch-clamp technique under visual control at room temperature (20-22 °C; Fig. 2*a*, *inset*). The Ringer solution had the following composition (in mM): 115 NaCl, 3 KCl, 10 HEPES free acid [*N*-(2-hydroxyethyl)piperazine-*N*′-(2-ethanesulfonic acid)], 0.6 $MgCl_2$, 0.6 $MgSO_4$, 1.5 $CaCl_2$, 10 glucose (osmolality 260 mOsm/Kg, buffered to pH = 7.6 with NaOH). All chemicals were purchased from Sigma (St. Louis, MO, USA).

The current amplitude (recorded employing an Axopatch 200B; Molecular Devices, Sunnyvale, CA, USA) elicited by repetitive -10 mV pulses was used to measure the seal resistance during cell-attached recording. Once the whole-cell recording configuration was obtained, the current transients produced by these repetitive pulses were used to measure the OS membrane resistance (R_m), the access resistance (R_a), and the membrane capacitance (C_m). Peptides (whose primary sequence are reported in Table 1) were applied and removed in ~50 ms by switching forth and back the OS from a stream of control solution [composition (in mM): 130 KCl, 1 $CaCl_2$ and 10 HEPES; osmolality 260 mOsm/Kg, buffered to pH=7.6

with KOH] to a stream of control solution containing the peptide under test (see paragraph 2.5). Patch pipettes were filled with control solution as well in order to drive the current just with the holding potential (V_h, that was typically -20 mV).

The stability of the recording was checked by routinely measuring R_m, R_a, and C_m in control perfusion solution, i.e. before and after each peptide application. Recordings were filtered at 2 kHz via an eight-pole Butterworth filter (VBF/8 Kemo, Beckenham, UK), sampled on-line at 5 kHz by a Digidata 1322A (Molecular Devices) connected to the SCSI port of a Pentium computer running the pClamp 9.0 software package (Molecular Devices), and stored on disk. Data were further low-pass filtered off-line at 200 or 500 Hz using a Gaussian filter, or by using the "running average" routine of SigmaPlot (version 8.0; Jandel Scientific, San Rafael, CA, USA), and analyzed using Clampfit (version 9.0; Molecular Devices). Figures and statistics were performed using SigmaPlot; results are given as means±SEM.

2.4 Peptide sequences and usage

The primary structure of all peptides studied in this chapter is reported in Table 1. CM15 and its scrambled version were dissolved in bi-distilled water, F50/5 and its analogs in methanol, and the peptide fragments of poliovirus 2B in dimethyl sulfoxide (DMSO), to get a 50, 100, 500, and 1000 µM stock solutions; an aliquot of one of these peptide stocks was dissolved in the perfusion solution to get a final peptide concentration of 0.25, 0.33, 0.1, 1, 2.5, 5 and 10 µM, and used within 30 min. Control experiments proved that the methanol and DMSO contamination of the perfusion solution (no larger than 10 nl/ml and 1 µl/ml, respectively) did not cause any non-specific membrane permeabilization.

Peptide	Sequence
F50/5	Ac-UPUAUAQUVUGLUPVUUQQ-Phol
[L-Glu(OMe) 18,19]	Ac-UPUAUAQUVUGLUPVUUEE-Phol
[L-Glu(OMe) 7,18,19]	Ac-UPUAUAEUVUGLUPVUUEE-Phol
CM15	Ac-KWKLFKKIGAVLKVL-NH_2
Scrambled CM15	Ac-KWKLKFKIGLVKLVAV-NH_2
Poliovirus 2B	GITNYIESLGAAFGSGFTQQISDKITELTNMVTSTITEKLLKNLI KIISSLVIITRNYEDTTTVLATLALLGCDASPWQWLRKKACDV LEIPYVIKQ

Table 1. Residue sequences of the peptides studied here. Ac, acetyl; Phol, phenylalaninol.

CM15 and its scrambled version were a generous gift of Dr. Feix of Department of Biophysics, Medical College of Wisconsin, Milwaukee, WI, USA; F50/5 and its analogs were synthesized in the lab of Dr. Toniolo of Department of Chemistry of the University of Padua, Italy (Peggion et al., 2004); the peptide fragments of poliovirus 2B were synthesized in the lab of Dr. Nieva of Unidad de Biofísica (CSIC-UPV/EHU), Universidad del País Vasco, Bilbao, Spain (Madan et al., 2007).

2.5 Fast perfusion system

After obtaining the whole-cell recording, the OS was aligned in front of a multibarreled perfusion pipette of a fast microperfusion system (Fig. 2*a*). The perfusion pipette was moved on a horizontal plane with a precision step motor, controlled by a user-friendly interface

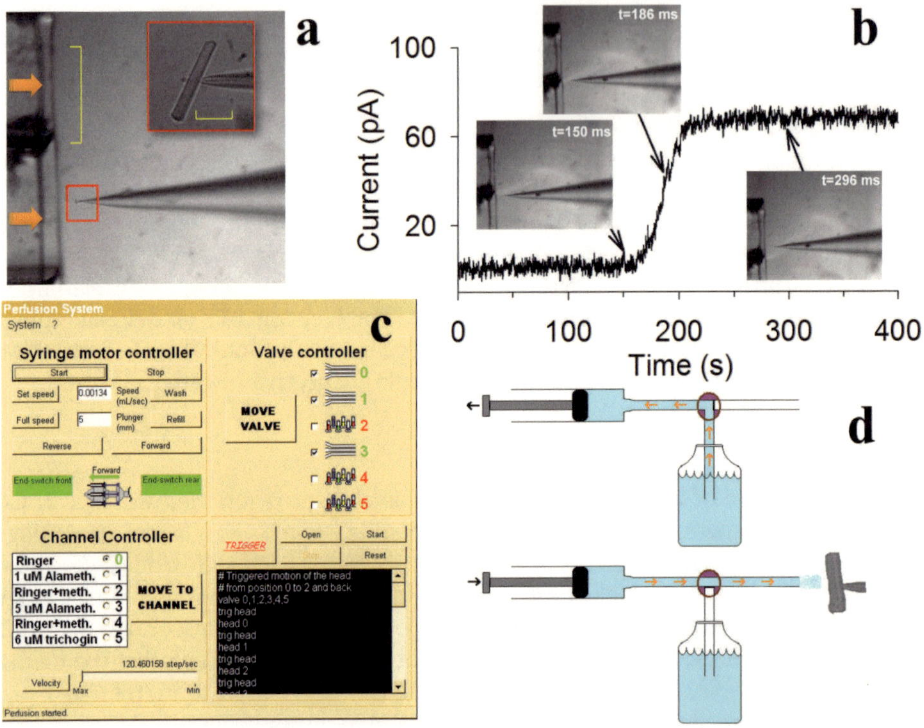

Fig. 2. The technique employed to investigate the permeabilization properties of the peptides inserted in a natural membrane. *a*, a whole-cell recorded OS (shown enlarged in the *inset*; scale bar is 20 µm) aligned in front of the perfusion pipette (scale bar is 500 µm; horizontal orange arrows denote perfusion flows). *b*, trace shows the current jump upon switching an open patch pipette filled with 130 mM K^++1 mM Ca^{2+} (9 MΩ, V_h=0 mV) from 65 mM choline+65 mM K^++1 mM Ca^{2+} to 130 mM K^++1 mM Ca^{2+}. The two latter solutions had a 0.6 mV junction potential, producing a current jump of 70 pA, lasting ~50 ms, that had the same kinetics of the solution change. This kinetics is also shown by the three still frames (650x494 pixels at 12 bits grey-scale resolution), extracted from a 30 frame/s movie synchronized with the voltage-clamp recording. The boundary separating the two solution streams was clearly visible, allowing a precise electrical and visual correlation of the solution change dynamics. *c*, user interface of the perfusion apparatus. *d*, scheme of one perfusion line (composed by a syringe, a three-way valve, a cylinder and one perfusion pipette barrel) and perfusion flow (in orange) during syringe refilling (*upper panel*) and during perfusion of an OS recorded in whole-cell (*lower panel*); drawing not in scale.

(Fig. 2*c*) running in a host computer, connected to the microperfusion system via the serial port. The perfusion pipette was constituted of up to six barrels (500 µm of side; two barrels of a four barrelled pipette are visible in Fig. 2*a*) made with precision, square glass capillaries glued together. Peptides were applied and removed in ~50 ms (see below) by moving the perfusion pipette so that to switch the whole-cell recorded OS back and forth from a stream of control perfusion solution (usually containing 130 mM of a monovalent cation and 1 mM Ca^{2+}; see Results) to a stream containing the peptide (dissolved in the same perfusion solution). This strategy allowed to assess the dynamics of the pore formation and the possible reversibility of the process. The temporal lag between the time in which the command (internal or triggered by an external device) moving the perfusion pipette was imparted, and the time in which the solution change effectively occurred, as well as the speed of the solution change, were occasionally measured (since they were very reproducible) as illustrated in Fig. 2*b*. At the end of experiment, the cell was blown off the patch pipette with a positive pressure pulse, and the odd perfusion lines were filled with a solution having 50% choline chloride and 50% KCl, while the even perfusion lines were filled with the patch pipette solution. The choline chloride had a different refraction index in respect to all the patch pipette solutions used, therefore the boundary separating two adjacent streams was clearly visible (see the three still frames of Fig. 2*b*). Since a junction potential was also developed between the two solutions, upon repeating the solution changes previously performed with the OS, a current jump was recorded by the open pipette (voltage-clamped at 0 mV) that had the same kinetics of the solution change (Fig. 2*b*). By comparing the electrical and the imaging recordings, it results that the solution change was completed within the 50 ms necessary to the OS to cross the boundary separating two adjacent streams.

The perfusion system could be also used to apply solutions to the cell of different ionic composition or containing a specific drug, so that to study, besides the current-to-voltage characteristics, the ion selectivity and blockade of the pore as well. To this aim, all perfusion solutions contained the peptide under study at the same concentration, so that to compare the effects of the test solutions at a constant membrane permeabilization.

The solutions flowing in the perfusion pipette were fed by means of precision syringes (Hamilton, Reno, NV, USA), whose piston was moved by a DC motor controlled via the computer interface (*Syringe motor controller* panel, Fig. 2*c*). The typical perfusion speed was 15 μl/min, therefore minimal amounts of peptide solution (<500 μl) were required to perform peptide applications lasting more than half an hour. The fine regulation of the perfusion flow speed and of the velocity of perfusion pipette horizontal movements (controlled by the *Velocity* button in the *Channel Controller* panel) allowed to perform fast solution changes on practically any isolated cell type, or even on small cell aggregates, without any significant change of the seal resistance. This system has been successfully tested on cells mechanically or enzymatically isolated from different amphibian, reptilian, fish, and mammalian tissues, on cultured cells and on giant unilamellar vesicles made with many different lipid mixtures. Depending upon the position of the six three-way solenoid valves (Fig. 2*d*; all tubing and valve components in contact with the solutions were made in teflon), each one independently controlled by the *Valve controller* panel (Fig. 2*c*), the solution contained in each syringe could be sent to the perfusion pipette (symbolized by the perfusion pipette in the *Valve controller* panel; Fig. 2*c*) or redirected to one of the six reservoirs (in which each solution was made, symbolized by a group of cylinders in the *Valve controller* panel). This allowed to save the solutions that were not used, or to avoid that, upon switching between two non-adjacent barrels, the OS was transiently exposed to a third undesired solution. Once emptied, the syringes could be refilled from the cylinders by clicking on the *Refill* button (in the *Syringe motor controller* panel, Fig. 2*c*). This command moved the six three-way valves in the cylinder position and backed up the DC motor at full speed until the syringes were filled: the motor was then stopped by an end-position switch; a second end-position switch stopped the motor when syringes were emptied (the status of the switches was signaled by the *End switch rear* and the *End switch front* indicators in the *Syringe motor controller* panel, that turned from green to red). When particularly precious solutions were used, all the perfusion lines were filled with control solution and the connectors to the perfusion pipette were detached and immersed in the vials containing the precious solutions. The syringe motor was then backed up and these solutions filled just the terminal portion of the tube between the valve and the perfusion pipette. The connectors were replaced and the motor was started in the forward direction for just enough time to fill the pipette barrels; at this point the perfusion pipette was brought in the recording chamber and the experiment was initiated. At the end of the experiment, all the perfusion lines could be washed by dipping all the tubes previously dipped in the cylinders (or in the vials), into a container filled with distilled water (or other washing solution, as methanol, ethanol, DMSO, etc.), and clicking the *Wash* button in the *Syringe motor controller* panel. This command activated the valves and moved the motor back and forth at full speed, so that the syringes were emptied through the perfusion pipette and refilled with the wash solution, a number of times set by the user in a sub-window opened by the *Wash* button (not shown). The timing and sequence of the solution changes (i.e. the direction, speed and travel of the perfusion pipette step motor), the syringe motor speed, start, stop and direction, and each valve position could be also automatically controlled by simple instructions, entered as a text code (visible in white characters in the black window of Fig. 2*c*). A set of instructions of arbitrary length could be manually executed in sequence by clicking on the *TRIGGER* button; each instruction could be executed after an arbitrary delay set by a particular instruction, that would therefore set the sequence timing. Every instruction could be also executed after receiving a trigger pulse from an external device, allowing, for example, the synchronization of the solution changes with the pClamp voltage protocols (as in the experiments shown in Fig. 5-9) or with an imaging system (as in the experiments of Fig. 2*b*). It was also possible to send out a trigger pulse, after executing a set of instruction of arbitrary length, to synchronize other external devices.

2.6 Pressure polish pipette

Patch pipettes were pulled in the conventional manner from 50 or 100 μl borosilicate glass microcaps (Drummond, Broomall, PA, USA), with a vertical puller (model PP-830, Narishige, Tokyo, Japan), and tightened into a pipette holder. The latter was clamped to the microscope stage, and a three-way valve allowed to connect the pipette to a pressurized air line (set to ~4 atm and filtered to 0.2 μm to avoid pipette clogging), or to vent it to air. The pipette holder was moved by means of the XY manipulator of the microscope stage, to center the pipette tip into the central bend of an "omega" shaped, glass-coated platinum filament (50 μm of diameter; Fig. 3*a* and *b*). This shape ensured the homogeneous softening of the pipette shank, when the filament was heated by passing a constant current through it. To avoid metal evaporation onto the pipette, the filament was uniformly glass-coated by dipping it in borosilicate glass powder when heated to yellow color. The filament was tin soldered to a copper holder (that functioned also as a heat sink, Fig. 3*a*) mechanically coupled to a micromanipulator, and electrically connected to a variable current generator via a "push-to-make" switch. To produce the adequate heat to soften the pipette tip (i.e. the filament was brought to reddish color), the current was typically set to ~1.2 A for filaments shaped as shown in Fig. 3*a*. The pressure polishing set-up was enclosed in a box, to protect the filament from air currents, that could strongly affect the filament temperature as well, and the entire set-up from dust. The optical field containing the filament and the pipette was viewed on an LCD monitor connected to a contrast-intensified CCD camera (VX 44, Till Photonics, Gräfelfing, Germany), that replaced the objective turret of a bright-field stereomicroscope (YS2-T, Nikon). The video signal was also digitized on-line (by Pinnacle Studio MovieBox DV, Avid, Burlington, MA, USA) and stored on a computer, to have a record of the pipette shaping process.

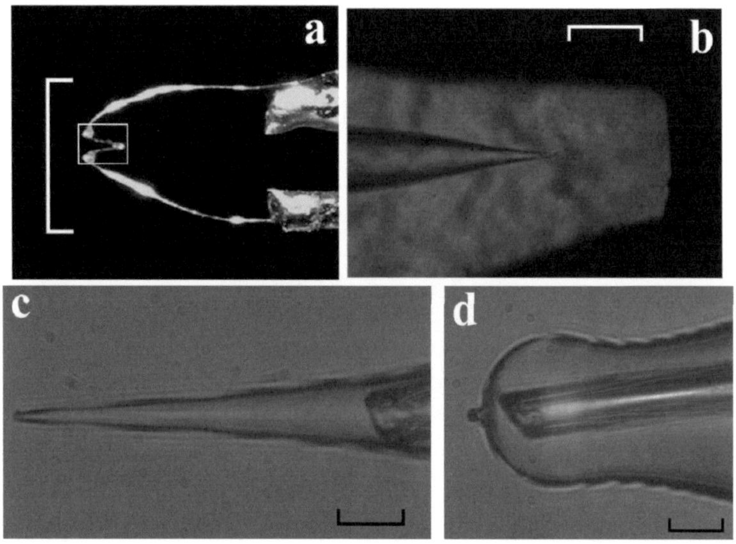

Fig. 3. Filament shape, pipette alignment, and controlled intracellular perfusion. *a*, the glass-coated platinum filament tin soldered to the holder; scale bar is 5 mm. The "omega" shaped region of the filament within the white box is enlarged in *b*, where it is also shown a pipette before pressure polishing, correctly aligned with the filament; scale bar: 100 μm. *c*, a pulled quartz tube is positioned inside the lumen of a conventional pipette as close as possible to its tip; *d*, the same tube is inserted in a pressure polished pipette. Scale bar is 20 μm in *c* and *d*.

3. Results and discussion

3.1 OS endogenous conductances and peptide characterization

The vertebrate OS possesses just two endogenous conductances: the main one is the light sensitive (or cGMP) channel, the other one is the Na^+:Ca^{2+},K^+ exchanger (reviewed in Rispoli, 1998). If the OS is illuminated, the light sensitive channels close, while the exchanger can be blocked if just one of the ion species transported by it (i.e. Na^+, Ca^{2+} or K^+) is removed from both sides of the membrane (Vedovato & Rispoli, 2007b). To simplify the interpretation of the experiments, patch pipettes were filled with the same perfusion solution (that typically contained 130 mM of KCl or 130 mM of NaCl) to ensure the current through the exogenous peptide pore was only driven by the holding potential (V_h, usually set to -20 mV). To preserve the membrane integrity during long recordings, it was necessary to include a physiological concentration of Ca^{2+} (1 mM) to the external solution (Vedovato & Rispoli, 2007a). Therefore, 1 mM Ca^{2+} was added to the intracellular solution as well, to ensure that the current was still entirely driven by V_h. Under these ionic conditions and under room lights (that will close all the light-regulated channels), the OS membrane resistance (R_m) was usually larger than 1 GΩ in the absence of the peptide, exhibiting a linear (ohmic) current-to-voltage characteristic (Fig. 4).

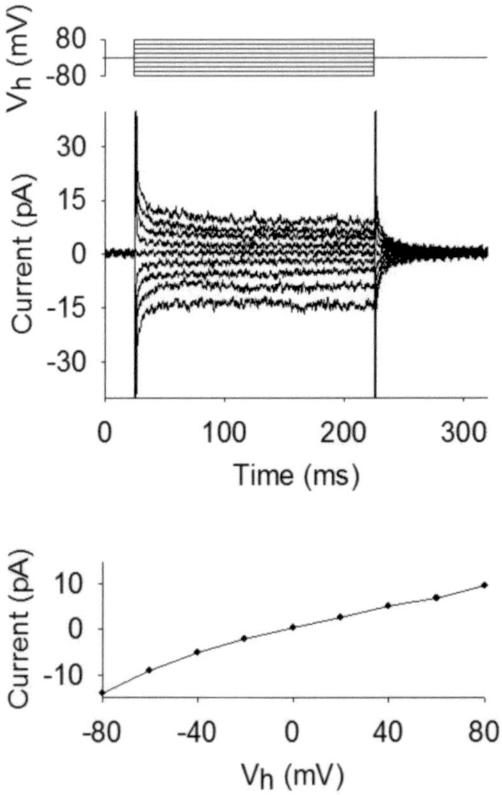

Fig. 4. Electrical properties of OS. *Central panel*: average whole-cell current recorded from a representative OS under room lights (pipette and external solution: 130 mM K^++1 mM Ca^{2+}), subjected to 5 s voltage steps from –80 mV to +80 mV in 10 mV increments (*top panel*) starting from V_h=0 mV and repeated 10 times; *lower panel*, the average current recorded for each voltage step shown in the *central panel* is plotted against the voltage step (the resulting current-to-voltage characteristic is linear, giving R_m~8 GΩ).

The high signal to noise ratio, given by the large value of R_m, allowed the detection of current signals down to the single channel level. The dynamics of the pore formation was tested by means of the following protocol (shown in Fig. 5*a*, 6*b*, 7*a*, 8*a*, 9*a* and *c*). With the isolated OS continuously held to V_h, R_m was measured before peptide perfusion by means of a brief -10 mV step (indicated with an asterisk in all figures); the peptide was then quickly applied (in about ~50 ms) using the fast perfusion system. Once the current had stabilized, the OS was finally returned to the control solution (without the peptide) to assess the possible recovery of the current, and R_m was again measured. In control solution, repetitive 10 mV pulses were routinely applied to check that R_a was unchanged, otherwise the cell was discarded.

In general, the waveform of the current induced by a peptide application and withdrawal can be described quantitatively by the following five kinetics parameters:

1. The activation delay (D_a), defined as the time lag between peptide application and the time in which the current deviates from its baseline (following peptide application), more than three times the noise average fluctuation (indicated by the arrows in Fig. 7*c* and *e*, 8*b*, and 9*b*);
2. The activation time constant (τ_a), defined as the time constant of the single exponential fit to current activation (Fig. 7*a* and *b*, *black* traces);
3. The current amplitude at steady-state (I_{max});
4. The deactivation delay (D_d), defined as the time lag between peptide removal and the time in which the current deviates from I_{max} more than three times the noise average fluctuation;
5. The deactivation time constant (τ_d) defined as the time constant of the single exponential fit to current deactivation (Fig. 7*a* and *b*, *black* traces).

To avoid errors produced by the noise, the above parameters were measured on the low-pass filtered or on the smoothed traces (see Methods).

3.2 Comparison between CM15 and F50/5 permeabilization properties

Application of CM15 at concentrations ≤1 μM gave no detectable macroscopic currents nor single channel events, but currents were routinely obtained at concentrations ≥2.5 μM (Fig. 5, 6*a*, 7*a*, *c*, and *d*). Occasionally, at early times of peptide application at low concentration, current waveform resembled (barely) single channel events (the clearest recording ever

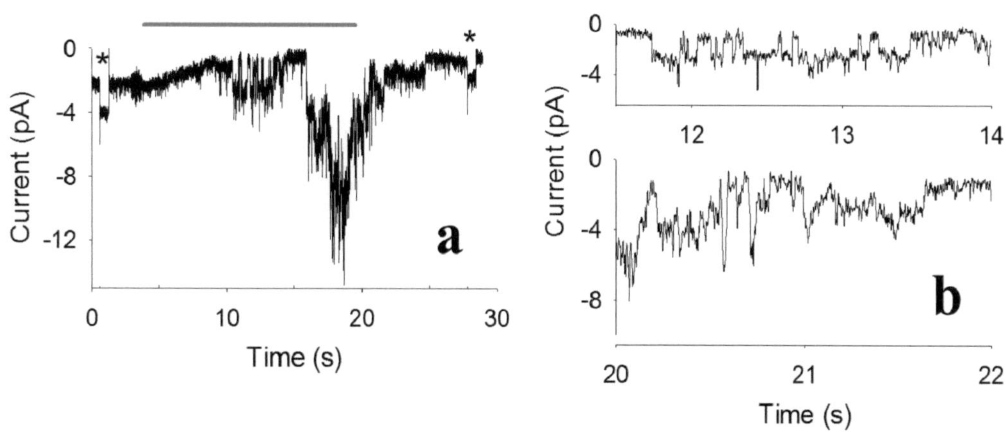

Fig. 5. Membrane permeabilization induced by CM15 at early times of recording at low concentrations. *a*, application for 16 s (*grey thick* line) and withdrawal of CM15 at concentration of 2.5 μM on an OS recorded in whole-cell configuration. The membrane resistance was checked before and after peptide application by means of -10 mV pulses, indicated by the asterisks; holding potential was -20 mV throughout the recording. *b*, enlargements of the recording of *a* 6.6 s after the perfusion onset (*upper panel*) and just after peptide removal from external solution (*lower panel*). To identify the temporal location of the two enlargements, timing of panels in *b* follows the one of panel *a*.

obtained is shown in Fig. 5). These events were followed by a noisy macroscopic current, that was expected to flow through several pores, which number was progressively increasing up to a steady-state number, as the peptide was continuously applied. It can be argued that the mechanism of pore formation by CM15 occurs according to a barrel and stave model: this requires that a certain number of monomers binds together once in the plasma membrane to form an ion conductive pore (Fig. 1). If the number of peptide

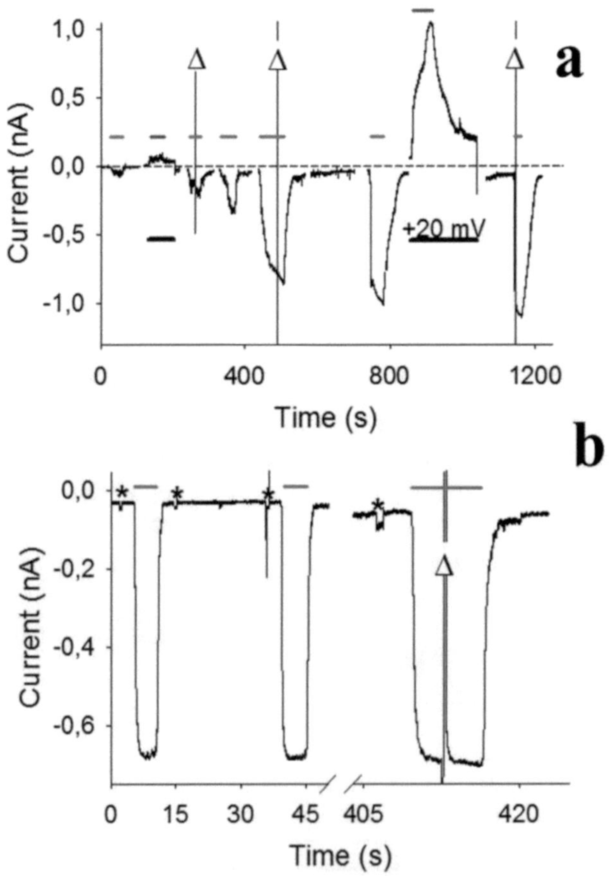

Fig. 6. Comparison between the currents elicited by repetitive applications of CM15 and F50/5. *a*, repetitive applications of CM15 at concentration of 5 μM (*thick grey* lines); V_h was -20 mV but at the *thick black* lines, where it was +20 mV; *dotted* line indicates the zero current at the beginning of recording (i.e. before any peptide application). *b*, repetitive applications of F50/5 at concentration of 1 μM at -20 mV. In both panels, the clipped vertical lines, indicated by the symbol Δ, are actually the response to quick voltage ramps (see text) used to construct the current-to-voltage relationships (shown in Fig. 7*d*); the membrane resistance was checked before and after peptide application by means of -10 mV pulses, indicated by the asterisks; V_h was -20 mV throughout the recording. The blanks in the recordings in control solution in *a* and *b* omit the repetitive voltage pulses used to measure R_m, C_m and R_a and/or the response to voltage ramps (a representative one is shown in Fig. 7*d*, *thick black* trace).

monomers inserted in the membrane is small, as it occurs at early times of peptide application at low concentration, the pores are formed and disaggregated frequently, producing single channel events. However, besides the rarity of the occurrence of these events at early time of the recordings, they were never detected just after peptide removal, as instead expected, since the number of peptide monomers inserted in the membrane should return to be small again. At difference with CM15, low concentrations of F50/5 (and of their analogs reported in Table 1) gave clear and sustained single channel events (Fig. 8*c*; Vedovato & Rispoli, 2007a). Incidentally, these channel events were produced by F50/5 at concentrations as low as 250 nM, showing a much larger efficiency in pore formation in eukaryotic cell membrane in respect to CM15.

The CM15 application at concentration of 5 μM gave consistently macroscopic currents (Fig. 6*a*) that developed exponentially to a relatively stable level (Fig. 7*a*). This level was not however maintained if the peptide application was lengthen, but current kept slowly increasing throughout the peptide application (Fig. 6*a* and 7*a*), at difference with sustained F50/5 perfusion (Fig. 7*b*), that produced very stable current. Moreover, repetitive CM15 applications produced currents of increasingly amplitude (Fig. 6*a*), but no such increase was ever observed with F50/5 (Fig. 6*b*). This again excludes that CM15 permeabilizes the plasma membrane according to a barrel and stave mechanism, since during continuous peptide application at high concentration, membrane peptides are expected to equilibrate with the ones externally perfused, giving a stable macroscopic current, as in the case of F50/5. Moreover, the current did not return to the zero level following peptide removal (Fig. 6*a* and 7*a*), but it recovered to a plateau level, where R_m was consequently smaller in respect to the one measured before the peptide application. The CM15 concentration was larger, and/or more and more applications were performed (i.e. the larger was the current induced by CM15), the larger was the plateau amplitude and the smaller was the R_m (measured during the plateau phase). This is again in contrast to F50/5, where current and R_m fully recovered (Fig. 6*b* and 7*b*) following peptide removal. In these experiments, the F50/5 and CM15 concentrations were selected to give currents smaller than 1 nA, since there is a voltage error induced by R_a (that was typically ~10 MΩ in the recordings considered in this chapter) that can be as high as 10 mV at 1 nA of current.

The kinetics parameters of current activation and deactivation defined in paragraph 3.1 can be measured from the fittings and the interpolations to the recordings as illustrated in Fig. 7*a*, *b*, *c* and *e*. However, these parameters cannot be unambiguously estimated in the case of CM15, since they depend by the current amplitude. For instance, from the first exposure to the last one of Fig. 6*a*, D_a progressively decreased from 2.6 s to 0.5 s and τ_d increased from 3.4 to 26.1 s, as the current increased from 50 pA to about 1 nA. To obtain a reproducible value of the kinetics parameters of the CM15-induced current, in order to compare them with the ones obtained with F50/5, CM15 recordings were selected to approximately give current amplitudes comparable to the ones induced by 1 μM F50/5, irrespective of the CM15 concentration and/or the recording time. A representative example of one of these recordings is shown in Fig. 7*a* (*thick grey* trace), that is compared to the much faster kinetics of F50/5 recordings (one example is in Fig. 7*b*, *thick grey* trace). This strong difference of permeabilization kinetics is better evidenced numerically by the 32-fold larger value of τ_a and τ_d of the CM15 recordings in respect to the F50/5 ones (Table 2), obtained by the monoexponential fittings to the activation and deactivation phases of both recordings

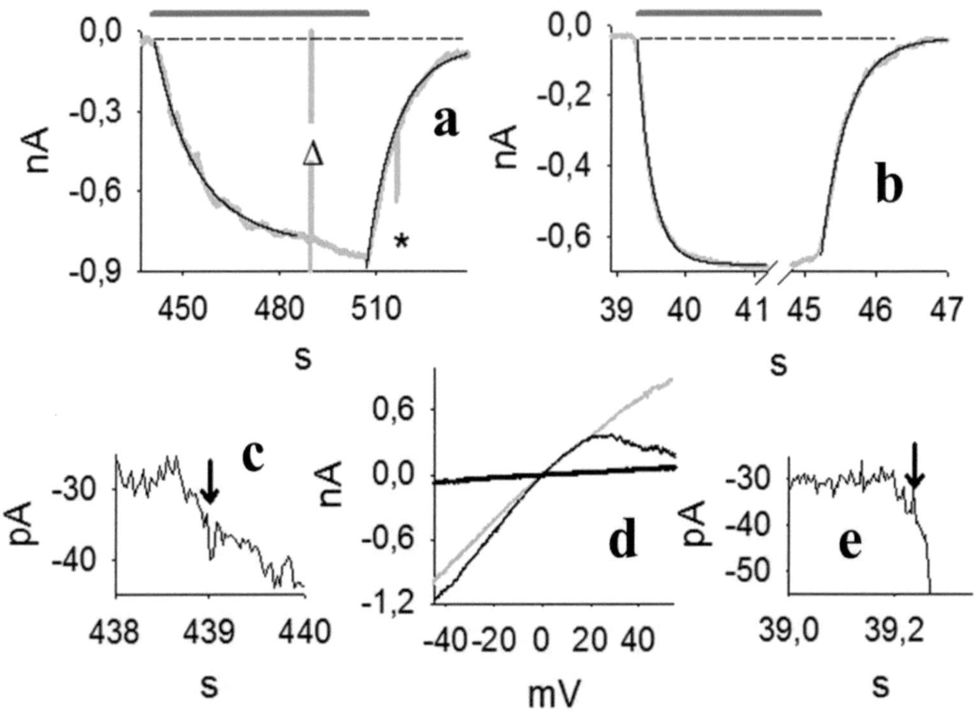

Fig. 7. Comparison of the kinetics parameters and the voltage dependence of the currents elicited by CM15 and F50/5. *a*, the fifth CM15 exposure of Fig. 6*a* (the time scale follows the one of Fig. 6*a*) is enlarged and the activation and deactivation of the current are fitted with monoexponential curves, which time constants are τ_a~14 s and τ_d ~9.1 s, respectively. The first two seconds of this trace is further enlarged in *c*, starting with the time in which the peptide was applied (at ~438 s from the beginning of recording of Fig. 6*a*), to show how D_a was measured (indicated by the arrow, D_a~1 s). *b*, the second F50/5 exposure of Fig. 6*b* (the time scale follows the one of Fig. 6*b*) is enlarged and the activation and deactivation of the current are fitted with monoexponential curves which time constants are τ_a~0.22 s and τ_d ~0.36 s, respectively. The first 350 ms of this trace is enlarged in *e*, starting with the time in which the peptide was applied (at ~39 s from the beginning of recording of Fig. 6*b*); D_a~0.21 s (indicated by the arrow). *d*, voltage dependence of the current elicited by CM15 (*thick grey* trace) and by F50/5 (*thin black* trace) subtracted from the leakage, that was measured by applying the same voltage ramp in the absence of peptide (*thick black* trace). D_d was measured for both CM15 and F50/5 with the same procedure shown in *c* and *e*. *Dotted* lines in *a* and *b* indicate the zero current before peptide application.

(shown in Fig. 7*a* and *b*, *black* traces). As expected, the average value of D_a (Table 2) measured from the CM15 recordings on a fast time scale (indicated by the arrow in Fig. 7*c*) was larger than D_a of F50/5 (Table 2; Fig. 7*e*). Moreover, D_d was not measurable in the case of F50/5 (since the current started to decrease as soon as the solution change was completed), while it was significant in the case of CM15 (Table 2). The dramatically slower kinetics parameters of CM15 in respect to F50/5 ones, indicate that it takes a longer time to

Parameter	CM15 (n=6)	F50/5 (n=10)	L-Glu(OMe)[7,18,19] (n=9)
D_a (s)	0.8±0.2	0.21±0.03	1.7±0.4
τ_a (s)	8.4±1.4	0.26±0.02	4.1±0.8
I_{max} (nA)	0.51±0.05	0.70±0.03	0.74±0.20
D_d (s)	0.6±0.2	0	0
τ_d (s)	10±2	0.31±0.02	1.9±0.6

Table 2. Kinetics parameters of current elicited by CM15, F50/5, and L-Glu(OMe)[7,18,19].

form pores with CM15 monomers than with F50/5 (i.e. D_a and τ_a of the CM15 are larger than the ones of F50/5). Moreover, once formed, the CM15 pores are more difficult to be disaggregated upon ceasing the peptide supply at the external solution (D_d and τ_d of CM15 are again larger than the corresponding values measured for F50/5). In general, as I_{max} and the plateau amplitude following peptide removal progressively increased with repetitive CM15 applications, D_a and τ_a progressively decreased while D_d and τ_d progressively increased. It is conceivable that the CM15 monomers build up in several regions of the OS membrane to a concentration not enough large to give rise to conductive pores: these peptides may then contribute in forming new pores once the CM15 is applied again extracellularly, giving rise to the observed acceleration of activation kinetics of the current (i.e. decreasing D_a and τ_a). As expected, the resulting progressive build up of the number of conductive channels upon repeating the CM15 applications (that produced the progressive increase of I_{max}), would slow down more and more the deactivation kinetics (i.e. increasing D_d and τ_d), since larger the number of channels (that are also more stable than the ones produced by F50/5), longer it takes to disaggregate them when ceasing the CM15 supply from the external solution. Collectively, all data presented so far exclude a barrel and stave model of pore formation of CM15, while they strongly support this model of permeation in the case of the F50/5. Moreover, the cell integrity and the substantial recovery of current, observed for CM15 applications and withdrawal at concentrations as high as 10 μM (data not shown), exclude the carpet mechanism of membrane permeabilization at concentrations smaller than 10 μM, because micellation would produce the irreversible disruption of the membrane and cell lyses. It is however conceivable that, at high concentrations, CM15 could irreversibly permeabilize the membrane, by producing micellation according to the mechanism described in Fig. 1*f*. It therefore can be concluded that CM15 permeabilizes the membrane according to a toroidal model of pore formation. This view is also supported by the voltage-independency of CM15 membrane permeabilization, in contrast to F50/5, that instead inserts in the OS membrane at negative voltages (Vedovato et al., 2007; Vedovato & Rispoli, 2007a; Fig. 7*d*). Indeed, CM15 application at +20 mV or -20 mV produced currents with similar I_{max} (and D_a, D_d, τ_a, τ_d as well; Fig. 6*a*). However, the latter protocol is not suitable to assess the precise voltage dependency of the current, since the current progressively increases with repetitive applications of CM15 (note that the maximal current amplitude at +20 mV in Fig. 6*a* is larger in respect to the one recorded at -20 mV). To circumvent this problem, rapid voltage ramps (slope: 0.25 mV/ms) were applied during CM15 perfusion at V_h=-20 mV, once the current stabilized for a period at least as long as the ramp (400 ms). To avoid the loss of voltage control due to R_a at extreme voltages (-60 and +40 mV, where current may became very large), cells were selected to have currents of about 100-200 pA at V_h=-20 mV. The response to the voltage ramp during CM15 perfusion was subtracted to the response to the same voltage ramp recorded in control conditions (i.e. in the absence of peptide; Fig. 7*d*, *thick black* trace) to obtain the current-to-voltage relationship corrected for the leakage (Fig. 7*d*, *thick grey* trace). Relationship was almost perfectly ohmic for physiological voltages, in all cells examined (n=6), in contrast with the strong inward rectification of the current-to-voltage relationship (corrected for the leakage) produced by F50/5 (Fig. 7*d*, *thin black* trace, n=16), obtained with the same voltage protocol used for CM15. Therefore, it is required a particular orientation of the F50/5 monomers (attained with the voltage) to form a conductive channel, while the CM15 monomers permeabilize the membrane regardless their orientation. This again supports the barrel and stave model for F50/5 and the toroidal model for CM15.

It can be argued that the pore formation according to the toroidal model could be ascribed to the relative abundance of lysine and/or leucine in respect to other aminoacids present in the CM15 sequence. Alternatively, the CM15-induced permeabilization might be due to some aspecific effect produced by the interaction of one or more of its amino acids with some membrane protein/s. These possibilities were however ruled out because a random sequence of the CM15 amino acids, (as the scrambled CM15; Table 1), was not able to produce any permeabilization, even for repetitive applications at 10 μM concentration (lasting up to 3 min; n=3 OS; data not shown). Therefore, it is concluded that CM15 must be lined up in a precise sequence to produce efficient membrane permeabilization.

3.3 F50/5 monomer assembling mechanism

The barrel and stave model of pore formation requires that a certain number of peptide monomers binds together to form a functioning pore. The hydrophilic glutammine residues at positions 7, 18 and 19 of F50/5 are supposed to face the pore lumen and form hydrogen-bonded rings, therefore they are expected to play a key role in channel formation. To examine the role of these residues in the pore formation, the amino acids in position 18 and 19, or all of three, were substituted with a glutamic acid, in which a methyl ester group was linked to the carboxyl function in the γ position. Two synthetic analogs, [L-Glu(OMe)^{18,19}] and [L-Glu(OMe)^{7,18,19}], were examined by applying them to the OS with the same protocol used to study the F50/5. Surprisingly, 1 μM concentration of both analogs produced a current as large that induced by F50/5, showing that these residue substitutions do not prevent pore formation. However, while all the kinetics parameters and the noise of current elicited by [L-Glu(OMe)^{18,19}] were undistinguishable by the ones of F50/5, D_a and τ_a of current elicited by [L-Glu(OMe)^{7,18,19}] were ~8-fold and ~16-fold, respectively, larger than those of F50/5 (Table 2): therefore, the kinetics of pore formation was fairly impaired in the latter analog, but not in the former, in respect to F50/5. Moreover, as the current developed, there was a much larger noise increase in respect to F50/5 (and [L-Glu(OMe)^{18,19}]), indicating that [L-Glu(OMe)^{7,18,19}] may form fewer channels, but with a larger single channel conductance, than those generated by F50/5. The latter peptide applied at 250 nM concentration on the OS produced single channel events of several amplitudes, being not a simple multiple of a fixed size. Analysis of traces displaying very few channels (as the one shown in Fig. 8*c*, *upper panel*) allowed to estimate the smallest average single channel size, that resulted of ~50±8 pS at -20 mV (3500 events averaged in 3 cells). As expected, the single channel events produced by [L-Glu(OMe)^{7,18,19}] at concentration of ~250 nM, were much larger than the F50/5 ones. Many different event amplitudes were produced by [L-

Glu(OMe)[7,18,19]], and some were so fast that they were cut by the patch-clamp amplifier, even when filtering as high as 2 KHz (Fig. 8c, *bottom panel*, *grey* trace). Since the majority of the events were anyway much slower, in order to compare the single channel amplitude of [L-Glu(OMe)[7,18,19]] with the one of F50/5, these very fast events were usually cancelled out

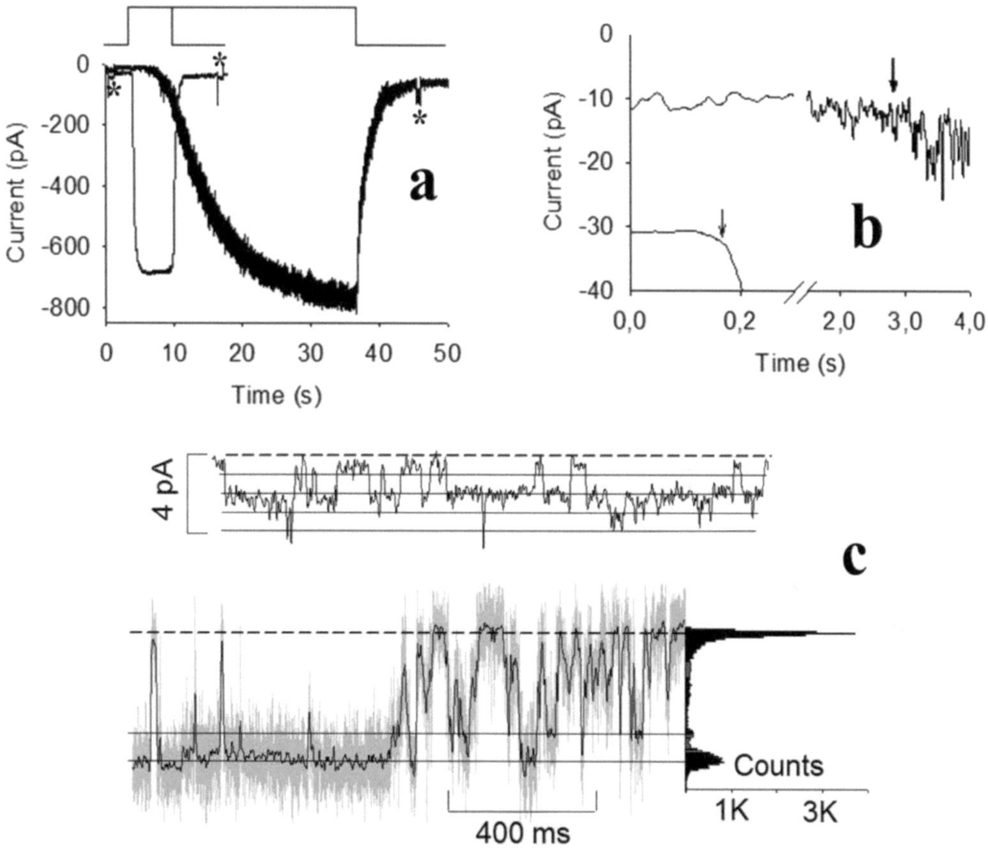

Fig. 8. Kinetics of OS membrane permeabilization induced by synthetic F50/5 and its [L-Glu(OMe)[7,18,19]] analog. *a*, *upper* traces, timing of application and withdrawal of peptides at 1 μM concentration; *lower* trace, the voltage-clamp, whole-cell current recordings from one OS perfused with F50/5 (*thin* trace) and with [L-Glu(OMe)[7,18,19]] (*thicker* trace). Asteriscs indicate the occurrence of the two -10 mV voltage pulses (superimposed to V_h=-20 mV), used to measure R_m, that was about 1 GΩ before, and 20 s after, peptide application in both OS. *b*, the whole-cell current recordings of *a* are smoothed to accurately measure the activation delay, indicated by the arrows, that is 0.17 s for F50/5 and 2.8 s for [L-Glu(OMe)[7,18,19]], computed from the solution change (that occurs at time 0 in this plot); the other kinetics parameters were, respectively: τ_a (0.21 s and 7.4 s), τ_d (0.29 s and 1.6 s), and I_{max} (700 pA and 760 pA at V_h=-20 mV; see Table 2). *c*, *upper panel*, single channel events produced by F50/5; *lower panel*, *left*, channel activity produced by [L-Glu(OMe)[7,18,19]] low-pass filtered at 2 KHz (*grey* trace) and at 100 Hz (superimposed *black* trace). *Right*, cumulative current amplitude distribution. Both recordings of *c* have the same scales and V_h=-20 mV; *continuous* and *dotted* lines indicate the most probable single channel amplitudes and 0 pA, respectively.

upon low pass filtering the traces at 100 Hz (Fig. 8*c*, *bottom panel*, *black* trace). The distribution of current amplitudes (Fig. 8*c*, *bottom panel*, *right*) showed that the most probable amplitude was 310±30 pS (820 events, n=3), but events of smaller sizes (as 260 pS) were recorded as well. For both peptides, current bursts occurred at very irregular intervals, making it impossible to systematically investigate interval duration, and measure key parameters as the open probability and the average open and closed time of the channel levels. The range of single channel amplitudes indicates that a channel is probably formed by several peptides assembled together, and various configurations of the peptide assembly are possible, depending upon the voltage, the peptide concentration and the duration of peptide application.

The slower kinetics of current activation formed by [L-Glu(OMe)^{7,18,19}] may simply reflect the smaller probability to assemble such large conductance channels. Consistently with this view, single channel events were recorded after ~20 s of [L-Glu(OMe)^{7,18,19}] application to the OS (V_h=-20 mV), instead of ~5 s as in the case of F50/5. Similarly to F50/5, the current fell to zero without any delay in respect to [L-Glu(OMe)^{7,18,19}] removal time, therefore this analog also does not stably integrate into the plasma membrane. Moreover, the current-to-voltage relationship of [L-Glu(OMe)^{7,18,19}] was almost identical to the F50/5 one (shown in Fig. 7*d*, *thin black* trace).

Incidentally, the fact that the current elicited by F50/5 and by its analogs fell from several hundreds of pA to the single channel level following a 4-fold reduction of peptide concentration, and that the peptide-elicited current fell to 0 within a few hundred of ms following peptide removal from the external solution (see, for example, Fig. 8*a* and *c*), indicate that the pore formation and disaggregation are very fast and cooperative events.

For all peptide tested, the D_a, I_{max}, τ_a, and τ_d recorded in symmetric Na^+ were very similar to those reported in Table 2 (i.e., recorded in symmetric K^+; data not shown).

All the above findings show that F50/5 and its analog [L-Glu(OMe)^{7,18,19}], although exhibiting a very different kinetics of pore formation, retain the poor cation selectivity and the voltage dependency, that are typical features of peptaibols. However, they cannot conclusively indicate whether the ion conduction occurs through a pore formed with several peptide monomers assembled together, or it occurs along the inner portion of a single α-helices. The three dimensional structure of [L-Glu(OMe)^{7,18,19}] at a resolution of 0.95 Å, resolved from the electron density maps obtained by using the synchrotron radiation, consisted of two crystallographically independent bent α-helices, just long enough to span the membrane, but that cannot conduct ions through the hole produced by their helical arrangement (Crisma et al., 2007). Therefore, to explain the above electrophysiological results, and in particular the large current obtained at 1 μM concentration of peptide and the occurrence of single channel events at low concentrations, it is concluded that the α-helices of F50/5 and its analogs assemble around a central pore that is able to conduct ions, according to the barrel and stave model (Fig. 1). The peptaibols constitute therefore the most ancestral form of ion channel, since all the voltage gated channels, the ligand gated channels and the gap junction found in all eukaryotic cells, are formed by several α-helices that pack together around a central ion conducting pore.

3.4 Viroporin permeabilization

The experimental strategy described above to study antibiotic peptides, turned out to be very powerful to investigate also the molecular mechanism of membrane permeabilization induced by viroporins. One of the most dangerous viroporins for human health is the 97-residues long poliovirus 2B, a non-structural protein required for effective viral replication, that has been implicated in cell membrane permeabilization during the late phases of infection. The experiments presented so far demonstrated that antimicrobial peptides comprising 15-20 residues are able to form membrane pores: it is then conceivable that a short domain of poliovirus 2B sufficed to generate membrane permeabilization. Therefore, it was assessed the pore-forming properties of three peptides about 20 amino acid -long of an overlapping library that spanned the complete viroporin sequence. No significant changes of current level and of R_m (Fig. 9*a*, *grey* trace) were induced by continuous perfusion of the 22-residues peptide at the beginning of poliovirus 2B sequence (7 out of 7 OS; Table 1, residues indicated in *red*) or at its end (5 out of 5 OS; Table 1, residues indicated in *green*) lasting up to 100 s, for peptide concentrations up to 1 µM. Since the permeabilization induced by peptaibols was voltage-dependent (Fig. 7*d*), V_h was changed in a range between -60 and +60 mV so to exclude that the lack of permeabilization was due to an inappropriate V_h (Fig. 9*a*, *grey* traces). Concentrations as little as 33 nM of the peptide spanning 2B residues 35–55, (indicated in *blue* in Table 1), sufficed instead to significantly permeabilize the OS, and 100 nM permeabilized the OS to a point that R_m decreased by three order of magnitude and became comparable to R_a (Fig. 9*a*, *black* trace and Fig. 9*c*; Table 3). This permeabilization process can be quantitatively described, besides by the R_m change, by the parameters D_a (indicated by the arrow in the *inset* of Fig. 9*b*) and I_{max} (see paragraph 3.1). However, τ_d and D_d cannot be measured, because, at difference with peptaibols and CM15, no significant current recovery was observed upon removing this peptide from the external solution.

Moreover, current increased during the peptide application following a curve that cannot be interpolated with a single (or even a multiple) exponential: therefore, as far as the viroporin is concerned, τ_a symbolizes the 10-90% activation time, defined as the time interval in which the peptide induced current goes from 10% to 90% of the steady-state amplitude. The kinetics parameters of poliovirus 2B permeabilization on OS (Table 3) were measured at V_h=-20 mV, so to compare them with the ones of the antibiotic peptides (Table 2). The current-to-voltage relationship with a voltage ramp from -60 to +60 mV, measured once the current was stable, was linear and it was similar to the one predicted by the Ohm's law with a resistance equal to R_m (22 MΩ in Fig. 9*d*).

This is consistent with the view that the peptide must permeabilize the eukaryotic cell with high efficacy, independently by the cell state (as its membrane potential). Larger the peptide concentration, shorter was D_a, being as small as 0.5 s at 1 µM. However, at such high peptide concentration, R_m became smaller than R_a, then R_m could not be measured and the steady-state amplitude was meaningless (these two parameters were therefore not reported in Table 3). The large values of D_a and τ_a indicate that this peptide must interact with the lipids for a long time, in respect to peptaibols and CM15, in order to permeabilize the membrane. This suggests that the membrane permeabilization occurs initially according to a toroidal model of pore formation; then, as more and more pores are formed, adjacent pores may restrict a micelle, that could detach and moves to the water phase, leaving a permanent hole behind, according to the carpet model of pore formation.

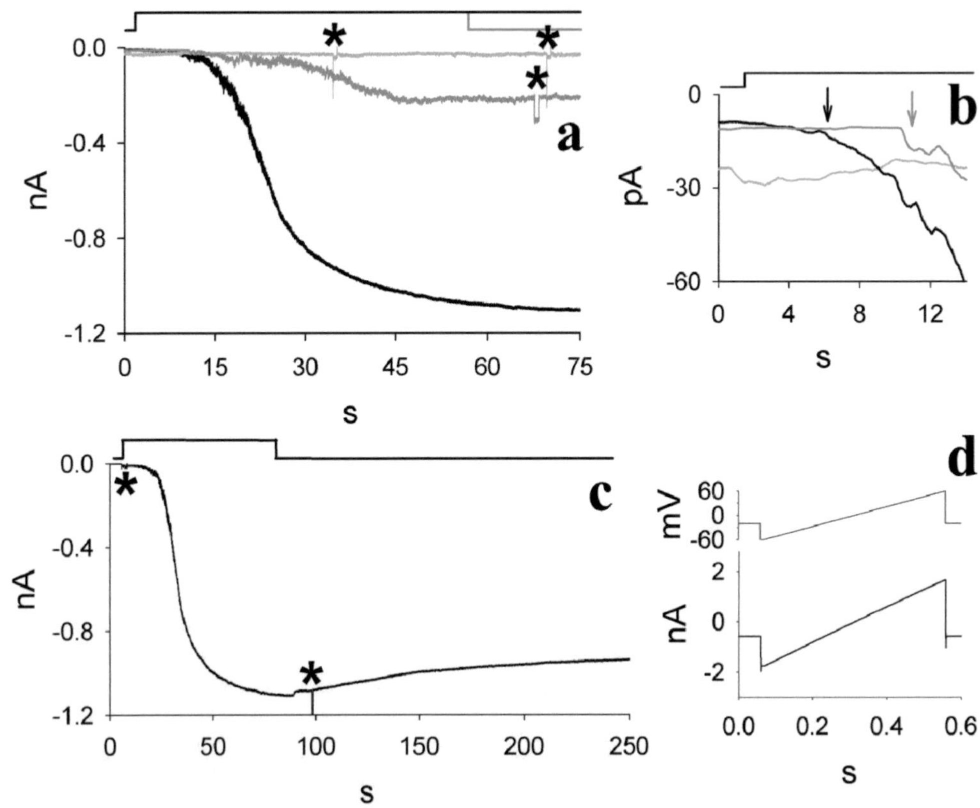

Fig. 9. Kinetics of OS membrane permeabilization induced by three domains of poliovirus 2B. *a*, current recorded during a 75 s application of: 1 µM of a domain at the beginning (indicated in *red* in Table 1) and at the end (indicated in *green*) of the viroporin sequence (*light grey* trace), 33 nM (*grey* trace), and 100 nM (*black* trace) of the central domain (*blue*) of 2B sequence (V_h=-20 mV throughout the recording). The timing of peptide application is indicated by the *black* line above the recording but the 33 nM peptide application, which is indicated in *grey*. *b*, same traces as *a*, but smoothed to accurately measure D_a (indicated by the arrow) and I_{max}, whose numerical values are reported in Table 2. R_m before peptide application was measured from the current amplitude at –20 mV (-9.5 pA, R_m= 2.1 GΩ, *black* trace; -27.0 pA, R_m=0.8 GΩ, *grey* trace; average: 1.8±0.9 GΩ, *n*=7). The R_m following peptide application was measured once the current attained steady-state by using short (0.74 s) and small (-10 mV) voltage pulses (indicated by the *asterisks* close to both traces) superimposed on V_h. At steady-state, R_m (22 MΩ) became comparable to R_a (12 MΩ) after the application of the central domain of poliovirus 2B (in *blue* in Table 1), whereas it did not change significantly after perfusion of 1 µM of either the other two domains (in *red* and in *green* in Table 1). *c*, current elicited by 100 nM of the central domain shown in *a* on a slower time scale. *d*, current (*lower* trace) elicited by a voltage ramp (*top* trace) from -60 mV to 60 mV (slope: 0.244 mV/ms; V_h=-20 mV) applied once the current attained the steady-state during perfusion with 33 nM of the central domain.

Conc. (nM)	D_a (s)	10-90% (s)	I_{max} (nA)	R_m pre(GΩ)	R_m post (MΩ)	R_a (MΩ)
33 (*n*=3)	9.9±2.6	31.8±7.2	0.22±0.08	1.3±0.3	110±40	18±7
100 (*n*=5)	5.2±0.7	25.0±1.7	1.0±0.3	2.0±0.8	29±14	17±8
1000 (*n*=3)	0.5±0.2	---	---	1.2±0.5	---	18±8

Table 3. Kinetics parameters of current elicited by poliovirus 2B at different concentrations.

3.5 Pressure polishing procedure

The permeabilizing domain of poliovirus 2B has been applied on the external side of the OS. However, the ideal experiment would be delivering this peptide intracellularly, with a controlled timing, via the patch pipette by a tube inserted in its lumen, that should be placed as close as possible to the pipette tip. However, the long tapered shape of the pipette shank make it very difficult to perfuse efficiently the cell with this strategy (Fig. 3*c*). In the Methods is illustrated a further improvement of an existing technique widening the patch pipette shank, through the calibrated combination of heat and air pressure, that allowed to insert quartz or plastic tubes in the pipette lumen very close to its tip. Once the pipette was properly aligned with the filament (Fig. 3*b*), the box was closed and the air pressure and the filament heating were turned on and off for an appropriate time (to attain the desired pipette shape), with the three-way valve and the push-to-make switch, respectively. The shaping process of the pipette shank and of its tip was precisely followed using a calibration grating superimposed on the LCD monitor. The pipette shank geometry and its tip opening diameter could be finely controlled by adjusting: 1) the relative position of the pipette in respect to the filament, 2) the current intensity passing through the filament, 3) the duration of this current flow, 4) the pressure intensity, and 5) the duration of the pressure application while the current was flowing in the filament. In order to standardize the technique, the most efficient strategy found was to optimize the parameters 1), 2), and 4) and to keep them fixed throughout the pressure polishing process, while the third and the fifth parameters were left to be adjusted each time in order to obtain the desired shank profile and tip opening diameter. Long pressure applications (while the current was flowing in the filament), were used to obtain very enlarged shanks. If the pipette opening resulted too large for a given pressure duration, then its tip was heated again in the absence of pressure until the desired size was achieved; if it was instead too small, then that pressure duration was applied to pipettes pulled with larger tips. Therefore, by carefully regulating the duration of pressure and heating application, the pipette shank was widened as desired, while the tip opening diameter could be increased, decreased or left unchanged.

Pressure polished pipettes increased the rate of molecular diffusion between the pipette and the cell interior, allowing to perfuse viroporins or their domains and small proteins, making it possible the "real time modulation" of a multitude of cellular processes as well, as the signal transduction cascades. However, proteogenic molecules stick to the glass tip to the point to prevent seal formation, therefore it is important to add them to the pipette solution to a later time, i.e. after the seal was attained. In general, it is important to apply to the cytosol a particular molecule under test not at the beginning of the whole-cell recording, but to a desired later time. This goal can be realized by inserting in the pipette lumen pulled quartz tubes (or tubes fabricated with plastics that can be easily softened with a Bunsen flame, as polyethylene or polypropylene), filled with a solution containing the molecule under test. The latter can be ejected into the pipette solution with a pulse of pressure applied to the tube lumen. These tubes could be inserted very close to the tip of pipettes with a particularly enlarged shank (i.e. very close to the cell cytosol; Fig. 3*d*). Therefore, the molecules under test would be applied to the cell cytosol with a much shorter delay in respect to conventional pipettes (Fig. 3*c*), delay that can be up to ~70 s for a small protein (having a diffusion coefficient of ~10^{-6} $cm^2 s^{-1}$; Benedusi et al., 2011).

Another great advantage of the pressure polished pipettes is the strong reduction of R_a, since all the electrical resistance of these pipettes is given to the tip opening diameter and not also to the pipette shank. The pressure polished pipettes were thoughtfully tested on frog OS, because this is the most critical cell system among the ones tested. Indeed, to have reliably recordings from frog OS, it is necessary to use pipettes with a very small tip diameter (yielding therefore high R_a if fabricated with conventional methods) to avoid to break the OS, because they are very fragile (Fig. 2*a*, *inset*) and they are often soiled with retina debris and vitreous humor. For pipettes shown in Fig. 3*d* it resulted a 4-fold reduction of R_a in respect to the conventional ones (Benedusi et al., 2011).

Surprisingly, the chosen borosilicate glass (see Methods) was extremely good at attaining tight seals on an extremely wide variety of cells (being isolated or inserted in small tissue fragment or slice), or on giant unilamellar vesicles of many different lipid composition, providing that the pipettes and the pressure polishing set-up were maintained dirt-free, the pipette was kept in the bath as little as possible (to avoid that cell debris soiled the tip while approaching the cell), and the membrane target was "pressure cleaned" by ejecting as fast as possible the pipette solution onto the cell. It was not therefore true, as commonly believed, that different types of glass work better on different cell types (Penner, 1995): when there was a failure at attaining a seal, it was because the patch pipette tip was soiled, and/or because the membrane target was contaminated with cell debris, connective tissue or other organic material, not because the glass type was inappropriate. The day-to-day variability in attaining the seal was therefore entirely due to the cell quality (soiled or too fragile to sustain the sealing process), not to some uncontrollable parameter in the pipette fabrication. Given the "sealing power" of the glass pipette mentioned above, the pressure polishing technique was optimized just for this glass type.

4. Conclusion

The controlled application and withdrawal of peptides on isolated photoreceptor rod outer segments, recorded with the whole-cell voltage-clamp technique, resulted a powerful method to investigate the dynamics of pore formation and disassembly of membrane-permeabilizing peptides (as antibiotic peptides and viroporins) under strict physiological conditions. Moreover, the patch pipettes fabricated with the pressure polished technique, and used as described above, allowed to reduce of at least 4-fold the access resistance, to efficiently incorporate exogenous molecules in the cytosol especially via pulled quartz or plastic perfusion tubes, and to seal virtually on any natural or artificial phospholipid bilayer. The main results presented in this chapter can be summarized as follows:

1. One μM of alamethicin F50/5 produced an inward rectifying current that activated exponentially with a time constant of ~300 ms to a steady-state amplitude of 0.7 nA within ~200 ms from peptide application; at low concentration the F50/5 produced single channel activity of most probable conductance of ~50 pS. The hydrophilic Gln residues at positions 7, 18 and 19 are not a key factor for pore formation, but [L-Glu(OMe)^{7,18,19}] analog produced larger pores (~300 pS), with a lower probability of formation than F50/5. All these peptides do not stably integrate into the plasma membrane and they form pores according to the barrel and stave model with a highly cooperative mechanism.
2. The cecropin-melittin hybrid peptide CM15 produced voltage-independent permeabilization of OS; repetitive peptide applications at concentrations >2 μM caused the progressive increase of the steady-state current amplitude. No clear single-channel events were detected at low peptide concentrations, while at concentrations as high as 10 μM the cell integrity was preserved and the permeabilization was substantially reversible. Collectively, these results indicate that CM15 inserts in the plasma membrane according to the toroidal mechanism of pore formation.
3. A peptide spanning residues 35-55 of the poliovirus 2B produced irreversible membrane permeabilization, probably according to the carpet model, producing voltage-independent current at concentrations as low as 33 nM.

All these simple peptides are therefore not only an ideal tool to develop new antimicrobial and antiviral drugs, but they are important to understand the biophysical properties of much more complex molecules as channels and transporter proteins. The large variability of their pore-forming properties make them a potential lead for the development of bioactive compounds.

5. Acknowledgment

Technical assistance was given by Andrea Margutti, while Anna Fasoli and Cristina Mantovani helped with analysis and preparation of the new experiments presented in this chapter. Financial support to G.R. included grants from the Ministero dell'Università e della Ricerca (MIUR), Roma (Italy; Project PRIN 2008), from the "Comitato dei sostenitori dell'Università di Ferrara" (Project FAR: Correnti ioniche nei peptidi antibiotici e virali e nelle cellule sensoriali e gangliari, Ferrara, Italy).

6. References

Aili D., & Stevens MM. (2010) Bioresponsive peptide-inorganic hybrid nanomaterials. *Chem Soc Rev.*, Vol. 39, No.9, (July 2010), pp.3358-3370, ISSN 0306-0012

Andreu D., Ubach J., Boman A., Wahlin B., Wade D., Merrifield R.B., & Boman H.G. (1992) Shortened cecropin A-melittin hybrids. Significant size reduction retains potent antibiotic activity. *FEBS Lett.*, Vol.296, No.2, (January 1992), pp.190-194, ISSN 0014-5793

Benedusi M., Aquila M., Milani A., & Rispoli G. (2011) A pressure polishing set-up to fabricate patch pipettes virtually sealing on any membrane, yielding low access resistance and efficient intracellular perfusion. *Eur Biophys J.*, Vol.40, No.4 (November 2011), pp.1215-1223, ISSN 0175-7571

Bockmann RA., Hac A., Heimburg T., & Grubmuller H. (2003) Effect of sodium chloride on a lipid bilayer. *Biophys J.*, Vol.85, No.3, (September 2003), pp.1647-1655, ISSN 0006-3495

Brogden K.A. (2005) Antimicrobial peptides: pore formers or metabolic inhibitors in bacteria? *Nat. Rev. Microbiol.*, Vol.3, No.3, (March 2005), pp. 238–250, ISSN 1740-1526

Carrasco L. (1995). Modification of membrane permeability by animal viruses. *Advan. Virus Res.*, Vol. 45, pp. 61–112, ISSN 0065-3527

Crisma M., Peggion C., Baldini C., Maclean EJ., Vedovato N., Rispoli G., & Toniolo C. (2007) Crystal structure of a spin-labeled, channel-forming alamethicin analogue. *Angew Chem Int Ed Engl.*, Vol. 46, No.12, (September 2011), pp.2047-2050, ISSN 1433-7851

Chen HM., Clayton AH., Wang W., & Sawyer WH. (2001) Kinetics of membrane lysis by custom lytic peptides and peptide orientations in membrane. *Eur J Biochem.*, Vol.268, No.6, (March 2001), pp. 1659-1669, ISSN 0014-2956

Gadsby DC. (2009) Ion channels versus ion pumps: the principal difference, in principle. *Nat Rev Mol Cell Biol.*, Vol. 10, No.5, (April 2009), pp.344-352, ISSN 1471-0072

Gonzalez ME., & Carrasco L. (2003) Viroporins. *FEBS Lett.*, Vol. 552, No.1, (September 2003), pp.28-34, ISSN 0014-5793

Goodman MB., & Lockery SR. (2000) Pressure polishing: a method for re-shaping patch pipettes during fire polishing. *J Neurosci Methods.*, Vol. 100, No.1-2, (July 2000), pp.13-15, ISSN 0165-0270

Johnson BE., Brown AL., & Goodman MB. (2008) Pressure-polishing pipettes for improved patch-clamp recording. *J Vis Exp*. doi: 10.3791/964.

Hille B. (2001) Ionic channels of excitable membranes. (Third edition), Sinauer Associates, Inc. Sunderland, MA (USA), ISBN 0-87893-321-2

Hoskin DW., & Ramamoorthy A. (2008) Studies on anticancer activities of antimicrobial peptides. *Biochim Biophys Acta.*, Vol. 1778, No. 2, (November 2007), pp.357-375, ISSN 0006-3002

Madan V., Sánchez-Martínez S., Vedovato N., Rispoli G., Carrasco L., & Nieva JL. (2007) Plasma membrane-porating domain in poliovirus 2B protein. A short peptide mimics viroporin activity. *J Mol Biol.*, Vol. 374, No.4, (December 2007), pp.951-964, ISSN 0022-2836

Milani A., Benedusi M., Aquila M., & Rispoli G. (2009) Pore forming properties of cecropin-melittin hybrid peptide in a natural membrane. *Molecules.*, Vol.14, No.12, (December 2009), pp.5179-5188, ISSN 1420-3049

Noshiro D., Asami K., & Futaki S. (2010) Metal-assisted channel stabilization: disposition of a single histidine on the N-terminus of alamethicin yields channels with extraordinarily long lifetimes. *Biophys J.*, Vol. 98, No.9, (May 2010), pp.1801-1808, ISSN 0006-3495

Papo N., & Shai Y.(2005) Host defense peptides as new weapons in cancer treatment. *Cell Mol Life Sci.*, Vol.62, No.7-8, (April 2005), pp.784-790, ISSN 1420-682X

Peggion C., Coin I., & Toniolo C. (2004) Total synthesis in solution of alamethicin F50/5 by an easily tunable segment condensation approach. *Biopolymers.*, Vol.76, No.6, pp. 485-493, ISSN 0006-3525

Penner R. (1995) A pratical guide to patch clamping. In: *Single-Channel Recording*, Sakmann B., & Neher E., pp. 3-30, Plenum Press, ISBN 978-0-306-44870-6, New York

Pusch M., & Neher E. (1988) Rates of diffusional exchange between small cells and a measuring patch pipette. *Pflugers Arch.*, Vol.411, No.2, (February 1988), pp.204-211, ISSN 0031-6768

Rispoli G., Sather W.A., & Detwiler P.B. (1993) Visual transduction in dialyzed detached rod outer segments from lizard retina. *Journal of Physiology.*, Vol.465, (June 1993), pp. 513-537, ISSN 0022-3751

Rispoli G., Navangione A., & Vellani V. (1995) Transport of K+ by Na(+)-Ca2+, K+ exchanger in isolated rods of lizard retina. *Biophys J.*, Vol.69, No.1, (July 1995), pp.74-83, ISSN 0006-3495

Rispoli G. (1998) Calcium regulation of phototransduction in vertebrate rod outer segments. *J Photochem Photobiol B.*, Vol. 44, No.1, (June 1998), pp. 1-20, ISSN 1011-1344

Saint N., Cadiou H., Bessin Y., & Molle G. (2002) Antibacterial peptide pleurocidin forms ion channels in planar lipid bilayers. *Biochim Biophys Acta.*, Vol.1564, No.2, (August 2002), pp.359-364, ISSN 0006-3002

Suchyna TM., Markin VS., & Sachs F. (2009) Biophysics and structure of the patch and the gigaseal. *Biophys J.*, Vol.97, No.3, (August 2009), pp. 738-747, ISSN 0006-3495

Toniolo C., Crisma M., Formaggio F., Peggion C., Epand R. F., & Epand R. M. (2001) Lipopeptaibols, a novel family of membrane active, antimicrobial peptides. *Cell. Mol. Life Sci.*, Vol. 58, No.9, (August 2001), pp.1179–1188, ISSN 1420-682X

Vedovato N., Baldini C., Toniolo C., & Rispoli G. (2007) Pore-forming properties of alamethicin F50/5 inserted in a biological membrane. *Chem Biodivers.*, Vol.4, No.6, (June 2007), pp.1338-1346, ISSN 1612-1872

Vedovato N., & Rispoli G.(2007a) A novel technique to study pore-forming peptides in a natural membrane. *Eur Biophys J.*, Vol.36, No.7, (September 2007), pp.771-778, ISSN 0175-7571

Vedovato N., & Rispoli G. (2007b) Modulation of the reaction cycle of the Na+:Ca2+, K+ exchanger. *Eur Biophys J.*, Vol.36, No.7, (September 2007), pp.787-793, ISSN 0175-7571

Wallace DP., Tomich JM., Eppler JW., Iwamoto T., Grantham JJ., & Sullivan LP. (2000) A synthetic channel-forming peptide induces Cl(-) secretion: modulation by Ca(2+)-dependent K(+) channels. *Biochim Biophys Acta.*, Vol.1464, No.1, (March 2000), pp.69-82, ISSN 0006-3002

Wilde AA. (2008) Channelopathies in children and adults. *Pacing Clin Electrophysiol.*Vol.31, No.1, (February 2008), pp.S41-45, ISSN 0147-8389